T0235305

Lecture Notes in Computer Science 12044

More information about this series at http://www.springer.com/series/7407

Roman Wyrzykowski · Ewa Deelman ·
Jack Dongarra · Konrad Karczewski (Eds.)

Parallel Processing and Applied Mathematics

13th International Conference, PPAM 2019
Bialystok, Poland, September 8–11, 2019
Revised Selected Papers, Part II

 Springer

Editors
Roman Wyrzykowski
Czestochowa University of Technology
Czestochowa, Poland

Ewa Deelman
University of Southern California
Marina del Rey, CA, USA

Jack Dongarra
University of Tennessee
Knoxville, TN, USA

Konrad Karczewski
Czestochowa University of Technology
Czestochowa, Poland

ISSN 0302-9743   ISSN 1611-3349   (electronic)
Lecture Notes in Computer Science
ISBN 978-3-030-43221-8   ISBN 978-3-030-43222-5   (eBook)
https://doi.org/10.1007/978-3-030-43222-5

LNCS Sublibrary: SL1 – Theoretical Computer Science and General Issues

This Springer imprint is published by the registered company Springer Nature Switzerland AG
The registered company address is: Gewerbestrasse 11, 6330 Cham, Switzerland

Preface

This volume comprises the proceedings of the 13th International Conference on Parallel Processing and Applied Mathematics (PPAM 2019), which was held in Białystok, Poland, September 8–11, 2019. It was organized by the Department of Computer and Information Science of the Częstochowa University of Technology together with Białystok University of Technology, under the patronage of the Committee of Informatics of the Polish Academy of Sciences, in technical cooperation with the IEEE Computer Society and IEEE Computational Intelligence Society. The main organizer was Roman Wyrzykowski.

PPAM is a biennial conference. 12 previous events have been held in different places in Poland since 1994, when the first PPAM took place in Częstochowa. Thus, the event in Białystok was an opportunity to celebrate the 25th anniversary of PPAM. The proceedings of the last nine conferences have been published by Springer in the Lecture Notes in Computer Science series (Nałęczów, 2001, vol. 2328; Częstochowa, 2003, vol. 3019; Poznań, 2005, vol. 3911; Gdańsk, 2007, vol. 4967; Wrocław, 2009, vols. 6067 and 6068; Toruń, 2011, vols. 7203 and 7204; Warsaw, 2013, vols. 8384 and 8385; Kraków, 2015, vols. 9573 and 9574; and Lublin, 2017, vols. 10777 and 10778).

The PPAM conferences have become an international forum for the exchange of the ideas between researchers involved in parallel and distributed computing, including theory and applications, as well as applied and computational mathematics. The focus of PPAM 2019 was on models, algorithms, and software tools which facilitate the efficient and convenient utilization of modern parallel and distributed computing architectures, as well as on large-scale applications, including artificial intelligence and machine learning problems.

This meeting gathered more than 170 participants from 26 countries. A strict refereeing process resulted in the acceptance of 91 contributed papers for publication in these conference proceedings. For regular tracks of the conference, 41 papers were selected from 89 submissions, thus resulting in an acceptance rate of about 46%.

The regular tracks covered important fields of parallel/distributed/cloud computing and applied mathematics such as:

- Numerical algorithms and parallel scientific computing, including parallel matrix factorizations
- Emerging HPC architectures
- GPU computing
- Parallel non-numerical algorithms
- Performance analysis in HPC systems
- Environments and frameworks for parallel/distributed/cloud computing
- Applications of parallel computing
- Soft computing with applications

The invited talks were presented by:

- David A. Bader from the New Jersey Institute of Technology (USA)
- Fabio Baruffa from the Intel Corporation
- Anne Benoit from ENS Lyon (France)
- Jack Dongarra from the University of Tennessee and ORNL (USA)
- Lin Gan from the Tsinghua University and National Supercomputing Center in Wuxi (China)
- Mary Hall from the University of Utah (USA)
- Torsten Hoefler from the ETH Zurich (Switzerland)
- Kate Keahey from the University of Chicago and Argonne National Lab (USA)
- Alexey Lastovetsky from the University College Dublin (Ireland)
- Miron Livny from the University of Wisconsin (USA)
- Satoshi Matsuoka from the Tokyo Institute of Technology (Japan)
- Bernd Mohr from the Jülich Supercomputing Centre (Germany)
- Manish Parashar from the Rutgers University (USA)
- Javier Setoain from ARM (UK)
- Leonel Sousa from the Technical University of Lisbon (Portugal)
- Jon Summers from the University of Lulea (Sweden)
- Manuel Ujaldon from the University of Malaga (Spain) and Nvidia
- Jeffrey Vetter from the Oak Ridge National Laboratory and Georgia Institute of Technology (USA)
- Tobias Weinzierl from the Durham University (UK)

Important and integral parts of the PPAM 2019 conference were the workshops:

- The 8th Workshop on Language-Based Parallel Programming Models (WLPP 2019), organized by Ami Marowka from the Bar-Ilan University (Israel)
- Workshop on Models, Algorithms and Methodologies for Hierarchical Parallelism in New HPC Systems, organized by Giulliano Laccetti and Marco Lapegna from the University of Naples Federico II (Italy), and Raffaele Montella from the University of Naples Parthenope (Italy)
- Workshop on Power and Energy Aspects of Computation (PEAC 2019), organized by Ariel Oleksiak from the Poznan Supercomputing and Networking Center (Poland) and Laurent Lefevre from Inria (France)
- Special Session on Tools for Energy Efficient Computing, organized by Tomas Kozubek and Lubomir Riha from the Technical University of Ostrava (Czech Republic), and Andrea Bartolini from the University of Bologna (Italy)
- Workshop on Scheduling for Parallel Computing (SPC 2019), organized by Maciej Drozdowski from the Poznań University of Technology (Poland)
- The Third Workshop on Applied High-Performance Numerical Algorithms in PDEs, organized by Piotr Krzyżanowski and Leszek Marcinkowski from the Warsaw University (Poland) and Talal Rahman from the Bergen University College (Norway)
- Minisymposium on HPC Applications in Physical Sciences, organized by Grzegorz Kamieniarz and Wojciech Florek from the A. Mickiewicz University in Poznań (Poland)

- Minisymposium on High Performance Computing Interval Methods, organized by Bartłomiej J. Kubica from the Warsaw University of Technology (Poland)
- Workshop on Complex Collective Systems, organized by Paweł Topa and Jarosław Wąs from the AGH University of Science and Technology in Kraków (Poland)

The PPAM 2019 meeting began with two tutorials:

- Modern GPU Computing, by Dominik Göddeke from the Stuttgart University (Germany), Robert Strzodka from the Heidelberg University (Germany), and Manuel Ujaldon from the University of Malaga (Spain) and Nvidia.
- Object Detection with Deep Learning: Performance Optimization of Neural Network Inference using the Intel OpenVINO Toolkit, by Evgenii Vasilev and Iosif Meyerov from the Lobachevsky State University of Nizhni Novgorod (Russia), and Nadezhda Kogteva and Anna Belova from Intel Corporation.

The PPAM Best Paper Award is awarded in recognition of the research paper quality, originality, and significance of the work in high performance computing (HPC). The PPAM Best Paper was first awarded at PPAM 2019 upon recommendation of the PPAM Chairs and Program Committee. For the main track, the PPAM 2019 winners were Evgeny Kuznetsov, Nikolay Kondratyuk, Mikhail Logunov, Vsevolod Nikolskiy, and Vladimir Stegailov from the National Research State University High School of Economics in Moscow and Russian Academy of Sciences (Russia), who submitted the paper "Performance and portability of state-of-art molecular dynamics software on modern GPUs." For workshops, the PPAM 2019 winners were Dominik Ernst, Georg Hager, and Gerhard Wellein from the Erlangen Regional Computing Center and Jonas Thies from the German Aerospace Center (Germany), who presented the paper "Performance Engineering for a Tall & Skinny Matrix Multiplication Kernel on GPUs."

A New Topic at PPAM 2019 was the Special Session on Tools for Energy Efficient Computing, focused on tools designed to improve the energy-efficiency of HPC applications running at scale.

With the steaming out of Moore's law and the end of Dennard's scaling, the pace dictated on the performance increase of HPC systems among generations has led to power constrained architectures and systems. In addition, the power consumption represents a significant cost factor in the overall HPC system economy. For these reasons, it is important to develop new tools and methodologies to measure and optimize the energy consumption of large scale high performance system installation. Due to the link between the energy consumption, power consumption, and execution time of the application executed by the final user, it is important for these tools and methodologies to consider all these aspects empowering the final user and the system administrator with the capability to find the best configuration given different high level objectives.

This special session provided a forum to discuss and present innovative solutions in following topics: (i) tools for fine grained power measurements and monitoring of HPC infrastructures, (ii) tools for hardware and system parameter tuning and its challenges in the HPC environment, (iii) tools and libraries for dynamic tuning of HPC applications at runtime, (iv) tools and methodology for identification of potential dynamic tuning for performance and energy, and (v) evaluation of applications in terms of runtime and energy savings.

These topics were covered by four presentations:

- Overview of application instrumentation for performance analysis and tuning (by O. Vysocky, L. Riha, and A. Bartolini) focused on automatizing the process of an application instrumentation which is an essential step for future optimization that leads to time and energy savings.
- Energy-efficiency tuning of the Lattice Boltzmann simulation using MERIC (by E. Calore, et al.) presents the impact of CPU core and uncore frequency dynamic tuning on energy savings, that reaches up to 24 % for this specific application.
- Evaluating the advantage of reactive MPI-aware power control policies (by D. Cesarini, C. Cavazzoni, and A. Bartolini) shows the COUNTDOWN library, that automatically down-scales the CPU core frequency during long-enough communication phases, with neither any modifications of the code nor complex application profiling.
- Application-aware power capping using Nornir (by D. De Sensi and M. Danelutto) presents how to combine DVFS and thread packing approaches to keep power consumption under a specified limit. This work shows that the proposed solution performs better than the state-of-the-art Intel RAPL power-capping approach for very low power budgets.

The organizers are indebted to PPAM 2019 sponsors, whose support was vital to the success of the conference. The main sponsor was the Intel Corporation and the other sponsors were byteLAKE and Gambit. We thank all the members of the International Program Committee and additional reviewers for their diligent work in refereeing the submitted papers. Finally, we thank to all of the local organizers from the Częstochowa University of Technology and the Białystok University of Technology, who helped us to run the event very smoothly. We are especially indebted to Łukasz Kuczyński, Marcin Woźniak, Tomasz Chmiel, Piotr Dzierżak, Grażyna Kołakowska, Urszula Kroczewska, and Ewa Szymczyk from the Częstochowa University of Technology; and to Marek Krętowski and Krzysztof Jurczuk from the Białystok University of Technology.

We hope that this volume will be of use to you. We would like everyone who reads it to feel welcome to attend the next conference, PPAM 2021, which will be held during September 12–15, 2021, in Gdańsk, the thousand-year old city on the Baltic coast and one of the largest academic centers in Poland.

January 2020   Roman Wyrzykowski
                                                                                      Jack Dongarra
                                                                                       Ewa Deelman
                                                                              Konrad Karczewski

Organization

Program Committee

Jan Węglarz (Honorary Chair)	Poznań University of Technology, Poland
Roman Wyrzykowski (Chair of Program Committee)	Częstochowa University of Technology, Poland
Ewa Deelman (Vice-chair of Program Committee)	University of Southern California, USA
Robert Adamski	Intel Corporation, Poland
Francisco Almeida	Universidad de La Laguna, Spain
Pedro Alonso	Universidad Politecnica de Valencia, Spain
Hartwig Anzt	University of Tennessee, USA
Peter Arbenz	ETH Zurich, Switzerland
Cevdet Aykanat	Bilkent University, Turkey
Marc Baboulin	University of Paris-Sud, France
Michael Bader	TU Munchen, Germany
Piotr Bała	ICM, Warsaw University, Poland
Krzysztof Banaś	AGH University of Science and Technology, Poland
Olivier Beaumont	Inria Bordeaux, France
Włodzimierz Bielecki	West Pomeranian University of Technology, Poland
Paolo Bientinesi	RWTH Aachen, Germany
Radim Blaheta	Institute of Geonics, Czech Academy of Sciences, Czech Republic
Jacek Błażewicz	Poznań University of Technology, Poland
Pascal Bouvry	University of Luxembourg, Luxembourg
Jerzy Brzeziński	Poznań University of Technology, Poland
Marian Bubak	AGH Kraków, Poland, and University of Amsterdam, The Netherlands
Tadeusz Burczyński	Polish Academy of Sciences, Poland
Christopher Carothers	Rensselaer Polytechnic Institute, USA
Jesus Carretero	Universidad Carlos III de Madrid, Spain
Andrea Clematis	IMATI-CNR, Italy
Pawel Czarnul	Gdańsk University of Technology, Poland
Zbigniew Czech	Silesia University of Technology, Poland
Jack Dongarra	University of Tennessee and ORNL, USA
Maciej Drozdowski	Poznań University of Technology, Poland
Mariusz Flasiński	Jagiellonian University, Poland
Tomas Fryza	Brno University of Technology, Czech Republic
Jose Daniel Garcia	Universidad Carlos III de Madrid, Spain

Norbert Meyer	PSNC, Poland
Iosif Meyerov	Lobachevsky State University of Nizhni Novgorod, Russia
Marek Michalewicz	ICM, Warsaw University, Poland
Ricardo Morla	INESC Porto, Portugal
Jarek Nabrzyski	University of Notre Dame, USA
Raymond Namyst	University of Bordeaux and Inria, France
Edoardo Di Napoli	Forschungszentrum Juelich, Germany
Gabriel Oksa	Slovak Academy of Sciences, Slovakia
Tomasz Olas	Częstochowa University of Technology, Poland
Ariel Oleksiak	PSNC, Poland
Marcin Paprzycki	IBS PAN and SWPS, Poland
Dana Petcu	West University of Timisoara, Romania
Jean-Marc Pierson	University Paul Sabatier, France
Loic Pottier	University of Southern California, USA
Radu Prodan	University of Innsbruck, Austria
Enrique S. Quintana-Ortí	Universitat Politecnica de Valencia, Spain
Omer Rana	Cardiff University, UK
Thomas Rauber	University of Bayreuth, Germany
Krzysztof Rojek	Częstochowa University of Technology, Poland
Witold Rudnicki	University of Białystok, Poland
Gudula Rünger	Chemnitz University of Technology, Germany
Leszek Rutkowski	Częstochowa University of Technology, Poland
Emmanuelle Saillard	Inria, France
Robert Schaefer	Institute of Computer Science, AGH, Poland
Olaf Schenk	Universita della Svizzera Italiana, Switzerland
Stanislav Sedukhin	University of Aizu, Japan
Franciszek Seredyński	Cardinal Stefan Wyszyński University in Warsaw, Poland
Happy Sithole	Centre for High Performance Computing, South Africa
Jurij Silc	Jozef Stefan Institute, Slovenia
Karolj Skala	Ruder Boskovic Institute, Croatia
Renata Słota	Institute of Computer Science, AGH, Poland
Masha Sosonkina	Old Dominion University, USA
Leonel Sousa	Technical University of Lisabon, Portugal
Vladimir Stegailov	Joint Institute for High Temperatures of RAS and MIPT/HSE, Russia
Przemysław Stpiczyński	Maria Curie-Skłodowska University, Poland
Reiji Suda	University of Tokio, Japan
Lukasz Szustak	Częstochowa University of Technology, Poland
Boleslaw Szymanski	Rensselaer Polytechnic Institute, USA
Domenico Talia	University of Calabria, Italy
Christian Terboven	RWTH Aachen, Germany
Andrei Tchernykh	CICESE Research Center, Mexico
Parimala Thulasiraman	University of Manitoba, Canada
Sivan Toledo	Tel-Aviv University, Israel

Victor Toporkov	National Research University MPEI, Russia
Roman Trobec	Jozef Stefan Institute, Slovenia
Giuseppe Trunfio	University of Sassari, Italy
Denis Trystram	Grenoble Institute of Technology, France
Marek Tudruj	Polish Academy of Sciences and Polish-Japanese Academy of Information Technology, Poland
Bora Ucar	École Normale Supérieure de Lyon, France
Marian Vajtersic	Salzburg University, Austria
Vladimir Voevodin	Moscow State University, Russia
Bogdan Wiszniewski	Gdańsk University of Technology, Poland
Andrzej Wyszogrodzki	Institute of Meteorology and Water Management, Poland
Ramin Yahyapour	University of Göttingen, GWDG, Germany
Jiangtao Yin	University of Massachusetts Amherst, USA
Krzysztof Zielinski	Institute of Computer Science, AGH, Poland
Julius Žilinskas	Vilnius University, Lithuania
Jarosław Żola	University of Buffalo, USA

Steering Committee

Jack Dongarra	University of Tennessee and ORNL, USA
Leszek Rutkowski	Częstochowa University of Technology, Poland
Boleslaw Szymanski	Rensselaer Polytechnic Institute, USA

Contents – Part II

Workshop on Power and Energy Aspects of Computations (PEAC 2019)

Special Session on Tools for Energy Efficient Computing

Workshop on Scheduling for Parallel Computing (SPC 2019)

Workshop on Applied High Performance Numerical Algorithms for PDEs

Minisymposium on HPC Applications in Physical Sciences

Minisymposium on High Performance Computing Interval Methods

Workshop on Complex Collective Systems

Contents – Part I

Emerging HPC Architectures

Performance Analysis and Scheduling in HPC Systems

Environments and Frameworks for Parallel/Distributed/Cloud Computing

Applications of Parallel Computing

Special Session on GPU Computing

Special Session on Parallel Matrix Factorizations

Workshop on Language-Based Parallel Programming Models (WLPP 2019)

Parallel Fully Vectorized *Marsa-LFIB4*: Algorithmic and Language-Based Optimization of Recursive Computations

Przemysław Stpiczyński[(✉)](iD)

Institute of Computer Science, Maria Curie–Skłodowska University,
Akademicka 9/519, 20-033 Lublin, Poland
przem@hektor.umcs.lublin.pl

Abstract. The aim of this paper is to present a new high-performance implementation of *Marsa-LFIB4* which is an example of high-quality multiple recursive pseudorandom number generators. We propose a new algorithmic approach that combines language-based vectorization techniques together with a new *divide-and-conquer* method that exploits a special sparse structure of the matrix obtained from the recursive formula that defines the generator. We also show how the use of *intrinsics* for Intel AVX2 and AVX512 vector extensions can improve the performance. Our new implementation achieves good performance on several multicore architectures and it is much more energy-efficient than simple SIMD-optimized implementations.

Keywords: Pseudorandom numbers · Recursive generators · Language-based vectorization · Intrinsics · Algorithmic approach · OpenMP

1 Introduction

Pseudorandom numbers are very important and pseudorandom number generators are often central parts of scientific applications such as simulations of physical systems using Monte Carlo methods. There are a lot of such generators with different properties [8]. Recursion-based generators have good statistical properties and they are commonly used [5,9,11,14,15]. *Marsa-LFIB4* [13] is a great example of such recursive generators. It is simple and it passed all empirical tests from TestU01 Library [12] and it was used in practical applications [10]. However, we do not know any of its high performance implementations.

The problem of effective implementation of pseudorandom number generators is very important from a practical point of view [1,3,16,17]. It is clear that an efficient implementation should utilize not only multiple cores of modern processor architectures but also exploit their vector extensions. It is crucial when we expect such implementations to achieve really high performance.

SPRNG Library [14] has been developed using *cycle division* or other parameterizing techniques like *block splitting* or *leapfrogging* [2,4]. Our proposed approach for developing not only parallel but also fully vectorized pseudorandom

© Springer Nature Switzerland AG 2020
R. Wyrzykowski et al. (Eds.): PPAM 2019, LNCS 12044, pp. 3–12, 2020.
https://doi.org/10.1007/978-3-030-43222-5_1

number generators is quite different. Instead of using rather complicated parallelization techniques [4], we rewrite recurrence relations as systems of linear equations and try to optimize them for multicore processors. Then statistical properties of such parallel generators are exactly the same as for corresponding sequential ones, thus there is no need to perform special statistical (and rather expensive) tests. Such systems can be solved using vectorized parallel algorithms with more efficient data layouts. Recently, this *algorithmic approach* has been successfully applied to develop new parallel versions of *Linear Congruential Generator* and *Lagged Fibonacci Generator* [19–21]. Unfortunately, in case of LFG, the number of operations required by the algorithm increases when the lag parameter increases. This is the reason why this approach cannot be directly applied for *Marsa-LFIB4*, where the lag is 256. In order to design a high-performance implementation of the generator, we propose a new algorithmic approach that combines language-based vectorization techniques together with a new *divide-and-conquer* parallel algorithm that can exploit the special sparse structure of the matrix obtained from the recursive formula that defines the generator. We also show how the use of *intrinsics* for Intel AVX2 and AVX512 vector extensions can improve the performance of the new implementation. Numerical experiments show that the new implementation achieves good performance on several multicore architectures. Moreover, it is much more energy-efficient than simple SIMD-optimized implementations.

2 SIMD Optimization of Marsa-LFIB4

A *multiple recursive generator* (MRG) of order k is defined by the linear recurrence of the form $x_i = (a_1 x_{i-1} + \ldots + a_k x_{i-k}) \bmod m$. It produces numbers from $\mathbb{Z}_m = \{0, 1, \ldots, m-1\}$. Usually m is a power of two, thus modulus operations to be computed by merely truncating all but the rightmost 32 bits. When we use "`unsigned int`" C/C++ data type, we can neglect "mod m". A simple example of MRG is *Lagged Fibonacci generator* $x_i = (x_{i-p_1} + x_{i-p_2}) \bmod m$, $0 < p_1 < p_2$. Another important high-quality recursive generator is *Marsa-LFIB4* [13] based on the following recursive equation

$$x_i = (x_{i-p_1} + x_{i-p_2} + x_{i-p_3} + x_{i-p_4}) \bmod 2^{32}, \qquad (1)$$

where $p_1 = 55$, $p_2 = 119$, $p_3 = 179$, and $p_4 = 256$. A simple implementation of (1) requires $3n$ arithmetic operations (additions) to generate a sequence of n numbers.

Figure 1 (left) shows a SIMD-optimized version of the algorithm. It utilizes the OpenMP `simd` directive that asks the compiler to make every possible effort to vectorize the loop [6]. The `safelen` clause indicates the maximum number of iterations per chunk. It is clear that it should be less than p_1, but the best performance can be achieved when the indicated value is a power of two, thus we use 32. It should be noticed that due to obvious data dependencies the loop from lines 6–7 cannot be automatically vectorized by the compiler, even if the highest optimization level is switched on.

```
1    void LFIB4(uint32_t n,uint32_t *x)        __m512i x1,x2;
2    {                                         for(uint32_t k=P4;k<n;k+=16){
3      // x[0..P4] contains a 'seed'            x1 = _mm512_load_si512(&x[k-P4]);
4      __assume_aligned(x,64);                  x2 = _mm512_loadu_si512(&x[k-P3]);
5      #pragma omp simd safelen(32)             x1 = _mm512_add_epi32(x1,x2);
6      for(uint32_t k=P4;k<n;k++)               x2 = _mm512_loadu_si512(&x[k-P2]);
7        x[k]=x[k-P1]+x[k-P2]                   x1 = _mm512_add_epi32(x1,x2);
8                   +x[k-P3]+x[k-P4];           x2 = _mm512_loadu_si512(&x[k-P1]);
9    }                                          x1 = _mm512_add_epi32(x1,x2);
10                                              _mm512_store_si512(&x[k],x1);
11                                            }
```

Fig. 1. Two SIMD-optimized sequential versions of *Marsa-LFIB4* using OpenMP `simd` directive (left) and AVX512 intrinsics (right)

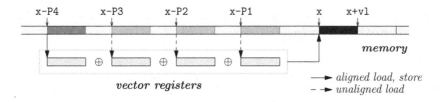

Fig. 2. Vectorization of *Marsa-LFIB4* using SIMD extensions

Figure 1 (right) shows another SIMD-optimized version using *intrinsics* for AVX512 instructions that take full advantage of Intel SIMD extensions. Intrinsics allow programmers to write constructs that look like C/C++ function calls corresponding to actual SIMD instructions. Such calls are replaced with assembly code inlined directly into programs. The general idea is presented in Fig. 2. The output data is produced as a sequence of vectors of length `vl`. Necessary previously computed numbers are loaded into vector registers and added using simple *vector-add* operations. Note that in each iteration we have one load of aligned data, three load operations of unaligned vectors (less efficient) and one store to aligned memory area.

The disadvantage of this solution is the lack of the code portability between different versions of vector extensions. Figure 3 (left) shows the intrinsic-based implementation of *Marsa-LFIB4* for Intel AVX2 256-bit extensions that is quite similar to AVX512 but $vl = 8$. The implementation for Intel KNC 512-bit extensions (Fig. 3, right) is more complicated because this architecture does not support simple load operations of unaligned vectors. Thus, we have to use more complicated technique that uses two special vector load operations.

3   New Algorithmic Approach

Recently, we have developed a new parallel approach that can be used to implement multiple recursive generators [19–21]. Unfortunately, in case of *Lagged Fibonacci generator*, the number of operations required by the algorithm increases when the

```
1   __m256i x1,x2;
2   for(uint32_t k=P4;k<n;k+=8){
3     x1=_mm256_load_si256(
4           (union __m256i*)&x[k-P4]);
5     x2=_mm256_loadu_si256(
6           (union __m256i*)&x[k-P3]);
7     x1=_mm256_add_epi32(x1,x2);
8     x2=_mm256_loadu_si256(
9           (union __m256i*)&x[k-P2]);
10    x1=_mm256_add_epi32(x1,x2);
11    x2=_mm256_loadu_si256(
12          (union __m256i*)&x[k-P1]);
13    x1=_mm256_add_epi32(x1,x2);
14    _mm256_store_si256(
15          (union __m256i*)&x[k],x1);
16  }
17
```

```
1   __m512i x1,x2;
2   for(uint32_t k=P4;k<n;k+=16){
3     x1=_mm512_load_si512(&x[k-P4]);
4     x2=_mm512_loadunpacklo_epi32(x2,&x[k-P3]);
5     x2=_mm512_loadunpackhi_epi32(x2,
6                          &x[k-P3]+16);
7     x1=_mm512_add_epi32(x1,x2);
8     x2=_mm512_loadunpacklo_epi32(x2,&x[k-P2]);
9     x2=_mm512_loadunpackhi_epi32(x2,
10                         &x[k-P2]+16);
11    x1=_mm512_add_epi32(x1,x2);
12    x2=_mm512_loadunpacklo_epi32(x2,&x[k-P1]);
13    x2=_mm512_loadunpackhi_epi32(x2,
14                         &x[k-P1]+16);
15    x1=_mm512_add_epi32(x1,x2);
16    _mm512_store_si512(&x[k],x1);
17  }
```

Fig. 3. Two SIMD-optimized sequential versions of *Marsa-LFIB4* using AVX2 (left) and 512-bit KNC intrinsics (right)

values of p_2 increases. This is the reason why this approach cannot be directly applied for *Marsa-LFIB4*, where $p_4 = 256$. In order to design a high-performance implementation of the generator, we will propose a new approach that will combine techniques presented in Sect. 2 with more efficient *divide-and-conquer* approach for solving linear recurrence systems (see Algorithm 1 in [18]).

Let $n = rs$, $s > 2p_4$. To find a sequence of pseudorandom numbers defined by (1) for a given seed $d_0, \ldots, d_{p_4-1}$, we have to solve the following system of linear equations

$$
\begin{bmatrix}
A_0 & & & \\
B & A & & \\
& \ddots & \ddots & \\
& & B & A
\end{bmatrix}
\begin{bmatrix}
\mathbf{x}_0 \\
\mathbf{x}_1 \\
\vdots \\
\mathbf{x}_{r-1}
\end{bmatrix}
=
\begin{bmatrix}
\mathbf{f} \\
0 \\
\vdots \\
0
\end{bmatrix},
\tag{2}
$$

where $\mathbf{f} = (d_0, \ldots, d_{p_4-1}, 0, \ldots, 0)^T \in \mathbb{Z}_m^s$, $\mathbf{x}_i = (x_{is}, \ldots, x_{(i+1)s-1})^T \in \mathbb{Z}_m^s$, and the matrices $A, A_0, B \in \mathbb{Z}_m^{s \times s}$ are as shown in Fig. 4. The block system of linear equations (2) can be rewritten as follows

$$
\begin{cases}
A_0 \mathbf{x}_0 = \mathbf{f} \\
B\mathbf{x}_{i-1} + A\mathbf{x}_i = 0, \quad i = 1, \ldots, r-1.
\end{cases}
\tag{3}
$$

Let $\mathbf{e}_k$ denotes the k-th unit vector from $\mathbb{Z}_m^s$, i.e. $\mathbf{e}_k = (0, \ldots, 0, 1, 0, \ldots, 0)^T$. It can be observed that non-zero columns of B satisfy

$$
-B_i = \begin{cases}
\mathbf{e}_{i-s-p_4}, & i = s-p_4, \ldots, s-1-p_3, \\
\mathbf{e}_{i-s-p_3} + \mathbf{e}_{i-s-p_4}, & i = s-p_3, \ldots, s-1-p_2, \\
\mathbf{e}_{i-s-p_2} + \mathbf{e}_{i-s-p_3} + \mathbf{e}_{i-s-p_4}, & i = s-p_2, \ldots, s-1-p_1, \\
\mathbf{e}_{i-s-p_1} + \mathbf{e}_{i-s-p_2} + \mathbf{e}_{i-s-p_3} + \mathbf{e}_{i-s-p_4}, & i = s-p_1, \ldots, s-1.
\end{cases}
$$

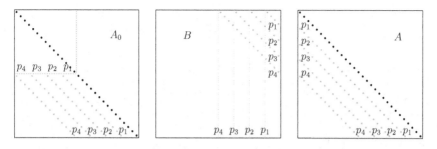

Fig. 4. Shapes of A_0, B, A. Black dots: 1, grey dots: -1, otherwise: 0

From (3) we have

$$\mathbf{x}_i = -A^{-1}B\mathbf{x}_{i-1} = \sum_{k=0}^{p_4-1} x_{is-k-1}A^{-1}(-B_{s-1-k}).$$

It is clear that $A^{-1}(\mathbf{e}_i + \mathbf{e}_j) = A^{-1}\mathbf{e}_i + A^{-1}\mathbf{e}_j$, thus

$$\mathbf{a}_0 = A^{-1}(-B_{s-p_4}) \tag{4}$$
$$\mathbf{b}_0 = A^{-1}(-B_{s-p_3}) = \mathbf{a}_0 + \mathbf{a}_{p_4-p_3} \tag{5}$$
$$\mathbf{c}_0 = A^{-1}(-B_{s-p_2}) = \mathbf{a}_0 + \mathbf{a}_{p_4-p_2} + \mathbf{a}_{p_3-p_2} \tag{6}$$
$$\mathbf{d}_0 = A^{-1}(-B_{s-p_1}) = \mathbf{a}_0 + \mathbf{a}_{p_4-p_1} + \mathbf{a}_{p_3-p_1} + \mathbf{a}_{p_2-p_1} \tag{7}$$

Moreover, all vectors $\mathbf{a}_k = A^{-1}(-B_{s-p_4+k})$, $k = 1, \ldots, p_4 - p_3 - 1$,

$$\mathbf{a}_k = (\underbrace{0, \ldots, 0}_{k}, a_0, \ldots, a_{s-1-k})^T, \tag{8}$$

where $\mathbf{a}_0 = (a_0, \ldots, a_{s-1})^T$. Similarly, $\mathbf{b}_k = A^{-1}(-B_{s-p_3+k})$, $k = 1, \ldots, p_3 - p_2 - 1$, $\mathbf{c}_k = A^{-1}(-B_{s-p_4+k})$, $k = 1, \ldots, p_2 - p_1 - 1$, $\mathbf{d}_k = A^{-1}(-B_{s-p_4+k})$, $k = 1, \ldots, p_1 - 1$, can be easily derived from (5–7) using simple shift operations (8). Finally, we get the following:

$$\mathbf{x}_i = \sum_{k=0}^{n_4-1} x_{is-p_4+k}\mathbf{a}_k + \sum_{k=0}^{n_3-1} x_{is-p_3+k}\mathbf{b}_k + \sum_{k=0}^{n_2-1} x_{is-p_2+k}\mathbf{c}_k + \sum_{k=0}^{n_1-1} x_{is-p_1+k}\mathbf{d}_k, \tag{9}$$

where $n_4 = p_4 - p_3$, $n_3 = p_3 - p_2$, $n_2 = p_2 - p_1$, $n_1 = p_1$. Note that $n_1 + n_2 + n_3 + n_4 = p_4$, thus each $\mathbf{x}_i$ is the sum of p_4 vectors.

The direct application of (9) allows to develop a new fully vectorized parallel algorithm. Unfortunately, the required number of operations is really huge

$$N_1(s, r) = 2p_4(s - p_4)(r - 1) = 512n - 2p_4^2 r - 2p_4 s + 2p_4^2.$$

In order to propose a more efficient method, let us observe that to find each vector $\mathbf{x}_i$, $i = 1, \ldots, r-1$, we need p_4 last entries of $\mathbf{x}_{i-1}$. Thus, we can apply (9) to find

---
Algorithm 1: Parallel Vectorized *Marsa-LFIB4*

Data: n, $x_0, \ldots, x_{p_4-1}$ – seed

Result: $x_{p_4}, \ldots, x_{n-1}$ – generated numbers

1 $r \leftarrow \#cores$ $\triangleright$ initially r should equal to the number of cores

2 **if** $s\%256 \neq 0$ **then**

3 $\quad\big|\quad s \leftarrow (s/256 + 1) * 256$ $\triangleright$ s should be a multiple of 256

4 $\quad\big|\quad r \leftarrow n/s$

5 **end**

   $\quad\triangleright$ now each $\mathbf{x}_i$ is aligned

6 apply (1) to find $\mathbf{x}_0$ and $\mathbf{a}_0$ $\triangleright$ using SIMD-optimized method

7 apply (5–7) to find $2p_4$ last entries of $\mathbf{b}_0, \mathbf{c}_0, \mathbf{d}_0$ $\triangleright$ using OpenMP **simd**

8 **for** $i = 1, \ldots, r - 1$ **do**

9 $\quad\big|\quad$ apply (9) to find p_4 last entries of $\mathbf{x}_i$ $\triangleright$ using OpenMP **simd**

10 **end**

11 **parallel for** $i = 1, \ldots, r - 1$ **do**

12 $\quad\big|\quad$ apply (1) to find $s - p_4$ first entries of $\mathbf{x}_i$ $\triangleright$ using SIMD-optimized method

13 **end**

14 apply (1) to find $x_{rs}, \ldots, x_{n-1}$

---

p_4 last entries of each $\mathbf{x}_i$ and then find in parallel $s - p_4$ first entries of each vector using the SIMD-optimized implementation from Sect. 2 (see Algorithm 1 for details). It can be easily verified that the total number of arithmetic operations required by the new algorithm is only

$$N_2(s,r) = 12p_4 r + 2p_4^2(r-1) + 3(s - p_4)(r+1)$$
$$= 3n + 3s + (9p_4 + 2p_4^2)r - 2p_4^2 - 3p_4. \tag{10}$$

The question is how to choose the values of the parameters r and s. It is clear that the total number of operations grows when the value of r grows. However, then the potential parallelism of the algorithm also grows. Therefore, the number of available cores can be used as the actual value of r. Then the value of s is calculated to ensure that each vector $\mathbf{x}_i$ is properly aligned (lines 2–5).

4   Results of Experiments

All experiments were carried out on two platforms. The first one was a server with two Intel Xeon E5-2670 v3 processors (totally 24 cores with hyperthreading, 2.3 GHz, 256-bit AVX2), 128GB RAM, with Intel Xeon Phi Coprocessor 7120P (KNC, 61 cores with multithreading, 1.238 GHz, 16GB RAM, 512-bit vector extensions). The next one was a server with Intel Xeon Phi 7210F (KNL, 64 cores, 1.3 GHz, AVX512), 128GB RAM. Both servers worked under Linux with Intel Parallel Studio version 2017. Experiments on Xeon Phi have been carried out using its native mode. We tested **Simple** (non optimized) implementation, two SIMD-optimized (non parallel) implementations of *Marsa-LFIB4* using the

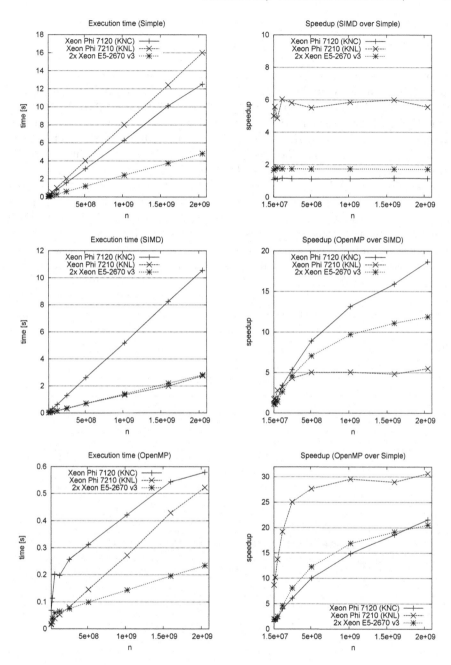

Fig. 5. Performance of the implementations of *Marsa-LFIB4*: execution time (left) and speedup (right)

OpenMP `simd` construct and intrinsics, respectively, and the parallel algorithm (i.e. `OpenMP`). Examples of the results are presented in Fig. 5.

As expected, the simple (non optimized) implementation achieved the worst performance. The intrinsic-based implementation is faster than the implementation based on the OpenMP `simd` construct on both Intel MIC architectures (up to 5% on KNL and up to 6% on KNC). In case of Xeon E5-2670 both implementations achieve the same performance. Thus, our parallel implementation (`OpenMP`) uses intrinsics. It should be noticed that on all platforms the best performance is achieved when only one thread per core is used. The use of SIMD extensions improves the performance of *Marsa-LFIB4* $5-6\times$ on KNL and about $1.8\times$ on CPU with AVX2. In case of KNL, the SIMD-optimized implementation is only 18% faster than `Simple`. The use of multiple cores results in a significant increase in performance. On Xeon E5-2670 the highest speedup relative to `SIMD` is up to 12, thus the efficiency is about 0.5. On KNC and KNL the efficiency of the use of multiple cores is worse, especially in case of KNL. However, on KNL we can observe the best speedup relative to `Simple` (about 31). In case of Xeon E5-2670 such speedup is about 21. It means that on this platform the efficiency of our parallel implementation relative to `Simple` is up to 88%. The low efficiency of using multiple cores on KNC and KNL is probably due to time overheads associated with the *fork-join* operation and the synchronization of multiple threads.

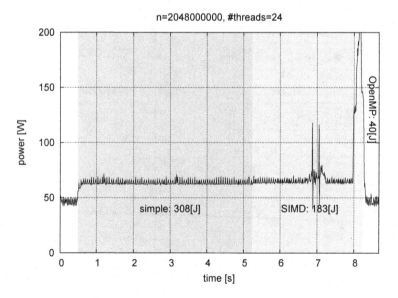

Fig. 6. Current power draw and total power consumption required by all considered implementations (`Simple`, `SIMD` and `OpenMP`)

Figure 6 presents the exemplary results of our experiments concerning the energy efficiency of the considered implementations. Data for this plot were collected on the server with two Intel Xeon E5-2670 v3 processors using Intel's Running Average Power Limit (RAPL) [7]. This interface enables to measure the power consumption for CPUs and DRAMs. This figure shows how the power consumption changes during the execution of the program which comprises calls to `Simple`, `SIMD` and `OpenMP+SIMD` implementations, respectively. Time ticks from 0 to 0.5 show the power consumption of CPUs and DRAMs when the system was idle. Then we can observe the power consumption of all three implementations. It is clear that current power draw during the execution of `OpenMP+SIMD` is much higher but it only lasts during a very short time. `Simple` needs $308J$ of energy, `SIMD` $183J$, while `OpenMP+SIMD` requires only $40J$. Thus, its power consumption is only 22% and 13% of `SIMD` and `Simple`, respectively.

5   Conclusions

We have shown that *Marsa-LFIB4* which is a fine example of linear recurrence computations can be efficiently implemented on modern multiprocessors with vector extensions using language-based tools together with the algorithmic approach. Using intrinsics instead of the simple `simd` construct increases performance slightly but also limits code portability. Our parallel SIMD-optimized implementation achieves good performance and it is much more energy efficient.

Acknowledgements. The use of computer resources installed at Maria Curie-Skłodowska University in Lublin and Czestochowa University of Technology is kindly acknowledged.

References

1. Aluru, S.: Lagged Fibonacci random number generators for distributed memory parallel computers. J. Parallel Distrib. Comput. **45**(1), 1–12 (1997). https://doi.org/10.1006/jpdc.1997.1363
2. Bauke, H., Mertens, S.: Random numbers for large-scale distributed Monte Carlo simulations. Phys. Rev. E **75**, 066701 (2007). https://doi.org/10.1103/PhysRevE.75.066701
3. Bisseling, R.H.: Parallel Scientific Computation. A structured Approach Using BSP and MPI. Oxford University Press, Oxford (2004)
4. Bradley, T., du Toit, J., Tong, R., Giles, M., Woodhams, P.: Parallelization techniques for random numbers generators. In: GPU Computing Gems, pp. 231–246. Gems Emerald Edition (2011)
5. Brent, R.P.: Uniform random number generators for supercomputers. In: Proceedings of the Fifth Australian Supercomputer Conference, pp. 95–104 (1992)
6. Jeffers, J., Reinders, J., Sodani, A.: Intel Xeon Phi Processor High-Performance Programming. Knights Landing Edition. Morgan Kaufman, Cambridge (2016)
7. Khan, K.N., Hirki, M., Niemi, T., Nurminen, J.K., Ou, Z.: RAPL in action experiences in using RAPL for power measurements. ACM Trans. Model. Perform. Eval. Comput. Syst. **3**(2), 9:1–9:26 (2018). https://doi.org/10.1145/3177754
8. Knuth, D.E.: The Art of Computer Programming, Volume II: Seminumerical Algorithms, 2nd edn. Addison-Wesley, Boston (1981)

9. Knuth, D.E.: MMIXware. LNCS, vol. 1750. Springer, Heidelberg (1999). https://doi.org/10.1007/3-540-46611-8
10. Łapa, K., Cpałka, K., Przybył, A., Grzanek, K.: Negative space-based population initialization algorithm (NSPIA). In: Rutkowski, L., Scherer, R., Korytkowski, M., Pedrycz, W., Tadeusiewicz, R., Zurada, J.M. (eds.) ICAISC 2018, Part I. LNCS (LNAI), vol. 10841, pp. 449–461. Springer, Cham (2018). https://doi.org/10.1007/978-3-319-91253-0_42
11. L'Ecuyer, P.: Good parameters and implementations for combined multiple recursive random number generators. Oper. Res. **47**(1), 159–164 (1999). https://doi.org/10.1287/opre.47.1.159
12. L'Ecuyer, P., Simard, R.J.: TestU01: AC library for empirical testing of random number generators. ACM Trans. Math. Softw. **33**(4), 22:1–22:40 (2007). https://doi.org/10.1145/1268776.1268777
13. Marsaglia, G.: Random numbers for C: The END? Posted to the electronic billboard sci.crypt.random-numbers (1999)
14. Mascagni, M., Srinivasan, A.: Algorithm 806: SPRNG: a scalable library for pseudorandom number generation. ACM Trans. Math. Softw. **26**(3), 436–461 (2000). https://doi.org/10.1145/358407.358427
15. Mascagni, M., Srinivasan, A.: Parameterizing parallel multiplicative lagged-Fibonacci generators. Parallel Comput. **30**(5–6), 899–916 (2004). https://doi.org/10.1016/j.parco.2004.06.001
16. Ökten, G., Willyard, M.: Parameterization based on randomized quasi-Monte Carlo methods. Parallel Comput. **36**(7), 415–422 (2010). https://doi.org/10.1016/j.parco.2010.03.003
17. Percus, O.E., Kalos, M.H.: Random number generators for MIMD parallel processors. J. Parallel Distrib. Comput. **6**(3), 477–497 (1989). https://doi.org/10.1016/0743-7315(89)90002-6
18. Stpiczyński, P.: Parallel algorithms for solving linear recurrence systems. In: Bougé, L., Cosnard, M., Robert, Y., Trystram, D. (eds.) CONPAR/VAPP -1992. LNCS, vol. 634, pp. 343–348. Springer, Heidelberg (1992). https://doi.org/10.1007/3-540-55895-0_428
19. Stpiczyński, P.: Vectorized algorithm for multidimensional Monte Carlo integration on modern GPU, CPU and MIC architectures. J. Supercomput. **74**(2), 936–952 (2018). https://doi.org/10.1007/s11227-017-2172-x
20. Stpiczyński, P., Szałkowski, D., Potiopa, J.: Parallel GPU-accelerated recursion-based generators of pseudorandom numbers. In: Proceedings of the Federated Conference on Computer Science and Information Systems, September 9–12, 2012, Wroclaw, Poland, pp. 571–578. IEEE Computer Society Press (2012). http://fedcsis.org/proceedings/2012/pliks/380.pdf
21. Szałkowski, D., Stpiczyński, P.: Using distributed memory parallel computers and GPU clusters for multidimensional Monte Carlo integration. Concurr. Comput. Pract. Exp. **27**(4), 923–936 (2015). https://doi.org/10.1002/cpe.3365

Studying the Performance
of Vector-Based Quicksort Algorithm

Ami Marowka$^{(\boxtimes)}$ (iD)

Parallel Research Lab, Jerusalem, Israel
amimar2@yahoo.com

Abstract. The performance of parallel algorithms is often inconsistent with their preliminary theoretical analyses. Indeed, the difference is increasing between the ability to theoretically predict the performance of a parallel algorithm and the results measured in practice. This is mainly due to the accelerated development of advanced parallel architectures, whereas there is still no agreed model for parallel computation, which has implications for the design of parallel algorithms.

In this study, we examined the practical performance of Cormen's Quicksort parallel algorithm. We determined the performance of the algorithm with different parallel programming approaches and examine the capacity of theoretical performance analyses of the algorithm for predicting the actual performance.

Keywords: Python · Quicksort · Performance modeling

1 Introduction

The runtime of a parallel algorithm is affected by many parallel overheads and most are not considered by the performance models employed, thereby resulting in a difference between the predicted performance and actual performance. These significant parallel overheads include barrier synchronization, spawning and destroying processes and threads, cache memory effects (e.g., cache misses and false sharing), TLB effects, memory bandwidth and latency issues, cache replacement policies and algorithms, the non-deterministic behavior of parallel algorithm, multi-level cache allocation policies (exclusive versus inclusive), and load unbalancing [1].

Cormen recently presented a vector-based parallel Quicksort algorithm for shared-memory multi-core processors [2] based on Blelloch's study of vector models [3,4]. Cormen described the design and analyzed the theoretic performance of the algorithm in detail, but did not implement the proposed algorithm on an actual multi-core machine or present actual performance analyses for any of the algorithm's functions. Therefore, the effectiveness of the algorithm was not evaluated in a current multi-core processor environment. The functions of the algorithm were presented as Python pseudo-code. Cormen claimed that:

"The Python code we use does not itself run in parallel. In order to run code in parallel with Python, we would have to lock into a particular Python library.

© Springer Nature Switzerland AG 2020
R. Wyrzykowski et al. (Eds.): PPAM 2019, LNCS 12044, pp. 13–24, 2020.
https://doi.org/10.1007/978-3-030-43222-5_2

The concepts behind parallel programming are more important than the exact means to achieve parallelism." We argue that this claim is not realistic with respect to parallel computing and programming. Indeed, the design of Cormen's parallel Quicksort algorithm and the analysis of its computational complexity are theoretically impressive, but we found that its efficiency cannot be demonstrated in actual tests. We argue that a parallel algorithm designed on the basis of sophisticated programming principles with inspiring theoretical computational complexity has no value when it cannot be efficiently implemented in practice in order to achieve linear performance and scalability improvements as the number of cores increases. The main contributions of this study are as follows.

- In the present study, we analyzed the performance of Cormen's parallel Quicksort algorithm based on implementations using Python shared-memory multiprocessing and multi-threading approaches. *It is important to emphasize that we implemented Cormen's algorithm in Python because the algorithm was originally presented in Python.*
- We compared the performance of Cormen's algorithm based on implementations of recursive parallel Quicksort using Python multi-processing and multi-threading approaches with Python built-in Quicksort implementations. We aimed *to show that the method used for achieving parallelism is not less important than the concepts behind parallel programming.*
- We compared the capacity for the theoretical prediction of an algorithm's performance with predictions based on a combination of theoretical and practical analyses.

The remainder of this paper is organized as follows. In Sect. 2, we describe the building blocks of Cormen's parallel Quicksort algorithm based on an example, and we demonstrate the computational complexity of the algorithm. In Sect. 3, we explain the implementation of the algorithm for different parallel programming models and assessments of their performance are presented. In Sect. 4, we examine the capacity of theoretical performance analyses of the algorithm for predicting the actual performance. In Sect. 5, we summarize the findings obtained in this study and give our conclusions.

2   Vector-Based Quicksort Algorithm

In the following, we briefly review the key functions of Cormen's algorithm and their associated data structures. Highly technical descriptions of the operating methods for the algorithm are presented but we do not provide exhaustive explanations of why these methods are applied. These descriptions provide the reader with a general idea of the design of the algorithm. A full description of the algorithm was given by Cormen [2]. The algorithm uses two main functions: *Reduction* and *Scan* . The Scan function is also known as *Prefix Sums.* The input for the Reduction function is an array of elements and its output is a scalar, where the value is the sum of the array elements, e.g., Reduction

Table 1. Demo of the main processing steps of Cormen's parallel Quicksort algorithm.

Steps	First iteration										
	index	0	1	2	3	4	5	6	7	8	9
	input	5	2	8	4	7	2	9	2	8	0
	seg	1	0	0	0	0	0	0	0	0	0
1	pivot	5	5	5	5	5	5	5	5	5	5
2	less	0	1	0	1	0	1	0	1	0	1
3	equal	1	0	0	0	0	0	0	0	0	0
4	greater	0	0	1	0	1	0	1	0	1	0
5	less scan	0	0	1	1	2	2	3	3	4	4
6	equal scan	0	1	1	1	1	1	1	1	1	1
7	greater scan	0	0	0	1	1	2	2	3	3	4
8	less reduce	5	5	5	5	5	5	5	5	5	5
9	equal reduce	1	1	1	1	1	1	1	1	1	1
10	index scan	0	0	0	0	0	0	0	0	0	0
11	less perm	0	0	1	1	2	2	3	3	4	4
12	equal perm	5	6	6	6	6	6	6	6	6	6
13	greater perm	6	6	6	7	7	8	8	9	9	10
14	perm	5	0	6	1	7	2	8	3	9	4
	partition	2	4	2	2	0	5	8	7	9	8
15	equal start	5	5	5	5	5	5	5	5	5	5
16	greater start	6	6	6	6	6	6	6	6	6	6
17	new seg	1	0	0	0	0	1	1	0	0	0
	Second partition										
	input	2	4	2	2	0	5	8	7	9	8
	seg	1	0	0	0	0	1	1	0	0	0
	pivot	2	2	2	2	2	5	8	8	8	8
	less	0	0	0	0	1	0	0	1	0	0
	equal	1	0	1	1	0	1	1	0	0	1
	greater	0	1	0	0	0	0	0	0	1	0
	less scan	0	0	0	0	0	0	0	0	1	1
	equal scan	0	1	1	2	3	0	0	1	1	1
	greater scan	0	0	1	1	1	0	0	0	0	1
	less reduce	1	1	1	1	1	0	1	1	1	1
	equal reduce	3	3	3	3	3	1	2	2	2	2
	index scan	0	0	0	0	0	5	6	6	6	6
	less perm	0	0	0	0	0	5	6	6	7	7
	equal perm	1	2	2	3	4	5	7	8	8	8
	greater perm	4	4	5	5	5	6	9	9	9	10
	perm	1	4	2	3	0	5	7	6	9	8
	partition	0	2	2	2	4	5	7	8	8	9

Table 2. The sequence of steps of a single iteration of Cormen's Algorithm and their relative average performance cost for Multi-Threading (MT) and Multi-Processing (MP) implementations for n = 100 K and p = 4.

	The *input* array is the array to be sorted. The *index* array marks the indexes of the input's elements. The *seg* array marks the beginning of each segment. The algorithm starts with one segment.	MT	MP
steps			
1	The first element of each segment of the input array is selected as the pivot of the segment. The pivot array holds these values.	40%	15%
2-4	The *less, equal* and *greater* arrays mark whether the corresponding element is less, equal or greater than the pivot respectively.	10%	3%
5-7	The *less scan, equal scan* and *greater scan* arrays hold the results of scan operations on the less, equal and greater arrays respectively.	11%	27%
8-9	The *less reduce* and *equal reduce* arrays hold the results of reduction operations on the less and equal arrays respectively.	20%	40%
10	The *index scan* holds for each segment the index value of its first element.	4%	9%
11-13	The *less perm, equal perm* and *greater perm* arrays are computed as follows: $less\_perm[i] = index\_scan[i] + less\_scan[i]$ $equal\_perm[i] = index\_scan[i] + less\_reduce[i] + equal\_scan[i]$ $greater\_perm[i] = index\_scan[i] + less\_reduce[i] + equal\_reduce[i] + greater\_scan[i]$	1%	1%
14	The *less perm, equal perm* and *greater perm* arrays are used to create the perm array. We do not show here how this is done. The perm array holds the corresponding indexes for the permutation operation on the input array. The partition array is the result of this permutation operation.	10%	4%
15-16	The *equal start* and *greater start* arrays are computed as follows: $equal\_start[i] = index\_scan[i] + less\_reduce[i]$ $greater\_start[i] = less\_reduce[i] + equal\_reduce[i]$	4%	1%
17	The *new seg* array is computed using the *equal start* and *greater start* arrays as follows: *if* $index[i] == index\_scan[i]$ *or* $index[i] == equal\_start[i]$ *or* $index[i] == greater\_start[i]$ *than* $new\ seg[i] = 1$ *else* $new\ seg[i] = 0$		

([1, 2, 3, 4]) = 10. The input for the Scan function is an array of elements and its output is an array of elements, where the values are computed as follows:

$$\text{output } [i] = \text{input } [0] + \text{input } [1] + ... + \text{input } [i\text{-}1],$$

e.g., Scan ([1, 2, 3, 4]) = [0, 1, 3, 6]. The algorithm is a vector-based and iterative algorithm. In each stage, the input array is divided into smaller partitions according to the value of the selected pivots. Cormen refers to these partitions as *Segments,* and the *Segmented Scan* and *Segmented Reduction* functions perform the reduction and scan operations on the different segments simultaneously. The algorithm is constructed in a modular manner, which allows most of the auxiliary functions in the algorithm to be performed using these two functions. Furthermore, the Segmented Reduction function is implemented by the Segmented Scan function with the help of a nice trick. It is important to note that the algorithm uses data parallelism, i.e., parallel processing is performed on the different arrays by dividing them equally between the different threads or processes. Moreover, each element of the input array is processed independently and without any dependency on its neighbor elements. This feature provides the algorithm with high scalability.

In order to illustrate the behavior of the algorithm, we use the example shown in Table 1, where the input comprises 10 non-sorted items. In this case, two iterations of the algorithm are sufficient to sort the input array. Table 2 shows the sequence of steps for a single iteration.

The *partition* and *new seg* arrays in the first iteration are the *input* and the *seg* arrays in the second iteration. The algorithm is ended after the partition array has been sorted in the second iteration. The two columns on the right in Table 2 show the average runtimes for each step in Cormen's algorithm with the Multi-threading (MT) and Multi-processing (MP) implementations, and for n = 100K and p = 4. We discuss these results in the next section. It is important to note that the full algorithm uses functions and other data structures that are not described in the present study.

Next, we analyze the computational complexity of the algorithm in a simplified manner. A more detailed analysis was given by Cormen [2]. As shown in Tables 1 and 2, the dominant functions during one partition step are the Segmented Scan and Segmented Reduction functions. These functions use simple recursive Reduction and Scan functions. Again, step 1 and steps 5–10 in Table 2 are implemented using the Segmented Scan function. Let us analyze the parallel runtime of the Reduction function. The parallel runtime of an operation usually comprises two factors: computational time + synchronization time. We assume that the Reduction operation is applied to an array of size n while using p processes. We also assume that $n \gg p$, which is what usually occurs in reality, such as in our performance benchmarks where $n \geq 100000$ and $p \leq 4$.

In the Reduction function, the input array is evenly divided between processes. In the first stage, each process performs a local sequential reduction operation on n/p elements. At the end of this stage, barrier synchronization is invoked in order to wait for all of the processes to complete their operations. This synchronization operation requires time β. The output from the first stage is an array of size p containing the local results from each process, which we refer to as *Reductions*.

Next, a reduction operation is performed on the elements of the Reductions array, where the Reduction function invokes the Simple Reduction function to perform the reduction operation on Reductions in parallel and recursively. Therefore, $log2(p))$ recursive steps are required and every recursion step ends with barrier synchronization. Hence, the parallel runtime for the Reduction operation is: $n/p + \beta + \beta log2(p)$. It can be shown that the parallel runtime of a scan operation is similar. The number of reduction and scan operations is constant, so it can be estimated that the parallel runtime of a partition step is: $O(n/p + \beta log2(p))$. However, is the first term dominated by the second term? We measured the actual values of these terms in our test environment for $n = 100000$ and $p = 4$, and found that $n/p = 0.01534$ s and $\beta log2(p) = 0.12693$ s in the case of the MP approach, and $n/p = 0.024177$ s and $\beta log2(p) = 0.01560$ s in the case of the MT approach. None of the terms differed significant compared with the others, so none can be dropped.

How many partition steps are required to accomplish the whole Quicksort algorithm? If we assume that in the average case, each partition step divides each segment into sub-segments of equal sizes, then $log2(n)$ recursive partition steps are required to complete the whole Quicksort algorithm. Therefore, the total cost of the parallel runtime for the algorithm is: $O(log2(n)(n/p + \beta log2(p)))$.

3   Implementations and Results

In the following, we describe our test environment, the methods used in this study, the implementation of the algorithms, and an analysis of the performance results.

3.1   Anaconda-Numba Python Environment

We used Numba [5] Python programming environment for developing and testing the algorithms. Numba offers a comprehensive, user-friendly solution for portable high performance computing. Numba is a NumPy-aware Python programming model and a Just-in-Time compiler based on source-code annotations. Numba uses the LLVM compiler for generating optimized machine code similar in performance to C. It was designed in mind for array-oriented and numerical code that supports CPUs, CUDA GPUs, and HSA APUs. Numba is in active development (the current version is 0.30.1). Numba is part of Anaconda Accelerate [9], which is available under a free license for academic users. It runs on top of Anaconda Python [10], which is a completely free package and environment manager for large scale data processing and scientific computing. It includes hundreds of open source packages to include the popular packages of NumPy [6], SciPy [7], Matplotlib [8], IPython, and Spider IDE. Our benchmarking environment includes Windows 10; Python 2.7; Anaconda version 4.2.13; Numba 0.30.1; Spyder 3.0; NumPy 1.11.1; and Intel Core-i7 3.4 GHz quad-core 2-hyper-threads processor.

3.2   Experimental Methods

All of the algorithms were developed in the same environment. Each algorithm was tuned to obtain the maximum performance. Side effects such as just-in-time compilations and cache effects were eliminated and amortized based on pre-runs where the time was not considered, where we ran the algorithms at least 10 times and calculated their average runtimes. *In order to avoid confounding the comparisons of the runtime results obtained with different models, numbers of processes, and input sizes, all of the tests were performed using the same pool of random inputs.*

Table 3 shows the runtime measurements for the sequential Quicksort algorithm, Cormen's vector-based parallel Quicksort algorithm, and the recursive parallel Quicksort algorithm. The Python parallel algorithms were implemented using two parallel programming models: MT and MP. The times shown in Table 3 are in seconds.

Table 3. Runtime results of sequential, recursive parallel and Cormen's Quicksort algorithms. Times shown are in seconds

Sequential quicksort							
	Multi-processing				Multi-threading		
100K	0.2988				0.0094		
1M	2.8700				0.0562		
4M	**10.900**				**0.1985**		
Parallel quicksort							
	Multi-processing				Multi-threading		
Number of threads	1	2	4	Input size	1	2	4	Input size
Cormen's	840	900	1196	100K	2.14	2.75	5.64	100K
Parallel	4578	4497	5180	1M	16.62	12.67	18.87	1M
Quicksort	17920	16128	**21590**	4M	55.61	46.73	**45.67**	4M
Speedup (4M)		1.11	0.83			1.19	1.21	
Recursive	0.54	0.39	0.54	100K	0.0250	0.0122	0.0123	100K
Parallel	12.72	7.08	5.79	1M	0.1843	0.1014	0.0721	1M
Quicksort	153.13	87.77	**72.26**	4M	0.1970	0.1317	**0.1187**	4M
Speedup (4M)		1.74	2.11			1.49	1.65	

The results in Tables 2 and 3 indicate the following.

- The Python MT algorithms performed better than the Python MP algorithms by one to three orders of magnitude. *These results clearly demonstratde that the choice of the method utilized to achieve parallelism was dramatically affected the performance of the algorithm.*
- The Cormen's parallel Quicksort algorithms had very low scalability.

- The Python recursive parallel Quicksort algorithms based on the MT and MP approaches exhibited low scalability as the input size increased. However, the performance achieved by the MT approach with an input size of 4M elements was better than that with the sequential approach. For example, the runtime of the sequential MT algorithm with an input size of 4M was 0.1985 s and the runtime of the parallel MT algorithm was 0.1187 s, which indicated a speed up of 1.67.
- The last two columns in Table 2 show the percentage of the total runtime for each step in Cormen's algorithm with the Python MT and MP implementations. These results showed that the performance costs of the different steps in the two implementations were not identical because they employed two different parallelism approaches. For example, the first step in the MT approach required 40% of the runtime, whereas it only required 15% in the MP approach.
- The results in Table 2 illustrated another phenomenon that was important for understanding the performance of the algorithm. The first step in the algorithm consumed 40% and 15% of the runtime with the MT and MP approaches, respectively. However, why does the simple operation involving filling all of the segments with their respective pivot values takes so much time? The answer is that this filling operation is implemented with the Segmented Scan function, which is not effective for this purpose. Thus, why is this function used? The main features of the design of the algorithm were explained in Sect. 2, which shows that the idea is to construct an algorithm based on two functions: Segmented Scan and Segmented Reduce. Theoretically, this approach is impressive but it is not efficient in practice, as clearly demonstrated by the results.

The observations presented above demonstrated that an apparently well-designed parallel algorithm in theory can obtain very disappointing performance when tested in practice. Moreover, the runtime results illustrated the differences in performance between various parallel programming models.

Is it possible to analyze the performance of a parallel algorithm by using a performance model to determine its actual performance? In the next section, we present the theoretical performance analysis of the algorithm proposed by Cormen and we examine its predictive ability in practice.

4   Performance Prediction

In order to assess the degree to which Cormen's parallel Quicksort algorithm performs better than the sequential Quicksort algorithm, Cormen theoretically analyzed the computational complexities of the *Scan* and *Reduction* functions. These theoretical analysis results were not supported by actual experiments. Next, we briefly review Cormen's theoretical analysis of the *Reduction* function and its ability to predict the performance in practice.

Cormen started his analysis by determining when the parallel execution time T_p will be less than the sequential execution time T_s:

$$T_p < T_s \tag{1}$$

This analysis considered the following two cases.

1. When $n \leq p$ and the reduce operation is performed by the *Simple Reduction* function.
2. When $n > p$ and the reduce operation is performed by the *Reduction* function.

The relationship between *Simple Reduction* and *Reduction* functions is explained in Sect. 2. Let us start with the case where $n \leq p$. In order to maintain the inequality $n \leq p$, the following inequality must be satisfied:

$$T_p = (t + h)log2(n) < t(n - 1) = T_s \tag{2}$$

where t is the time of one plus operation and h is the time overhead per recursive call, which includes barrier synchronization and all other computations excluding the plus operation. After rearranging the above inequality, we obtain:

$$n^{1+h/t} < 2^{n-1} \tag{3}$$

What does the inequality (3) indicate? The inequality allows us to find the input size at the crossover point given the ratio between the parallel overhead time (h) and the computation time (t) as defined above, and vice versa. Thus, any input size greater than the input size at the crossover point ensures that the parallel computation time of the *Simple Reduction* will be less than the sequential computation time of the reduce operation.

Now, let us examine inequality (3) in practice. First, let us assume that our parallel system is a 6-core system, i.e., $p = n = 6$. In order to maintain inequality (3), our system must comply with the condition that $h/t < 0.934$. Therefore, in our 6-core system, the parallel overhead time of the *Simple Reduction* algorithm must be less than the computation time of one plus operation. This result is excessively optimistic compared with the overhead costs in real systems. In reality, the parallel overhead costs are significantly greater than the computational time of simple arithmetic operations.

Now, let us look at inequality (3) from the opposite perspective. Let us assume that $h/t = 20$. Then, according to inequality (3), the parallel time required for *Simple Reduction* will be less than the sequential time for $n \geq 154$. Thus, because the *Simple Reduction* function is invoked for $n \leq p$ and $n \geq 154$, then p must hold as $p \geq 154$. A typical parallel system of 154 cores or more will yield a ratio of the overhead time relative to the computational time of greater than 20, as defined above. We measured the values of h/t in our parallel system for $n = p = 4$ and found that these values were not constant and they varied among the development environments. The measured values were 105 and 18132 for the Python MT and MP approaches, respectively, which are higher than the optimistic assumption obtained from inequality (3).

We measured the performance of the *Simple Reduction* function for $n = p = 4$. Table 4 shows the runtime of the *Simple Reduction* function with the Python MT and MP approaches, and the performance of sequential reduction and the built-in Python reduction function $np.sum$. The performance of the recursive parallel *Simple Reduction* function was slower than that of the sequential algorithm by two and five orders of magnitude with the MT and MP approaches, respectively.

Table 4. Sequential and parallel runtimes of the **simple reduction** function for Python MT and MP approaches and for input size n = 4. Times shown are in **micro seconds**.

Multi-processing			Multi-threading		
Sequential	np.sum	Parallel	Sequential	np.sum	Parallel
1.61	13.58	130991	1.79	21.7	181

Now, let's examine the case where $n > p$. Here, we first divide the n elements into p partitions, and each process performs a local reduction operation on one partition of n/p elements. In the second stage, a reduction operation is performed on the local results using the *Simple Reduction* function that was analyzed in the previous case. Therefore, the following inequality must be satisfied:

$$T_p = (tn/p + \beta + (t + h)log2(p) < t(n - 1) = T_s \tag{4}$$

After rearranging the above inequality and taking in account that $h > \beta$ we get:

$$n/p + h/t + (1 + h/t)log2(p) < n - 1 \tag{5}$$

Now, let us examine this inequality (5) in practice.

First, let us consider the performance of the *Reduction* function. Table 5 shows the runtime results for the *Reduction* function with the MT and MP approaches for one and four processes/threads, and the performance of sequential reduction and the built-in Python reduction function $np.sum$ for various input sizes. The performance of the MT approach was up to three orders of magnitude better than that of the MP approach. These results highlight an important issue.

In the analysis given above, we defined the sequential runtime as $t(n-1) = T_s$. The sequential runtime is a reference time that determines whether the parallel runtime obtained improves the performance. However, what time do we use as the practical reference time? We could use the runtime of the parallel algorithm when using one process/thread, the time of the sequential algorithm that uses the same data structures as the parallel algorithm, or the time of the fastest sequential algorithm.

Next, we consider the results in Table 5 and focus on the measurements obtained with the MP approach. If the reference time is selected as the runtime of a single process or the sequential runtime, the performance of the parallel algorithm improved as the input size increased compared with the performance of the sequential algorithms. Table 5 shows that for four processes and an input size of 4M elements, a speedup of 1.86 was achieved when the reference time was set as the runtime for a single process. However, if we select the runtime of $np.sum$ as the reference time, then the parallel reduction algorithm only matched the runtime for $np.sum$ with a very large input, if at all.

Table 5. Sequential and parallel runtimes of the **reduction** function for MP and MT approaches and for various input sizes. Times shown are in **micro seconds**.

Input size	Multi-processing			
	Sequential	np.sum	1 process	4 processes
100000	9913	78	153200	292000
1000000	125300	589	362300	358600
4000000	503088	2023	1067100	573500
Input size	Multi-threading			
	Sequential	np.sum	1 thread	4 threads
100000	24	66	425	1607
1000000	519	583	1272	1501
4000000	1612	1970	3761	2807

Now, let us consider the measurements obtained with the MT approach. If the reference time is selected as that for one of the sequential algorithms, then the parallel reduction algorithm obtained very poor performance compared with the sequential algorithms. However, if we select the runtime of a single process as the reference, then a speedup of 1.33 was achieved with four threads and an input size of 4M elements.

Thus, which sequential algorithm should be selected as a reference?

In general, if we want to examine the scalability of a parallel system, the runtime of the parallel algorithm for a single thread/process is selected as the reference time. However, if we are interested in examining the speed of our parallel algorithm, then we select the best sequential algorithm as a reference algorithm.

Now, let us reconsider inequality (5). Suppose that $p = 4$ and $h/t = 20$. The inequality shows that the parallel reduction function will perform better than the sequential reduction function for $n > 84$. However, as shown by the results shown in Table 5, the parallel algorithm actually performed better when $n > 4M$ elements and $n > 1M$ elements with the MT and MP approaches, respectively, and when the comparison was performed against the runtime for a single process/thread. If the runtime of one of the other sequential algorithms is selected as a reference, then much larger input values will be required.

Thus, what is the source of the difference in the predicted theoretical performance and the actual performance? The answer is that the difference is due to the deficiencies of the performance model employed, i.e., inequalities (3) and (5). In particular, these inequalities lack expressions for all of the parallel overheads, and thus they are not considered when analyzing the theoretical performance. As shown by our experiments, these overheads significantly affect the performance of a parallel system. Table 5 shows that the differences are substantial between the sequential algorithm runtimes and the parallel runtimes obtained for a single thread/process. These results showed that considerable overheads are incurred simply by activating the parallelism mechanisms and without parallelizing anything.

5   Conclusions

In this study, we investigated the vector-based Quicksort algorithm proposed by Cormen. In particular, we implemented this algorithm in an advanced Python environment using two parallel programming models: process-based and thread-based parallel programming models. We compared the performance of these algorithms with those of recursive parallel algorithms and serial algorithms.

Our analysis of the runtime results indicated that Cormen's algorithm did not exhibit good scalability. In addition, the parallel programming model selected for the implementation of the algorithm significantly affected the performance of the algorithm. We also examined the predictive ability of the theoretical performance model and found a large difference between the predicted performance and the actual performance derived from parallel overhead sources that are not considered in the theoretical model.

References

1. Marowka, A.: Pitfalls and issues of many-core programming. Adv. Comput. **79**, 71–117 (2010)
2. Cormen, T.H.: Chapter 9: Parallel computing in a Python-based computer science course. In: Prasad, S.K., et al. (eds.) Topics in Parallel and Distributed Computing: Introducing Concurrency in Undergraduate Courses. Morgan Kaufmann (2015)
3. Blelloch, G.E.: Scan primitives and parallel vector models. Ph.D. dissertation, Massachusetts Institute of Technology (1988)
4. Blelloch, G.E.: Vector Models for Data-Parallel Computing. The MIT Press, Cambridge (1990)
5. Numba. http://numba.pydata.org/
6. Numpy. http://www.numpy.org/
7. Scipy. http://www.scipy.org/
8. Matplotlib. http://matplotlib.org/
9. Anaconda Accelerate. https://docs.continuum.io/accelerate/
10. Anaconda Python. https://www.continuum.io/downloads

Parallel Tiled Cache and Energy Efficient Code for Zuker's RNA Folding

Marek Palkowski[✉] and Wlodzimierz Bielecki

Faculty of Computer Science and Information Systems, West Pomeranian University
of Technology in Szczecin, Zolnierska 49, 71210 Szczecin, Poland
{mpalkowski,wbielecki}@wi.zut.edu.pl,
http://www.wi.zut.edu.pl

Abstract. In this paper, we consider Zuker's RNA folding algorithm, which is a challenging dynamic programming task to optimize because it is resource intensive and has a large number of non-uniform dependences. We apply a previously published approach, proposed by us, to automatically tile and parallelize each loop in the Zuker RNA Folding loop nest, which is within the polyhedral model. First, for each loop nest statement, rectangular tiles are formed within the iteration space of the Zuker loop nest. Then, those tiles are corrected to honor all dependences exposed for the original loop nest. Correction is based on applying the exact transitive closure of a dependence graph. We implemented our approach as a part of the source-to-source TRACO compiler. We compare code performance and energy consumption with those obtained with the state-of-the-art PluTo compiler based on the affine transformation framework as well as with those generated by means of the cache-efficient manual method *Transpose*. Experiments were carried out on a modern multi-core processor to achieve the significant locality improvement and energy saving for generated code.

Keywords: RNA folding · High-performance computing · Zuker algorithm · Loop tiling · Energy consumption

1 Introduction

Dynamic programming (DP) recurrences have been one of the most ongoing approaches to sequence analysis and structure prediction in biology. However, achieving good code performance is limited due to memory latency and bandwidth on modern multi-core platforms.

Fortunately, DP algorithms involve mathematical computations, which are easily implemented as affine control loop nests [2,4], thus, the iteration space can be represented by the polyhedral model for optimizing their locality and parallelism. It provides a powerful theoretical framework that can analyze regular loop programs with static dependences [10].

Loop transformations such as tiling for improving locality group loop nest statement instances in the loop nest iteration space into smaller blocks (tiles)

© Springer Nature Switzerland AG 2020
R. Wyrzykowski et al. (Eds.): PPAM 2019, LNCS 12044, pp. 25–34, 2020.
https://doi.org/10.1007/978-3-030-43222-5_3

allowing reuse when the block fits in local memory. To our best knowledge, well-known tiling techniques are based on linear or affine transformations [1,2,4,20].

In this paper, we focus on the automatic code locality improvement of RNA folding realized with Zuker's algorithm [22]. It predicts the structure using more detailed and accurate energy models [5] and entails a large number of non-uniform loop dependences. Therefore, the algorithm belongs to the non-serial polyadic dynamic programming (NPDP) class. The NPDP irregular loop dependence patterns prevent applying commonly-know polyhedral optimization techniques [12].

Recently, we introduced an algorithm to tile affine arbitrary nested loops. It is based on the transitive closure of program dependence graphs [12]. First, rectangular tiles are formed, then they are corrected to establish tiling validity and a cycle-free inter-tile dependence graph by means of the transitive closure of loop nest dependence graphs. The approach is able to tile non-fully permutable loops, which are exposed for NPDP algorithms. To parallelize corrected tiles, we use the ISL scheduler [18]. The tiling strategy is implemented within the TRACO compiler[1].

In the experimental study, we compare generated code performance with that obtained with related cache-efficient strategies and analyze the energy consumption on a modern multi-core computer.

The rest of the paper is organized as follows. Section 2 explores related approaches. Section 3 presents Zuker's algorithm. Section 4 discusses optimization ways to accelerate Zuker's code. In Sect. 5, we report experimental results to validate our claims. Finally, conclusions are presented in Sect. 6.

2   Related Work

In recent years, many research groups have been doing research in the area of manual and automatic accelerating NPDP algorithms used in bioinformatics for multi-core processors, graphics accelerators, and FPGAs [5,7–10,17]. In this paper, we consider implementations dedicated to reducing memory access latency and the cache bandwidth on modern multi-core CPUs.

GTfold [9] is a well-known optimized multi-core implementation of RNA secondary structure prediction algorithms. It optimizes the memory layout of the arrays to improve spatial locality. However, GTfold does not perform tiling to improve temporal locality.

Li et al. [6] suggested a manual cache efficient version of simplified Nussinov's recurrence by using the lower triangle of a dynamic programming table to store the transpose of computed values in the upper triangle. Li's modifications accelerate rapidly code execution because reading values in a row is more cache efficient than reading values in a column. Diagonal scanning of statements exposes parallelism in Li's code.

Zhao et al. improved the *Transpose* method to generate energy-efficient code [21]. Their benchmarking shows that depending on a studied computational

---

[1] traco.sourceforge.net.

platform and programming language, either *ByRow* or *ByBox* gives minimal run time and energy consumption. As a result, using the same amount of memory, the algorithms proposed by Zhao et al. can solve problems up to 40% larger than those solvable by *Transpose*. However, the authors do not present how to generate multi-threaded code.

The state-of-the-art source-to-source PluTo compiler [1] is able to tile RNA folding loop nests in an automatic way. It forms and applies affine transformations to generate tiled code within the polyhedral model. However, PluTo fails to generate tiles of the maximal dimension for NPDP codes [12] because the tile dimensionality is limited to the number of linearly independent solutions to the space/time partition constraints.

Mullapudi and Bondhugula presented dynamic tiling for Zuker's optimal RNA secondary structure prediction [10]. An iterative tiling for dynamic scheduling is calculated by means of reduction chains. Operations along each chain can be reordered to eliminate cycles in an inter-tile dependence graph. But this technique is not able to generate static tiled code.

Wonnacott et al. introduced 3-d tiling of "mostly-tileable" loop nests of RNA secondary-structure prediction codes in paper [19]. However, the authors demonstrated how to generate only serial tiled code for the Nussinov loop nest.

In our previous work, we have accelerated Nussinov's RNA folding loop nest [12] by means of the exact transitive closure of loop dependence graphs. Paper [15] presents locality improvements for Zuker's recurrence without any parallelism. Papers [13,14] consider the tile correction application for the minimum cost polygon triangulation and Smith-Waterman alignment codes, respectively.

3   Zuker's Algorithm

Zuker's algorithm is executed in two steps. First, it calculates the minimal free energy of the input RNA sequence on recurrence relations as shown in the formulas below. Then, it performs a trace-back to recover the secondary structure with the base pairs. The first step consumes almost all of the total execution time. Thus, optimization of computing energy matrices is crucial to improve code performance.

Zuker defines two energy matrices, $W(i,j)$ and $V(i,j)$, where $\mathcal{O}(n^2)$ pairs (i,j) satisfying $1 \leq i \leq N; i \leq j \leq N$, and N is the length of a sequence. $W(i,j)$ is the total free energy of sub-sequence defined with values of i and j, $V(i,j)$ is defined as the total free energy of the sub-sequence starting with i and ending with j if i and j pairs, otherwise $V(i,j) = \infty$.

The main recursion of Zuker's algorithm for all i, j with $1 \leq i < j \leq N$, where N is the length of a sequence, is the following.

$$
W(i,j) = \begin{cases}
W(i+1,j) & (1) \\
W(i,j-1) & (2) \\
V(i,j) & (3) \\
\min_{i<k<j}\{W(i,k) + W(k+1,j)\} & (4)
\end{cases}
$$

Below, we present the computation of V.

$$V(i,j) = \begin{cases} eH(i,j) & (5) \\ V(i+1,j-1) + eS(i,j) & (6) \\ \min_{\substack{i \le i' \le j' \le j \\ 2 < i'-i+j-j' < d}} \{V(i',j') + eL(i,j,i',j')\} & (7) \\ \min_{i < k < j-1} \{W(i+1,k) + W(k+1,j-1)\} & (8) \end{cases}$$

eH (hairpin loop), eS (stacking) and eL (internal loop) are the structure elements of energy contributions in the Zuker algorithm.

The computation of Eqs. 1, 2, 3, 5, 6 takes $\mathcal{O}(n^2)$ steps. Equations 4 and 8 requires $\mathcal{O}(n^3)$ steps. The time complexity of a direct implementation of this algorithm is $\mathcal{O}(n^4)$ because we need $\mathcal{O}(n^4)$ operations to compute Eq. 7. This formulation as a computational kernel involves float arrays and operations.

The computation domain and dependences for Zuker's recurrence cell (i,j) is similar but more complex than Nussinov's recurrence dependence pattern. Long-range (non-local) dependences of cell (i,j) are generated within Eqs. 3, 4 and 8. The $V(i',j')$ element computation in Eq. 3 is spread within the triangle whose area is limited to several dozens or hundreds of cells in nature. The other equations present short-range (local) dependences.

Listing 1 shows the affine loop nest for finding the minimums of the V and W energy matrices.

Listing 1. Zuker's recurrence loop nest

```
for (i = N-1; i >= 0; i--){
 for (j = i+1; j < N; j++) {
  for (k = i+1; k < j; k++){
   for(m=k+1; m <j; m++){
    if(k-i + j - m > 2 && k-i + j - m < 30)
      V[i][j] = MIN(V[k][m] + EL(i,j,k,m), V[i][j]);   // Eq. 3
   }
   W[i][j] = MIN ( MIN(W[i][k], W[k+1][j]), W[i][j]);   // Eq. 8
   if(k < j-1)
      V[i][j] = MIN(W[i+1][k] + W[k+1][j-1], V[i][j]);  // Eq. 4
  }
  V[i][j] = MIN( MIN (V[i+1][j-1] + ES(i,j), EH(i,j), V[i][j]);
                                                  // Eq. 1,2
  W[i][j] = MIN( MIN ( MIN ( W[i+1][j], W[i][j-1]), V[i][j]), W[i][j]);
                                                  // Eq. 5,6,7
 }
}
```

4   Optimizing the Zuker Loop Nest

The Zuker loop nest has similar dependence patterns to those of the Nussinov one. It uses also only the upper-right triangles of the energy arrays. It is possible to parallelize the second outermost loop nest with the statement execution in the diagonal order (Listing 2). No data exchange between threads is needed. All threads synchronize before moving to the next diagonal [6].

Li et al. [6] proposed the Transpose method to improve locality of this code. In Eqs. 4 and 8, there are not cache-efficient column reading of the W array, $W[k+1][j]$ and $W[k+1][j-1]$, respectively. The transpose method changes these array accesses to the row reading and add the following statement $W[col][row] = W[row][col]$ to make a transposed copy of the cells in the lower-left triangle.

Li's algorithm uses all elements in the W array, i.e., the algorithm keeps and writes the double number of cells in the memory.

Listing 2. Zuker's recurrence loop nest after applying *Transpose*

```
for(diag=2; diag<=N-1; diag++)
#pragma omp parallel for shared(diag) private(col, row, k, m)
  for(row=0; row<=N-diag-1; row++){
    col = diag+row;
    for(k=row; k<col; k++){
      for(m=k+1; m<col; m++)
        if(k-row + col - m > 2 && k-row + col - m < 30 )
          V[row][col] = MIN(V[k][m] + EFL[row][col], V[row][col]);
      W[row][col] += MIN ( MIN(W[row][k], W[col][k+1]), W[row][col]);
      if(k < col-1)
        V[row][col] = MIN(W[row+1][k] + W[col-1][k+1], V[row][col]);
    }
    V[row][col] = MIN( MIN (V[row+1][col-1], EHF[row][col]), V[row][col]);
    W[row][col] = MIN( MIN ( MIN ( W[row+1][col], W[col][col-1]),
                                                V[row][col]), W[row][col]);
    W[col][row] = W[row][col];
}
```

Optimizing compilers allow us to automatically produce parallel tiled code using the serial code (Listing 1) as the input. We use the two source-to-source tools in an experimental study: Traco and PluTo.

The PluTo compiler parallelizes code by means of the loop skewing method found with the PluTo scheduler algorithm based on the affine transformation framework (ATF). The compiler is unable to tile the third loop k [10]. Because the third loop k is innermost and not tiled, locality improvement of generated tiled code is limited.

The Traco compiler is able to tile all loops in this nest. It uses a tile correction algorithm based on the transitive closure of program dependence graphs [12]. First, rectangular tiles are formed, then they are corrected to establish tiling validity. Tile statement instances, which are dependence destinations, are moved to lexicographically greater tiles containing dependence sources. Next, tiled loops are skewed to generate parallel code. To our best knowledge, tile correction is the only method that allows us to tile all loops in the nest and generate static parallel tiled code for Zuker's recurrence.

Figure 1 exhibits an example pattern of two uniform dependences $i, j \rightarrow i, j+1$ and $i, j \rightarrow i+1, j-1$, respectively, with the dependece vectors $[0, 1]$ and $[1, -1]$ in the $\mathcal{O}(n^2)$ domain. Statement instances in tiles are executed in the serial order. Bold arrows represent inter-tile dependences.

The second negative element -1 in the dependence vector $[1, -1]$ disables rectangular tiling. Dark gray statement instances are "problematic" because they are the dependence destinations whose sources belong to lexicographically greater tiles. There are cycles in the inter-tile dependence graph.

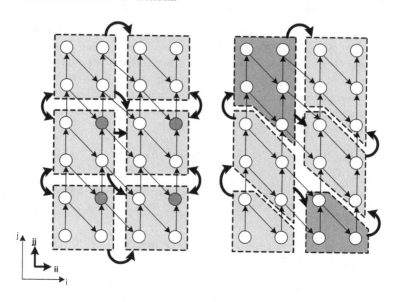

Fig. 1. An example dependence pattern and its tile correction.

Transitive closure allows us to find all invalid tiles and statement instances that make the corresponding tiles invalid. Then such statement instances are moved from original tiles to the lexicographically greatest tiles including the corresponding dependence sources.

It is worth noting that the computation of point (i, j) in the (i, j, k) domain depends on the points in the domain from the i-th row and the points from the j-th column. Dependences among neighboring points (connected with paths of length 1) are uniform, while ones among points connected with paths of length 2 or more for each value of k are non-uniform. Both uniform and non-uniform dependences involve original tile correction along those axes corresponding loop iterators (i, j, k, m), for which the corresponding elements of distance vectors are negative. For the Zuker loop nest, distance vectors have negative elements in the third and fourth positions (k and m, respectively). Tile correction along axes k and m allows for tiling loops k and m that improves code locality. Techniques based on affine transformations do not allow for tiling loops k and m.

Tile correction changes the shape of original tiles and reduces the number of inter-tile dependences (from 13 to 7 in the example depicted in Fig. 1). The inter-tile dependence graph is acyclic. If the calculation of the transitive closure of a dependence graph is possible, TRACO is able to tile all loops in the nest. Details of the tile correction algorithm are presented in paper [15]. For the Zuker loop nest, TRACO calculates the exact transitive closure of a dependence graph.

Dark gray tiles in Fig. 1 are independent, so they can be scheduled to the same schedule time. Corrected valid tiles can be scheduled with an arbitrary technique. In other words, tile correction and scheduling are independent steps in opposite to the PluTo strategy.

To generate parallel code on the tile level, we apply the ISL scheduler [18], which is as a variation of the PluTo scheduler with Feautrier's one as fallback. The ISL scheduler is improved with unscaling, domain compression, incremental scheduling, grouping and coalescing.

The schedule returned with the ISL scheduler corresponds to the loop skewing transformation, which can be represented with the following map of tile identifiers $(ii, jj, ...) \rightarrow (ii + jj, jj, ...)$. TRACO applies this schedule to generate target parallel tiled code with the OpenMP pragmas *parallel for* [11] placed before the second loop in the nest, which scans values of tile identifier jj.

Table 1. Execution time in seconds

Size	Serial	TRACO	PluTo	Transpose
1000	21.1	3.7	4.5	3.7
1500	102.0	23.9	27.0	18.3
2000	317.8	45.2	63.8	56.5
2500	773.0	105.6	153.8	138.8
3000	1595.2	211.9	325.8	285.9
3500	2946.1	391.9	619.2	526.9
4000	4923.8	664.9	1074.5	894.2
4500	7819.0	1030.8	1683.5	1413.7
5000	11474.4	1561.6	2602.9	2148.8

5   Experimental Study

We implemented the Zuker loop nest and its transposed version [6] in C and generated parallel tiled code by means of the Traco and PluTo compilers[2]. Then, we examined code performance on an Intel i7-8700 processor (3.2 GHz, 4.6 GHz in turbo, 6 cores, 12 threads, 12 MB Cache). Codes were compiled using the GCC version 7.3.0 with the O3 option. We used the "perf" [16] software to measure energy consumption through the RAPL interface [3].

Randomly generated RNA sequences, used by us for carrying out experiments, are comparable to real similarly sized sequences [12].

For generated tiled code, we empirically discover that the best tile size is 16 × 16 × 16 × 16. All studied codes were parallelized by applying the OpenMP *parallel* directives and *for* directives [11] with the dynamic schedule of loop iterations and the chunk size equal to 1 as the most efficient work-sharing.

Table 1 presents execution time (in seconds) for serial code and parallel code generated with TRACO, PluTo, and the Transpose technique for all 12 available

---

[2] The source codes are available at the repository https://github.com/markpal/zuker. The tiled codes are too long to present in this paper.

i7 threads. The tiled code generated with TRACO demonstrates the best speed-up 7.35 for the longest studied sequence. Li's code is more cache-efficient than the PluTo code.

Table 2 presents energy consumption measured at the processor and its components (cores and cache) in kilojoules. The TRACO code is also the most energy-efficient program. The serial implementation requires from 3 to 7 times larger energy (it engages only one CPU thread). The rest optimized codes also reduce energy consumption, but to a lesser extent.

Table 2. Energy usage on CPU cores and cache (kJ).

Size	Serial	TRACO	PluTo	Transpose
1000	0.574	0.253	0.299	0.246
1500	2.720	1.535	1.752	1.189
2000	8.448	2.939	4.140	3.672
2500	20.072	6.855	9.981	9.0230
3000	41.730	13.756	21.014	18.574
3500	78.405	25.442	39.953	34.220
4000	134.566	43.130	68.842	58.076
4500	214.407	66.869	107.940	91.806
5000	319.190	98.319	166.379	139.540

6 Conclusion

In this paper, we presented how to generate parallel cache and energy efficient code for the Zuker loop nest, which is one of the most sophisticated algorithms for folding single RNAs. We applied the transitive closure of a dependence graph to improve locality of the RNA folding code and compared its performance and energy consumption with those obtained with the *Transpose* and affine transformation methods. We observed higher performance and lower energy consumption at the cores of obtained code in comparison with those of code generated with the techniques mentioned above.

In future work, we plan to study more complex polyhedral implementations of Zuker's recurrence and compare them with related implementations included in sophisticated bioinformatics packages. We intend to develop novel more efficient tiling strategies dedicated to irregular dependence patterns.

References

1. Bondhugula, U., et al.: A practical automatic polyhedral parallelizer and locality optimizer. SIGPLAN Not. **43**(6), 101–113 (2008)
2. Griebl, M.: Automatic parallelization of loop programs for distributed memory architectures (2004)
3. Intel Corporation: Intel® 64 and IA-32 Architectures Software Developer's Manual, Volume 3B: System Programming Guide, Part 2, September 2016. https://www.intel.com/content/dam/www/public/us/en/documents/manuals/64-ia-32-architectures-software-developer-vol-3b-part-2-manual.pdf
4. Irigoin, F., Triolet, R.: Supernode partitioning. In: Proceedings of the 15th ACM SIGPLAN-SIGACT Symposium on Principles of Programming Languages, POPL 1988, pp. 319–329. ACM, New York (1988)
5. Jacob, A.C., Buhler, J.D., Chamberlain, R.D.: Rapid RNA folding: analysis and acceleration of the Zuker recurrence. In: 2010 18th IEEE Annual International Symposium on Field-Programmable Custom Computing Machines, pp. 87–94, May 2010
6. Li, J., Ranka, S., Sahni, S.: Multicore and GPU algorithms for Nussinov RNA folding. BMC Bioinformatics **15**(8) (2014). S1. http://dx.doi.org/10.1186/1471-2105-15-S8-S1
7. Liu, L., Wang, M., Jiang, J., Li, R., Yang, G.: Efficient nonserial polyadic dynamic programming on the cell processor. In: IPDPS Workshops, Anchorage, Alaska, pp. 460–471. IEEE (2011)
8. Markham, N.R., Zuker, M.: UNAFold, pp. 3–31. Humana Press, Totowa (2008)
9. Mathuriya, A., Bader, D.A., Heitsch, C.E., Harvey, S.C.: GTfold: a scalable multicore code for RNA secondary structure prediction. In: Proceedings of the 2009 ACM Symposium on Applied Computing, SAC 2009, pp. 981–988. ACM, New York (2009)
10. Mullapudi, R.T., Bondhugula, U.: Tiling for dynamic scheduling. In: Rajopadhye, S., Verdoolaege, S. (eds.) Proceedings of the 4th International Workshop on Polyhedral Compilation Techniques, Vienna, Austria, January 2014
11. OpenMP Architecture Review Board: OpenMP application program interface version 4.0 (2012). http://www.openmp.org/mp-documents/OpenMP4.0RC1_final.pdf
12. Palkowski, M., Bielecki, W.: Parallel tiled Nussinov RNA folding loop nest generated using both dependence graph transitive closure and loop skewing. BMC Bioinformatics **18**(1), 290 (2017)
13. Palkowski, M., Bielecki, W.: Accelerating minimum cost polygon triangulation code with the TRACO compiler. In: Communication Papers of the 2018 Federated Conference on Computer Science and Information Systems, FedCSIS 2018, Poznań, Poland, 9–12 September 2018, pp. 111–114 (2018)
14. Palkowski, M., Bielecki, W.: Parallel tiled codes implementing the Smith-Waterman alignment algorithm for two and three sequences. J. Comput. Biol. **25**(10), 1106–1119 (2018)
15. Palkowski, M., Bielecki, W.: A practical approach to tiling Zuker's RNA folding using the transitive closure of loop dependence graphs. In: Świątek, J., Borzemski, L., Wilimowska, Z. (eds.) ISAT 2017. AISC, vol. 656, pp. 200–209. Springer, Cham (2018). https://doi.org/10.1007/978-3-319-67229-8_18
16. de Melo, A.C.: The new linux 'perf' tools. Technical report, Linux Kongress, Georg Simon Ohm University Nuremberg, Germany (2010)

17. Tan, G., Feng, S., Sun, N.: Locality and parallelism optimization for dynamic programming algorithm in bioinformatics. In: SC 2006 Conference, Proceedings of the ACM/IEEE, pp. 41–41 (2006)
18. Verdoolaege, S.: Integer set library - manual. Technical report (2011). www.kotnet. org/~skimo//isl/manual.pdf
19. Wonnacott, D.G., Strout, M.M.: On the scalability of loop tiling techniques. In: Proceedings of the 3rd International Workshop on Polyhedral Compilation Techniques (IMPACT), January 2013
20. Xue, J.: Loop Tiling for Parallelism. Kluwer Academic Publishers, Norwell (2000)
21. Zhao, C., Sahni, S.: Cache and energy efficient algorithms for Nussinov's RNA folding. BMC Bioinformatics 18(15), 518 (2017)
22. Zuker, M., Stiegler, P.: Optimal computer folding of large RNA sequences using thermodynamics and auxiliary information. Nucleic Acids Res. 9(1), 133–148 (1981)

Examining Performance Portability with Kokkos for an Ewald Sum Coulomb Solver

Rene Halver[1], Jan H. Meinke[1], and Godehard Sutmann[1,2](✉)

[1] Jülich Supercomputing Centre, Institute for Advanced Simulation,
Forschungszentrum Jülich, 52425 Jülich, Germany
{r.halver,j.meinke,g.sutmann}@fz-juelich.de
[2] ICAMS, Ruhr-Universität Bochum, 44801 Bochum, Germany

Abstract. We have implemented the computation of Coulomb interactions in particle systems using the performance portable C++ framework Kokkos. Coulomb interactions are evaluated with an Ewald-sum-based solver, where the interactions are split into long- and short-range contributions. The short-range contributions are calculated using pairwise contributions of particles while long-range interactions are calculated using Fourier sums. We evaluate the performance portability of the implementation on Intel CPUs, including Intel Xeon Phi, and Nvidia GPUs.

Keywords: Programming model · Accelerator · Performance modeling · Long-range interaction

1 Introduction

The development of modern computer architectures shows a clear trend towards increased complexity and heterogeneity. This increases the complexity of efficient code development for multiple architectures that takes advantage of all available components. GPUs, for example, are powerful processors available in cell phones as well as supercomputers that usually require their own programming model. As a matter of fact GPUs have become more and more important as a source of computing power in supercomputers, as can be seen in the increase of systems using GPUs in the Top500 list [17] over the past ten years. While in November 2008 there was no system that included GPUs, in November 2013 there were 39 systems, and in the current list from November 2018 126 systems contained GPUs. There are many ways of programming GPUs but unfortunately few are even function portable without large changes to the source code [10]. This matter becomes even worse if we want to write code for different kind of accelerators, e.g., Intel's Xeon Phi series.

In the domain of particle simulation methods of complex systems, electrostatic interactions represent a class of algorithms of high computational complexity. This arises as a result of pair-wise interactions between all particles

© Springer Nature Switzerland AG 2020
R. Wyrzykowski et al. (Eds.): PPAM 2019, LNCS 12044, pp. 35–45, 2020.
https://doi.org/10.1007/978-3-030-43222-5_4

in a system, which basically scale as $\mathcal{O}(N^2)$. More efficient methods can be reduced to $\mathcal{O}(N \log(N))$ or even $\mathcal{O}(N)$ but come with a large implementation effort [3,6,13,15,19]. In the present paper we consider the Ewald summation method, which is suitable for particle systems in three dimensions under periodic boundary conditions and which can be optimized by proper choice of parameters to $\mathcal{O}(N^{3/2})$ (there are also formulations for one- or two-dimensional systems, which we do not consider here). The basic structure of the Ewald summation is sufficiently transparent and not too complex, allowing an analysis of the operational count and providing insight into the procedure to measure performance portability.

Performance portable approaches have been supported recently by the US department of Energy and have resulted in frameworks like Kokkos [1,5] or Raja [2,4], which offer C++ software abstractions for code execution and memory management.

In this paper, we compare the performance of an Ewald sum implemented in Kokkos on various Intel CPUs including Intel Xeon Phi Knights Landing and Nvidia GPUs. We start with a quick overview of Kokkos and its main features (Sect. 3). Then we introduce the problem of a system of electric charges with periodic boundary conditions and show how the Ewald sum can be used to calculate it efficiently (Sect. 2). In Sect. 4 we establish a base line for the achievable performance. Afterwards we present our implementation and show our performance benchmarks (Sect. 5).

2   Calculating Long-range Interactions with Periodic Boundaries

When computing energies and forces in particle systems composed of N particles, which are dominated by long range interactions, each particle i gets partial contributions of each other particle $j \in [1, N]$. Long range interactions arise when the potential energy function $\phi(r)$ decays slower than $1/r^d$, where d is the dimension of the system and r is the distance from a point in space to a particle. Here, we consider electrostatic potentials created by point charges, for which the potential energy at a point r in free space is given by $\phi(r) = q_j/|r - r_j|$ which leads to a total electrostatic energy $U = 1/2 \sum_{i,j} q_i \phi(r_{ij})$. When simulating bulk systems, the number of particles in a simulation is always small, compared with laboratory samples and therefore, in order to avoid surface effects, periodic boundary conditions are often applied [9] and the electrostatic potential energy at particle position $\mathbf{r}_i$ can formally be written as

$$\phi(\mathbf{r}_i) = \sum_{\mathbf{n}} {}^\dagger \sum_{j=1}^{N} \frac{q_j}{\|\mathbf{r}_{ij} + \mathbf{n}L\|_2}$$

where $\mathbf{n} \in \mathbb{Z}^3$ is a so called lattice vector, L the length of the (cubic) system and "†" indicates that $j \neq i$ for $\|\mathbf{n}\|_2 = 0$. This sum cannot be evaluated by a straightforward summation rule, since (i) the first sum is formally over an infinite number of lattice vectors; and (ii) the lattice sum is conditionally convergent, i.e., the result depends on the order of summation. Ewald proposed [7] a way to overcome the conditional convergence by subdividing the expression into a short range and a long range part, which is introduced via a splitting function, $f(r)$, which decays to zero within a finite range. A function $u(r) = 1/r$ can then be rewritten as $u(r) = f(r)/r + (1 - f(r))/r$. The first term is short range (since it decays to zero), while the second one is long range (since asymptotically it decays as $1/r$). This reformulation has the advantage that it can be transformed into an unconditionally convergent sum for a proper choice of f. Originally, $f(r) = \mathrm{erfc}(\alpha r)$ was chosen, where α is a splitting parameter, controlling the width of the short range part. The long range part can be elegantly computed in Fourier space, which leads to [9]

$$\phi(\mathbf{r}_i) = \sum_{j=1}^{N}\sum_{\mathbf{n}}^{\dagger} q_j \frac{\mathrm{erfc}(\alpha\|\mathbf{r}_{ij} + \mathbf{n}L\|)}{\|\mathbf{r}_{ij} + \mathbf{n}L\|_2} + \frac{4\pi}{L}\sum_{|\mathbf{k}|\neq 0}\sum_{j=1}^{N}\frac{q_j}{|\mathbf{k}|^2}e^{-\frac{|\mathbf{k}|^2}{4\alpha^2}}e^{i\mathbf{k}\mathbf{r}_{ij}} - q_i\frac{2\alpha}{\sqrt{\pi}} \quad (1)$$

The last term corresponds to a correction for particle i, which also appears in the k-space summation (second term). For practical computations the infinite sums (over $\mathbf{n}$ and $\mathbf{k}$) have to be approximated. For large arguments $\mathrm{erfc}(x)$ decays as a Gaussian, as it does the k-space summation. Therefore, both sums can be limited to a finite range of values, which still allows for control of approximation error. In most cases, due to the spherical symmetry of $\mathrm{erfc}(x)$ and a fast decay, the first sum can be restricted to contributions within a spherical region of radius R_c. Furthermore, it can be shown that via a proper set of parameters [8,16], the computational complexity is reduced from $\mathcal{O}(N^2)$ to $\mathcal{O}(N^{3/2})$.

3   Kokkos at a Glance

Kokkos uses C++ to provide an abstraction of parallel algorithms, their execution and memory spaces. The basic algorithms include parallel_for, parallel_reduce, and parallel_scan. Each of these algorithms can be executed in different execution spaces, for example, using an OpenMP execution space on the CPU or a CUDA execution space on an Nvidia GPU.

CPUs and GPUs use different approaches to vectorization. CPUs use a single instruction multiple data paradigm. GPUs use a single instruction multiple threads paradigm. These two approaches lead to different preferred memory layouts. To accommodate different memory layouts and memory locations Kokkos introduces so called *Views*.

A View is a thin wrapper around the data. It knows its dimensionality, its sizes, its layout, and its memory space. Kokkos::View<**double**∗> v(n);, for example, initializes a one dimensional array of doubles of size n in the default execution space, which can be set at compile time. A View can be mirrored on the host side. In GPU computing it is not uncommon to initialize data on the host, transfer them to the GPU, perform computations, and transfer the results back. A mirrored View can do just that. Any transfer between a View and its

mirror needs to be done explicitly using Kokkos::deep_copy. The code shown in
Listing 1.1 creates a 1d View and a host mirror of it, fills the mirrored View with
random numbers, copies the numbers to the View, sums them up in parallel, and
gets the result.

Listing 1.1. Reduction using Kokkos. This program can be executed using OpenMP
on a CPU or on a GPU. Second level curly brackets are needed to ensure deallocation
of views before calling Kokkos::finalize.

```cpp
#include <random>
#include <Kokkos_Core.hpp>

int main(int argc, char* argv[]) {
  Kokkos::initialize(argc, argv);
  {
    std::default_random_engine generator;
    std::uniform_real_distribution<double> uniform_dist(0, 1);
    auto uniform = [&]{return uniform_dist(generator);};
    int n = 1024;
    double sum = 0;
    // Create a view in the default execution space
    Kokkos::View<double*> v("v", n);
    // Create a mirror of v in host memory
    auto h_v = Kokkos::create_mirror_view(v);
    for(int i = 0; i < n; ++i) h_v(i) = uniform();
    // Copy data from host to device if necessary
    Kokkos::deep_copy(v, h_v);
    // Parallel reduction in default execution space
    Kokkos::parallel_reduce(n, KOKKOS_LAMBDA(int i, double&
        localSum){
      localSum += v(i);
    }, sum);
    std::cout << "The_average_value_of_the_elements_of_v_is_"
        << (sum / n) << ".\n";
  }
  Kokkos::finalize();
}
```

If the program is compiled for OpenMP, the mirror view becomes an alias
and the deep copy does not have to do anything, but if the program is compiled
for CUDA, the original View lives on the GPU and the deep copy transfers the
data from the CPU to the GPU.

On GPUs neighboring threads should access consecutive memory. Thread i
should access a[i] and thread i+1 a[i+1], but on a CPU this prevents *vectorization* and can introduce an unnecessary dependency. If a[i] and a[i+1]
belong to the same cache line and thread i writes to a[i] it invalidates the entire

cache line. If thread `i+1` wants to access `a[i+1]` it first needs to read the entire cache line again. This effect is called false sharing [18]. So for CPUs a single thread should deal with a chunk of data. The effects due to different memory layout requirements become even more pronounced for multi-dimensional data. Note that in Kokkos, the left-most index is assumed to be the one over which parallelization is performed.

4 Achievable Performance

To determine how well our implementation takes advantage of the available hardware, we need to know what the hardware is capable of. Theoretical peak performance is *not* a good measure of the performance that is achievable for a particular algorithm. If the calculation is dominated by square roots or exponentials, for example, it does not matter how quickly a compute device can calculate multiplications and additions. To estimate the number of cycles needed for the Ewald summation (Eq. 1), we use vendor information and mini benchmarks.

We first initialize an array of elements to some range of values and then loop over this array applying the operation in question one to a few times. The idea is to access data from cache or registers to minimize the effect of memory bandwidth and latency. We check that vectorized versions of the functions are used where available. The important operations are multiplication, division, square roots, exponentials (exp), sine (sin) and cosine (cos), and the error function (erfc). Table 1 lists the duration of an operation in cycles and its inverse (throughput per cycle) for each device. For operations for which we found information from the vendors, the values are listed in parenthesis as well.

In the following sections we look at the number of instructions performed by the Ewald solver.

4.1 Ewald Solver

The Ewald solver consists of a k-space (Fourier space) and a real-space part. Let N be the number of particles in the central cell and N_k be the number of wave vectors.

Real-Space Contributions. To calculate a single particle-particle interaction energy, we first need to calculate the distance between the particles (c.f. first term of Eq. 1). In our implementation the central cell is large enough that we do not need to add additional image cells and thus do not have contributions of the type nL. A distance calculation consists of 3 subtraction, 2 multiply-adds, 1 multiplication and a square root (sqrt). For the particles within a cutoff radius defined by a tunable parameter α, we then calculate the error function (erfc) of the distance, divide by it, and multiply the result by the charge of particle j. All these partial results need to be added up for each particle i and the result is multiplied by the charge of particle i and a constant. Finally, the potential energy of all particles needs to be summed up to get the total energy. This leads

Table 1. Average duration in cycles per core or streaming multiprocessor (SMs) for additions, multiplication, (fused) multiply-add, division, square roots, exponentials, sin and cos, and the error function. All values are approximate. The number of cycles is calculated as the number of cycles per vector instruction divided by the width of the vector. A division using AVX512 instruction on Skylake-X, for example, takes 16 cycles and does 8 division in parallel during those 16 cycles. We therefore have 2 cycles per division. The numbers in parenthesis are from [11,14]. The line for Skylake (zmm) shows results when the compiler was asked for a high usage of zmm registers. On the CPU architectures, we used a single core for the measurements. On the GPUs, we used all SMs and divided the throughput by the number of SMs.

	add	mul	(f)ma	div	sqrt	exp	sin	cos	erfc
Skylake	$\left(\frac{1}{16}\right)$	$\left(\frac{1}{16}\right)$	$\left(\frac{1}{16}\right)$	2.0 (2)	3.06 (3)	12.5	11.58	12.63	16.77
Skylake (zmm)	$\left(\frac{1}{16}\right)$	$\left(\frac{1}{16}\right)$	$\left(\frac{1}{16}\right)$	0.96	1.20	6.32	6.51	6.76	9.25
Haswell	$\left(\frac{1}{8}\right)$	$\left(\frac{1}{8}\right)$	$\left(\frac{1}{8}\right)$	3.97	3.97	2.66	2.84	3.35	5.85
KNL	$\left(\frac{1}{16}\right)$	$\left(\frac{1}{16}\right)$	$\left(\frac{1}{16}\right)$	1.12	2.04	3.82	6.94	7.09	9.20
Kepler	$\left(\frac{1}{64}\right)$	$\left(\frac{1}{64}\right)$	$\left(\frac{1}{64}\right)$	0.15	0.21	0.37	0.55	0.55	1.25
Volta	$\left(\frac{1}{32}\right)$	$\left(\frac{1}{32}\right)$	$\left(\frac{1}{32}\right)$	0.30	0.31	0.57	0.84	0.84	2.04

to a total of $N^2((3 \text{ sub} + 2 \text{ multiply-add} + 1 \text{ mul} + 1 \text{ sqrt}) + V_f (2 \text{ mul} + 1 \text{ div} + 1 \text{ erfc})) + N\text{multiply-adds}$, where V_f is the fraction of the total volume within the cutoff radius. For Skylake (zmm) this becomes

$$(1.58 + 10.34 V_f)N^2 + N/16 \tag{2}$$

cycles. On a Volta GPU we need

$$(0.50 + 2.40 V_f)N^2 + N/32 \tag{3}$$

cycles.

K-Space Contributions. The second term of Eq. 1 contains 2 nested sums. The outer sum is over $N_k = (2k_{\text{int}} + 1)^3$, where k_{int} is the integer ceiling of k_{max} and k_{max} is determined by the required precision and the factor α mentioned in the Sect. 2. It requires the calculation of the square of the length of the k-vector (2 multiply-add, 1 multiplication), which is used twice. Only wave vectors with a length less than k_{max} are included for the remaining calculations. There are 3 divisions, 7 multiplication, 2 multiply-adds, and 1 exponential. The argument of the inner sum includes the dot product between $\boldsymbol{k}$ and $\boldsymbol{r}_i$(2 multiply-add, 1 multiplication). The exponential of the complex argument is calculated using 1 sin and 1 cos. This is then multiplied by q_i and summed up (1 multiply-add). The argument of the inner sum is executed $N N_k V_f$ times, where $V_f = \frac{\frac{4\pi}{3}k_{\text{max}}^3}{N_k}$. In total this becomes $N_k(3 \text{ (sub,multiply-add,mul)} + V_f(9 \text{ (sub,multiply-add,mul)} + 3 \text{ div} + 1 \text{ exp} + N(\sin + \cos + 3 \text{ (sub,multiply-add,mul)})))$. For Skylake (zmm) this becomes

$$(0.19 + (9.76 + 13.46N)V_f)N_k \tag{4}$$

cycles. On a Volta GPU we need

$$(0.09 + (1.75 + 1.77N)V_f)N_k \tag{5}$$

cycles.

5   Results

The program was benchmarked on five different architectures: three different Intel CPUs and two different Nvidia GPUs. The benchmarks were performed on the JURECA and JUWELS clusters at the Jülich Supercomputing Centre [12]. On JURECA the tests were run on (i) a CPU compute node, equipped with two Intel Xeon E5-2680 v3 Haswell CPUs, (ii) a GPU node equipped with two NVIDIA K80 (Kepler) cards, of which only a single one was used, and (iii) a booster node consisting of a single Xeon-Phi 7250-F Knights Landing (KNL) processor. The nodes used on JUWELS are (i) a CPU node containing two Intel Xeon Platinum 8168 Skylake-X (SKX) processors and (ii) a GPU node with four NVIDIA V100 (Volta) cards, of which again only one is used for the benchmarks.

For each benchmark the same source code was used, containing only minor adjustments concerning the used ExecutionSpace and MemorySpace, depending on the use of (i) a GPU architecture and (ii) the use of the host_mirror mechanic of Kokkos. The possibility to change the memory layout is also included. For the CPU benchmark runs a complete node was used, i.e., two processors of Haswell and Skylake and one KNL processor, while for the GPU benchmarks only a single GPU was used, i.e., 'half' a K80 and a single Volta V100 card. Therefore the presented runtimes are per-node runtimes, not per processor runtimes.

For the benchmarks a cubic NaCl crystal was simulated, for which the exact solution to the Coulomb potential is known, so that the accuracy of the computed solution could be compared to the exact solution. During the benchmark the size of the crystal was increased by increasing the edge length L of the crystal, thereby increasing the number of particles by L^3. Due to the nature of the system, the contribution of the Fourier-space is much smaller than the real-space contribution, due to screening effects. This does not decrease the computational demand of the algorithm if a given accuracy has to be achieved.

In the optimal case the Ewald solver shows an complexity of $\mathcal{O}(N^{3/2})$, which depends on an optimal choice of the splitting parameter α, the real-space cut-off radius r_c and the k-space cut-off k_{max}. Due to the implementation of the real-space computation, which is basically a direct solver of complexity $\mathcal{O}(N^2)$, in our results it can be seen that for larger systems sizes the resulting runtimes behave more like $\mathcal{O}(N^2)$ than $\mathcal{O}(N^{3/2})$ (see Fig. 1). The figure also shows the expected relations of runtime to architecture, as the more powerful architectures shows faster runtimes than the less powerful ones. Another detail that can be seen is that the GPUs show the same scaling behavior as the CPUs, with the Volta card resulting in the shortest runtimes of all architectures.

In order to achieve some more insight into the performance portability between the same types of architectures, i.e., CPU-CPU, GPU-GPU, and across

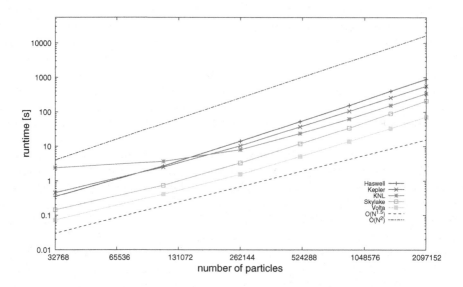

Fig. 1. Comparison of the runtime for different systems sizes on all tested architectures. For reference two guide lines showing a complexity of $\mathcal{O}(N^{3/2})$ and $\mathcal{O}(N^2)$ are shown.

Table 2. Nominal peak performance data for each of the architectures used in the benchmarks

Architecture	Note	Nominal peak performance [TFlops/s]
Haswell	Complete node (two processors)	0.9
Kepler	Single GPU (half a K80 card)	0.945
KNL	One processor	3.05
Skylake	Complete node (two processors)	4.1
Volta	Single GPU	7.8

types of architectures, i.e., CPU-GPU, we consider the peak-performance normalized runtime on the architectures. The runtime can be expressed by the number of operations divided by a fraction γ, where γ is a measure for the proximity to maximum performance, $t_{run} = \frac{N_{instruc}}{\gamma P_{peak}}; \gamma \in [0, 1]$. This can be rewritten as

$$t_{run} \cdot P_{peak} = N_{instruct}/\gamma. \tag{6}$$

To compare the performance portability of the implementation the runtimes need to be compared between the different architectures. Assuming that on each architecture a comparable number of instructions are executed for a given simulation, one can assume that the product of runtime t_{run} and the nominal peak performance P_{peak} will be equal across all platforms, if the reached relative performance γ is equal (Eq. 6).

For a qualitative comparison based on this thought, the runtimes are multiplied with the nominal peak performance for each of the architectures given in

Table 2. All peak performance data is with regard to double precision compu-tations, which are used in the code. The resulting plot (Fig. 2) shows that the normalized number of instructions computed for each of the different architectures is similar. It can also be seen that the lines for the same type of architecture (CPU, GPU) are nearly identical to each other, indicating on a qualitative level, that the achieved relative performance is similar within a given type of architecture (with the exception of KNL).

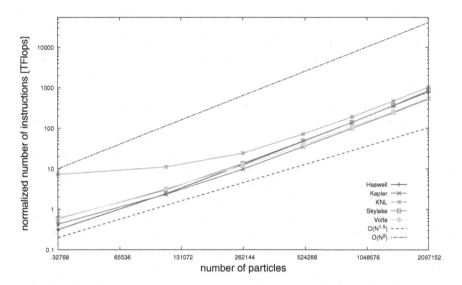

Fig. 2. Runtime on each architecture normalized by the nominal peak performance of the architecture. If the achieved performance on two architectures relative to the nominal peak performance is equal, the lines should be overlapping.

No significant difference could be measured between the variants of the code using the Kokkos host_mirror functionality and using the option with unified memory. While this was expected on CPU architectures, it was interesting to see that required hard copies of data between devices where handled in a very efficient way on GPU architectures independent of the GPU model.

We can do a quantitative analysis of the performance based on the results from Sect. 4. As an example, for $N = 128^3$ we obtain an optimal runtime of 56.44 s based on Eqs. 2 and 4 for the Skylake system. The measured runtime is 206.74 s. This corresponds to a relative performance γ_{SKX} of 0.27. For Volta we get an optimal runtime of 18.61 s and a measured runtime of 71.56 s resulting in a relative performance γ_{SKX} of 0.26. While the scaling with the nominal peak performance P_{peak} (Fig. 2) suggests that γ_{Volta} is larger than γ_{SKX}, above analysis shows that they are nearly equal.

6   Discussion and Conclusions

The performed benchmarks indicate that it is possible to write a performance portable Ewald solver code with Kokkos that can utilize different architectures without the requirement of intensive code adaptations for each of the architectures. Of course it might be possible to write more efficient code specialized for certain architectures, but this kind of code would lose the advantage of versatility concerning architectures it could usefully run on. The quantitative analysis shows that the expected runtime for Skylake is three times longer than for Volta, which our measurement confirm. On the other hand, the nominal peak performance predicts only a factor of two leading to the discrepancy with Fig. 2.

It is noticeable that the performance of the KNL is worse than the performance of the other architectures when using smaller number of particles. This could be related to a massive overhead in the administration of threads, as each thread might not be fully utilized due to the smaller amount of work for each thread. For larger system sizes, it can be seen that the KNL behaves comparable to the other architectures, with regard to the scaling behavior.

Implementing the Ewald solver with Kokkos was slightly more difficult than implementing the code with OpenMP, as the correct usage of the corresponding *parallel_for* and *parallel_reduce* constructs is a bit more intricate than the usage of OpenMP pragmas. The advantage is that they can be used on GPUs as well if certain restrictions regarding memory access are obeyed. As can be seen from our benchmarks this can be done with nearly no loss of relative performance on the different architectures.

Our first results indicate that implementations of algorithms based on Kokkos on a given architecture allows a simplified way of porting to other architectures without a redesign of code (e.g., porting an efficient code for GPUs from standard C++ to CUDA). This allows for an easier transition to other (future) architectures and to investigate and utilise this hardware in an earlier stage of their availability.

For the future it needs to be examined, if the $\mathcal{O}(N^{3/2})$ complexity can be achieved for the Kokkos implementation, e.g., by implementing nearest-neighbor lists for the real-space contribution computation. Also, it would be beneficial to implement more advanced Coulomb solvers, like PME, P3M or the fast multipole method, with Kokkos to see if the solvers can also be used performance portable. With regard to Kokkos features, it will also be investigated how large the impact is of choosing an unsuitable memory layout for a given architecture.

References

1. https://github.com/kokkos/kokkos
2. https://github.com/LLNL/RAJAPerf
3. Arnold, A., et al.: Comparison of scalable fast methods for long-range interactions. Phys. Rev. E **88**, 063308 (2013)

4. Beckingsale, D., Hornung, R., Scogland, T., Vargas, A.: Performance portable C++ programming with RAJA. In: Proceedings of the 24th Symposium on Principles and Practice of Parallel Programming, PPoPP 2019, pp. 455–456. ACM, New York (2019). https://doi.org/10.1145/3293883.3302577
5. Carter Edwards, H., Trott, C.R., Sunderland, D.: Kokkos: enabling manycore performance portability through polymorphic memory access patterns. J. Parallel Distrib. Comput. **74**(12), 3202–3216 (2014). https://doi.org/10.1016/j.jpdc.2014.07.003
6. Deserno, M., Holm, C.: How to mesh up Ewald sums. I. A theoretical and numerical comparison of various particle mesh routines. J. Chem. Phys. **109**, 7678 (1998)
7. Ewald, P.P.: Die Berechnung optischer und elektrostatischer Gitterpotentiale. Annalen der Physik **369**(3), 253–287 (1921). https://doi.org/10.1002/andp.19213690304
8. Fincham, D.: Optimisation of the Ewald sum for large systems. Mol. Simul. **13**(1), 1–9 (1994). https://doi.org/10.1080/08927029408022180
9. Frenkel, D., Smit, B.: Understanding Molecular Simulation: From Algorithms to Applications, 2nd edn. Academic Press, San Diego (2001)
10. Halver, R., Homberg, W., Sutmann, G.: Function portability of molecular dynamics on heterogeneous parallel architectures with OpenCL. J. Supercomput. **74**(4), 1522–1533 (2018). https://doi.org/10.1007/s11227-017-2232-2
11. Intel: Intel® 64 and IA-32 Architectures Optimization Reference Manual, April 2019
12. JSC: Forschungszentrum Jülich - Jülich Supercomputing Centre (JSC) (2019). https://www.fz-juelich.de/ias/jsc/
13. Luty, B., Davis, M., Tironi, I., van Gunsteren, W.: A comparison of particle-particle, particle-mesh and Ewald methods for calculating electrostatic interactions in periodic molecular systems. Mol. Simul. **14**, 11–20 (1994)
14. NVIDIA: CUDA C Programming Guide, March 2019. http://docs.nvidia.com/cuda/cuda-c-programming-guide/index.html
15. Pollock, E.L., Glosli, J.: Comments on P^3M, FMM, and the Ewald method for large periodic Coulombic systems. Comput. Phys. Commun. **95**, 93–110 (1996)
16. Sutmann, G.: Molecular dynamics - vision and reality. In: Grotendorst, J., Blügel, S., John von Neumann-Institut für Computing (eds.) Computational Nanoscience: Do It Yourself! Winter School, 14–22 February 2006, Forschungszentrum Jülich, Germany; Lecture Notes. No. 31 in NIC Series, NIC-Secretariat, Research Centre Jülich, Jülich (2006). oCLC: 181556319
17. Top500: TOP500 Supercomputer Sites. https://www.top500.org/
18. Torrellas, J., Lam, H.S., Hennessy, J.L.: False sharing and spatial locality in multiprocessor caches. IEEE Trans. Comput. **43**(6), 651–663 (1994). https://doi.org/10.1109/12.286299
19. Toukmaji, A.Y., Board Jr., J.A.: Ewald summation techniques in perspective: a survey. Comput. Phys. Commun. **95**, 73–92 (1996)

Efficient cuDNN-Compatible Convolution-Pooling on the GPU

Shunsuke Suita[1], Takahiro Nishimura[1], Hiroki Tokura[1], Koji Nakano[1(✉)],
Yasuaki Ito[1], Akihiko Kasagi[2], and Tsuguchika Tabaru[2]

[1] Department of Information Engineering, Hiroshima University, 1-4-1 Kagamiyama,
Higashi-Hiroshima 739-8527, Japan
`nakano@cs.hiroshima-u.ac.jp`
[2] Fujitsu Laboratories Ltd., 4-1-1 Kamikodanaka, Nakahara-ku, Kawasaki,
Kanagawa 211-8588, Japan

Abstract. The main contribution of this paper is to show efficient implementations of the convolution-pooling in the GPU, in which the pooling follows the multiple convolution. Since the multiple convolution and the pooling operations are performed alternately in earlier stages of many Convolutional Neural Networks (CNNs), it is very important to accelerate the convolution-pooling. Our new GPU implementation uses two techniques, (1) convolution interchange with direct sum, and (2) conversion to matrix multiplication. By these techniques, the computational and memory access cost are reduced. Further the convolution interchange is converted to matrix multiplication, which can be computed by cuBLAS very efficiently. Experimental results using Telsa V100 GPU show that our new GPU implementation compatible with cuDNN for the convolution-pooling is at least 1.34 times faster than the multiple convolution and then the pooling by cuDNN, the most popular library of primitives to implement the CNNs in the GPU.

Keywords: Deep learning · Neural Networks · Convolution · Average pooling · GPU

1 Introduction

The GPU (Graphics Processing Unit) is a specialized circuit designed to accelerate computation for building and manipulating images [4,5,9,13,15]. Latest GPUs are designed for general purpose computing and can perform computation in applications traditionally handled by the CPU. Hence, GPUs have recently attracted the attention of many application developers. NVIDIA provides a parallel computing architecture called *CUDA* (Compute Unified Device Architecture) [10], the computing engine for NVIDIA GPUs. CUDA gives developers access to the virtual instruction set and memory of the parallel computational elements in NVIDIA GPUs. Application programs running on GPUs can be developed using CUDA C programming language. Further, NVIDIA provides

© Springer Nature Switzerland AG 2020
R. Wyrzykowski et al. (Eds.): PPAM 2019, LNCS 12044, pp. 46–58, 2020.
https://doi.org/10.1007/978-3-030-43222-5_5

several libraries of primitives to accelerate application programs. For example, cuBLAS [11], a linear algebra library including matrix computations, is optimized for each of GPU architecture generations, such as Kepler, Maxwell, Pascal, Volta, and Turing. So, we can attain the best performance for operations of linear algebra using cuBLAS, and it makes no sense to develop them using CUDA C language by ourselves in most cases.

GPUs have been used for accelerating machine learning by Deep Neural Networks (DNNs). NVIDIA provides cuDNN [2,12], a GPU-accelerated library of primitives for DNNs such as the convolution and the pooling. Developers can use cuDNN APIs to implement DNN operations in GPUs. Further, popular machine learning frame works such as TensorFlow, CNTK, PyTorch, and Caffe2 call cuDNN APIs to accelerate operations of DNN using GPUs. Hence, it is very important to improve library calls of cuDNN. The main purpose of this paper is to provide an efficient cuDNN-compatible GPU implementation for the convolution-pooling, in which the pooling follows the convolution as illustrated in Fig. 1. Since the convolution and the pooling are performed alternately in earlier stages of a Convolutional Neural Network (CNN), a kind of DNN for images, training and inference of CNNs can be accelerated.

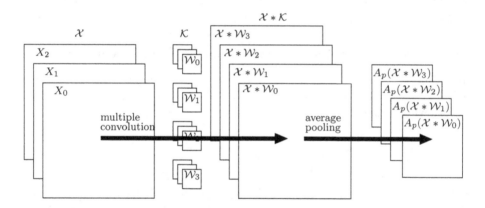

Fig. 1. The convolution-pooling for $I = 3$ input channels and $R = 4$ output channels

Our new GPU implementation for the convolution-pooling uses two techniques, (1) convolution interchange with direct sum, and (2) conversion to matrix multiplication. In (1), the direct sum operation is performed before the convolution to obtain the same results. The computational and memory access cost are reduced by this technique. To further accelerate the convolution-pooling, the computation in (1) is converted to equivalent matrix multiplication, which can be computed by cuBLAS very efficiently. Our experimental results using Telsa V100 GPU show that our new GPU implementation for the convolution-pooling is at least 1.34 times faster than the multiple convolution and then the pooling by cuDNN APIs. From the theoretical analysis, our convolution-pooling algorithm reduces the computational cost of the convolution-pooling. Thus, our acceleration technique is applicable for any architecture if regular memory access performed by it does not have large memory access penalty.

There are a lot of approaches to accelerate the operations in the DNN [1,8,14, 16]. In [7], the convolution interchange technique to accelerate the convolution-pooling has been presented. They use the Summed Area Table (SAT) of the input channels to reduce the computational cost, which is inverse proportional to the pooling size. However, in most DNNs, the pooling of size only 2×2 is used. We have used the direct sum computation, which is more efficient than the SAT when the pool size is small. In addition, we have used the matrix multiplication conversion for the convolution interchange technique for further acceleration. We have used two techniques above and evaluate the performance on Tesla V100 GPU.

2   Convolution-Pooling in the CNN

The main purpose of this section is to explain the details of convolution-pooling, in which pooling operation follows convolution operation and show *the computational cost*, which is the number of arithmetic operations such as addition and multiplication.

2.1   Convolution-Pooling and Straightforward Implementation

Let X and W be matrices of size $n \times n$ and $k \times k$, respectively. *The convolution of X and W* denoted by $X * W$ is a $(n - k + 1) \times (n - k + 1)$ matrix defined by the following formula:

$$(X * W)[i, j] = \sum_{i'=0}^{k-1} \sum_{j'=0}^{k-1} X[i + i', j + j']W[i', j'] \qquad (0 \le i, j \le n - k) \quad (1)$$

Sometimes *zero padding operation*, which expands the size of X or W by padding zero elements, is performed before the convolution to obtain an $n \times n$ resulting matrix. Usually, in the area of image processing and machine learning, $n \gg k$ holds and matrices X and W are called *a channel* and *a kernel*, respectively. For a set $\mathcal{X} = \{X_0, X_1, \ldots, X_{I-1}\}$ of I channels and a set $\mathcal{W} = \{W_0, W_1, \ldots, W_{I-1}\}$ of I kernels, we write $\mathcal{X} * \mathcal{W}$ to denote the element-wise sum of the pairwise convolutions, that is,

$$(\mathcal{X} * \mathcal{W})[i, j] = \sum_{l=0}^{I-1} (X_l * W_l)[i, j] \qquad (0 \le i, j \le n - k) \quad (2)$$

Suppose that a set $\mathcal{X}$ of I channels and R sets $\mathcal{K} = \{\mathcal{W}_0, \mathcal{W}_1, \ldots, \mathcal{W}_{R-1}\}$ of I kernels each are given. *The multiple convolution* is a task to compute R products

$$\mathcal{X} * \mathcal{K} = \{\mathcal{X} * \mathcal{W}_0, \mathcal{X} * \mathcal{W}_1, \ldots, \mathcal{X} * \mathcal{W}_{R-1}\}.$$

Clearly, the total computational cost of $\mathcal{X} * \mathcal{K}$ is $O(n^2k^2IR)$. The reader should refer to Fig. 1 illustrating multiple convolution for $I = 3$ input channels and $R = 4$ output channels.

The (average) pooling of a matrix is a down-sampling by dividing an input matrix into blocks, and computing the average of each block. More specifically, for an $n \times n$ matrix X, the resulting matrix $A_p(X)$ of the average pooling is an $\frac{n}{p} \times \frac{n}{p}$ matrix such that

$$A_p(X)[i,j] = \sum_{i'=pi}^{pi+p-1} \sum_{j'=pj}^{pj+p-1} X[i',j']/p^2 \qquad (0 \le i,j \le \tfrac{n}{p} - 1) \qquad (3)$$

where $p \times p$ is *the pooling size*. Since the sum of p^2 input elements is computed for each element of the resulting $\frac{n}{p} \times \frac{n}{p}$ matrix, the computational cost is $p^2 \times (\frac{n}{p})^2 = O(n^2)$.

In the CNN, it is often the case that the pooling follows the multiple convolution as illustrated in Fig. 1. We call these computations combined *the convolution-pooling*, which is a task to output

$$A_p(\mathcal{X} * \mathcal{K}) = \{A_p(\mathcal{X} * \mathcal{W}_0), A_p(\mathcal{X} * \mathcal{W}_1), \ldots, A_p(\mathcal{X} * \mathcal{W}_{R-1})\}$$

Clearly, the total computational cost to obtain these R matrices is $(O(n^2k^2I) + O(n^2)) \cdot R = O(n^2k^2IR)$, and we have,

Lemma 1. *The convolution-pooling can be done in $O(n^2k^2IR)$ computational cost.*

We will show that the computational cost can be reduced to $O(\frac{n^2k^2IR}{p^2})$ later.

2.2 Fused Kernel Implementation of Convolution-Pooling Layer

The convolution operation is associative, that is, $(X * Y) * Z = X * (Y * Z)$ holds for any matrices X, Y, and Z. We will show that, using this associative law, the convolution-pooling can be implemented by the convolution with the down-sampling.

Let S_p be a down-sampling operation to pick one element in each $p \times p$ block of a matrix X. More specifically, $S_p(X)$ of size $\frac{n}{p} \times \frac{n}{p}$ is defined as follows:

$$S_p(X)[i,j] = X[pi,pj] \qquad (0 \le i,j \le \tfrac{n}{p} - 1).$$

Let α_p be a kernel of size $p \times p$ with every element taking value $\frac{1}{p^2}$. Clearly, the convolution $X * \alpha_p$ corresponds to the average filter for X. Hence, the resulting matrix $A_p(X)$ of the average pooling for X can be computed by evaluating $S_p(X * \alpha_p)$, that is, $A_p(X) = S_p(X * \alpha_p)$ always holds. Thus, each resulting matrix of convolution-pooling can be obtained by the following formula:

$$A_p(\mathcal{X} * \mathcal{W}_r) = S_p(\sum_{l=0}^{I-1}(X_l * (W_{r,l} * \alpha_p))) \qquad (0 \le r \le R - 1), \qquad (4)$$

where $W_{r,l}$ denotes the l-th kernel in W_r. We can think that each *fused kernel* $W_{r,l}*\alpha_p$ is a fixed matrix of size $(k+p-1)\times(k+p-1)$. After that, $X_l*(W_{r,l}*\alpha_p)$ is computed. However, it is not necessary to compute all matrix elements of $X_l*(W_{r,l}*\alpha_p)$, because the down-sampling S_p is performed; Only one element in every $p\times p$ block is necessary. Thus, we have,

Lemma 2. *The convolution-pooling by fused kernels can be done in* $O(\frac{n^2(k+p)^2 IR}{p^2})$ *computational cost.*

The computational cost is not better than that of the straightforward implementation shown for Lemma 1. However, since convolution operation is performed only once, fused kernel implementation can be faster from the practical point of view.

2.3   Convolution Interchange for the Convolution-Pooling

This section shows the convolution interchange technique to implement the convolution-pooling, and it runs in only $O(\frac{n^2k^2 IR}{p^2})$ computational cost by computing the summed area table (Fig. 2, [7]). We then go on to show that our direct sum technique for the convolution interchange for further acceleration.

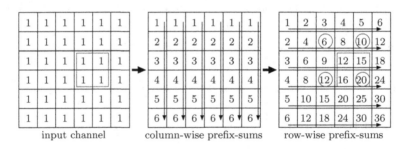

Fig. 2. The summed area table (SAT) computed by column-wise prefix-sums and then by row-wise prefix-sums

Since convolution operation is associative and commutative, we can rewrite formula (4) as follows:

$$A_p(\mathcal{X}*W_r) = S_p \sum_{l=0}^{I-1}((X_l*\alpha_p)*W_{r,l}) \qquad (0 \le r \le R-1). \qquad (5)$$

Clearly, each $X_l*\alpha_p$ can be computed in $O(n^2p^2)$ computational cost. If we use *the summed area table (SAT)* as presented in [7], the computational cost can be reduced to $O(n^2)$. For an $n\times n$ matrix X, the SAT $S(X)$ is defined as follows:

$$S(X)[i,j] = \sum_{i'=0}^{i}\sum_{j'=0}^{j} X[i',j'] \qquad (6)$$

It is known that $S(X)$ can be obtained by computing the column-wise prefix-sums of X and then computing the row-wise prefix-sums [3,6]. Hence, $S(X)$ can be computed in $O(n^2)$ computational cost. The sum of any rectangular block can be computed by four elements of $S(X)$. For example, the sum of any $p \times p$ block of X can be computed by four elements of $S(X)$ as follows:

$$\sum_{i'=i}^{i+p-1} \sum_{j'=i}^{j+p-1} X[i',j'] = S(X)[i+p-1, j+p-1]$$
$$+ S(X)[i-1, j-1] - S(X)[i-1][j+p-1] - S(X)[i+p-1][j-1] \quad (7)$$

For example, in Fig. 2, the sum of elements in the red square can be computed by four elements with blue circle such that $20 + 6 - 10 - 12 = 4$. Thus, each element of $X_l * \alpha_p$ can be computed by $O(1)$ computational cost by computing the sum of each $p \times p$ region and dividing it by p^2, and so the total computational cost to obtain all $X_l * \alpha_p$ for all l ($0 \leq l \leq I - 1$) is $O(n^2 I)$. After that, each element of $S_p(\sum_{l=0}^{I-1}((X_l * \alpha_p) * W_l))$ is computed in $O(k^2 I)$ computational cost. Since we have $\frac{n^2}{p^2}$ elements, $S_p(\sum_{l=0}^{I-1}((X_l * \alpha_p) * W_l))$ can be computed in $\frac{n^2}{p^2} \cdot O(k^2 I) = O(\frac{n^2 k^2 I}{p^2})$ computational cost. Since the convolution-pooling is performed for R sets of I kernels each, we have,

Theorem 1. *The convolution-pooling by the convolution interchange can be completed in $O(\frac{n^2 k^2 IR}{p^2} + n^2 I)$ computational cost.*

Clearly, if $k\sqrt{R} \geq p$, then the computational cost is $O(\frac{n^2 k^2 IR}{p^2})$. Actually, p is smaller than both k and R in practical implementations of CNNs. Further, in the CNN, most pooling operation is performed with parameter $p = 2$. If this is the case, it makes no sense to compute the SAT to obtain $X_l * \alpha_p$. By computing the sum of each neighboring pair in row direction, and then by computing the sum of each neighboring pair in row direction, we can obtain the sum of every 2×2 block as illustrated in Fig. 3. By dividing each sum by 4, we can obtain $X_l * \alpha_2$ in $O(n^2 I)$. For later reference, we call this computation *direct sum*. After computing the direct sum of each input channel, we can compute $S_p \sum_{l=0}^{I-1}((X_l * \alpha_2) * W_{r,l})$ to complete the convolution-pooling in the same way.

3   Matrix Multiplication Conversion for Convolution-Pooling

This section explains the matrix multiplication conversion technique known as im2col, which is implemented in Python, MATLAB, cuDNN, etc. It converts input channels combined are converted in a single matrix and kernel set combined is also converted in a single matrix so that the product of them equals to the result of the multiple convolution. We apply this technique to the convolution-pooling, and use cuBLAS to multiply two matrices.

We first show that the multiple convolution represented as formula (2) can be computed by a matrix multiplication. First, I input channels $\mathcal{X} =$

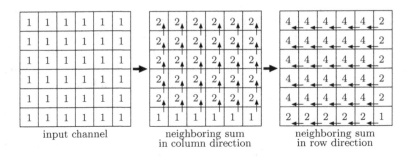

Fig. 3. The direct-sum computation to obtain $X_l * \alpha_2$

$\{X_0, X_1, \ldots, X_{I-1}\}$ and R kernel sets $\mathcal{K} = \{\mathcal{W}_0, \mathcal{W}_1, \ldots, \mathcal{W}_{R-1}\}$ are converted into two matrices $D(\mathcal{X})$ and $V(\mathcal{K})$ of size $(n - k + 1)^2 \times k^2 I$ and $k^2 I \times R$ as illustrated in Fig. 4. The matrix $D(\mathcal{X})$ has $k^2 I$ columns such that consecutive k^2 columns are copied from an input channel. Each row in consecutive k^2 columns corresponds to a $k \times k$ block. For example, green blocks of X_0 and X_1 are arranged in the top row of $D(\mathcal{X})$. Hence, $D(\mathcal{X})$ has $(n - k + 1)^2$ rows. Each column of the matrix $V(\mathcal{K})$ corresponds to a kernel set and the value of I kernels in a kernel set are copied in the corresponding column as illustrated in Fig. 4. From the figure, the reader should have no difficulty to confirm that the product of $D(\mathcal{X})$ and $V(\mathcal{K})$ is equal to the values of R output channels $\mathcal{X} * \mathcal{W}_i$ for all i ($0 \leq i \leq R-1$). Thus, the multiple convolution can be obtained by the product of $D(\mathcal{X})$ and $V(\mathcal{K})$.

The computational cost for generating $D(\mathcal{X})$ is $O((n - k + 1)^2 \times k^2 I) \leq O(n^2 k^2 I)$. Also, that for $V(\mathcal{K})$ is $O(k^2 IR)$. Their product can be computed in $O((n - k + 1)^2 \cdot k^2 I \cdot R) \leq O(n^2 k^2 IR)$. Hence, the total computing cost is $O(n^2 k^2 IR)$.

We will show that the same technique can be used for the convolution interchange, which computes $S_p \sum_{l=0}^{I-1} ((X_l * \alpha_p) * W_l)$. Suppose that $\mathcal{X} * \alpha_p = \{X_0 * \alpha_p, X_1 * \alpha_p, \ldots, X_{I-1} * \alpha_p\}$ and R kernel sets $\mathcal{K} = \{\mathcal{W}_0, \mathcal{W}_1, \ldots, \mathcal{W}_{R-1}\}$ are given. Clearly, by the product of two matrices $D(X_l * \alpha_p)$ and $V(\mathcal{K})$, we can obtain $\sum_{l=0}^{I-1} ((X_l * \alpha_p) * W_{r,l})$. Since we need the down-sample $S_p \sum_{l=0}^{I-1} ((X_l * \alpha_p) * W_{r,l})$, which is obtained by selecting one element from every $p \times p$ block of $\sum_{l=0}^{I-1} ((X_l * \alpha_p) * W_{r,l})$, we can remove unnecessary rows from $D(X_l * \alpha_p)$ to obtain $S_p \sum_{l=0}^{I-1} ((X_l * \alpha_p) * W_{r,l})$. Let $D_p(X_l * \alpha_p)$ denote the matrix obtained by this down-sampling such that one out of every p^2 rows in $D(X_l * \alpha_p)$ is picked appropriately. The product of $D_p(X_l * \alpha_p)$ and $V(\mathcal{K})$ can be computed in $O(\frac{n^2 k^2 IR}{p^2})$ and so the total computational cost is $O(\frac{n^2 k^2 IR}{p^2} + n^2 I)$.

Also, the same technique can be used for fused kernels. Let $\mathcal{W}'_i = \{W_0 * \alpha_p, W_1 * \alpha_p, \ldots, W_{I-1} * \alpha_p\}$ be a set of I fused kernels, and $\mathcal{K}' = \{\mathcal{W}'_0, \mathcal{W}'_1, \ldots, \mathcal{W}'_{R-1}\}$ be R sets of I fused kernels. The convolution-pooling by fused kernels can be computed by the product of $D_p(\mathcal{X})$ and $V(\mathcal{K}')$. The total computational cost is $O(\frac{n^2 (k+p)^2 IR}{p^2})$.

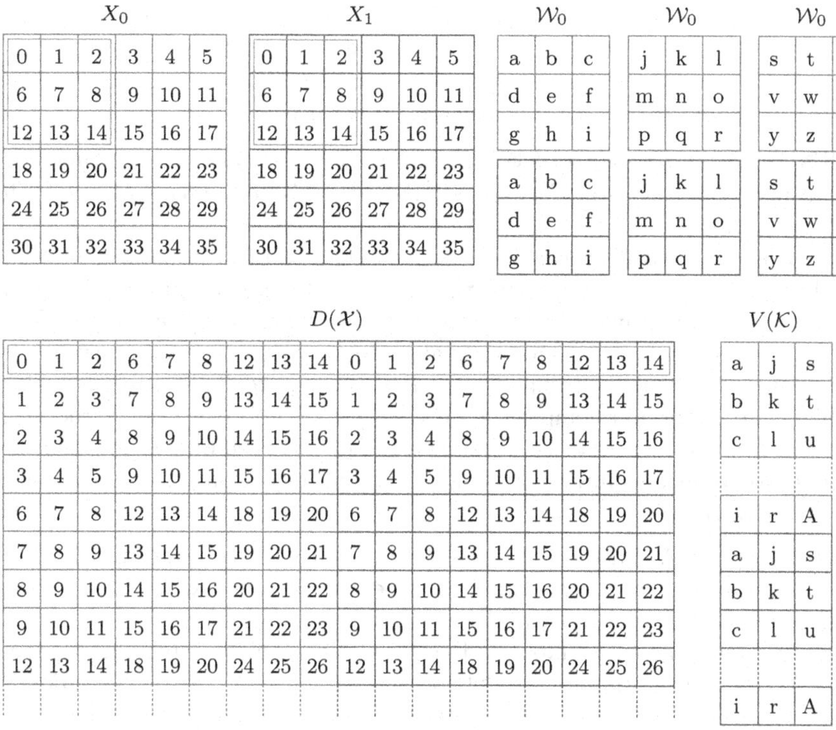

X_0

0	1	2	3	4	5
6	7	8	9	10	11
12	13	14	15	16	17
18	19	20	21	22	23
24	25	26	27	28	29
30	31	32	33	34	35

X_1

0	1	2	3	4	5
6	7	8	9	10	11
12	13	14	15	16	17
18	19	20	21	22	23
24	25	26	27	28	29
30	31	32	33	34	35

$\mathcal{W}_0$

a	b	c
d	e	f
g	h	i
a	b	c
d	e	f
g	h	i

$\mathcal{W}_0$

j	k	l
m	n	o
p	q	r
j	k	l
m	n	o
p	q	r

$\mathcal{W}_0$

s	t	u
v	w	x
y	z	A
s	t	u
v	w	x
y	z	A

$D(\mathcal{X})$

0	1	2	6	7	8	12	13	14	0	1	2	6	7	8	12	13	14
1	2	3	7	8	9	13	14	15	1	2	3	7	8	9	13	14	15
2	3	4	8	9	10	14	15	16	2	3	4	8	9	10	14	15	16
3	4	5	9	10	11	15	16	17	3	4	5	9	10	11	15	16	17
6	7	8	12	13	14	18	19	20	6	7	8	12	13	14	18	19	20
7	8	9	13	14	15	19	20	21	7	8	9	13	14	15	19	20	21
8	9	10	14	15	16	20	21	22	8	9	10	14	15	16	20	21	22
9	10	11	15	16	17	21	22	23	9	10	11	15	16	17	21	22	23
12	13	14	18	19	20	24	25	26	12	13	14	18	19	20	24	25	26

$V(\mathcal{K})$

a	j	s
b	k	t
c	l	u
i	r	A
a	j	s
b	k	t
c	l	u
i	r	A

Fig. 4. The multiple convolution by matrix multiplication for $I = 2$ input channels $\mathcal{X} = \{X_0, X_1\}$ with $R = 3$ kernel sets $\mathcal{K} = \{\mathcal{W}_0, \mathcal{W}_1, \mathcal{W}_2\}$.

4 GPU Implementations

In this section, we explain the details of five GPU implementations, cuDNN(naive), cuDNN(fused), cuBLAS(fused), cuDNN(direct), and cuBLAS(direct) to compute the convolution-pooling. We call parameters of the multiple convolution such as data type (double, float, half), the size $n \times n$ and the number I of channels, the size $k \times k$ and the number I of kernels, and the batch size B, *the configuration of the multiple convolution*.

cuDNN(naive): The multiple convolution $\mathcal{X} * \mathcal{K}$ is computed by the cuDNN and then the pooling $A_p(\mathcal{X} * \mathcal{K})$ is computed by cuDNN. The convolution of cuDNN can have several options of convolution algorithms.

We use the function call cudnnGetConvolutionForwardAlgorithm() returns the best algorithm for the configuration of the multiple convolution. After that, we first execute cudnnGetConvolutionForwardWorkspaceSize() with the best selected algorithm and the configuration which allocates memory space in the global memory for multiple convolution computation by cuDNN. We call cudnnConvolutionForward() with the best selected algorithm and the configuration to perform the multiple convolution. Finally, we cal cudnnPoolingForward() with the configuration to perform the pooling. When we evaluate the running time of cuDNN(naive), the time for cudnnConvolutionForward() and cudnnPooling-Forward() is included. The running time for cudnnGetConvolutionForwardAlgorithm() is excluded because it is executed only once.

cuDNN(fused): All fused kernels $W_{r,l} * \alpha_p$ are computed in advance and then the multiple convolution $S_p \sum_{l=0}^{I-1} (X_l * (W_{r,l} * \alpha_p))$ is computed for each r-th kernel set by cuDNN.
We first compute the fused kernel $W_{r,l} * \alpha_p$ for a kernel set by our implementation in an obvious way. Similarly, cudnnGetConvolutionForwardAlgorithm() is called to obtain the best algorithm for the configuration. We then executes cudnnGetConvolutionForwardWorkspaceSize() to allocate the global work memory space, and cudnnConvolutionForward() with stride p to compute the multiple convolution. Since the fused filter computation is executed once for the same kernel set, and cudnnGetConvolutionForwardWorkspaceSize() is called only once, the running time of cudnnConvolutionForward() is used for evaluating the performance of cuDNN(fused).

cuBLAS(fused): A matrix $V(\mathcal{K}')$ is generated from the resulting values of $W_{r,l} * \alpha_p$ in advance, a matrix $D_p(\mathcal{X})$ is generated by our CUDA C program, and then the product $D_p(\mathcal{X}) \cdot V(\mathcal{K}')$ is computed by cuBLAS.
We first compute the fused filter and convert it to the corresponding matrix $V(\mathcal{K}')$. We then covert input channels of each channel set to the corresponding matrix $D_p(\mathcal{X})$ by our CUDA C program. Since we have B channel sets, the corresponding B matrices are concatenated into one large matrix. Finally, we execute cublasSgemmStridedBatched() to complete the convolution-pooling. Since the $V(\mathcal{K}')$ is computed only once, the running time of the computation of the corresponding matrix $D_p(\mathcal{X})$ and cublasSgemmStridedBatched() are evaluated.

cuDNN(direct): Each $X_l * \alpha_p$ is computed by the direct sum using our CUDA C program and then the multiple convolution $S_p \sum_{l=0}^{I-1} ((X_l * \alpha_p) * W_{r,l})$ is computed by cuDNN.
Each $X_l * \alpha_p$ is computed by the direct sum using our CUDA C program. Similarly to cuDNN(naive), the best algorithm is obtained by calling cudnnGetConvolutionForwardAlgorithm() for the configuration of the multiple convolution. We then executes cudnnGetConvolutionForwardWorkspaceSize() to allocate the global work memory space, and cudnnConvolutionForward() with stride p to compute the multiple convolution. Since cudnnGetConvolutionForwardWorkspaceSize() is executed only once, the running time of the computation of $X_l * \alpha_p$ by our CUDA C program and cudnnConvolutionForward() are used to evaluate the performance.

cuBLAS(direct): Each $X_l * \alpha_p$ is computed by the direct sum and then $D_p(X_l * \alpha_p)$ and $V(\mathcal{K})$ are generated by our CUDA C program.
The product of $D_p(X_l * \alpha_p)$ and $V(\mathcal{K})$ are computed by cuBLAS. We first convert kernels to the corresponding matrix $V(\mathcal{K})$ by our CUDA C program. We then compute Each $X_l * \alpha_p$ is computed by the direct sum and convert it to the

corresponding matrix $D_p(X_l * \alpha_p)$ by our CUDA C program. We execute cublasS-gemmStridedBatched() to compute the product of $D_p(X_l * \alpha_p)$ and $V(\mathcal{K})$. Since $V(\mathcal{K})$ for a kernel set $\mathcal{K}$ is computed only once, the running time of the computation of $D_p(X_l * \alpha_p)$ and cublasSgemmStridedBatched() are evaluated.

If developers implement the convolution-pooling using cuDNN as it is, they will use cuDNN(naive) implementation. Further, if they use DNN frame works such as Chainer, PyTorch, and TensorFlow, the convolution-pooling is executed on the GPU as cuDNN(naive). Thus, the performance of cuDNN(naive) approximates that using DNN frameworks. If developers know the fused kernel technique, they may use cuDNN(fused) to implement the convolution-pooling. Both cuDNN(direct) and cuBLAS(direct) use the convolution interchange and the direct sum. Their difference is to use cuDNN or cuBLAS to compute the convolution.

Also, please note that the convolution performed for a multiple channel sets called *batch* at the same time in most DNNs. More specifically, let B denote the size of batch, *i.e.* the number of channel sets. The multiple convolution of a batch of size B performs the convolution for B channel sets of I channels each with respect to a single kernel set of I kernels. We should evaluate the performance of the running time of the convolution for a batch.

5   Experimental Results

Table 1 shows the running time of the convolution-pooling by cuDNN(naive), cuDNN(fused), cuBLAS(fused), cuDNN(direct), and cuBLAS(direct) for input channel size from 8×8 to 64×64 and for the number of input/output channels from 32/32 to 512/512. The data type is a 32-bit single precision floating point number. We have used NVIDIA Tesla V100 with cuDNN v7.1.4 and cuBLAS v9.0. Since kernels of size 3×3 and the pooling for 2×2 are used in the most DNNs, we use these parameters for the experiments. The running time is evaluated for 64 sets of the multiple convolution, thus, it corresponds to batch size 64 in the DNN. The running time in the table is the average of 100 computations. In the table, the best running time of the five implementations for each parameter set is highlighted. It also shows the speedups of the best result of cuDNN(direct) and cuBLAS(direct) over cuDNN(naive), and that of over the best result of cuDNN(fused) and cuBLAS(fused). From the table, we can see that either cuDNN(direct) or cuBLAS(direct) is always faster than cuDNN(naive) for each case. Also, they are faster than cuDNN(fused) and cuBLAS(fused) in most cases. They are slower for few cases but the difference is quite small. The maximum speedup of 9.49 is achieved for 128 input/output channels of size 64×64, because cuDNN(naive) does not select an appropriate algorithm and takes a lot of time for the multiple convolution.

Unfortunately, the best algorithm of the five differs depending on configurations. Usually, DNNs have many layers with different configurations. We can choose the best one for each layer to minimize the total computing time.

Table 1. The running time (ms) of the convolution-pooling with 3×3 kernels and 2×2 pooling for I input channels and R output channels for batch size 64

Input channel size: 8×8					
channels I/R	32/32	64/64	128/128	256/256	512/512
cuDNN(naive)	0.104	0.137	0.206	0.651	1.87
cuDNN(fused)	0.105	0.153	0.259	0.425	0.961
cuBLAS(fused)	0.103	0.141	0.281	0.929	3.17
cuDNN(direct)	0.107	0.133	0.340	**0.341**	**0.701**
cuBLAS(direct)	**0.0473**	**0.0739**	**0.154**	0.479	1.92
Speed-up: cuDNN(naive)	2.20	1.85	1.34	1.91	2.67
Speed-up: fused	2.18	1.91	1.68	1.25	1.37
Input channel size: 16×16					
channels I/R	32/32	64/64	128/128	256/256	512/512
cuDNN(naive)	0.186	0.262	0.663	1.87	6.51
cuDNN(fused)	0.102	0.155	0.299	0.824	3.18
cuBLAS(fused)	0.110	0.174	0.335	0.922	3.34
cuDNN(direct)	0.112	0.146	0.342	0.616	1.92
cuBLAS(direct)	**0.0488**	**0.101**	**0.192**	**0.577**	**1.89**
Speed-up: cuDNN(naive)	3.81	2.59	3.45	3.24	3.44
Speed-up: fused	2.09	1.53	1.56	1.43	1.68
Input channel size: 32×32					
channels I/R	32/32	64/64	128/128	256/256	512/512
cuDNN(naive)	0.247	0.42	2.03	5.98	20.9
cuDNN(fused)	**0.156**	0.254	0.836	2.82	10.7
cuBLAS(fused)	0.294	0.538	1.05	3.3	12.5
cuDNN(direct)	0.163	**0.255**	0.626	**1.85**	7.25
cuBLAS(direct)	0.164	0.315	**0.610**	**1.85**	**7.02**
Speed-up: cuDNN(naive)	1.52	1.65	3.33	3.23	2.98
Speed-up: fused	0.957	0.996	1.37	1.52	1.52
Input channel size: 64×64					
channels I/R	32/32	64/64	128/128	256/256	512/512
cuDNN(naive)	0.936	1.85	20.6	13.9	47
cuDNN(fused)	**0.327**	1.19	4.03	13.5	49.3
cuBLAS(fused)	0.989	1.96	4.06	12.8	44.3
cuDNN(direct)	0.355	**0.685**	**2.17**	7.03	**24.2**
cuBLAS(direct)	0.548	1.12	2.28	**7.01**	25.0
Speed-up: cuDNN(naive)	2.64	2.70	9.49	1.98	1.94
Speed-up: fused	0.921	1.74	1.86	1.83	1.83

6   Conclusion

We have presented new GPU implementations for the convolution-pooling based on convolution interchange with direct sum. Experimental results using Tesla V100 GPU show that our new GPU implementation compatible with cuDNN for the convolution-pooling is at least 1.34 times faster than the multiple convolution and then the pooling by cuDNN.

References

1. Cheng, Y., Wang, D., Zhou, P., Zhang, T.: A survey of model compression and acceleration for deep neural networks. CoRR abs/1710.09282, October 2017
2. Chetlur, S., Woolley, C., Vandermersch, P., Cohen, J., Tran, J., Catanzaro, B., Shelhamer, E.: cuDNN: efficient primitives for deep learning. CoRR abs/1410.0759, August 2014
3. Emoto, Y., Funasaka, S., Tokura, H., Honda, T., Nakano, K., Ito, Y.: An optimal parallel algorithm for computing the summed area table on the GPU. In: Proceedings of International Parallel and Distributed Processing Symposium Workshops, pp. 763–772, February 2018
4. Honda, T., Yamamoto, S., Honda, H., Nakano, K., Ito, Y.: Simple and fast parallel algorithms for the Voronoi map and the Euclidean distance map, with GPU implementations. In: Proceedings of International Conference on Parallel Processing, pp. 362–371, August 2017
5. Hwu, W.W.: GPU Computing Gems, Emerald edn. Morgan Kaufmann, Burlington (2011)
6. Kasagi, A., Nakano, K., Ito, Y.: Parallel algorithms for the summed area table on the asynchronous hierarchical memory machine, with GPU implementations. In: Proceedings of International Conference on Parallel Processing (ICPP), pp. 251–260, September 2014
7. Kasagi, A., Tabaru, T., Tamura, H.: Fast algorithm using summed area tables with unified layer performing convolution and average pooling. In: Proceedings of International Workshop on Machine Learning for Signal Processing, September 2017
8. Li, C., Yang, Y., Feng, M., Chakradhar, S., Zhou, H.: Optimizing memory efficiency for deep convolutional neural networks on GPUs. In: Proceedings of the International Conference for High Performance Computing, Networking, Storage and Analysis, November 2016
9. Matsumura, N., Tokura, H., Kuroda, Y., Ito, Y., Nakano, K.: Tile art image generation using conditional generative adversarial networks. In: Proceedings of International Symposium on Computing and Networking Workshops, pp. 209–215 (2018)
10. NVIDIA Corporation: NVIDIA CUDA C programming guide version 4.0 (2011)
11. NVIDIA Corporation: CUBLAS LIBRARY user guide, February 2019. https://docs.nvidia.com/cuda/cublas/index.html
12. NVIDIA Corporation: CUDNN developer guide, February 2019. https://docs.nvidia.com/deeplearning/sdk/cudnn-developer-guide/index.html
13. Ogawa, K., Ito, Y., Nakano, K.: Efficient Canny edge detection using a GPU. In: Proceedings of International Conference on Networking and Computing, pp. 279–280. IEEE CS Press, November 2010

14. Sze, V., Chen, Y.H., Yang, T.J., Emer, J.S.: Efficient processing of deep neural networks: a tutorial and survey. Proc. IEEE **105**(12), 2295–2329 (2017)
15. Takeuchi, Y., Takafuji, D., Ito, Y., Nakano, K.: ASCII art generation using the local exhaustive search on the GPU. In: Proceedings of International Symposium on Computing and Networking, pp. 194–200, December 2013
16. Zhang, Q., Zhang, M., Chen, T., Sun, Z., Ma, Y., Yu, B.: Recent advances in convolutional neural network acceleration. Neurocomputing **323**, 37–51 (2019)

Reactive Task Migration for Hybrid MPI+OpenMP Applications

Jannis Klinkenberg[1]([✉]) [ID], Philipp Samfass[2] [ID], Michael Bader[2],
Christian Terboven[1], and Matthias S. Müller[1]

[1] Chair for High Performance Computing, IT Center, RWTH Aachen University,
Aachen, Germany
{j.klinkenberg,terboven,mueller}@itc.rwth-aachen.de
[2] Department of Informatics, Technical University of Munich, Garching, Germany
{samfass,bader}@in.tum.de

Abstract. Many applications in high performance computing are designed based on underlying performance and execution models. While these models could successfully be employed in the past for balancing load within and between compute nodes, modern software and hardware increasingly make performance predictability difficult if not impossible. Consequently, balancing computational load becomes much more difficult. Aiming to tackle these challenges in search for a general solution, we present a novel library for fine-granular task-based reactive load balancing in distributed memory based on MPI and OpenMP. With our approach, individual migratable tasks can be executed on any MPI rank. The actual executing rank is determined at run time based on online performance data. We evaluate our approach under an enforced power cap and under enforced clock frequency changes for a synthetic benchmark and show its robustness for work-induced imbalances for a realistic application. Our experiments demonstrate speedups of up to 1.31X.

Keywords: Reactivity · Task migration · Hybrid MPI+OpenMP ·
Load balancing · Tasking

1 Introduction and Related Work

Over the past decades, most scientific applications have been developed under the assumption of a homogenous execution environment where every compute node – and even every single core – in a larger cloud or High Performance Computing (HPC) system has a constant equal speed. Therefore, executing the same work on every node should require the same computation time. In the past, this execution model was shown to be highly accurate and efficient for balancing computational load. However, as both hardware and software become increasingly complex, this model might no longer be sufficient on current and future systems.

© Springer Nature Switzerland AG 2020
R. Wyrzykowski et al. (Eds.): PPAM 2019, LNCS 12044, pp. 59–71, 2020.
https://doi.org/10.1007/978-3-030-43222-5_6

Today's architectures already exhibit run time variations, e.g., with dynamic voltage and frequency scaling (DVFS), sophisticated memory hierarchies comprising caches, DRAM, NVRAM and HBM or features like Intel's Turbo Boost [1,3]. Further, CPU power efficiency variations arising from the manufacturing process can lead to performance variations in presence of an enforced power cap [7]. Another source of *dynamic variability* stems from modern numerics in simulation applications such as particle simulations or iterative codes employing adaptive mesh refinement (AMR) where the workload distributed across processing units changes over time causing load imbalances both in shared and distributed memory. In the ADER-DG numerical scheme with a-posteriori limiting [15], additional computation work arises dynamically in regions where the solution is not considered to be physically admissible.

Consequently, the assumption that the execution time can be accurately predicted does no longer apply. In order to prevent load imbalances and performance declines resulting from *performance variability* both in hardware and software, we believe that it is necessary that applications are able to dynamically react on the changing execution conditions.

The literature describes several approaches to mitigate effects of load imbalance. Shared memory runtime systems such as Cilk [2], TBB [12] and several OpenMP [10] implementations apply work stealing to dynamically balance load between threads in shared-memory only. Other distributed runtime systems such as Charm++ [8] enable work redistribution in distributed memory. However, they act to the best of our knowledge typically on a rather coarse-grained level and require defined synchronization points where load migration is triggered.

Producer-consumer patterns or global repartitioning of work or data (e.g., [11]) are common application-level load balancing approaches. Both strategies typically induce high overhead for message and data transfer between processes. Further, while global repartitioning of work – usually done at global synchronization points – was an effective technique to ensure proper balance in the past, it is based on a cost model to predict future execution time. Such a cost model is doomed to fail in increasingly complex hardware and software environments.

To mitigate these shortcomings of traditional predictive load balancing, we present a library for fine-grained reactive load balancing of task-parallel MPI+X applications that allows reactive load balancing within and across process boundaries. Further, as our goal is not to create a completely new programming language or paradigm, our library rather builds on top of the established standards MPI and OpenMP and provides an incremental solution to support the large amount of existing codes developed in C, C++ and Fortran. In our previous work [13], we carried out a feasibility study of our reactive approach in the PDE framework sam(oa)^2. In this work, we extend and improve our concept as well as generalize and modularize it to make it available to other MPI-parallel applications. Consequently, this paper makes the following contributions:

1. We present the first conceptual generalization of reactive load balancing to arbitrary MPI-parallel task-based applications, detailing both requirements and limitations associated with it.

2. We present our library implementation based on MPI+OpenMP that allows an incremental integration into existing task-based applications with minimal programming efforts. Further, we have a deeper look at implementation extensions and decisions that have changed compared to our feasibility study.
3. To demonstrate the effectiveness and scalability of our approach we conduct a systematic evaluation using a comprehensible synthetic benchmark comparing different implementation decisions as well as the sam(oa)^2 framework [9].

The remainder of this paper is structured as follows. Section 2 introduces the fundamental concept and requirements for a reactive hybrid task-based load balancing solution. In Sect. 3, we describe our implementation and different design choices. An experimental evaluation is carried out in Sect. 4 before we conclude and discuss future work in Sect. 5.

2 Reactive Load Balancing

This section details our concept of fine-granular task-based load balancing in both shared and distributed memory. We review fundamental assumptions and objectives with respect to the generalizability of our approach. Further, we identify three essential components: a *task-based execution environment, self introspection* and an *analysis* component and discuss their requirements and implications for a general solution.

2.1 Assumptions and Objectives

Our guiding underlying observation is that any imbalances (both predictable as well as unpredictable imbalances) manifest in increased waiting times at global MPI synchronization points. In case an imminent imbalance is detected, our approach attempts to quickly/immediately migrate tasks to underloaded processes, thus replacing the aforementioned waiting times with useful computation. A key assumption is that tasks represent *basic units of work* without any side effects (e.g., accessing global variables inside the task) that can be executed on the local or on a remote process. As these tasks are candidates for being executed remotely, we call them *migratable*.

In our previous work [13], we identified the following key objectives of a distributed work stealing implementation:

1. **Reactivity:** Since load imbalances can result from dynamically changing execution conditions or computational load on a very short time scale, it is necessary to detect these changes as quickly as possible.
2. **Smart decision making:** Relying on permanently collected introspection data an implementation has to identify an emerging imbalance and decide whether to migrate tasks or not. Further, it has to select adequate victims to migrate tasks to. However, inaccurate or incorrect decisions can result in a performance decline.

3. **Hiding overhead:** Compared to work stealing in shared memory, migrating tasks in distributed memory induces additional overhead as task-related information and data needs to be transferred over the network. Consequently, it is desired to migrate tasks as soon as possible to sufficiently overlap communication with computation and hide any migration overhead.
4. **Ease of integration:** Augmenting existing applications with task migration should not require extensive programming efforts or code modifications.
5. **Generalization and modularity:** Although the objective is to create a generally applicable solution that can be integrated into arbitrary applications, it might be desired and profitable to customize introspection/load specification or migration strategy in order to incorporate domain and application knowledge. Nevertheless, an implementation should provide a default behavior.

2.2   Execution Environment for Migratable Tasks

An execution environment for migratable tasks needs to satisfy the following requirements. First, it has to provide means to create migratable tasks by specifying an *action* to perform as well as *data items* accessed by the task. To be able to migrate tasks via inter-process communication the specification for a *data item* must contain a reference to the corresponding data, its size, as this information might not be available automatically (e.g., when using native pointers in C/C++), and a type $t \in \{input, output\}$ that indicates whether the corresponding item is only used within the task (*input*) or whether it is updated and needs to be available for subsequent operations (*output*).

To trigger the execution of queued tasks synchronization is required, similar to `taskwait` or `barrier` in OpenMP. However, this synchronization is not allowed to terminate until all tasks (of all processes) and outstanding communication is finished. An implementation can then decide at run time to either execute a task locally or migrate the task to another process. Contrary to other approaches that perform a redistribution in separate defined phases, it is desired to detect impending imbalances and take appropriate counter measures as soon as possible to overlap data transfers and communication with calculation, i.e., the execution of other tasks.

After a migrated task has been executed on a remote process, *data items* specified as *output* will be sent back to the original process that created the task. This process allows an incremental integration into existing applications that use the resulting data for subsequent calculations or communication such as a halo exchange, effectively preventing a complex change of communication partners.

It is recommended to perform both a thread-parallel task creation and task execution to apply load balancing within a process and exploit a large degree of shared memory concurrency.

2.3   Introspection and Analysis

A reactive solution requires to quickly detect changing run conditions, hardware behavior or unequal workload distribution. Hence, each process continuously monitors its execution condition and characteristics. As suitable characteristics can range from coarse-grained information to fine-grained metrics (e.g., time measurements or hardware performance counters) and might depend on the application, one question is:

Question 1. *What is an appropriate general load metric that can be used for arbitrary applications?*

Complemented with an analysis component that consolidates collected per-process introspection data to a coherent global view this procedure lays the foundations for identifying dynamically changing execution conditions and predicting imbalances between processes. Based on the result of the analysis the implementation can decide to trigger task migrations in order to mitigate upcoming imbalances. Yet, an implementation also needs to address the following questions (implementation details are discussed in Sect. 3):

Question 2. *Based on provided introspection/load data, what is a good default strategy to decide whether to migrate tasks and when to stop migrating?*

Question 3. *How to select proper victims for task migration?*

3   Implementation

We implemented our reactive load balancing concept in a MPI+OpenMP-parallel library allowing existing codes to use our proposed solution with only minimal code modifications.

3.1   A Migratable Task Paradigm

In contrast to our previous application-level prototype [13] where we used a pull-oriented work stealing approach, we now follow a push-oriented mechanism, where *migratable* tasks are offloaded from overloaded to underloaded processes. While this is logically only an inversion of responsibility, it saves some communication overhead: in the pull-oriented variant, a handshake between a stealing rank and the selected stealing victim was required. Further, offloading allows us to leverage OpenMP's target offloading infrastructure making a first step towards an extension of OpenMP's programming model. Conceptually, instead of offloading tasks to an accelerator, tasks are offloaded to MPI processes in our approach. However, while the decision where to execute the offloaded task is already made at task creation for classic OpenMP offloading, we need to defer that decision to runtime as we strive to reactively balance load.

We implemented a custom libomptarget plugin in the LLVM OpenMP runtime that calls our library at task creation. Combined with the clang compiler,

Listing 1.1. Example of a synthetic dense matrix multiplication code creating migratable tasks in parallel with the OpenMP target construct or API

```
1    // function that performs MxM
2    void compute_matrix_matrix(double *A, double *B, double *C, int mat_size);
3
4    int main()
5    {
6      void* lit_size = *(void**)(&size); // pointer literal representing value of size
7      #pragma omp parallel
8      {
9        #pragma omp for nowait
10       for(int i=0; i<num_tasks; i++) {
11         double *A = matrices_a[i];
12         double *B = matrices_b[i];
13         double *C = matrices_c[i];
14
15 #if USE_OPENMP_TARGET_CONSTRUCT
16         #pragma omp target map(tofrom: C[0:size*size]) map(to: A[0:size*size], B[0:size*size])
17         compute_matrix_matrix(A, B, C, size);
18 #else // API approach
19         map_data_entry_t* args = new map_data_entry_t[4];
20         args[0] = map_data_entry_create(A, size*size*sizeof(double), MAPTYPE_INPUT);
21         args[1] = map_data_entry_create(B, size*size*sizeof(double), MAPTYPE_INPUT);
22         args[2] = map_data_entry_create(C, size*size*sizeof(double), MAPTYPE_INPUT | MAPTYPE_OUTPUT);
23         args[3] = map_data_entry_create(lit_size, sizeof(void*), MAPTYPE_INPUT | MAPTYPE_LITERAL);
24
25         add_task((void *)&compute_matrix_matrix, 4, args);
26 #endif
27       }
28
29       // trigger execution (In background: introspection + task migration)
30       distributed_taskwait();
31     }
32 }
```

this enables us to fully specify a migratable task using the `#pragma omp target` directive and its data environment using the associated `map` clause. The compiler takes care of creating a task entry function and generates appropriate calls to the custom plugin, where both a reference to the task entry function (*action*) and the task's data environment (*data items*) are then forwarded to our library. As usually the same hybrid binary is executed by all ranks, an offset from the start of the loaded binary to the corresponding task entry function can be used to determine the correct entry point on a remote rank in case a task is offloaded.

While creating *migratable* tasks using OpenMP's target offloading construct is our preferred choice, we found that there is a lack of compiler support for this variant, specifically for Fortran compilers. Therefore, we additionally implemented an API (C and Fortran available) that allows to create *migratable* tasks by manually specifying a reference to the task entry function and the task's data environment. An example code snippet for both approaches is shown in Listing 1.1.

3.2   Communication Infrastructure

We implemented a communication infrastructure to handle task migration as well as introspection and continuous global distribution of online load information. All communication is fully non-blocking using a dedicated communication thread on a separate core per rank with the desire to overlap communication and computation. We found that using a dedicated core is essential to guarantee sufficient progression of MPI messages and achieving our objective of fine-granular reactivity. Similar findings have been reported in [4,6]. In contrast to predictive load balancing, there is no mutual a-priori agreement on the task migration pattern. In fact, the reactive nature of our approach demands that an overloaded rank can very quickly migrate tasks which requires responsiveness on the sender and victim rank. Even employing hyper-threading where a physical core is shared by the communication thread and another application thread is not a viable option here: there is in-determinism in thread scheduling by the OS and the hyper-thread would compete for resources with a computation thread, thus creating additional imbalances and degrading reactivity.

3.3 Task Execution and Termination Detection

As mentioned in Sect. 2.2 an implementation requires a synchronization point that ensures that all tasks of all ranks (i.e., local and migrated tasks executed on a remote rank) and all outstanding communication (i.e., transferring results back to the original rank) have been completed. We implemented a `distributed_taskwait` function (see Listing 1.1 line 30) where each rank participates in completing the created tasks and communication before terminating. Although there are more efficient solutions for global termination detection like proposed by Dinan et al. [5], we already exchange load information continuously. Hence, we append the number of outstanding operations per rank to the corresponding messages, achieving a termination detection almost for free. The `distributed_taskwait` routine triggers the execution of queued tasks and activates the communication thread that handles task migration and load exchange. As it is desired to overlap communication as much as possible, our implementation prioritizes the execution of incoming migrated tasks before working on local tasks.

3.4 Making Effective Load Balancing Decisions

To build a generally applicable responsive load balancing solution three relevant questions have been pointed out in Sect. 2.3 that we address with our implementation.

What is an appropriate general load metric that can be used for arbitrary applications? A suitable introspection metric that precisely reflects the load or run condition of a rank is the key for a good outcome. This metric might highly depend on the hardware, application and domain knowledge confronting us with two conflicting goals. On one hand it might be desired to incorporate such domain or application knowledge. On the other hand we are seeking for an appropriate default metric that can be used for arbitrary applications. Since most tasking codes apply over-decomposition we selected the *number of tasks per rank* as a general load metric. While it is easy to determine and induces low overhead compared to more sophisticated calculations, it might not work well for tasks with varying complexity or size. However, our library also provides a *tools interface* that enables the user to customize introspection, load specification and migration strategy including victim selection.

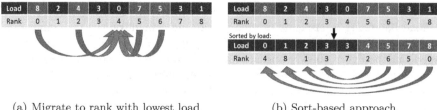

(a) Migrate to rank with lowest load   (b) Sort-based approach

Fig. 1. Potential choices for identifying proper task migration victims assuming that only ranks with load higher than average (red) migrate tasks (Color figure online)

What is a good default strategy to decide whether to migrate tasks and when to stop migrating? The migration strategy is a sensitive component of this approach. Although our solution targets to compensate small imbalances, migrating a tasks comes with an additional communication cost that has to be considered. An imbalance between sender and victim rank has to be large enough in order to amortize the task migration. Thus, our default strategy only migrates tasks if the imbalance is larger than a configurable absolute or relative *threshold*. Further, migration should not be performed too late, as this would prevent full overlap. As a result, threads would only wait for completing outstanding communication. It is critical to determine when to stop migrating tasks. Our strategy only migrates if the *number of outstanding local tasks per rank* is greater than a configurable value, e.g. *number of threads*. Thus, we ensure to keep all threads busy while preventing idle times and overhead caused by late migrations. All thresholds can be set via environment variables.

How to select proper victims for task migration? Our previous prototype selected the rank with the highest load as victim for task stealing. However, as now task migration decisions are made on each rank separately[1] in a short time frame only pushing tasks from the rank with maximum load might not be sufficient and pushing tasks to the rank with the lowest load might lead to contention or could result in load imbalances again. Our library applies a sort-based approach to identify proper victims aiming to achieve a good load balance while avoiding contention as illustrated in Fig. 1. Ranks are sorted by load. After that, in case the imbalance between the current rank and the corresponding counterpart exceeds the configured *threshold* this rank is selected as victim.

4   Experimental Evaluation

In this section, we evaluate our generalized approach and implementation decisions against hardware variability (Sect. 4.1) and work-induced imbalances (Sect. 4.2).

---

[1] Migration decision are made on a each rank separately based on per rank load information that has been exchanged before. Consequently, this step does not require any additional two-sided or collective communication.

All tests are conducted on the HPC production system of RWTH Aachen University CLAIX that is equipped with an Intel Omni-Path interconnect and dual-socket Intel Xeon E5-2650v4 (codename "Broadwell") processor nodes with a TDP of 105 W and 24 cores in total running at 2.2 GHz.[2] Our library as well as benchmarks are compiled with Intel C/C++ or Intel Fortran Compiler 19.0.1 and Intel MPI 2018.4. Hybrid MPI+OpenMP application runs are performed using a single rank per node where OpenMP threads are pinned to cores using OMP_PLACES and OMP_PROC_BIND. In order to exploit the shared-memory parallelism of the nodes using OpenMP we differentiate between the following two situations:

1. **Runs without task migration:** Applications are executed with n_{cores} OpenMP threads acting as a baseline.
2. **Runs with task migration:** As a separate communication thread is running on the last core, applications are executed with $n_{cores} - 1$ threads.

4.1   Robustness Against Hardware Variations

In order to evaluate the robustness against dynamic variability caused by hardware we use a synthetic hybrid matrix multiplication benchmark, where each rank has to perform a configurable number of dense matrix multiplications $C = A * B$ with the same computational complexity (see Listing 1.1). To ensure an adequate execution time and sufficiently large tasks, every rank has to solve 2400 multiplications with a matrix size of $S = 600 \times 600$. In an ideal scenario where all nodes and CPUs have the same speed and efficiency, all ranks are expected to finish the calculation in the same time. However, to demonstrate the effectiveness when working with dynamic imbalances or show effects of varying hardware efficiencies we run experiments under an enforced power cap or by adapting CPU core frequencies.

Experiment 1: Power Capping. In this experiment, we run the aforementioned benchmark on 4 nodes with a version that solely employs regular OpenMP tasks to balance load in shared memory and a version featuring our task migration approach that is also capable of balancing load in distributed memory. Every run is conducted 10 times under enforced power caps ranging from 40 W to 105 W (no powercap). Resulting mean values and standard deviations are depicted in Fig. 2. However, it should be noted that the results highly depend on the energy efficiency of the selected compute nodes. For our tests we included a compute node known to have a lower energy efficiency.

Empirical tests have shown that the average power draw for such an application run is around 90 W. If the selected power cap is close to or larger than 90 W task migration will not result in much improvement but is also not suffering

---

[2] Although we planned to conduct the tests on our new Intel Xeon Skylake processors, this partition was still in the process of getting into production at the time of creating the paper.

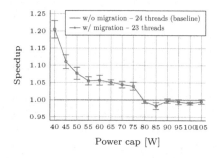
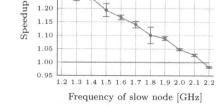

Fig. 2. Speedup under enforced power cap

Fig. 3. Speedup with a single slow node

much from overhead. With lower thresholds effects from varying hardware efficiencies become visible and task migration can help mitigate arising imbalances. This leads to improvements of 4% to 20% depending on the selected power cap. As an example, an execution of the task migration version with a power cap of 60 W took on average 17.42 s. Investigating a single execution showed that during this time frame the dedicated communication threads that are also responsible for continuous self introspection (e.g. load of the corresponding rank) performed 1,267,978 load exchanges. Based on that information it was able to dynamically detect imbalances between ranks at run time leading to a migration of 148 out of 4,800 tasks between the participating ranks.

Experiment 2: Varying Core Frequencies. To provoke imbalances we run the same setups as in experiment 1. However, we are not setting any power cap but use *likwid-setFrequencies* [14] to reduce the core frequency of a single node whereas the other nodes run with the default frequency of 2.2 GHz. We conduct tests varying the frequency of the single slow node from 1.2 GHz to 2.1 GHz. Results are shown in Fig. 3. As expected, setting no power cap leads to a slight performance decline due to loosing one core for communication purposes. With a power cap close to the base frequency of the other nodes there is only a marginal speedup. With larger frequency differences, e.g. with 1.2 GHz, task migration achieves a speedup up to 1.31X.

4.2 Robustness Against Work-Induced Imbalances

While our load balancing approach is targeted at treating unpredictable imbalances, we also evaluated whether it can be used to improve performance of work-imbalanced codes. As a realistic application, we use the sam(oa)^2 framework for parallel adaptive mesh refinement [9] to simulate the Tohoku tsunami in 2011. To provoke work imbalances, we disabled the application-level load balancing and benchmark against our reactive load balancing library instead. Figure 4 presents the degree of work imbalance that changes during the simulation time for a run with and without reactive load balancing on 32 nodes. As illustrated,

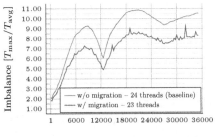

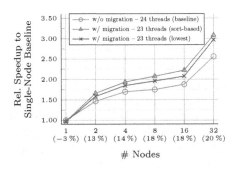

Fig. 4. Degree of work imbalance over time due to AMR for the Tohoku scenario using 32 nodes

Fig. 5. Strong scaling results for the Tohoku scenario with and without reactive load balancing. Speedup compared to the base line is depicted below corresponding number of nodes.

task migration can help to reduce the emerging imbalance but is not capable of completely eliminating it. Figure 5 shows the strong scaling results on up to 32 nodes where we tested the victim selection strategies depicted in Fig. 1. We find that using reactive load balancing improves scalability despite using one core less for computation. The sort-based victim selection outperforms the strategy where the rank with the lowest load is selected as a victim. Speedups of up to $1.20X$ are obtained with our approach relative to the baseline.

5   Conclusion and Future Work

In this paper, we presented a library for reactive load balancing across process boundaries for hybrid task-parallel applications. We demonstrated how continuous introspection, a (possibly user-defined) migration strategy and a task-based execution environment interplay in order to effectively balance the load in distributed memory at run time. Our results show performance improvements up to $1.31X$ for hardware-induced imbalances and $1.20X$ for work-induced imbalances using a realistic application with AMR. Our approach is minimal invasive in that it builds upon the established programming models MPI and OpenMP and requires little code modifications, facilitating the integration into existing MPI+OpenMP applications. The approach is designed to tackle fine granular and unpredictable imbalances by temporarily migrating tasks to other processes while any communication overhead is hidden. We demonstrated that task migration is even capable of reducing higher work imbalances (see Fig. 4). However, for those kinds of applications and scenarios we recommended to combine a domain decomposition based load balancing scheme with reactive task migration: the computational domain could be repartitioned every x iteration steps to account for large load imbalances while our reactive approach allows to target emerging fine-granular imbalances in between.

There are multiple natural directions for future work. Our present model assumes that tasks are independent from each other. A more general approach would allow for dependencies between tasks, rendering the decision making of when and where to offload tasks more complicated. It is an ongoing research question how fine-granular tasking in distributed memory with dependencies can be implemented effectively.

Moreover, we are currently exploring another reactive load balancing mechanism that makes use of task replication. Keeping tasks replicated on multiple MPI ranks and deciding at run time which rank computes a replicated task would allow us to further boost reactivity and help mitigate potential issues arising when too many tasks have been migrated to a victim rank.

Finally, we strive to broaden the class of applications benefiting from our approach.

Acknowledgements. Some of the experiments were performed with computing resources granted by JARA-HPC from RWTH Aachen University under projects jara0001 and nova0027. Parts of this work were funded by the German Federal Ministry of Education and Research (BMBF) under grant numbers 01IH16004B and 01IH16004C (Project Chameleon).

References

1. Acun, B., Miller, P., Kale, L.V.: Variation among processors under Turbo Boost in HPC systems. In: Proceedings of the 2016 International Conference on Supercomputing, ICS 2016, pp. 6:1–6:12. ACM, New York (2016). https://doi.org/10.1145/2925426.2926289

2. Blumofe, R.D., Joerg, C.F., Kuszmaul, B.C., Leiserson, C.E., Randall, K.H., Zhou, Y.: Cilk: an efficient multithreaded runtime system. In: Proceedings of the Fifth ACM SIGPLAN Symposium on Principles and Practice of Parallel Programming, PPOPP 1995, pp. 207–216. ACM, New York (1995). https://doi.org/10.1145/209936.209958

3. Charles, J., Jassi, P., Ananth, N.S., Sadat, A., Fedorova, A.: Evaluation of the Intel® Core™ i7 Turbo Boost feature. In: Proceedings of the 2009 IEEE International Symposium on Workload Characterization (IISWC), IISWC 2009, pp. 188–197. IEEE Computer Society, Washington, DC (2009). https://doi.org/10.1109/IISWC.2009.5306782

4. Denis, A., Jaeger, J., Taboada, H.: Progress thread placement for overlapping MPI non-blocking collectives using simultaneous multi-threading. In: Mencagli, G., et al. (eds.) Euro-Par 2018. LNCS, vol. 11339, pp. 123–133. Springer, Cham (2019). https://doi.org/10.1007/978-3-030-10549-5_10

5. Dinan, J., Larkins, D.B., Sadayappan, P., Krishnamoorthy, S., Nieplocha, J.: Scalable work stealing. In: Proceedings of the Conference on High Performance Computing Networking, Storage and Analysis, SC 2009, pp. 1–11, November 2009. https://doi.org/10.1145/1654059.1654113

6. Hoefler, T., Lumsdaine, A.: Message progression in parallel computing - to thread or not to thread? In: Proceedings - IEEE International Conference on Cluster Computing, ICCC. Proceeding, pp. 213–222, September 2008. https://doi.org/10.1109/CLUSTR.2008.4663774

7. Inadomi, Y., et al.: Analyzing and mitigating the impact of manufacturing variability in power-constrained supercomputing. In: Proceedings of the International Conference for High Performance Computing, Networking, Storage and Analysis, SC 2015, pp. 78:1–78:12. ACM, New York (2015). https://doi.org/10.1145/2807591.2807638

8. Kale, L.V., Krishnan, S.: CHARM++: a portable concurrent object oriented system based on C++. SIGPLAN Not. **28**(10), 91–108 (1993). https://doi.org/10.1145/167962.165874

9. Meister, O., Rahnema, K., Bader, M.: Parallel memory-efficient adaptive mesh refinement on structured triangular meshes with billions of grid cells. ACM Trans. Math. Softw. **43**(3), 1–27 (2016). https://doi.org/10.1145/2947668

10. OpenMP Architecture Review Board: OpenMP Application Program Interface, Version 5.0, November 2018. http://www.openmp.org/

11. Pinar, A., Aykanat, C.: Fast optimal load balancing algorithms for 1D partitioning. J. Parallel Distri. Comput. **64**(8), 974–996 (2004). https://doi.org/10.1016/j.jpdc.2004.05.003

12. Reinders, J.: Intel Threading Building Blocks, 1st edn. O'Reilly & Associates Inc., Sebastopol (2007)

13. Samfass, P., Klinkenberg, J., Bader, M.: Hybrid MPI+OpenMP reactive work stealing in distributed memory in the PDE framework sam(oa)^2. In: 2018 IEEE International Conference on Cluster Computing (CLUSTER), CLUSTER 2018, pp. 337–347. IEEE, September 2018. https://doi.org/10.1109/CLUSTER.2018.00051

14. Treibig, J., Hager, G., Wellein, G.: LIKWID: a lightweight performance-oriented tool suite for x86 multicore environments. In: Proceedings of PSTI 2010, The First International Workshop on Parallel Software Tools and Tool Infrastructures, San Diego CA (2010)

15. Zanotti, O., Fambri, F., Dumbser, M., Hidalgo, A.: Space–time adaptive ADER discontinuous Galerkin finite element schemes with a posteriori sub-cell finite volume limiting. Comput. Fluids **118**, 204–224 (2015). https://doi.org/10.1016/j.compfluid.2015.06.020, http://www.sciencedirect.com/science/article/pii/S0045793015002030

Workshop on Models Algorithms and Methodologies for Hybrid Parallelism in New HPC Systems

Ab-initio Functional Decomposition of Kalman Filter: A Feasibility Analysis on Constrained Least Squares Problems

Luisa D'Amore[✉], Rosalba Cacciapuoti, and Valeria Mele

University of Naples Federico II, Naples, Italy
{luisa.damore,rosalba.cacciapuoti,valeria.mele}@unina.it

Abstract. The standard formulation of Kalman Filter (KF) becomes computationally intractable for solving large scale state space estimation problems as in ocean/weather forecasting due to matrix storage and inversion requirements. We introduce an innovative mathematical/numerical formulation of KF using Domain Decomposition (DD) approach. The proposed DD approach partitions ab-initio the whole KF computational method giving rise to local KF methods that can be solved independently. We present its feasibility analysis using the constrained least square model underlying variational Data Dssimilation problems. Results confirm that the accuracy of solutions of local KF methods are not impaired by DD approach.

Keywords: Kalman Filter · Domain Decomposition · Data Assimilation · Constrained Least Square Problem

1 Introduction and Related Works

Kalman Filter (KF) dates back to 1960, when Kalman [19] provided a recursive algorithm to compute the solution of a (linear) data filtering and prediction problem. It is also known as linear quadratic estimation algorithm that infers parameters of interest from indirect, inaccurate, or uncertain observations (Data Assimilation, DA). During the last years, DA reached a widespread interests at many federal research institutes as well as at many universities [NCAR (National Center for Atmospheric Research), NCEP (National Centers for Environmental Prediction), DWD (Deutscher Wetterdienst), Met Office with University of Reading and Imperial College of London in UK, JMA (Japan Meteorological Agency), CMC (Canadian Association of Management Consultants) and the CMCC (EuroMediterranean Center for Climate Changes)].

In the past years KF has become a main component in satellite navigation, economics, or telecommunications and in the validation of the mathematical models used in meteorology, climatology, geophysics, geology and hydrology. Its main strength is its recursive property: new measurements can be processed as they arrive. Nevertheless, the standard formulation of the KF becomes computationally intractable for solving large scale state space estimation problems

© Springer Nature Switzerland AG 2020
R. Wyrzykowski et al. (Eds.): PPAM 2019, LNCS 12044, pp. 75–92, 2020.
https://doi.org/10.1007/978-3-030-43222-5_7

due to matrix storage and inversion requirements. So, several variants have been proposed to reduce the computational complexity, designed on the basis of a reduction in the order of the system model (usually the approximation is performed trough the use of the Empirical Orthogonal Functions (EOF)) [18,21], or based on the Ensemble methods where a prediction of the error at a future time is computed by integrating each ensemble state independently by the model. The integrations are typically performed until observations are available. At this time, the information from the observations and the ensemble are combined by performing an analysis step based on KF [15]. However, the choice of the dimension of the reduced-state space or of the ensemble size giving an accurate approximation of KF still remains a delicate question [2].

We present a new Domain Decomposition (DD) framework suitable for using KF in large scale applications. As case study, we consider Constrained Least Squares (CLS) problem, as this is the prototype model of variational DA applications (Sect. 2.2). In Sect. 3 we first show how to compute CLS solution by using KF (we refer to this method as KF-CLS method); in Sect. 4, we introduce DD into CLS problem. As a consequence we get a certain number of local problems we call DD-CLS problems. We note that DD-CLS problems are defined by adding an overlapping operator to the variational model describing CLS problem. This is a sort of regularization approach needed in order to guarantee the matching of local solutions on the overlapping domains. A regularization parameter balances the weight given to the overlapping operator with respect to the CLS problem.

Then, by applying KF for concurrently solving local DD-CLS sub problems, we get the so called KF-DD-CLS method. Main contribution of the present work is to prove, both theoretically (see Theorem 3) and experimentally (see Validation results), that KF-DD-CLS method is equivalent to DD-KF-CLS method, i.e. the innovative DD method we propose in the present work (see Theorem 2), which arises decomposing ab-initio KF method once it is used for solving CLS problem, i.e. decomposing ab-initio KF-CLS method.

In Fig. 1 we give a schematic picture of these approaches showing how they arise. We see that KF-DD-CLS is obtained by following the path on the right, while on the left we get DD-KF-CLS. Finally, experiments are reported in Sect. 5 and conclusions are given in Sect. 6. It is worth noting that results we present are concerned not so much with parallel efficiency as with the capability to solve this problem by using such a full decomposition. We are currently working on designing the related parallel algorithm. We hope that these findings encourage readers to further extend the framework according to their specific application's requirements.

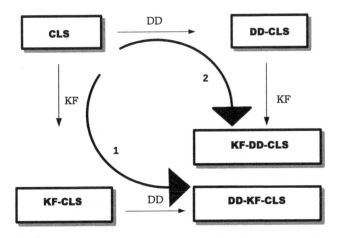

Fig. 1. Schematic description of DD framework we introduce in KF method solving CLS problems.

2 Preliminaries

2.1 Kalman Filter Method

Given $x_0 \in \mathbb{R}^n$, let $x(t) \in \mathbb{R}^n$, $\forall t \in [0, T]$, denote the state of a dynamic system governed by the mathematical model $\mathcal{M}_{t,t+\Delta t}[x(t)]$, $\Delta t > 0$:

$$\begin{cases} x(t + \Delta t) = \mathcal{M}_{t,t+\Delta t}(x(t)), & \forall t, t + \Delta t \in [0, T] \\ x(0) \quad\quad = x_0 \end{cases}, \tag{1}$$

and let:

$$y(t + \Delta t) = \mathcal{H}_{t+\Delta t}[x(t + \Delta t)], \tag{2}$$

denote observations where $\mathcal{H}_{t+\Delta t}$ is observations mapping. Chosen $r \in \mathbb{N}$, we consider $r + 2$ points in $[0, T]$ and $\Delta t = \frac{T}{r+1}$.

Let $\{t_k\}_{k=0,1,...,r+1}$ be a discretization of $[0, T]$, where $t_k = k\Delta t$, we will use the following discrete formulation of KF method [22]:

- $x_k \equiv x(t_k) \in \mathbb{R}^n$: state of system at time t_k, for $k = 0, 1, ..., r + 1$;
- $\widehat{x}_0 \equiv x_0$: state estimate at time $t_0 \equiv 0$;
- $\widehat{x}_k$: the state estimate at time t_k, for $k = 1, ..., r + 1$;
- $M_{k,k+1} \in \mathbb{R}^{n \times n}$: discretization of a linear approximation of $\mathcal{M}_{t_k,t_{k+1}}$, for $k = 0, 1, ..., r$;
- $H_k \in \mathbb{R}^{m \times n}$: discretization of a linear approximation of $\mathcal{H}_t$ with $m > n$, for $k = 0, 1, ..., r + 1$;
- $w_k \in \mathbb{R}^n$ and $v_k \in \mathbb{R}^m$: model and observation errors with normal distribution and zero mean such that $E[w_k v_i^T] = 0$, for $i, k = 0, 1, ..., r + 1$, where $E[\cdot]$ denotes the expected value;

– $Q_k \in \mathbb{R}^{n \times n}$ and $R_k \in \mathbb{R}^{m \times m}$: covariance matrices of the errors on the model and on the observations, respectively i.e.

$$Q_k := E[w_k w_k^T] \quad R_k := E[v_k v_k^T] \quad \forall\, k = 0, 1, ..., r+1. \tag{3}$$

These matrices are symmetric and positive definite.

KF Method: KF method consists in calculating an estimate $\widehat{x}_{k+1}$, at time t_{k+1}, of

$$x_{k+1} = M_{k,k+1} x_k + w_k, \quad \forall k = 0, 1, ..., r \tag{4}$$

which is

$$y_{k+1} = H_{k+1} x_{k+1} + v_{k+1}, \quad \forall k = 0, 1, ..., r. \tag{5}$$

KF Procedure: Given $\widehat{x}_0 \in \mathbb{R}^n$ and $P_0 = O \in \mathbb{R}^{n \times n}$ a null matrix, for each $k = 0, 1, ..., r$ KF procedure is composed by:

– Predicted phase.
  • Compute predicted state estimate:

$$x_{k+1} = M_{k,k+1} \widehat{x}_k; \tag{6}$$

  • Compute predicted error covariance matrix:

$$P_{k+1} = M_{k,k+1} P_k M_{k,k+1}^T + Q_k. \tag{7}$$

– Corrector phase.
  • Compute Kalman gain:

$$K_{k+1} = P_{k+1} H_{k+1}^T (H_{k+1} P_{k+1} H_{k+1}^T + R_{k+1})^{-1}. \tag{8}$$

  • Compute Kalman covariance matrix:

$$P_{k+1} = (I - K_{k+1} H_{k+1}) P_{k+1}, \tag{9}$$

  • Compute Kalman state estimate:

$$\widehat{x}_{k+1} = x_{k+1} + K_{k+1}(y_{k+1} - H_{k+1} x_{k+1}). \tag{10}$$

2.2   The Constrained Least Squares (CLS) Problem

Let

$$H_0 x_0 = y_0, \quad H_0 \in \mathbb{R}^{m_0 \times n}, \quad y_0 \in \mathbb{R}^{m_0}, \quad x_0 \in \mathbb{R}^n \tag{11}$$

be an overdetermined linear system, where $rank(H_0) = n > 0$, $m_0 > n$, Given $H_1 \in \mathbb{R}^{m_1 \times n}$, $y_1 \in \mathbb{R}^{m_1}$, $x_1 \in \mathbb{R}^n$, $x \in \mathbb{R}^n$, let us consider the system

$$S: \quad Ax = b \tag{12}$$

where

$$A = \begin{bmatrix} H_0 \\ H_1 \end{bmatrix} \in \mathbb{R}^{(m_0+m_1) \times n}, \quad b = \begin{bmatrix} y_0 \\ y_1 \end{bmatrix} \in \mathbb{R}^{m_0+m_1}, \tag{13}$$

and $m_1 > 0$. Let $R_0 \in \mathbb{R}^{m_0 \times m_0}$, $R_1 \in \mathbb{R}^{m_1 \times m_1}$ be the weight matrices and $R = diag(R_0, R_1) \in \mathbb{R}^{(m_0+m_1) \times (m_0+m_1)}$. CLS problem is

$$CLS: \quad \widehat{x} = argmin_{x \in \mathbb{R}^n} J(x) \tag{14}$$

with

$$J(x) = ||Ax - b||_R^2 = ||H_0 x - y_0||_{R_0}^2 + ||H_1 x - y_1||_{R_1}^2, \tag{15}$$

where $\widehat{x}$ is

$$(A^T RA)\widehat{x} = A^T Rb \Rightarrow \widehat{x} = (A^T RA)^{-1} A^T Rb \tag{16}$$

or in summation form

$$\widehat{x} = (H_0^T R_0 H_0 + H_1^T R_1 H_1)^{-1} (H_0^T R_0 y_0 + H_1^T R_1 y_1). \tag{17}$$

We refer to $\widehat{x}$ as solution in least squares sense of system in (12).

3 KF-CLS Method: KF Method Solving CLS Problems

We prove that solution of CLS problem in (14), can be obtained by applying KF to S in (12). To this end, regarding (12) as an inverse ill posed problem [7,9,12], we rewrite KF as a Variational problem, the so called VAR-KF formulation, obtained minimizing the sum of the weighted Euclidean norm $|| \cdot ||_{Q_k}$ of the model error $w_k = x_{k+1} - M_{k,k+1} x_k$ and the weighted Euclidean norm $|| \cdot ||_{R_{k+1}}$ of the observation error $v_{k+1} = y_{k+1} - H_{k+1} x_{k+1}$.

VAR-KF Method: Var KF method consists in computing or each $k = 0, 1, ..., r$

$$\begin{aligned} \widehat{x}_{k+1} &= argmin_{x_{k+1} \in \mathbb{R}^n} J_{k+1}(x_{k+1}) \\ &= argmin_{x_{k+1} \in \mathbb{R}^n} \left\{ ||x_{k+1} - M_{k,k+1}\widehat{x}_k||_{Q_k}^2 + ||y_{k+1} - H_{k+1}x_{k+1}||_{R_{k+1}}^2 \right\}. \end{aligned} \tag{18}$$

Then, by using linear algebra results we prove that:

Proposition 1 *(KF-CLS). Let S be the overdetermined linear system in (12) with $A \in \mathbb{R}^{(m_0+m_1) \times n}$, $b \in \mathbb{R}^{m_0+m_1}$ defined in (13). Let us consider*

- *for $k = 0, 1$, $H_k \in \mathbb{R}^{m_k \times n}$ and $y_k \in \mathbb{R}^{m_k}$ with $m_0 > n$ and $m_1 > 0$;*
- *$M = (H_0^T R_0 H_0)^{-1} H_0^T R_0 \in \mathbb{R}^{n \times m_0}$ with $R_0 \in \mathbb{R}^{m_0 \times m_0}$, $R_1 \in \mathbb{R}^{m_1 \times m_1}$ and $R = diag(R_0, R_1) \in \mathbb{R}^{(m_0+m_1) \times (m_0+m_1)}$ weight matrices;*
- *$\widehat{x}_0 = M y_0 \in \mathbb{R}^n$, solution in least squares sense of system in (11);*
- *$\widehat{x} = (A^T RA)^{-1} A^T Rb \in \mathbb{R}^n$ solution in least squares sense of S in (12).*

We pose:

$$M_{0,1} \equiv I_{n,n} \in \mathbb{R}^{n \times n},$$
$$Q_0 \equiv O_{n,n} \in \mathbb{R}^{n \times n}, \qquad (19)$$
$$P_0 \equiv (H_0^T R_0 H_0)^{-1} \in \mathbb{R}^{n \times n};$$

where $I_{n,n}$ is the identity matrix and $O_{n,n}$ is the null matrix, then by using KF procedure 2.1, for $k = 0$, we obtain KF estimate $\widehat{x}_1$ in (10) such that

$$\widehat{x} \equiv \widehat{x}_1. \qquad (20)$$

Proof. Solution of CLS problem (14), i.e. $\widehat{x}$, can be obtained solving the linear system

$$(A^T R A)\widehat{x} = A^T R b, \qquad (21)$$

or in summation form

$$(H_0^T R_0 H_0 + H_1^T R_1 H_1)\widehat{x} = H_0^T R_0 y_0 + H_1^T R_1 y_1. \qquad (22)$$

Consider $\widehat{x}_0 \in \mathbb{R}^n$, solution of the normal equations obtained by considering the matrix H_0 i.e. solution of the following system:

$$(H_0^T R_0 H_0)\widehat{x}_0 = H_0^T R_0 y_0, \qquad (23)$$

that can be written as

$$\widehat{x}_0 = (H_0^T R_0 H_0)^{-1} H_0^T R_0 y_0. \qquad (24)$$

We define

$$P_0 = (H_0^T R_0 H_0)^{-1}, \quad P_1 = (H_0^T R_0 H_0 + H_1^T R_1 H_1)^{-1} \qquad (25)$$

and we write $\widehat{x}_0$ as

$$\widehat{x}_0 \equiv M y_0 = (H_0^T R_0 H_0)^{-1} H_0^T R_0 y_0 = P_0 H_0^T R_0 y_0, \qquad (26)$$

so that we have

$$H_0^T R_0 y_0 = P_0^{-1} \widehat{x}_0. \qquad (27)$$

We write a recursive expression for P_1^{-1} and obtain P_0^{-1} as follows:

$$P_1^{-1} = P_0^{-1} + (H_1^T R_1 H_1) \; \Rightarrow \; P_0^{-1} = P_1^{-1} - (H_1^T R_1 H_1), \qquad (28)$$

so from (27) and (28), we have

$$H_0^T R_0 y_0 = (P_1^{-1} - (H_1^T R_1 H_1))\widehat{x}_0, \qquad (29)$$

and from (22), (25) and (29), $\widehat{x}$ can be rewritten as follows

$$
\begin{aligned}
\widehat{x} &= (A^T R A)^{-1} A^T R b \\
&= (H_0^T R_0 H_0 + H_1^T R_1 H_1)^{-1}(H_0^T R_0 y_0 + H_1^T R_1 y_1) \\
&= P_1 \left[(H_0^T R_0 H_0)\widehat{x}_0 + H_1^T R_1 y_1) \right] \\
&= P_1 \left[(P_1^{-1} - (H_1^T R_1 H_1)) \widehat{x}_0 + H_1^T R_1 y_1 \right].
\end{aligned}
\qquad (30)
$$

so, from (30) we have

$$\widehat{x} = \widehat{x}_0 + P_1 H_1^T R_1 (y_1 - H_1 \widehat{x}_0). \qquad (31)$$

Defining

$$K_1 = P_1 H_1^T R_1, \tag{32}$$

then $\widehat{x}$ in (31) can be rewritten as follows

$$\widehat{x} = \widehat{x}_0 + K_1(y_1 - H_1 \widehat{x}_0). \tag{33}$$

In particular

$$P_1^{-1} = P_0^{-1} + (H_1^T R_1 H_1)$$

and using the Sherman-Morrison-Woodbury formula[1] we get

$$P_1 = P_0 - P_0 H_1^T (R_1 + H_1 P_0 H_1^T)^{-1} H_1 P_0. \tag{34}$$

We note that $K_1 \in \mathbb{R}^{n \times m_1}$ in (32) can be rewritten as

$$K_1 = P_0 H_1^T (R_1 + H_1 P_0 H_1^T)^{-1}, \tag{35}$$

which coincides for $k = 0$ with Kalman gain in (8), from the hypothesis we have $M_{0,1} = I_{n,n}$ this means that predicted estimate in (6) is $x_1 = \widehat{x}_0$. So, we get

$$\widehat{x} \equiv \widehat{x}_1, \tag{36}$$

where $\widehat{x}_1$ is Kalman estimate in (10) for $k = 0$.

By adding $(r + 1) \cdot m$ equations, with $r \geq 0$, $m > 0$, to system in (11) and posing, for $k = 0, 1, ..., r$,

$$\begin{aligned} M_{k,k+1} &:= I_{n,n} \in \mathbb{R}^{n \times n}, \\ Q_k &:= O_{n,n} \in \mathbb{R}^{n \times n}, \\ P_0 &:= (H_0^T R_0 H_0)^{-1} \in \mathbb{R}^{n \times n} \end{aligned} \tag{37}$$

and $R \in \mathbb{R}^{(r+2) \cdot m \times (r+2) \cdot m}$ the weight matrix, KF procedure 2.1 can be applied to solve the overdetermined system

$$Mz = p \tag{38}$$

where

$$M = \begin{bmatrix} H_0 \\ H_1 \\ \vdots \\ H_{r+1} \end{bmatrix} \in \mathbb{R}^{(r+2) \cdot m \times n}; \ p = \begin{bmatrix} y_0 \\ y_1 \\ \vdots \\ y_{r+1} \end{bmatrix} \in \mathbb{R}^{(r+2) \cdot m}, \ z \in \mathbb{R}^n \tag{39}$$

and $H_k \in \mathbb{R}^{m \times n}$, $y_k \in \mathbb{R}^m$, where $m_0 \equiv m$ and . This means, as proved in Proposition 1, that the KF estimate $\widehat{z}_{r+1}$ at step $k = r$ coincides with $\widehat{z} = (M^T R M)^{-1} M^T R p$, i.e. the solution in least squares sense of (38).

---

[1] Let $A \in \mathbb{R}^{n \times n}$, $U \in \mathbb{R}^{n \times k}$, $V \in \mathbb{R}^{k \times n}$, $R \in \mathbb{R}^{k \times k}$ and $B = A + URV$. Then, $B^{-1} = A^{-1} - A^{-1}U(R + VA^{-1}U)^{-1}VA^{-1}$.

4   DD Approaches

We apply DD approach solving system S in (12). We refer to this problem as DD-CLS problem.

Definition 1 *(Reduction of matrices).* *Let $B = [B^1\ B^2\ ...\ B^n] \in \mathbb{R}^{m \times n}$ be a matrix with $m, n \geq 1$ and B^j the $j - th$ column of B and $I_j = \{1, ..., j\}$ and $I_{i,j} = \{i, ..., j\}$ for $i = 1, ..., n - 1; j = 2, ..., n$, and $i < j$ for every (i, j). The reduction of B to the set I_j is:*

$$|_{I_j} : B \in \mathbb{R}^{m \times n} \to B|_{I_j} = [B^1\ B^2\ ...\ B^j] \in \mathbb{R}^{m \times j}, \quad j = 2, ..., n, \quad (40)$$

and to $I_{i,j}$

$$|_{I_{i,j}} : B \in \mathbb{R}^{m \times n} \to B|_{I_{i,j}} = [B^i\ B^{i+1}\ ...\ B^j] \in \mathbb{R}^{m \times j - i}, \quad i = 1, ..., n - 1, j > i, \quad (41)$$

where $B|_{I_j}$ and $B|_{I_{i,j}}$ denote the reduction of B to I_j and $I_{i,j}$, respectively.

Definition 2 *(Reduction of vectors).* *Let $w = [w_t\ w_{t+1}\ ...\ w_n]^T \in \mathbb{R}^s$ be a vector with $t \geq 1$, $n > 0$, $s = n - t$ and $I_{1,r} = \{1, ..., r\}$, $r > n$ and $n > t$. The extension of w to I_r is:*

$$EO_{I_r} : w \in \mathbb{R}^s \to EO_{I_r}(w) = [\bar{w}_1\ \bar{w}_2\ ...\ \bar{w}_r]^T \in \mathbb{R}^r, \quad (42)$$

where for $i = 1, ..., r$

$$\bar{w}_i = \begin{cases} w_i & \text{if } t \leq i \leq n \\ 0 & \text{if } i > n \text{ and } i < t \end{cases} \quad (43)\ ,$$

We now introduce the reduction of J in (15).

Definition 3 *(Reduction of functionals).* *Let us consider $A \in \mathbb{R}^{(m_0+m_1) \times n}$, $b \in \mathbb{R}^{m_0+m_1}$, the matrix and the vector defined in (13), $I_1 = \{1, ..., n_1\}$, $I_2 = \{1, ..., n_2\}$ with $n_1, n_2 > 0$ and the vectors $x \in \mathbb{R}^n$. Let*

$$J|_{(I_i,I_j)} : (x|_{I_i}, x|_{I_j}) \longmapsto J|_{(I_i,I_j)}(x|_{I_i}, x|_{I_j}) \quad \forall i, j = 1, 2 \quad (44)$$

denote the reduction of J defined in (15). It is defined as

$$J|_{(I_i,I_j)}(x|_{I_i}, x|_{I_j}) = ||H_0|_{I_i} x|_{I_i} - (y_0 + H_0|_{I_j} x|_{I_j})||^2_{R_0} + ||H_1|_{I_i} x|_{I_i} - (y_1 + H_1|_{I_j} x|_{I_j})||^2_{R_1}, \quad (45)$$

for $i, j = 1, 2$.

For simplicity of notations we let $J_{i,j} \equiv J|_{(I_i,I_j)}$ with $i, j = 1, 2$.

4.1   DD-CLS Problems: DD of CLS Problems

Definition 4 *(DD-CLS problem).* *Let S be the overdetermined linear system in (12) and $A \in \mathbb{R}^{(m_0+m_1) \times n}$, $b \in \mathbb{R}^{m_0+m_1}$ the matrix and the vector defined in (13) and $R_0 \in \mathbb{R}^{m_0 \times m_0}$, $R_1 \in \mathbb{R}^{m_1 \times m_1}$, $R = diag(R_0, R_1) \in \mathbb{R}^{(m_0+m_1) \times (m_0+m_1)}$ be the weight matrices with $m_0 > n$ and $m_1 > 0$. Let us consider the index set of columns of A, $I = \{1, ..., n\}$. DD-CLS problem consists in:*

- *DD step:*
  - *decomposition of I into*

$$I_1 = \{1, ..., n_1\}, \quad I_2 = \{n_1 - s + 1, ..., n\}, \tag{46}$$

  where $s \geq 0$ is the number of indexes in common, $|I_1| = n_1 > 0$, $|I_2| = n_2 > 0$, and the overlap sets

$$I_{1,2} = \{n_1 - s + 1, ..., n_1\}, \tag{47}$$

  If $s = 0$, then decomposition of I is without overlap, i.e. $I_1 \cap I_2 = \emptyset$ and $I_{1,2} \neq \emptyset$, instead if $s > 0$ decomposition of I is with overlap, i.e. $I_1 \cap I_2 \neq \emptyset$ and $I_{1,2} = \emptyset$.
  - *reduction of A to I_1 and I_2 defined in (46)*

$$A_1 = A|_{I_1} \in \mathbb{R}^{(m_0+m_1) \times n_1}, \quad A_2 = A|_{I_2} \in \mathbb{R}^{(m_0+m_1) \times n_2}, \tag{48}$$

- *DD-CLS step: given $x_2^0 \in \mathbb{R}^{n_2}$, according to the ASM (Alternating Schwarz Method) in [16], DD approach consists in solving for $n = 0, 1, 2, ...$ the following overdetermined linear systems:*

$$S_1^{n+1}: \quad A_1 x_1^{n+1} = b - A_2 x_2^n \quad S_2^{n+1}: \quad A_2 x_2^{n+1} = b - A_1 x_1^{n+1}. \tag{49}$$

This means to solve

$$\begin{aligned} P_1^{n+1}: \quad \widehat{x}_1^{n+1} &= argmin_{x_1^{n+1} \in \mathbb{R}^{n_1}} J_1(x_1^{n+1}, x_2^n) \\ &= argmin_{x_1^{n+1} \in \mathbb{R}^{n_1}} \left[J|_{(I_1,I_2)}(x_1^{n+1}, x_2^n) + \mu \cdot \mathcal{O}_{1,2}(x_1^{n+1}, x_2^n) \right] \end{aligned} \tag{50}$$

$$\begin{aligned} P_2^{n+1}: \quad \widehat{x}_2^{n+1} &= argmin_{x_2^{n+1} \in \mathbb{R}^{n_2}} J_2(x_2^{n+1}, x_1^{n+1}) \\ &= argmin_{x_2^{n+1} \in \mathbb{R}^{n_2}} \left[J|_{(I_2,I_1)}(x_2^{n+1}, x_1^{n+1}) + \mu \cdot \mathcal{O}_{1,2}(x_2^{n+1}, x_1^{n+1}) \right] \end{aligned} \tag{51}$$

where I_i is defined in (46) and $J|_{I_i,I_j}$ is defined in (45), $\mathcal{O}_{1,2}$ is the overlapping operator and $\mu > 0$ is the regularization parameter.

Remark 1. If decomposition of I is without overlap (i.e. $s = 0$) then $\widehat{x}_1^{n+1} \in \mathbb{R}^{n_1}$ and $\widehat{x}_2^{n+1} \in \mathbb{R}^{n_2}$ can be written in terms of normal equations as follows

$$\begin{aligned} \widetilde{S}_1^{n+1}: \quad (A_1^T R A_1)\widehat{x}_1^{n+1} = A_1^T R(b - A_2 x_2^n) &\Rightarrow \widehat{x}_1^{n+1} = (A_1^T R A_1)^{-1} A_1^T R b_1^n \\ \widetilde{S}_2^{n+1}: \quad (A_2^T R A_2)\widehat{x}_2^{n+1} = A_2^T R(b - A_1 x_1^{n+1}) &\Rightarrow \widehat{x}_2^{n+1} = (A_2^T R A_2)^{-1} A_2^T R b_2^{n+1}, \end{aligned} \tag{52}$$

where $b_1^n = b - A_2 x_2^n$ and $b_2^{n+1} = b - A_1 x_1^{n+1}$.

Remark 2. Regarding the overlapping operator $\mathcal{O}_{1,2}$, we consider $x_1 \in \mathbb{R}^{n_1}$ and $x_2 \in \mathbb{R}^{n_2}$, for $i = 1, 2$ and we pose

$$\mathcal{O}_{1,2}(x_i, x_j) = ||EO_{I_i}(x_i|_{I_{1,2}}) - EO_{I_i}(x_j|_{I_{1,2}})||, \ i, j = 1, 2 \tag{53}$$

with $EO_{I_i} x_1|_{I_{1,2}}$, $EO_{I_i} x_2|_{I_{1,2}}$ be extension to I_i, of reduction to $I_{1,2}$ in (47) of $x_1 \in \mathbb{R}^{n_1}$ and $x_2 \in \mathbb{R}^{n_2}$, respectively. The overlapping operator $\mathcal{O}_{1,2}$ represents the exchange of data on the overlap set $I_{1,2}$ in (47).

We note that $\tilde{S}_1^{n+1}$ and $\tilde{S}_2^{n+1}$ in (52) can be obtained by applying Jacobi method to normal equations in (16).

Proposition 2. *DD approach (49) applied to S in (12) is equivalent to Block Jacobi method to normal equations in (16).*

Remark 3. In particular, we get the sequences $\{x^{n+1}\}_{n \in \mathbb{N}_0}$:

$$x^{n+1} = \begin{cases} \widehat{x}_1^{n+1}|_{I_1 \setminus I_{1,2}} & on \ I_1 \setminus I_{1,2} \\ \frac{\mu}{2}(\widehat{x}_2^{n+1}|_{I_{1,2}} + \widehat{x}_1^{n+1}|_{I_{1,2}}) & on \ I_{1,2} \\ \widehat{x}_2^{n+1}|_{I_2 \setminus I_{1,2}} & on \ I_2 \setminus I_{1,2} \end{cases}, \tag{54}$$

with the sets I_1, I_2 be defined in (46) and $I_{1,2}$ in (47).

We note that if decomposition of I is without overlap, if the matrix $D - 2A^T A$ where $D = diag(A_1^T R A_1, A_2^T R A_2)$ is symmetric and definite positive, the convergence of DD method is guaranteed, i.e.

$$\lim_{n \to \infty} x^{n+1} = \widehat{x}, \tag{55}$$

where $\widehat{x}$ is the solution of CLS problem in (14).

4.2   KF-DD-CLS Method: KF Method Solving DD-CLS Problems

Let $A \in \mathbb{R}^{(m_0+m_1) \times n}$ and $b \in \mathbb{R}^{(m_0+m_1)}$ be as in (13) and $R \in \mathbb{R}^{(m_0+m_1) \times (m_0+m_1)}$, with $m_0 > n$ and $m_1 > 0$. We aim to find an estimate of $\widehat{x} = (A^T R A)^{-1} A^T R b \in \mathbb{R}^n$, i.e. solution in least squares sense of S in (12), by using DD-CLS and KF procedure. To this end, for $n = 0, 1, 2, ...$, we prove that $\widehat{x}_1^{n+1} \in \mathbb{R}^{n_1}$, $\widehat{x}_2^{n+1} \in \mathbb{R}^{n_2}$, solutions of P_1^{n+1} and P_2^{n+1} in (50) and (51) respectively are equal to KF estimates $\widehat{x}_{1,1}^{n+1}$, $\widehat{x}_{2,1}^{n+1}$ obtained by applying KF procedure to CLS problems P_1^{n+1}, P_2^{n+1}. We refer to this as KF-DD-CLS method.

To apply KF method P_1^{n+1} and P_2^{n+1} in (50) and (51), we need $\widehat{x}_{1,0} \in \mathbb{R}^{n_1}$ and $\widehat{x}_{2,0} \in \mathbb{R}^{n_2}$ i.e. reduction of $\widehat{x}_0 = (H_0^T R H_0)^{-1} H_0^T R y_0$ that can be calculated as follows:

1. decomposition without overlap (i.e. $s = 0$).

Theorem 1. *Let us consider $H_0 \in \mathbb{R}^{m_0 \times n}$, $y_0 \in \mathbb{R}^{m_0}$ and $R_0 \in \mathbb{R}^{m_0 \times m_0}$, with $m_0 > n$, $m_1 > 0$, system $H_0 x_0 = y_0$ and I. Let us consider the following steps:*

– *decomposition of I:*

$$I_1 = \{1, ..., n_1\} \quad and \quad I_2 = \{n_1 + 1, ..., n\}. \tag{56}$$

– *reduction of H_0:*

$$H_{1,0} = H_0|_{I_1} \in \mathbb{R}^{m_0 \times n_1} \quad and \quad H_{2,0} = H_0|_{I_2} \in \mathbb{R}^{m_0 \times n_2}. \tag{57}$$

- *computation of* $P_{H_{i,0}} \in \mathbb{R}^{m_0 \times m_0}$:

$$P_{H_{i,0}} = R_0 - R_0 H_{i,0}(H_{i,0}^T R_0 H_{i,0})^{-1} H_{i,0}^T R_0, \quad i=1,2. \tag{58}$$

- *computation of* $x_1 \in \mathbb{R}^{n_1}$ *and* $x_2 \in \mathbb{R}^{n-n_1}$:

$$x_1 = (H_{1,0}^T P_{H_{2,0}} H_{1,0})^{-1} H_{1,0}^T P_{H_{2,0}} y_0, \quad x_2 = (H_{2,0}^T P_{H_{1,0}} H_{2,0})^{-1} H_{2,0}^T P_{H_{1,0}} y_0. \tag{59}$$

Then $\hat{x}_0 = (H_0^T R_0 H_0)^{-1} H_0^T R_0 y_0 \in \mathbb{R}^n$, *which is the solution in least squares sense of* $H_0 x_0 = y_0$, *is obtained as:*

$$\hat{x}_0|_{I_1} = x_1, \quad \hat{x}_0|_{I_2} = x_2, \tag{60}$$

where $\hat{x}_0|_{I_i}$ *is the reduction of* $\hat{x}_0$ *to respective sets* I_i, *for* $i = 1, 2$.

Proof. We consider

$$(H_0^T R_0^{-1} H_0)\hat{x}_0 = H_0^T R_0^{-1} y_0, \tag{61}$$

that can be written

$$[H_{1,0} \ H_{2,0}]^T R_0^{-1}[H_{1,0} \ H_{2,0}] \begin{bmatrix} \hat{x}_0|_{I_1} \\ \hat{x}_0|_{I_2} \end{bmatrix} = [H_{1,0} \ H_{2,0}]^T R_0^{-1} y_0, \tag{62}$$

where $H_{1,0}$, $H_{2,0}$ are defined in (57). We get two linear systems

$$(H_{1,0}^T R_0^{-1} H_{1,0})\hat{x}_0|_{I_1} = H_{1,0}^T R_0^{-1}(y_0 - H_{2,0}\hat{x}_0|_{I_2}) \ \rightarrow$$
$$\hat{x}_0|_{I_1} = (H_{1,0}^T R_0^{-1} H_{1,0})^{-1} H_{1,0}^T R_0^{-1}(y_0 - H_{2,0}\hat{x}_0|_{I_2})$$
$$(H_{2,0}^T R_0^{-1} H_{2,0})\hat{x}_0|_{I_2} = H_{2,0}^T R_0^{-1}(y_0 - H_{1,0}\hat{x}_0|_{I_1}) \ \rightarrow$$
$$\hat{x}_0|_{I_2} = (H_{2,0}^T R_0^{-1} H_{2,0})^{-1} H_{2,0}^T R_0^{-1}(y_0 - H_{1,0}\hat{x}_0|_{I_1}). \tag{63}$$

Below we get $\hat{x}_0|_{I_1}$ as follows:

$$(H_{1,0}^T R_0^{-1} H_{1,0})\hat{x}_0|_{I_1} = H_{1,0}^T R_0^{-1}(y_0 - H_{2,0}(H_{2,0}^T R_0^{-1} H_{2,0})^{-1} H_{2,0}^T R_0^{-1}(y_0 - H_{1,0}\hat{x}_0|_{I_1})),$$

that can be written

$$H_{1,0}^T (R_0^{-1} H_{1,0} - R_0^{-1} H_{2,0}(H_{2,0}^T R_0^{-1} H_{2,0})^{-1} H_{2,0}^T R_0^{-1} H_{1,0})\hat{x}_0|_{I_1} = H_{1,0}^T R_0^{-1}$$
$$\cdot (y_0 - H_{2,0}(H_{2,0}^T R_0^{-1} H_{2,0})^{-1} H_{2,0}^T R_0^{-1} y_0) \tag{64}$$

or

$$(H_{1,0}^T P_{H_{2,0}} H_{1,0})\hat{x}_0|_{I_1} = H_{1,0}^T P_{H_{2,0}} y_0, \tag{65}$$

where $P_{H_{2,0}}$ is defined in (58). So, we have

$$\hat{x}_0|_{I_1} = (H_{1,0}^T P_{H_{2,0}} H_{1,0})^{-1} H_{1,0}^T P_{H_{2,0}} y_0, \tag{66}$$

we obtain the thesis in (59) and we get $\hat{x}_0|_{I_2}$ in the same way.

2. decomposition with overlap i.e. $s \neq 0$ and overlap set $I_{1,2}$ in (47).

$$\widehat{x}_0|_{I_1} = x_1 \in \mathbb{R}^{n_1} \quad \widehat{x}_0|_{I_2} = \begin{cases} x_1|_{I_{1,2}} \in \mathbb{R}^s \text{ on } I_{1,2} \\ x_2 \in \mathbb{R}^{n_2-s} \text{ on } I_2 \setminus I_{1,2} \end{cases} \in \mathbb{R}^{n_2}, \qquad (67)$$

where $x_1 \in \mathbb{R}^{n_1}$ and $x_2 \in \mathbb{R}^{n_2}$ are defined in (59) and $n_1 = |I_1|$, $n_2 = |I_2|$, $s = |I_{1,2}|$.

Next theorem formally states the mathematical framework of DD approach, which is main contribution of the present work.

Theorem 2 (*DD-KF-CLS*). *Let us consider the overdetermined linear system in (12), $H_0 \in \mathbb{R}^{m_0 \times n}$, $R_0 \in \mathbb{R}^{m_0 \times m_0}$, $M = (H_0^T R_0 H_0)^{-1} H_0^T R_0 \in \mathbb{R}^{n \times m_0}$ and $y_0 \in \mathbb{R}^{m_0}$ and $\widehat{x}_0 = M y_0 \in \mathbb{R}^n$, $m_0 > n$, $m_1 > 0$. DD-KF-CLS procedure is composed by the following steps.*

- *DD step. decomposition of $I = \{1, ..., n\}$, i.e. the columns index set of $A \in \mathbb{R}^{(m_0+m_1) \times n}$ of S in (12) into*

$$I_1 = \{1, ..., n_1\} \quad and \quad I_2 = \{n_1 + 1, ..., n\}, \qquad (68)$$

  with $|I_1| = n_1$ and $|I_2| = n_2$.
- *KF-CLS step. Computation of $\widehat{x}_{1,0} \equiv \widehat{x}_0|_{I_1} \in \mathbb{R}^{n_1}$ and $\widehat{x}_{2,0} \equiv \widehat{x}_0|_{I_2} \in \mathbb{R}^{n_2}$, reduction of $\widehat{x}_0 \in \mathbb{R}^n$ as in (60) to I_1 and I_2 in (68).*

Given $\widehat{x}_{2,0}^0 \in \mathbb{R}^{n_2}$, we consider

$$\begin{aligned} M_{0,1}^1 &= I_{n,n} \in \mathbb{R}^{n \times n}, & M_{0,1}^2 &= I_{n,n} \in \mathbb{R}^{n \times n}, \\ Q_0^1 &= O_{n,n} \in \mathbb{R}^{n \times n}, & Q_0^2 &= O_{n,n} \in \mathbb{R}^{n \times n} \\ P_{1,0} &= (H_0^T|_{I_1} R_0 H_0|_{I_1})^{-1} \in \mathbb{R}^{n_1 \times n_1} & P_{2,0} &= (H_0^T|_{I_2} R_0 H_0|_{I_2})^{-1} \end{aligned} \qquad (69)$$

with $I_{n,n}$ identity matrix and $O_{n,n}$ null matrix.

For each $n = 0, 1, 2, ...$ and for $k = 0$, applying KF procedure 2.1 to P_1^{n+1} and P_2^{n+1} in (50) and (51), we obtain KF estimates $\widehat{x}_{1,1}^{n+1} \in \mathbb{R}^{n_1}$ and $\widehat{x}_{2,1}^{n+1} \in \mathbb{R}^{n_2}$ such that

$$\begin{aligned} \widehat{x}_{1,1}^{n+1} &= \widehat{x}_1^{n+1} \\ \widehat{x}_{2,1}^{n+1} &= \widehat{x}_2^{n+1} \end{aligned} \qquad (70)$$

where $\widehat{x}_1^{n+1}$, $\widehat{x}_2^{n+1}$ are solutions in least squares sense of systems $\tilde{S}_1^{n+1}$ and $\tilde{S}_2^{n+1}$ defined in (52).

Proof. We apply KF in 2.1 to P_1^{n+1}, P_2^{n+1} in (50), (51). For $k = 0$, we obtained as predicted estimates

$$\begin{cases} x_{1,1} = M_{0,1}^1 \widehat{x}_{1,0} = \widehat{x}_{1,0} \\ x_{2,1} = M_{0,1}^2 \widehat{x}_{2,0} = \widehat{x}_{2,0} \end{cases}, \qquad (71)$$

predicted covariance matrices

$$\begin{cases} P_{1,1} = M_{0,1}^1 P_{1,0} (M_{0,1}^1)^T + Q_0^1 = P_{1,0} \\ P_{2,1} = M_{0,1}^2 P_{2,0} (M_{0,1}^2)^T + Q_0^2 = P_{2,0} \end{cases}, \qquad (72)$$

Kalman gains

$$\begin{cases} K_{1,1} = P_{1,1}H_1|_{I_1}^T (H_1|_{I_1} P_{1,1}H_1^T|_{I_1} + R_1)^{-1}, \\ K_{2,1} = P_{2,1}H_1|_{I_2}^T (H_1|_{I_2} P_{2,1}H_1^T|_{I_2} + R_1)^{-1} \end{cases} \tag{73}$$

Kalman covariance matrices

$$\begin{cases} P_{1,1} = (I - K_{1,1}H_1|_{I_1})P_{1,1}, \\ P_{2,1} = (I - K_{2,1}H_1|_{I_2})P_{2,1}, \end{cases} \tag{74}$$

and matrices

$$\begin{cases} \tilde{K}_{1,1} = P_{1,1}H_0|_{I_1}^T R_0^{-1} \\ \tilde{K}_{2,1} = P_{2,1}H_0|_{I_2}^T R_0^{-1} \end{cases} \tag{75}$$

So, for $n = 0, 1, 2, \ldots$ Kalman estimates are

$$\begin{aligned} \widehat{x}_{1,1}^{n+1} &= \widehat{x}_{1,0} + K_{1,1}\left[(y_1 - H_1|_{I_2}\widehat{x}_{2,1}^n) - H_1|_{I_1}\widehat{x}_{1,0}\right] + \mathcal{S}_{I_1 \leftrightarrow I_2}(\widehat{x}_{2,1}^n) \\ \widehat{x}_{2,1}^{n+1} &= \widehat{x}_{2,0} + K_{2,1}\left[(y_1 - H_1|_{I_1}\widehat{x}_{1,1}^{n+1}) - H_0|_{I_2}\widehat{x}_{2,0}\right] + \mathcal{S}_{I_1 \leftrightarrow I_2}(\widehat{x}_{1,1}^{n+1}) \end{aligned} \tag{76}$$

where for $i, j = 1, 2$, $\mathcal{S}_{I_1 \leftrightarrow I_2}(\widehat{x}_{i,1}^n) := \tilde{K}_{i,j}\left[H_0|_{I_i}(\widehat{x}_{i,0} - \widehat{x}_{i,1}^n)\right]$, represents the exchange of data between the sets I_1, I_2.

For each $n = 0, 1, 2, \ldots$ and from Proposition 1 applied to P_1^{n+1} and P_2^{n+1} in (50), (51) we obtain:

$$\widehat{x}_{1,1}^{n+1} \equiv \widehat{x}_1^{n+1} \qquad \widehat{x}_{2,1}^{n+1} \equiv \widehat{x}_2^{n+1}, \tag{77}$$

therefore we get the thesis.

Remark 4. Decomposition of I with overlap i.e. $I_1 = \{1, \ldots, n_1\}$ *and* $I_2 = \{n_1 - s + 1, \ldots, n\}$, and $I_{1,2} = \{n_1 - s + 1, \ldots, n_1\}$ with $s \neq 0$ is similarly obtained by considering initial estimates as in (67). Furthermore, at each $n = 0, 1, 2, \ldots$, we add operator $\mathcal{O}_{1,2}$ to P_1^{n+1} and P_2^{n+1} in (50) and (51). It means that it is

$$\begin{aligned} \widehat{x}_{1,1}^{n+1} &\equiv \widehat{x}_{1,1}^{n+1} + P_{1,1}\mu\nabla\mathcal{O}_{1,2}(EO_{I_1}(\widehat{x}_{1,1}^n|_{I_{1,2}}), EO_{I_1}(\widehat{x}_{2,1}^n|_{I_{1,2}})) \\ \widehat{x}_{2,1}^{n+1} &\equiv \widehat{x}_{2,1}^{n+1} + P_{2,1}\mu\nabla\mathcal{O}_{1,2}(EO_{I_2}(\widehat{x}_{1,1}^{n+1}|_{I_{1,2}}), EO_{I_1}(\widehat{x}_{2,1}^{n+1}|_{I_{1,2}})) \end{aligned} \tag{78}$$

where $\nabla\mathcal{O}_{1,2}(EO_{I_i}(\widehat{x}_{i,1}^n|_{I_{1,2}}), EO_{I_i}(\widehat{x}_{j,1}^n|_{I_{1,2}})) = [EO_{I_1}(\widehat{x}_{j,1}^n|_{I_{1,2}}) - EO_{I_1}(\widehat{x}_{i,1}^n|_{I_{1,2}})]$, with $EO_{I_i}(\widehat{x}_{1,0}^n|_{\tilde{I}_{1,2}})$, $EO_{I_i}(\widehat{x}_{2,0}^n|_{\tilde{I}_{1,2}})$, $i, j = 1, 2$, be extensions to I_i of reduction to $\tilde{I}_{1,2}$ of $\widehat{x}_{1,0}^n$ and $\widehat{x}_{2,0}^n$ and μ regularization parameter.

Finally, last result proves that KF-DD-CLS is equivalent to DD-KF-CLS method. To this end, we prove that these methods provides the same solutions.

Theorem 3. *Let S be the overdetermined linear system in (12), $A \in \mathbb{R}^{(m_0+m_1) \times n}$, $b \in \mathbb{R}^{m_0+m_1}$ defined in (13) with $m_0 > n$ and $m_1 > 0$. Let $\widehat{x}_1 \in \mathbb{R}^n$ be Kalman estimate in (10) of $\widehat{x}$, i.e. solution of P in (14) and let $\widehat{x}_{1,1}^{n+1}$, $\widehat{x}_{2,1}^{n+1}$ in (78) be Kalman estimates of P_1^{n+1} and P_2^{n+1} given in (50), (51). Then, it holds that*

$$\widehat{x}_{1,1}^{n+1} \to \widehat{x}_1|_{I_1}, \qquad \widehat{x}_{2,1}^{n+1} \to \widehat{x}_1|_{I_2}. \tag{79}$$

Proof. We consider decomposition of I, the indexes set of the columns of matrix A, into I_1 and I_2 as in (68). From the convergence of $\{x^{n+1}\}_{n \in \mathbb{N}_0}$ to $\widehat{x}$ in (55), where $\widehat{x}$ is the solution in least squares sense of S in (12), we get

$$x^{n+1} \to \widehat{x} \Rightarrow \begin{cases} \widehat{x}_1^{n+1} \to \widehat{x}|_{I_1} \ on \ I_1 \\ \widehat{x}_2^{n+1} \to \widehat{x}|_{I_2} \ on \ I_2 \end{cases} \tag{80}$$

with $\widehat{x}_1^{n+1}$, $\widehat{x}_2^{n+1}$ solutions of $\tilde{S}_1^{n+1}$, $\tilde{S}_2^{n+1}$ in (52) and $\widehat{x}|_{I_1}$, $\widehat{x}|_{I_2}$ reduction of $\widehat{x}$ to I_1, I_2, respectively. For $i = 1, 2$, using Proposition 2 we have $\widehat{x}_{i,1}^{n+1} \equiv \widehat{x}_i^{n+1}$ and using Proposition 1 it follows that $\widehat{x}|_{I_i} \equiv \widehat{x}_1|_{I_i}$, so we get the thesis. The case with overlap is similarly obtained.

5 Validation Results

We perform validation analysis of the proposed approach. We underline that results we present are concerned not so much with parallel efficiency as with the trustworthy and usability of the proposed approach to solve this problem. Simulation results, implemented using MATLABR2018b on a laptop with 1.6 GHz CPU and 4 GB of memory, are described in details essentially to ensure their reproducibility.

We consider:

- $H_0 \in \mathbb{R}^{11 \times 6}$: random matrix;
- $H_1 \equiv h^T \in \mathbb{R}^{1 \times 6}$: random vector;
- $y_0 \in \mathbb{R}^{11}$: random vector;
- $y_1 \in \mathbb{R}$: a random constant;
- $b = [y_0, y_1] \in \mathbb{R}^{12}$ the vector in (12);
- $R_0 = 0.5 \cdot I$: weight matrix, with $I \in \mathbb{R}^{11 \times 11}$ identity matrix, $R_1 = 0.5$ and $R = diag(R_0, R_1) \in \mathbb{R}^{12 \times 12}$ weight matrix.

We calculate:

- $\widehat{x}_0 \in \mathbb{R}^6$: solution of normal equations in (11);
- $\widehat{x} \in \mathbb{R}^6$: solution of normal equations in (16) obtained by using Conjugate Gradient method;
- $\widehat{x}_1 \in \mathbb{R}^6$ Kalman estimate as in (10) at step $k = 1$.

We apply DD approach to CLS problem in (14) by using:

- $nmax = 50$: maximum number of iterations;
- $tol = 10^{-6}$: tolerance;
- $\widehat{x} \in \mathbb{R}^6$: solution of normal equations in (16) by Conjugate Gradient method.

Decomposition of $I = \{1, 2, 3, 4, 5, 6\}$ without overlap i.e.

- $I_1 = \{1, 2, 3, 4\}, I_2 = \{5, 6\}$;
- $n \equiv |I| = 6$,
- $n_1 \equiv |I_1| = 4$,
- $n_2 \equiv |I_2| = 2$;

- $\widehat{x}_{1,0} \equiv \widehat{x}_0|_{I_1} \in \mathbb{R}^{n_1}$,
- $\widehat{x}_{2,0} \equiv \widehat{x}_0|_{I_2} \in \mathbb{R}^{n_2}$: as in (67) with $\widehat{x}_0$ solution in least squares sense of (11);
- for $i = 1, 2$, $\widehat{x}_{i,1}^0 \equiv zeros(n_i) \in \mathbb{R}^{n_i}$, where $zeros(n_i)$ is the null vector;
- for $n = 1, 2, ..., nmax$, $\widehat{x}_{1,1}^{n+1} \in \mathbb{R}^4$, $\widehat{x}_{2,1}^{n+1} \in \mathbb{R}^2$: Kalman estimates;
- $\|r^{n+1}\| < tol$: stopping criterion, where $r^{n+1} := (A^T R A)x^{n+1} - A^T R b$ is the residual $n + 1$ (16);
- ns: number of iterations needed to stop iterative procedure.

x^{n+1}, i.e. DD solution, is:

$$x^{n+1} = \begin{cases} \widehat{x}_{1,1}^{n+1} & on \; I_1 \\ \widehat{x}_{2,1}^{n+1} & on \; I_2 \end{cases}. \tag{81}$$

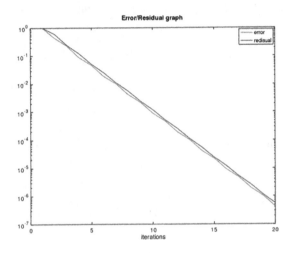

Fig. 2. Tolerance $tol = 10^{-6}$ is exceeded at $ns = 20$.

In Fig. 2 we see that residual norm exceeds $tol = 10^{-6}$ in correspondence of $ns = 20$. In particular, we note that the order of magnitude of $error = \|\widehat{x} - x^{ns}\| \approx 6.2668 \times 10^{-7}$ is the same of $\|r^{ns}\| \approx 6.6801 \times 10^{-7}$. In Table 1, we report values of $error$ and the relative number of iterations (ns).

Decomposition of $I = \{1, 2, 3, 4, 5, 6\}$ in I_1 and I_2 with overlap, for $s = 1, 2, 3$:

- $I_1 = \{1, 2, 3, 4\}$, $I_2 = \{4 - s, ..., n\}$ and $I_{1,2} = \{4 - s, ..., 4\}$;
- $n \equiv |I| = 6$,
- $n_1 \equiv |I_1| = 4$,
- $n_2 \equiv n + s - n_1 \equiv |I_2| = 2 + s$;
- for $i = 1, 2$, $\widehat{x}_{i,1}^0 \equiv zeros(n_i) \in \mathbb{R}^{n_i}$, where $zeros(n_i)$ is the null vector;

– for $n = 1, 2, ..., nmax$, we compute $\widehat{x}_{1,1}^{n+1} \in \mathbb{R}^4$, $\widehat{x}_{2,1}^{n+1} \in \mathbb{R}^2$: Kalman estimates as in (78) [6, 13, 14, 20].

DD estimate $x_s^{n+1} \in \mathbb{R}^9$ is obtained as

$$
x_s^{n+1} = \begin{cases} \widehat{x}_{1,1}^{n+1}|_{I_1 \setminus I_{1,2}} & on\ I_1 \setminus I_{1,2} \\ \frac{\mu}{2}(\widehat{x}_1^{n+1}|_{I_{1,2}} + \widehat{x}_2^{n+1}|_{I_{1,2}}) & on\ I_{1,2} \\ \widehat{x}_{2,1}^{n+1}|_{I_2 \setminus I_{1,2}} & on\ I_2 \setminus I_{1,2} \end{cases}, \tag{82}
$$

with $\mu \equiv 1$ regularization parameter; $\|r_s^{n+1}\| < tol$ stopping criterion, where $r^{n+1} := (A^T R A)x_s^{n+1} - A^T R b$ is the residual at $n + 1$ of (16); ns_s is the corresponding iteration. As expected and shown in Table 2, the size of the overlapping set impacts the convergence behaviour of the algorithm.

Table 1. Values of $error = \|\widehat{x} - x^{ns}\|\|$ for different values of tol.

tol	ns	error
10^{-6}	20	$6.4037e - 07$
10^{-9}	29	$4.8394e - 10$
10^{-14}	33	$6.7045e - 15$

Table 2. Values of $error_s = \|\widehat{x} - x_s^{ns}\|$ for $tol = 10^{-6}$.

ns_s	$error_s$	s
17	$7.2526e - 07$	1
15	$5.1744e - 07$	2
22	$7.2741e - 07$	3

6   Conclusions and Future Work

The present work is placed in the context of a research activity devoted to the development of scalable algorithms for using Data Assimilation in large scale applications [3, 5, 10, 11]. Main purpose of this article is to describe a mathematical framework for using a DD-based approach for KF method that is both relatively easy to implement and computationally efficient. We provide the mathematical framework of this method. The key point of the present work is to formulate and prove the necessary results that underpin this framework by considering, let us say, a first-level decomposition. Nevertheless, such configuration should be considered as a part of a multilevel DD scheme designed according to the features of the application and of the computing environment. Innovation mainly goes from the introduction - ab initio, i.e. on the numerical/mathematical model - of a decomposition approach [1]. Hence, results we presented were concerned not so much with parallel efficiency as with the capability to solve this problem by using such a full decomposition. We are currently working on designing the related parallel algorithm. We hope that these findings encourage readers to further extend the framework according to their specific application's requirements.

Results confirm that the accuracy of local solutions of the forecast model and hence of local KF estimates, are not impaired by DD approach. Just to prove the reliability of the proposed approach, for simplicity of notations, we considered a simplified configuration of domain decomposition. We are working on the design and development of the related parallel algorithm by considering a more general configuration which could be used in concrete scenarios. Indeed, DD configuration may depend both on the particular application and on the mapping on the available parallel computing environment. It is worth noting that one-level Schwarz DD methods are known to not often scale to a large number of processors, that is, we need a multi-level DD methods. Such a scalable approach is the subject of ongoing research [3].

References

1. Antonelli, L., Carracciuolo, L., Ceccarelli, M., D'Amore, L., Murli, A.: Total variation regularization for edge preserving 3D SPECT imaging in high performance computing environments. In: Sloot, P.M.A., Hoekstra, A.G., Tan, C.J.K., Dongarra, J.J. (eds.) ICCS 2002. LNCS, vol. 2330, pp. 171–180. Springer, Heidelberg (2002). https://doi.org/10.1007/3-540-46080-2_18

2. Arcucci, R., D'Amore, L., Pistoia, J., Toumi, R., Murli, A.: On the variational Data Assimilation problem solving and sensitivity analysis. J. Comput. Phys. **335**, 311–326 (2017)

3. Arcucci, R., D'Amore, L., Carracciuolo, L., Scotti, G., Laccetti, G.: A decomposition of the Tikhonov Regularization functional oriented to exploit hybrid multilevel parallelism. Int. J. Parallel Prog. **45**, 1214–1235 (2017). https://doi.org/10.1007/s10766-016-0460-3. ISSN 0885-7458

4. Arcucci, R., D'Amore, L., Carracciuolo, L.: On the problem-decomposition of scalable 4D-Var Data Assimilation model. In: Proceedings of the 2015 International Conference on High Performance Computing and Simulation, HPCS 2015, 2 September 2015, 13th International Conference on High Performance Computing and Simulation, HPCS 2015, Amsterdam, Netherlands, 20 July 2015 through 24 July 2015, pp. 589–594 (2015)

5. Arcucci, R., D'Amore, L., Celestino, S., Laccetti, G., Murli, A.: A scalable numerical algorithm for solving Tikhonov Regularization problems. In: Wyrzykowski, R., Deelman, E., Dongarra, J., Karczewski, K., Kitowski, J., Wiatr, K. (eds.) PPAM 2015. LNCS, vol. 9574, pp. 45–54. Springer, Cham (2016). https://doi.org/10.1007/978-3-319-32152-3_5

6. Bertero, M., et al.: MedIGrid: a medical imaging application for computational grids. In: Proceedings International Parallel and Distributed Processing Symposium (2003). https://doi.org/10.1109/IPDPS.2003.1213457

7. D'Amore, L., Campagna, R., Mele, V., Murli, A., Rizzardi, M.: ReLaTIve. An Ansi C90 software package for the Real Laplace Transform Inversion. Numer. Algorithms **63**(1), 187–211 (2013). https://doi.org/10.1007/s11075-012-9636-0

8. D'Amore, L., Mele, V., Laccetti, G., Murli, A.: Mathematical approach to the performance evaluation of matrix multiply algorithm. In: Wyrzykowski, R., Deelman, E., Dongarra, J., Karczewski, K., Kitowski, J., Wiatr, K. (eds.) PPAM 2015. LNCS, vol. 9574, pp. 25–34. Springer, Cham (2016). https://doi.org/10.1007/978-3-319-32152-3_3

9. D'Amore, L., Campagna, R., Mele, V., Murli, A.: Algorithm 946. ReLIADiff - a C++ software package for Real Laplace transform inversion based on automatic differentiation. ACM Trans. Math. Softw. **40**(4), 31:1–31:20 (2014). article 31. https://doi.org/10.1145/2616971

10. D'Amore, L., Cacciapuoti, R.: A note on domain decomposition approaches for solving 3D variational data assimilation models. Ricerche mat. (2019). https://doi.org/10.1007/s11587-019-00432-4

11. D'Amore, L., Arcucci, R., Carracciuolo, L., Murli, A.: A scalable approach for variational data assimilation. J. Sci. Comput. **61**, 239–257 (2014). https://doi.org/10.1007/s10915-014-9824-2. ISSN 0885-7474

12. D'Amore, L., Campagna, R., Galletti, A., Marcellino, L., Murli, A.: A smoothing spline that approximates Laplace transform functions only known on measurements on the real axis. Inverse Prob. **28**(2) (2012)

13. D'Amore, L., Laccetti, G., Romano, D., Scotti, G., Murli, A.: Towards a parallel component in a GPU–CUDA environment: a case study with the L-BFGS Harwell routine. Int. J. Comput. Math. **92**(1) (2015). https://doi.org/10.1080/00207160.2014.899589

14. D'Amore, L., Mele, V., Romano, D., Laccetti, G., Romano, D.: A multilevel algebraic approach for performance analysis of parallel algorithms. Comput. Inform. **38**(4) (2019). https://doi.org/10.31577/cai_2019_4_817

15. Evensen, G.: The Ensemble Kalman Filter: theoretical formulation and practical implementation. Ocean Dynam. **53**, 343–367 (2003)

16. Gander, M.J.: Schwarz methods over the course of time. ETNA **31**, 228–255 (2008)

17. Gander, W.: Least squares with a quadratic constraint. Numer. Math. **36**, 291–307 (1980)

18. Hannachi, A., Jolliffe, I.T., Stephenson, D.B.: Empirical orthogonal functions and related techniques in atmospheric science: a review. Int. J. Climatol. **1152**, 1119–1152 (2007)

19. Kalman, R.E.: A new approach to linear filtering and prediction problems. Trans. ASME J. Basic Eng. **82**, 35–45 (1960)

20. Murli, A., D'Amore, L., Laccetti, G., Gregoretti, F., Oliva, G.: A multi-grained distributed implementation of the parallel Block Conjugate Gradient algorithm. Concur. Comput. Pract. Exp. **22**(15), 2053–2072 (2010)

21. Rozier, D., Birol, F., Cosme, E., Brasseur, P., Brankart, J.M., Verron, J.: A reduced-order Kalman filter for data assimilation in physical oceanography. SIAM Rev. **49**(3), 449–465 (2007)

22. Sorenson, H.W.: Least square estimation:from Gauss to Kalman. IEEE Spectr. **7**, 63–68 (1970)

Performance Analysis of a Parallel Denoising Algorithm on Intel Xeon Computer System

Ivan Lirkov$^{(\boxtimes)}$ ⓘ

Institute of Information and Communication Technologies,
Bulgarian Academy of Sciences,
Acad. G. Bonchev, bl. 25A, 1113 Sofia, Bulgaria
ivan@parallel.bas.bg
http://parallel.bas.bg/~ivan/

Abstract. This paper presents an experimental performance study of a parallel implementation of the Poissonian image restoration algorithm. Hybrid parallelization based on MPI and OpenMP standards is investigated. The implementation is tested for high-resolution radiographic images on a supercomputer using Intel Xeon processors as well as Intel Xeon Phi coprocessors. The experimental results show an essential improvement when running experiments for a variety of problem sizes and number of threads.

Keywords: Anscombe transform · Image restoration · Parallel algorithm · Intel Xeon Phi coprocessor

1 Introduction

Accurate 3D Computed Tomography (CT) reconstruction of microstructures has numerous applications and is crucial for future realistic numerical simulations of the material's macro characteristics. It is also a quite complicated task, due to the presence of noise in the image. For example, directly segmenting the noisy 3D CT image is not reliable for porous data where standard algorithms may not be able to reconstruct even up to 50% of the material voxel data, thus important quantities (e.g., absolute porosity, average pore size, size and shape of individual pores) which determine its properties are completely miscomputed.

Typically, the CT data consist of thousands of 2D radiographic images. Optimal feature extraction from 2D radiographic image data is a vital process from an application point of view, since the image edges and singularities contain most of the important information about the structure and the properties of the scanned object. On the other hand, it is a theoretically challenging task and an ongoing research field due to the usually poor quality of the analyzed data. In radiography data acquisition is obtained by counting particles that gives rise to Poisson-dominated noisy output. Poisson noise is non-additive and exhibits

© Springer Nature Switzerland AG 2020
R. Wyrzykowski et al. (Eds.): PPAM 2019, LNCS 12044, pp. 93–100, 2020.
https://doi.org/10.1007/978-3-030-43222-5_8

mean/variance relationship, thus nonlinear filters are necessary to be applied for its successful removal. Using variance-stabilizing transformations, such as the Anscombe transform, the Poisson noise can be approximated by a Gaussian one, for which classical denoising filters can be used.

We consider an algorithm which solves an Anscombe-transformed constrained optimization problem, based on Least Squares techniques [5]. It allows for complete splitting of the pixel data and allows for their independent treatment within each iteration. Furthermore, it was experimentally observed that the convergence rate of the algorithm heavily depends on both the image size and the choice of input parameters, making the sequential realization of the algorithm impractical for large-scale industrial images. On the other hand, the CT data consist of thousands of high resolution (e.g. size 1446×1840) 2D radiographic images. Thus, the sequential implementation of the algorithm cannot be used for real-time 3D volume reconstruction.

The proposed algorithm is taken from [5] and can be written as follows:

Algorithm 1

Initialization:
$$\mathbf{u}^{(0)}, \boldsymbol{\zeta}^{(0)}, \left(\mathbf{p}_j^{(0)}\right)_{1 \leq j \leq 3} = \left(\bar{\mathbf{p}}_j^{(0)}\right)_{1 \leq j \leq 3}, \ \rho > 0, \ \sigma > 0, \ \rho\sigma < 1/9.$$
For $k = 0, 1, \ldots$ repeat until a stopping criterion is reached

1. $\mathbf{u}^{(k+1)} = \max\left\{\min\left\{\left(\mathbf{u}^{(k)} - \sigma\rho\left(H^*\bar{\mathbf{p}}_1^{(k)} + \nabla^*\bar{\mathbf{p}}_2^{(k)}\right)\right), \nu\mathbf{1}_n\right\}, 0\right\}$

2. $\boldsymbol{\zeta}^{(k+1)} = P_{V_n}\left(\boldsymbol{\zeta}^{(k)} - \sigma\rho\bar{\mathbf{p}}_3^{(k)}\right)$

3. $(v_{1,i}, \eta_i) = P_{\mathrm{epi}\varphi_i}\left(p_{1,i}^{(k)} + \left(H\mathbf{u}^{(k+1)}\right)_i + 3/8, \ p_{3,i}^{(k)} + \zeta_i^{(k+1)}\right), \quad i = 1, \ldots, n$

4. $\mathbf{v}_2 = \mathbf{p}_2^{(k)} + \nabla\mathbf{u}^{(k+1)}$

5. $\mathbf{p}_1^{(k+1)} = \mathbf{p}_1^{(k)} + H\mathbf{u}^{(k+1)} + 3/8 - \mathbf{v}_1$

6. $\mathbf{p}_2^{(k+1)} = \mathbf{v}_2 - \mathrm{prox}_{\sigma^{-1}\|\cdot\|_{2,1}}(\mathbf{v}_2)$

7. $\mathbf{p}_3^{(k+1)} = \mathbf{p}_3^{(k)} + \boldsymbol{\zeta}^{(k+1)} - \boldsymbol{\eta}$

8. $\bar{\mathbf{p}}_j^{(k+1)} = \mathbf{p}_j^{(k+1)} + \left(\mathbf{p}_j^{(k+1)} - \mathbf{p}_j^{(k)}\right), \quad j = 1, 2, 3.$

In [4] we have developed a hybrid parallel code based on the MPI and OpenMP standards [1–3,6,7]. It is motivated by the need to maximize the parallel efficiency of the implementation of the proposed algorithm.

The remainder of the paper is organized as follows. We introduce the experimental setup in the Sect. 2. A set of numerical tests are presented in the Sect. 3. Finally, some conclusions and next steps in our work are included in the last section.

2   Experimental Setup

Let us now report on the experiments performed with the parallel implementation of the algorithm. A portable parallel code was designed in C. As outlined above, the hybrid parallelization is based on joint application of the MPI and the OpenMP standards.

Table 1. Execution time (in seconds) for 100000 iterations of the algorithm using only processors of a single node of the Avitohol.

M	N	k					
		1	2	4	8	16	32
723	920	5520.07	4371.80	2294.13	1288.48	849.77	557.19
1446	1840	21547.00	18157.96	9400.75	5299.85	3576.14	3052.74
1840	1446	21057.86	18033.88	9389.83	5310.77	3581.53	2528.93

In our experiments, times were collected using the MPI provided timer, and we report the average time from multiple runs. In what follows, we report the average elapsed time T_p (in seconds), when using m MPI processes and k OpenMP threads per MPI process. During the numerical experiments, we have tested the parallel algorithm on one node for the number of OpenMP threads from one to the maximal available number of threads. On many nodes we tested the algorithm for the number of OpenMP threads varying from the number of cores per node to the maximal available number of threads. Let us denote the global number of threads by p. Then, we report the parallel speed-up $S_p = T_1/T_p$ (T_1 is the average elapsed time of the same algorithm using one MPI process and one thread) and the parallel efficiency $E_p = S_p/p$.

We have tested the parallel algorithm on images obtained from the Tomograph XTH 225 Compact industrial CT scanning. The images have size 723×920 or 1446×1840 pixels. Also, in order to study the performance of the developed algorithm, we applied it to a "transposed image," with size 1840×1446 pixels. In the tables the size of the image is denoted by $M \times N$.

The parallel code has been tested on cluster computer system Avitohol, at the Advanced Computing and Data Centre of the Institute of Information and Communication Technologies of the Bulgarian Academy of Sciences.

The computer system Avitohol is constructed with HP Cluster Platform SL250S GEN8. It has 150 servers, and two 8-core Intel Xeon E5-2650 v2 8 C processors and two Intel Xeon Phi 7120P coprocessors per node. Each processor runs at 2.6 GHz. Processors within each node share 64 GB of memory. Each Intel Xeon Phi has 61 cores, runs at 1.238 GHz, and has 16 GB of memory. Nodes are interconnected with a high-speed InfiniBand FDR network (see also http://www.hpc.acad.bg/). We used the Intel C compiler, and compiled the code using the following options: "-O3 -qopenmp" for the processors and "-O3 -qopenmp -mmic" for the coprocessors. Intel MPI was used to execute the code on the Avitohol computer system.

3 Experimental Results

Tables 1 and 2 present times collected on the Avitohol using only Intel Xeon processors. Table 1 shows that using only processors on one node the best results are obtained using 32 OpenMP threads. We gain from the effect of hyper-threading for all image sizes used in this set of experiments.

Table 2. Execution time (in seconds) for 100000 iterations of the algorithm using only processors on many nodes of the Avitohol.

M	N	Nodes					
		2	3	4	5	6	8
		k = 16					
723	920	384.70	233.90	**151.42**	**117.97**	**99.86**	**77.60**
1446	1840	1764.39	1162.12	864.54	699.46	550.16	397.40
1840	1446	1785.73	1160.34	862.35	**688.89**	**546.60**	**393.59**
		k = 32					
723	920	**246.15**	**207.84**	184.13	174.62	192.41	191.02
1446	1840	**1248.47**	**1137.24**	**815.63**	**671.48**	**523.21**	**330.31**
1840	1446	**1619.59**	**1139.02**	**851.25**	733.85	594.25	431.20

The execution time on up to eight nodes (again using only processors) is presented in Table 2. Here, a slightly different results are observed. We used bold numbers to mark the better performance varying the number of OpenMP threads for the same number of nodes. The results show that for up to three nodes we gain from the effect of hyper-threading for all image sizes. For images with size 1446×1840 the same holds true also for the number of nodes up to eight. For images with size 723×920 and 1840×1446 increasing the number of nodes the better performance is obtained using 16 OpenMP threads per node.

To provide an insight into performance of the parallel algorithm using only processors of the Avitohol, the obtained speed-up is reported in Table 3. It can be seen that for small images the obtained parallel efficiency is slightly better.

Table 3. Speed-up using only processors.

M	N	Cores									
		2	4	8	16	32	48	64	80	96	128
723	920	1.26	2.41	4.29	9.91	22.45	26.59	36.49	46.81	55.33	71.23
1446	1840	1.19	2.29	4.07	7.06	17.28	18.97	26.69	32.13	41.18	65.31
1840	1446	1.17	2.25	3.97	8.34	13.03	18.53	24.74	30.63	38.60	53.61

Tables 4 and 5 present times collected on the Avitohol using only Intel Xeon Phi coprocessors. As it was expected, the best performance on one coprocessor is obtained using the maximal available number of threads ($k = 244$). The same is true also for big images on up to 16 coprocessors. For small images there are exceptions—the execution of the code on four and eight nodes. In this cases the performance using 200 or 240 OpenMP threads is better. Comparing the results in Tables 1, 2, and 5 one can see that for big images the algorithm run faster

Table 4. Execution time (in seconds) for 100000 iterations of the algorithm using only one coprocessor of the Avitohol.

M	N	k					
		1	8	60	120	240	244
723	920	45176.10	8863.91	1340.72	822.56	550.51	549.88
1446	1840	205412.67	37072.82	6219.42	3818.24	2614.77	2599.64
1840	1446	208701.67	69456.75	9654.39	5043.79	2754.97	2719.17

Table 5. Execution time (in seconds) for 100000 iterations of the algorithm using only coprocessors of the Avitohol ($k = 244$).

M	N	Nodes						
		1	2	3	4	5	6	8
		$k = 200$						
723	920	357.33	224.31	179.89	159.44	148.92	138.12	**126.95**
1446	1840	1516.01	765.40	541.50	436.50	379.98	340.17	291.88
1840	1446	1608.39	759.03	532.22	434.51	373.03	347.94	286.84
		$k = 240$						
723	920	325.86	208.04	169.10	**153.52**	145.89	137.04	128.17
1446	1840	1350.51	681.63	493.32	405.82	355.23	320.43	277.93
1840	1446	1363.66	667.78	479.42	386.42	335.38	314.51	256.52
		$k = 244$						
723	920	**321.86**	**207.80**	**168.87**	153.55	**145.73**	**137.03**	128.13
1446	1840	**1336.17**	**674.21**	**488.44**	**402.27**	**352.23**	**317.99**	**277.76**
1840	1446	**1338.11**	**664.77**	**479.10**	**384.34**	**335.10**	**311.14**	**257.27**

Table 6. Speed-up using only coprocessors.

M	N	p					
		8	60	120	240	244	488
723	920	5.10	33.70	54.92	82.19	82.27	140.50
1446	1840	5.54	33.03	53.80	78.56	79.02	153.73
1840	1446	3.00	21.62	41.38	75.75	76.75	155.97
		976	1464	1952	2440	2928	3904
723	920	215.37	263.83	294.13	305.02	328.53	352.59
1446	1840	305.00	420.93	511.25	583.50	646.32	739.54
1840	1446	313.95	436.77	543.02	622.81	670.76	811.22

using only coprocessors on the same number of nodes. For small images this is true only on up to three nodes.

Table 6 shows the obtained speed-up of the parallel algorithm using only coprocessors of the Avitohol. Here, one can see that the obtained parallel efficiency is better for "transposed" images.

Table 7 shows the average execution time collected on the Avitohol using Intel Xeon processors as well as Intel Xeon Phi coprocessors. Comparing results in Tables 1, 2, and 7 it can be seen that for the small images there is an improvement in the performance only using one to four nodes. For the large images the algorithm has from two to three times better performance using both processors and coprocessors compared to the performance using only processors on up to eight nodes.

Table 7. Execution time (in seconds) for 100000 iterations of the algorithm using processors and coprocessors of the Avitohol.

M	N	Nodes						
		1	2	3	4	5	6	8
723	920	260.04	167.25	137.54	123.14	118.85	121.29	110.98
1446	1840	1110.95	554.87	394.63	318.25	273.65	267.12	215.83
1840	1446	1091.79	561.22	409.55	310.96	254.32	264.09	201.53

Table 8. The number of threads used to obtain the execution time in Table 7. We use the following notation: $m_c \times k_c + m_\varphi \times k_\varphi$ means m_c MPI processes on processors, k_c OpenMP threads on processors, m_φ MPI processes on coprocessors, k_φ OpenMP threads on coprocessors.

M	N	Nodes						
		1	2	3	4	5	6	8
723	920	1 × 16 + 2 × 244	2 × 16 + 4 × 240	6 × 8 + 6 × 240	8 × 8 + 8 × 240	10 × 8 + 10 × 244	12 × 8 + 12 × 244	16 × 8 + 16 × 244
1446	1840	1 × 32 + 2 × 244	2 × 16 + 4 × 240	3 × 16 + 6 × 244	4 × 16 + 8 × 244	5 × 16 + 10 × 244	6 × 16 + 12 × 244	16 × 8 + 16 × 240
1840	1446	1 × 32 + 2 × 244	2 × 16 + 4 × 240	3 × 16 + 6 × 244	4 × 16 + 8 × 244	5 × 16 + 10 × 244	6 × 16 + 12 × 240	16 × 8 + 16 × 240

Table 8 presents the number of threads used to obtain the execution time in Table 7. The last Table shows that for all image sizes and number of nodes the better results are obtained using 240 or 244 threads on Intel Xeon Phi. In general, for all images the best performance is observed using 16 OpenMP threads on processors.

Finally, we compare the performance of the parallel algorithm using only processors, only coprocessors, and using both processors and coprocessors. The average time obtained on up to eight nodes for different image sizes is shown in Fig. 1.

Execution time

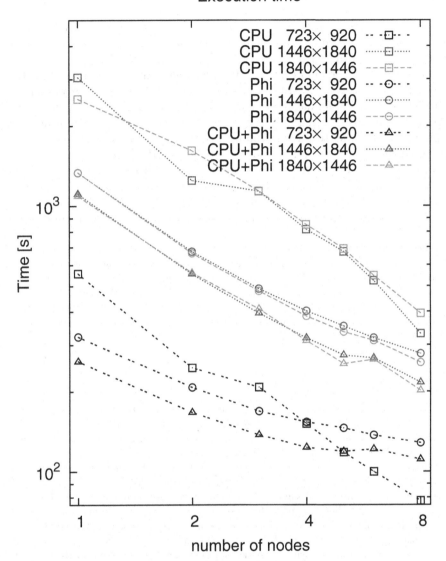

Fig. 1. Execution time for various image sizes.

4   Concluding Remarks and Future Work

We have studied the efficiency of the parallel implementation of an image restoration algorithm based on Least Squares techniques. Numerical experiments are conducted on a supercomputer using Intel Xeon processors as well as Intel Xeon Phi coprocessors. As it was expected, for big images the best results using only processors are obtained using the maximal available number of threads. Also, for both sizes of the images, the best results on coprocessors are obtained using the maximal available number of threads. Finally, the experimental results show an essential improvement when running experiments using processors as well as coprocessors for a variety of image sizes and number of threads.

In the current version of our parallel implementation, the image is divided into strips. The sizes of all strips are almost the same. During the experiments it was seen that for small images on number of nodes larger than three our parallel algorithm runs faster using only processors compared to the results using only coprocessors. In order to tune the implementation and to have good performance we tested the algorithm running various number of MPI processes on processors while using one MPI process per coprocessor. As a next step, in order to achieve better load balance on the hybrid architecture, we have to make further changes in the MPI code. In this way we can avoid or at least minimize the delay caused by different load in the Intel Xeon processors and Intel Xeon Phi coprocessors.

Acknowledgments. We acknowledge the provided access to the e-infrastructure of the National centre for high performance and distributed computing. This research was partially supported by grant KP-06-N27/6 from the Bulgarian NSF. This work has been accomplished with the partial support by the Grant No. BG05M2OP001-1.001-0003, financed by the Science and Education for Smart Growth Operational Program (2014–2020) and co-financed by the European Union through the European structural and Investment funds.

References

1. Chandra, R., Menon, R., Dagum, L., Kohr, D., Maydan, D., McDonald, J.: Parallel Programming in OpenMP. Morgan Kaufmann, San Francisco (2000)
2. Chapman, B., Jost, G., Van Der Pas, R.: Using OpenMP: Portable Shared Memory Parallel Programming, vol. 10. MIT Press, Cambridge (2008)
3. Gropp, W., Lusk, E., Skjellum, A.: Using MPI: Portable Parallel Programming with the Message-Passing Interface. MIT Press, Cambridge (2014)
4. Harizanov, S., Lirkov, I., Georgiev, K., Paprzycki, M., Ganzha, M.: Performance analysis of a parallel algorithm for restoring large-scale CT images. J. Comput. Appl. Math. **310**, 104–114 (2017). https://doi.org/10.1016/j.cam.2016.07.001
5. Harizanov, S., Pesquet, J.-C., Steidl, G.: Epigraphical projection for solving least squares anscombe transformed constrained optimization problems. In: Kuijper, A., Bredies, K., Pock, T., Bischof, H. (eds.) SSVM 2013. LNCS, vol. 7893, pp. 125–136. Springer, Heidelberg (2013). https://doi.org/10.1007/978-3-642-38267-3_11
6. Snir, M., Otto, S., Huss-Lederman, S., Walker, D., Dongarra, J.: MPI: The Complete Reference. Scientific and Engineering Computation Series. The MIT Press, Cambridge (1997). Second printing
7. Walker, D., Dongarra, J.: MPI: a standard message passing interface. Supercomputer **63**, 56–68 (1996)

An Adaptive Strategy for Dynamic Data Clustering with the K-Means Algorithm

Marco Lapegna[1](✉)[iD], Valeria Mele[1][iD], and Diego Romano[2][iD]

[1] Department of Mathematics and Applications, University of Naples Federico II, Naples, Italy
{marco.lapegna,valeria.mele}@unina.it
[2] Institute for High Performance Computing and Networking (ICAR), National Research Council (CNR), Naples, Italy
diego.romano@cnr.it

Abstract. K-means algorithm is one of the most widely used methods in data mining and statistical data analysis to partition several objects in K distinct groups, called clusters, on the basis of their similarities. The main problem of this algorithm is that it requires the number of clusters as an input data, but in the real life it is very difficult to fix in advance such value. For such reason, several modified K-means algorithms are proposed where the number of clusters is defined at run time, increasing it in a iterative procedure until a given cluster quality metric is satisfied. In order to face the high computational cost of this approach we propose an adaptive procedure, where at each iteration two new clusters are created, splitting only the one with the worst value of the quality metric.

Keywords: Modified k-means clustering · Adaptive algorithm · Unsupervised learning · Data mining

1 Introduction

The data clustering problem has been addressed by researchers in many disciplines, and it has several different applications in the scientific world, from biological research, to finance, marketing, logistic, robotics, mathematical and statistical analysis, image processing, identifying patterns, and the classifications of medical tests (e.g. [2,27]).

Thus, we can say that clustering algorithms are today one of the most important tools in exploratory data analysis and one the most important data mining methodology. They can be seen as unsupervised classification approaches whose main goal is to group similar data with in the same cluster according to precise metrics.

There are several surveys, reviews and comparative study about clustering applications and techniques, written in the last twenty years, that one can refer to get an overall picture of the clustering approaches in their evolution to the

© Springer Nature Switzerland AG 2020
R. Wyrzykowski et al. (Eds.): PPAM 2019, LNCS 12044, pp. 101–110, 2020.
https://doi.org/10.1007/978-3-030-43222-5_9

current state of art (e.g. [14,20,26,27]). Different surveys give often different tax-onomies, but all of them name the K-means as one the most popular clustering algorithm. K-means algorithms is the best known Squared Error-based cluster-ing approach [27] because of its simplicity, ability to deal with large number of attributes, and providing good quality clusters with the $N * K * d$ computa-tional complexity where N is the number of elements in data space, K is count of clusters to be identified, and d is the number of attributes/dimensions [11]. Results of K-Means clustering depends on cluster center initialization and it is not able to provide globally optimum results. For different data sets, diverse versions of K-Means clustering must be chosen, and many modified version of the K-means algorithm has been proposed in the last years [1,13,14] to decrease the complexity or increase the solution quality [23,25].

The present work joins this research trend. More precisely it describes an adaptive K-means algorithm for dynamic clustering where the number of clusters K is unknown in advance and it is aimed to realize a trade-off between the algorithm performance and a global quality index for the clusters.

The rest of the paper is organized as follow: in Sect. 2 we introduce an adap-tive K-means algorithm, in Sect. 3 we report some implementation details, in Sect. 4 we show the results obtained from several experiments we have done to validate the new algorithm, and in Sect. 4 we summarize the work.

2   An Adaptive K-Means Algorithm

The K-means algorithm is a procedure aimed to define K clusters where each of them contains at least one element and each element belong to one cluster only. A formal description of the procedure follows.

Given a set of N vectors $S = \{\mathbf{s}_n : \mathbf{s}_n \in \mathbf{R}^d \quad n = 1,..,N\}$ in the d-dimensional space, and an integer K, the K-means algorithm collects the items of S in the K subgroups of a partition $\mathcal{P}_K = \{C_k : C_k \subset S \quad k = 1,..,K\}$ of S, such that $\bigcup C_k = S$ and $C_{k1} \bigcap C_{k2} = \emptyset$ with $k1 \neq k2$, on the basis of their similarity. Usually the similarity between two objects is measured by means the Euclidean norm or some other metric. Its traditional description is then based on the following steps:

Step 1. Assign randomly the N elements $s_n \in S$ to K arbitrary subgroups C_k each of them with N_k items

Step 2. Compute the centers $\mathbf{c}_k$ of the C_k with the following vector operation:

$$\mathbf{c}_k = \frac{1}{N_k} \sum_{\mathbf{s}_n \in C_k} \mathbf{s}_n \quad k = 1,..,K \tag{1}$$

Step 3. $\forall \mathbf{s}_n \in S$ find the cluster $C_{\overline{k}}$ minimizing the Euclidean distance from the center of the cluster, that is:

$$\mathbf{s}_n \in C_{\overline{k}} \Leftrightarrow ||\mathbf{s}_n - \mathbf{c}_{\overline{k}}||_2 = \min_{k=1,..,K} ||\mathbf{s}_n - \mathbf{c}_k||_2 \tag{2}$$

Step 4. Reassign $\mathbf{s}_n$ to the new cluster $C_{\overline{k}}$
Step 5. Repeat steps 2 – 4 until there is no change.

One of the major flaws of this algorithm is the need to fix the number of clusters K before the execution. Mainly with large dimensions d and number of elements N is almost impossible to define a suitable K. If it is too large similar items will be put in different clusters. On the other hand, if K is too small, there is the risk that dissimilar items will be grouped in the same cluster.

Furthermore, several studies have shown that the previous algorithm does not produce an analytic solution, and the result strongly depends on the initial assignment of the elements to the clusters [24]. For such reasons the algorithm is executed several times with different vale of K, and some quality index is used to choose a "good solution". To this aim several indices have been introduced in the literature (see for example [11]).

As an example consider the root-mean-square standard deviation (RMSSTD) index:

$$R_{RMSSTD} = \left[\frac{\sum_{k=1}^{K} \sum_{\mathbf{s}_n \in C_k} \|\mathbf{s}_n - \mathbf{c}_k\|_2^2}{d(N-K)} \right]^{1/2} \tag{3}$$

that measures the homogeneity of the clusters quality. The RMSSTD quality index decreases when the number of clusters K increases, until a fair homogeneity is reached, so that the optimal number of clusters is then the value of K at which the RMSSTD starts to grow. On the basis of these considerations we can design the iterative procedure described in the Algorithm 1 that increases the number of clusters at each step.

This strategy repeatedly tests several partitioning configurations with different values of K and it is possible to implement it only if the Computational Cost (CC) of the kernels is not too large.

Algorithm 1. iterative K-means algorithm

---

1) Set the number of clusters $K = 0$
2) **repeat**
     2.1) Increase the number of clusters $K = K + 1$
     2.2) Assign randomly the N elements $s_n \in S$ to arbitrary
          K clusters C_k, each of them with N_k items
     2.3) **repeat**
          2.3.1) Compute the centers $\mathbf{c}_k$ of the C_k as in (1)
          2.3.2) For each $\mathbf{s}_n \in S$ find the cluster $C_{\overline{k}}$ minimizing the
                 Euclidean distance from $\mathbf{c}_k$ as in (2)
          2.3.3) Reassign the elements $\mathbf{s}_n$ to the new clusters
          until (no change in the reassignment)
     2.4) update RMSSTD as in (3)
   until (RMSSTD starts to grow or its reduction is smaller
         than a given threshold)

In order to analyze the computational cost of this procedure, we concentrate our attention on the most expensive steps in the inner iterative structure 2.3. At this regard we observe that the computation of the centers $\mathbf{c}_k$ of the clusters (step 2.3.1) based on the (1), requires

$$CC_{2.3.1} = \sum_{k=1}^{K} dN_k = Nd \quad \text{f.p. operations}$$

whereas the search of the cluster for each element $\mathbf{s_n}$ (step 2.3.2) based on the (2) requires

$$CC_{2.3.2} = NKd \quad \text{f.p. operations}$$

The cost of the step 2.3.3 is strongly dependent on how the elements are distributed in the K clusters C_k at the step 2.2. An unsuitable initial assignment can result in a huge number of movement of the elements $\mathbf{s}_n$ among the clusters C_k, in order to satisfy the stopping criterion of the iterative structure 2.3.

Our method is then designed to reduce the movements of the elements among the clusters, with the aim of achieving a trade-off between a good distribution with a reasonable computational cost.

The main idea of the proposed method is to use the partition $\mathcal{P}_{K-1}$ of the elements already defined in the previous iteration, working only on the clusters with the more dissimilar elements and avoiding to starting over with a random distribution in step 2.2. To this aim, let consider the standard deviation of the elements $\mathbf{s}_k$ in the cluster C_k:

$$\sigma_k = \sqrt{\frac{1}{N_k - 1} \sum_{n=1}^{N_k} (\mathbf{s}_n - \mathbf{c}_k)^2}$$

The value of σ_k can be used to measure the similarity of the elements in C_k. Smaller the value σ_k is, closer to the center $\mathbf{c}_k$ are the elements of C_k, and the cluster is composed by similar elements. For a such reason our strategy, in the step 2.2, defines the new partition $\mathcal{P}_K$ by splitting in two subset C_λ and C_μ only the cluster C_{K-1}^* with the largest standard deviation in the previous iteration. When $K = 1$ the partition $\mathcal{P}_1$ is defined by only 1 cluster $C_1 \equiv S$. More precisely:

$$\begin{aligned} K = 1 & \quad \mathcal{P}_1 = \{C_1\} \quad \text{where} \quad C_1 \equiv S \\ K > 1 & \quad \mathcal{P}_K = \mathcal{P}_{K-1} - \{C_{K-1}^*\} \cup \{C_\lambda, C_\mu\} \end{aligned} \quad (4)$$

This strategy is based on the assumption that, at a given iteration K, very similar items have been already grouped in compact clusters with small values for σ_k at the previous iteration $K - 1$, which therefore does not require an assignment to a new cluster. At this regard, it is interesting to note that the idea of reorganizing the elements of a partition, according the value of a given quality index computed at run time, is quite common in many procedures called *adaptive* algorithms. For example, with regard to the iterative algorithms for the numerical integration, many strategies are known for the refinement of the integration domain only where a large discretization error is estimated [7, 15–19].

From what has been said, we propose an adaptive modified K-mean algorithm as follow:

Algorithm 2. adaptive K-means algorithm
1) Set the number of clusters $K = 0$
2) **repeat**
2.1) Increase the number of clusters $K = K + 1$
2.2) find the cluster C_{K-1}^* with the largest standard deviation
2.3) define the new partition of clusters $\mathcal{P}_K$ as in (3)
2.4) **repeat**
2.4.1) Compute the centers $\mathbf{c}_k$ of the C_k as in (1)
2.4.2) For each $\mathbf{s}_n \in S$ find the cluster $C_{\overline{k}}$ minimizing the
Euclidean distance from $\mathbf{c}_k$ as in (2)
2.4.3) Reassign the elements $\mathbf{s}_n$ to the new clusters
until (no change in the reassignment)
2.5) update RMSSTD as in (3)
until (RMSSTD starts to grow or its reduction is smaller
than a given threshold)

3 Implementation Details

Following there are some implementation details regarding our adaptive K-means algorithm (see also Fig. 1).

All the elements $\mathbf{s}_n \in S$ are stored, row by row, in a $N \times d$ array. In order to improve the computational cost, our method does not change the order of the rows of the array, when the elements must be moved from a cluster to another one, in the step 2.4.3. In such step, the composition of each cluster is then defined by means of contiguous items in a array PT, pointing to the rows of S representing the elements of the cluster. All the displacements of elements among clusters required in the step 2.4.3, are then implemented by exchanging only the pointers in the array PT.

In order to identify the contiguous items of the array PT pointing to a given cluster C_k, a suitable data structure is defined: a Cluster Descriptor (CD_k) that contains

- the cluster identifier (k)
- the pointer to the first elements of the cluster in the array PT (F_k)
- the number of elements of the cluster (N_k)
- the center of the cluster ($\mathbf{c}_k$)
- the standard deviation of the elements of the cluster (σ_k)

Finally, the access to cluster descriptors CD_k is provided by a Cluster Table (CT), that is a pointers array whose k-th element refers to the cluster descriptor CD_k of the cluster C_k.

This data organization allows quick and efficient access to all clusters information required in Algorithm 2.

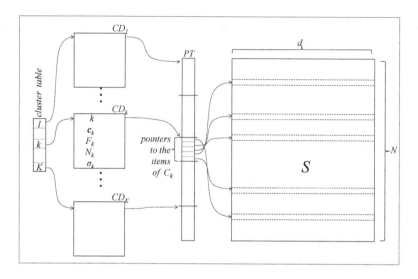

Fig. 1. Data structures organization for the K-means adaptive algorithm

4   Experimental Results

To test our method, from the accuracy and the efficiency points of view, several experiments have been conducted on a system equipped with a quad-core Intel i5-4460S (Haswell) CPU running a 2.9 Ghz and 16 Gbytes of main memory. The algorithms have been implemented in C language under Linux operating system.

First of all we report the results on the Iris data set [9] from the UCI Machine Learning Repository [8]. This is probably the earliest and the most commonly used data set in the literature of pattern recognition. It is a quite small set containing only $N = 150$ instances of iris flowers, divided into $K = 3$ classes of the same dimension $N_k = 50$ elements. The items are described on the basis of $d = 4$ attributes: petal's and sepal's width and length. Our experiments are aimed to measure the ability of Algorithm 2 to separate the items in three distinct sets and to compare the results with those obtained from Algorithm 1.

Figure 2 shows the scatter plots of each couple of features of the data set produced by Algorithm 2 setting 3 iterations as an input. The different marks represent the elements of the three clusters. An identical figure has been obtained with Algorithm 1, so that the two algorithms produce the same classification of the items. This is confirmed also by Table 1 reporting the number of elements N_k and the standard deviation σ_k of each cluster.

A more meaningful experiment was conducted using the Letter Recognition data set [10] from the UCI Machine Learning Repository. This is a large data set based on $N = 20,000$ unique items, each of them representing the black and withe image of an uppercase letter. The character images are based on 20 different fonts where each letter have been randomly distorted to produce an item of the data set. Each item was converted into $d = 16$ numerical attributes

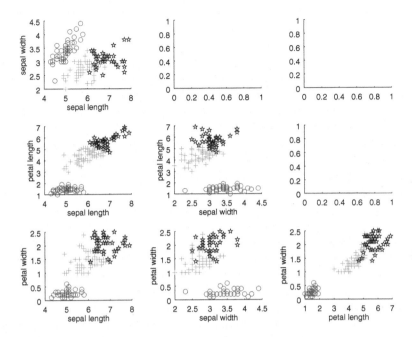

Fig. 2. Scatter plots of each couple of features of the data set produced by Algorithm 1 and Algorithm 2.

(statistical moments and edge counts) which were then scaled to fit into a range of integer values from 0 through 15. The classification task for this data set is considered especially challenging because of the wide diversity among the different fonts and because of the primitive nature of the attributes.

Table 1. Results comparison between Algorithm 1 and Algorithm 2 on the Iris dataset.

	Algorithm 1		Algorithm 2	
	N_k	σ_k	N_k	σ_k
C_1	50	0.26	50	0.26
C_2	61	0.30	61	0.30
C_3	39	0.34	39	0.34

The first set of experiment is aimed to compare the performance of our proposed Algorithm 2 with the basic Algorithm 1. In Table 2 we report the number of items $\mathbf{s}_n$ displaced from a cluster to another one in step 2.4.3 (*disp*) and the total elapsed time in second (*time*) for the generation of $K = 26$ clusters, i.e. one for each letter of the English alphabet. As expected, Algorithm 2 shows better performance than Algorithm 1, as its does not move items already grouped in compact clusters with a smaller standard deviation.

Table 2. Performance comparison between Algorithm 1 and Algorithm 2.

Algorithm 1		Algorithm 2	
disp	*time*	*disp*	*time*
1001349	55.9	130801	23.8

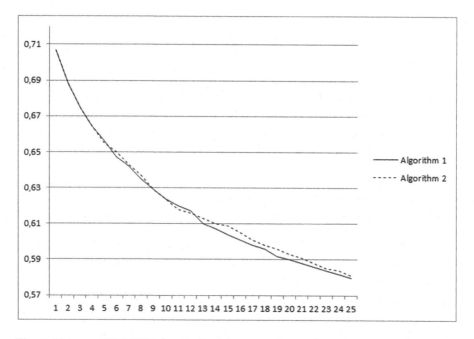

Fig. 3. Values of RMSSTD (vertical axis) versus the number of iterations (horizontal axis) for Algorithm 1 (solid line) and Algorithm 2 (dashed line).

A second set of experiments is aimed to measure the quality of the solution computed by Algorithm 2. To this aim we use the value of root-mean-square standard deviation RMSSTD, given by (3), as a measure of the global quality index of the generated clusters. Figure 3 shows the value of R_{RMSSTD} versus the number of generated clusters K for both Algorithm 1 and Algorithm 2. These values differ by less 1%, confirming that Algorithm 2 generates a partition very similar to the one computed by the traditional K-means algorithm.

5   Conclusions

In this paper we introduced an adaptive approach to improve the efficiency of dynamic data clustering with the K-means algorithm. The main drawback of this well known algorithm is that the number of clusters K should be fixed as input data, but in several real cases it is very difficult to define such a value in advance. Several algorithms attempts to overcame this problem by finding the optimal value at run time by increasing the number of clusters until some stopping criterion is satisfied, but the computational cost can be very expensive, because the need to reallocate the node of the set at each iteration. Our work addresses this aspect, avoiding the displacement of similar items already grouped into compact clusters, characterized by small values of the standard deviation. The achieved results are very promising, with accuracy similar to traditional approaches that redistribute all the items of the set at each step, and with a much lower computational cost. Future work can focus on the implementation of Algorithm 2 on advanced computing environment such as parallel/distributed computers, GPU based systems or low power devices [4,5,12,21,22] with special attention to the issues regarding the fault tolerance and the performance [3,6].

References

1. Abubaker, M., Ashour, W.M.: Efficient data clustering algorithms: improvements over Kmeans. Int. J. Intell. Syst. Appl. **5**, 37–49 (2013)
2. Aggarwal, C.C., Reddy, C.K.: Data Clustering, Algorithms and Applications. Chapman and Hall/CRC, London (2013)
3. Caruso, P., Laccetti, G., Lapegna, M.: A performance contract system in a grid enabling, component based programming environment. In: Sloot, P.M.A., Hoekstra, A.G., Priol, T., Reinefeld, A., Bubak, M. (eds.) EGC 2005. LNCS, vol. 3470, pp. 982–992. Springer, Heidelberg (2005). https://doi.org/10.1007/11508380_100
4. D'Ambra, P., Danelutto, M., diSerafino, D., Lapegna, M.: Advanced environments for parallel and distributed applications: a view of the current status. Parallel Comput. **28**, 1637–1662 (2002)
5. D'Ambra, P., Danelutto, M., diSerafino, D., Lapegna, M.: Integrating MPI-based numerical software into an advanced parallel computing environment. In: Proceedings of the Eleventh Euromicro Conference on Parallel Distributed and Network-based Processing, Clematis ed., pp. 283–291. IEEE (2003)
6. D'Amore, L., Mele, V., Laccetti, G., Murli, A.: Mathematical approach to the performance evaluation of matrix multiply algorithm. In: Wyrzykowski, R., Deelman, E., Dongarra, J., Karczewski, K., Kitowski, J., Wiatr, K. (eds.) PPAM 2015. LNCS, vol. 9574, pp. 25–34. Springer, Cham (2016). https://doi.org/10.1007/978-3-319-32152-3_3
7. D'Apuzzo, M., Lapegna, M., Murli, A.: Scalability and load balancing in adaptive algorithms for multidimensional integration. Parallel Comput. **23**, 1199–1210 (1997)
8. Dua, D., Graff, C.: UCI Machine Learning Repository. University of California, School of Information and Computer Science, Irvine, CA (2017). http://archive.ics.uci.edu/ml
9. Duda, R., Hart, P.E.: Pattern Classification and Scene Analysis. Wiley, Hoboken (1973). (Q327.D83)
10. Frey, P.W., Slate, D.J.: Letter recognition using holland-style adaptive classifiers. Mach. Learn. **6**, 161–182 (1991)
11. Gan, D.G., Ma, C., Wu, J.: Data Clustering: Theory, Algorithms, and Applications. ASA-SIAM Series on Statistics and Applied Probability. SIAM, Philadelphia (2007)
12. Gregoretti, F., Laccetti, G., Murli, A., Oliva, G., Scafuri, U.: MGF: a grid-enabled MPI library. Future Gener. Comput. Syst. **24**, 158–165 (2008)

13. Huang, Z.X.: Extensions to the K-means algorithm for clustering large datasets with categorical values. Data Min. Knowl. Disc. **2**, 283–304 (1998)
14. Joshi, A., Kaur, R.: A review: comparative study of various clustering techniques in data mining. Int. J. Adv. Res. Comput. Sci. Softw. Eng. **3**, 55–57 (2013)
15. Laccetti, G., Lapegna, M., Mele, V., Montella, R.: An adaptive algorithm for high-dimensional integrals on heterogeneous CPU-GPU systems. Concurr. Comput. Pract. Exp. **31**, cpe4945 (2018)
16. Laccetti, G., Lapegna, M., Mele, V., Romano, D., Murli, A.: A double adaptive algorithm for multidimensional integration on multicore based HPC Systems. Int. J. Parallel Program. **40**, 397–409 (2012)
17. Laccetti, G., Lapegna, M., Mele, V., Romano, D.: A study on adaptive algorithms for numerical quadrature on heterogeneous GPU and multicore based systems. In: Wyrzykowski, R., Dongarra, J., Karczewski, K., Waśniewski, J. (eds.) PPAM 2013. LNCS, vol. 8384, pp. 704–713. Springer, Heidelberg (2014). https://doi.org/10.1007/978-3-642-55224-3_66
18. Laccetti, G., Lapegna, M., Mele, V.: A loosely coordinated model for heap-based priority queues in multicore environments. Int. J. Parallel Prog. **44**, 901–921 (2016)
19. Lapegna, M.: A global adaptive quadrature for the approximate computation of multidimensional integrals on a distributed memory multiprocessor. Concurr. Pract. Exp. **4**, 413–426 (1992)
20. Patibandla, R.S.M.L., Veeranjaneyulu, N.: Survey on clustering algorithms for unstructured data. In: Bhateja, V., Coello Coello, C.A., Satapathy, S.C., Pattnaik, P.K. (eds.) Intelligent Engineering Informatics. AISC, vol. 695, pp. 421–429. Springer, Singapore (2018). https://doi.org/10.1007/978-981-10-7566-7_41
21. Marcellino, L., et al.: Using GPGPU accelerated interpolation algorithms for marine bathymetry processing with on-premises and cloud based computational resources. In: Wyrzykowski, R., Dongarra, J., Deelman, E., Karczewski, K. (eds.) PPAM 2017. LNCS, vol. 10778, pp. 14–24. Springer, Cham (2018). https://doi.org/10.1007/978-3-319-78054-2_2
22. Montella, R., et al.: Accelerating linux and android applications on low-power devices through remote GPGPU offloading. Concurr. Comput.: Pract. Exp. **29**, cpe.4950 (2017)
23. Pelleg, D., Moore. A.W.: X-means: extending k-means with efficient estimation of the number of clusters. In: Proceedings of the 17th International Conference on Machine Learning, pp. 727–734. Morgan Kaufmann (2000)
24. Pena, J.M., Lozano, J.A., Larranaga, P.: An empirical comparison of four initialization methods for the K-means algorithm. Pattern Recognit. Lett. **20**, 1027–1040 (1999)
25. Shindler, M., Wong, A., Meyerson, A.: Fast and accurate k-means for large datasets. In: Shawe-Taylor, J., Zemel, R.S., Bartlett, P.L., Pereira, F.C.N., Weinberger, K.Q. (eds.) Proceedings of 25th Annual Conference on Neural Information Processing Systems, pp. 2375–2383 (2011)
26. Xu, D., Tian, Y.: A comprehensive survey of clustering algorithms. Ann. Data Sci. **2**(2), 165–193 (2015). https://doi.org/10.1007/s40745-015-0040-1
27. Xu, R., Wunsch, D.: Survey of clustering algorithms. Trans. Neural Netw. **16**, 645–678 (2005)

Security and Storage Issues in Internet of Floating Things Edge-Cloud Data Movement

Raffaele Montella[1]([⊠]), Diana Di Luccio[1], Sokol Kosta[2], Aniello Castiglione[1], and Antonio Maratea[1]

[1] Department of Science and Technologies,
University of Naples Parthenope,
Naples, Italy
{raffaele.montella,diana.diluccio,
aniello.castiglione,antonio.maratea}@uniparthenope.it
[2] Department of Electronic Systems,
Aalborg University, Copenhagen, Denmark
sok@es.aau.dk

Abstract. Sensors and actuators became first class citizens in technologically pervasive urban environments. However, the full potential of data crowdsourcing is still unexploited in marine coastal areas, due to the challenging operational conditions, extremely unstable network connectivity and security issues in data movement. In this paper, we present the latest specification of our DYNAMO Transfer Protocol (DTP), a platform-independent data mover framework specifically designed for the Internet of Floating Things applications, where data collected on board of professional or leisure vessels are stored locally and then moved from the edge to the cloud. We evaluate the performance achieved by the DTP in data movement in a controlled environment.

Keywords: Internet of Floating Things · Data crowdsourcing · Data movement · Security · Cloud database

1 Introduction

The rise of the Internet of Things and the computational resource elasticity provided by the Cloud [6,33] are changing the human lifestyle [9]. The crowdsourcing paradigm [14] has been widely adopted in diverse contexts to solve large problems by engaging many human workers to solve manageable sub-problems [12]. When the problem involves data acquisition, management and analysis, it is referred to as *data crowdsourcing* [12]. Nowadays, data crowdsourcing is one of the most impacting technology raised as first-class citizen in the data science landscape [32], thanks to the flywheel effect generated by the availability of distributed human-carried sensor network – commonly referred as mobile computing, the reliable connection infrastructure provided by cellular and Wi-Fi

© Springer Nature Switzerland AG 2020
R. Wyrzykowski et al. (Eds.): PPAM 2019, LNCS 12044, pp. 111–120, 2020.
https://doi.org/10.1007/978-3-030-43222-5_10

networks, and the elastic computing and storage resources. Nevertheless, data crowdsourcing potentially gains more importance in environments where the use of conventional data acquisition methodologies [2,17,20,27] are expansive or unfeasible and the satellite data do not reach the adequate resolution and quality, mostly when approaching the coast [3,4]. The coastal areas host most part of human population, are fundamental for the global economy, and, above all, are one of the most sensitive environments to climate changes [26]: extreme weather events could impact negatively on human activities in a dramatic way [7,30].

Although the embryo of a distributed data collecting has been already designed at the early stage of the grid computing epic [10,34], unfortunately, data crowdsourcing marine applications are limited by the availability of stable, reliable, and cheap data connections. On the other hand, the measurement of seafloor bottom depth (bathymetry) via data crowdsourcing is common in both scientific [29] and business projects. The application of this technique to the measurement of weather and sea state parameters has previously been limited to ferries, freight carriers, professional vessels, and cruise ships. In a previous work, we developed FairWind, a smart, cloud-enabled, multi-functional navigation system for leisure and professional vessels [22,25]. In this paper, we introduce DYNAMO, an infrastructure for collecting marine environmental data from a distributed sensor network carried by leisure vessels [21,24]. DYNAMO could be considered as an improvement and evolution of FairWind, strongly leveraging on SignalK (http://signalk.org) as a common interchange format for marine data, but more focused on data logging and management.

The rest of this paper is organized as follows: Sect. 2 contains a synthetic description of similar solutions focusing on diverse and different data transfer protocols. Section 3 contains the most of the novel contribution of this paper with a detailed description of the DYNAMO Transfer Protocol, detailing on the security and storage issues. Section 4 describes the preliminary evaluation in an experimental controlled environment. Finally, Sect. 5 concludes and outlines future directions.

2   Related Work

The Bundle Protocol (BP) [28] has been proposed by the Delay Tolerant Networking Research Group (DTNRG) of the Internet Research Task Force (IRTF). The idea of this protocol is to group data in bundles in order to store and forward them when the networking is available. The main capabilities of the BP include: (i) custody-based re-transmission; (ii) ability to cope with intermittent connectivity; (iii) ability to take advantage of scheduled, predicted, and opportunistic connectivity; (iv) late binding of overlay network endpoint identifiers to constituent internet addresses. Even though BP is the only acknowledged data transfer protocol for DTNs as of today, and the best reference point for new proposals in this field, it is not designed for IoT devices and for their communication with cloud infrastructures.

The two widely used IoT application protocols that represent the current state of the art, are Message Queuing Telemetry Transport (MQTT) [15] and

Constrained Application Protocol (CoAP) [5]. MQTT is an internet application protocol for extreme environments. MQTT today is an OASIS standard, widely used for every kind of IoT application, including cloud data transfer. It is a publish/subscribe model working on top of TCP, ensuring the reliability of its approach also thanks to its small bandwidth footprint and a low loss rate in unstable networking.

CoAP is a modern standard specialized application protocol for constrained devices. It leverages on a REST model: servers allow resources access under a URL and clients use them through GET, POST, PUT, and DELETE methods. This makes CoAP integration with already different software straightforward, but working on UDP in order to maximize the efficiency.

A common middleware supporting both MQTT and CoAP and providing a common programming interface has been also implemented [31].

Nevertheless, MQTT and CoAP are both not resilient, since they are not able to elastically change the transmission rate in dependence of the bandwidth, and although both have a lightweight footprint, they do not compress the payload. The security is guaranteed by the transport layer not ensuring the firewall and proxy friendship.

3   Design

3.1   Vessel Side and Cloud Side Security

In [25], we already described the idea of a data transfer framework designed for vessel data logging and transferring to the cloud with an adaptive algorithm devoted to the maximization of the bandwidth usage leveraging on concurrent requests. In this work, we completely redesigned the vessel side in order to match a higher level of security avoiding any form of key exchange. The behaviour of the SignalK data logger on the vessel side is conceptually described by the block diagram shown in the Fig. 1.

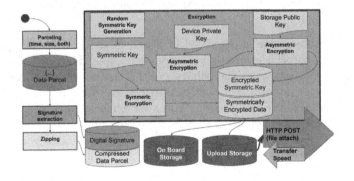

Fig. 1. The DYNAMO SignalK logger on the vessel side.

The SignalK data updates are grouped in parcels, as described in [21], and stored as text files in a temporary outbound folder. Each time a new parcel is available, the signature is extracted using the RSA-SHA 256 bit algorithm, encrypted using the vessel 1024 bit private key and finally encoded in base64 [11]. A 32 byte long symmetric encryption key is generated randomly. This key is encrypted with the cloud-side public key using the RSA PLCS1 OAEP padding. The encrypted symmetric key is finally encoded in base 64. A cipher key is generated using the SHA256 hash algorithm applied to the previously generated symmetric key. A secure pseudo random initialization vector is generated and then used to encrypt the compressed data parcel. Finally, the initialization vector and the encrypted symmetric key are prepended and the encrypted data parcel is stored in an upload folder. The DYNAMO data transfer framework mediates the vessel and the cloud sides in order to maximize the bandwidth usage. A NodeJS working implementation of a SignalK DYNAMO logger is available as open source (Apache 2.0 license) (https://github.com/OpenFairWind/signalk-dynamo-logger).

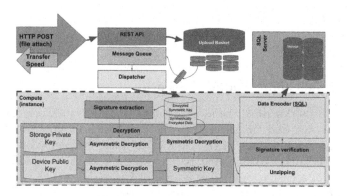

Fig. 2. *Cloud storage side* (consumer) data pipeline. The symmetric key is decrypted with the consumer's private key, and then is used to decrypt the data. The data are then uncompressed, the signature is verified, and the data are stored in a SQL database.

The cloud storage side is described in Fig. 2. Each time an encrypted and compressed data parcel is received, it is stored in an upload basket and enqueued to a message queue manager. The behavior of this component drastically affects the overall storage performance, which makes it a critical point on this kind of near-realtime applications. In order to be as much as possible independent from the practical implementation choices, we used a plug-in approach enabling the DYNAMO cloud storage administrator to change the message queue manager and its policies in order to match the specific application and, above all, the available storage and computational resources. For production scenarios the use of enterprise level components as RabbitMQ (https://www.rabbitmq.com) and Redis (https://redis.io) are recommended, but for performance evaluation we implemented our own message queue manager in order to make the metric measurements and the control over the used resources more effective [16]. The message queue manager dispatches the encrypted and compressed data parcels on available computational resources. Here the encoded encrypted symmetric key is extracted and decrypted using the cloud side private key in order to enforce the non repudiability of the data. Each data parcel contains a list of SignalK updates. The symmetric key is used to decrypt the compressed data parcel using the initialization vector leveraging on the AES algorithm using the CBC mode. Then, the data parcel is unzipped and the encrypted digital signature extracted and decrypted using the vessel public key. Finally, the decrypted digital signature is used for verification in order to enforce the integrity, then the updates are added to the update list. Once the update list is fully consistent, the storage in database process begins. A Python working implementation of a SignalK DYNAMO cloud storage is available as open source (Apache 2.0 license - https://github.com/OpenFairWind/dynamo-storage).

3.2   Cloud Side Storage

At the cost of a time-consuming, careful and proper data schematization, the relational solution offers many advantages over the plain NoSQL one, namely: (i) a highly expressive manipulation language (SQL); (ii) minimal redundancy; (iii) the possibility to enforce sophisticated integrity constraints; (iv) indexing, materialization, partitioning and all the arsenal of physical optimization to improve performance. In this case, the main challenge for designing an E/R diagram is the unknown schema of the received data, that concern different aspects of the vessel navigation gathered in real time and in a semi-structured form. Furthermore, all the measured quantities evolve over time and are sampled at irregular intervals, due to the possibly harsh environmental conditions and the consequent loss of signals.

The proposed solution is a *star schema*, with a strong entity at the center representing the VESSEL (the transmitters) and a variable number of weak entities at its side that hold the data relative to each variable to be stored, arranged time-wise (the TIMESTAMP of the measurements is the weak key for all the weak entities). For example, a variable like "position" is measured and transmitted ideally with a 5 to 10 Hz frequency and stored in a table named POSITION whose primary key is the combination of VESSEL-ID and TIMESTAMP and whose attributes are LATITUDE and LONGITUDE; a variable like "destination" is measured and transmitted with a given frequency and stored in a table named DESTINATION whose primary key is the combination of VESSEL-ID and TIMESTAMP and whose attributes are characteristics of the destination, like DESTINATION-COMMON-NAME. In both tables, the VESSEL-ID is also a foreign key connecting the weak entity to the strong entity (the VESSEL table). Once data relative to POSITION (or to DESTINATION) arrive with attached the TIMESTAMP of the measurements, they are stored in the corresponding tables. This schema naturally partitions the load on the tables, allowing parallel inserts.

The schema is built incrementally and dynamically as new variables arrive from the floating things, becoming new tables. After a *boot* period during which

many new tables (one for each unknown variable) are created, the schema sta-
bilizes itself and variable addition becomes very rare. All consecutive measurements of the same variable are stored in the corresponding table sorted by time
of measurement (TIMESTAMP). This schema does not require the *a priori* definition of the number or type of required variables, allows enforcing of integrity
constraints, holds the time series of each variable and can be indexed and tuned
via standard SQL to improve the access time.

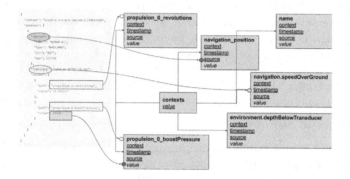

Fig. 3. SQL data encoder.

In the Internet of Floating Things, all transmitters use the GPS time, hence
they are all synchronized, however, depending on the application, a time quantization is normally necessary to have a meaningful match of the tuples representing
the measurements of the variables on the various tables. If we define ϵ the minimum time interval considered relevant for the application (for example $\epsilon = 1s$), all
values of a given variable in the same time windows (1 s) should be averaged and
only their average value (or a more robust index [18]) should be considered, with
a timestamp truncated to the second. Quantization can be performed adding to
each table a column QUANT representing the time as the number of ϵ unit of times
(seconds) passed since a reference date and then grouping.

In general, there are two possible choices: (a) storing the raw data at maximum time resolution and performing the quantization *after* inserting the data in
the database, querying and joining them with GROUP BY clauses on the QUANT
column, eventually saved in a materialized view; (b) using the GROUP BY clauses
on the QUANT column to quantize and join the data *before* inserting them in the
database, averaging their values.

The advantage of solution (a) is that the granularity can be changed at the
application level, while in solution (b) the raw data are lost and the data can only
be rolled up. On the other hand, solution (a) requires more space and more write
operations with respect to solution (b), that is lighter and may give a sufficient
precision in most real use cases. In the following we adopted solution (a).

4   Evaluation

In order to produce a preliminary evaluation of the implemented DYNAMO cloud storage, we set up an experiment using the HPC cluster *PurpleJeans* available at the Department of Science and Technologies of the University of Naples Parthenope as controlled environment. The cluster is devoted to Machine Learning and Data Science researches. Although the cluster is provided by a 16 NVidia V100 CUDA enabled GPGPU devices partition, we used the multicore intensive computing partition powered by 4 computing nodes equipped by 2 Intel Xeon 16-Core 5218 2,3 Ghz CPUs providing 32 computing cores per node and supported by 192 Giga Bytes of RAM. The computing nodes are connected to the front-end with an Infiniband Mellanox CX4 VPI SinglePort FDR IB 56 Gb/s ×16. The file system is shared using the Ethernet over the Infiniband protocol. The cluster supports Docker on both front-end and computing nodes. The total amount of storage is about 65 Terabytes. We configured the DYNAMO Cloud Storage using a custom message queue manager, dispatching the execution of the decryption and decompressing tasks on Docker-deployed DYNAMO cloud storage compute instances on the computing nodes. Using Docker, we deployed a single instance of PostgreSQL/PostGIS SQL database server on the front-end (Fig. 4). We simulated a workload with 5182 encrypted and compressed data parcels acquired during a real vessel navigation sequentially enqueued to the message queue manager. The used dataset produces 21 tables in the SQL database. We performed the overall wall clock measurement varying the number of deployed DYNAMO cloud storage compute instances on the computing nodes. All containers share the same Docker volume and they can interact with the SQL database server instance. The preliminary results obtained performing

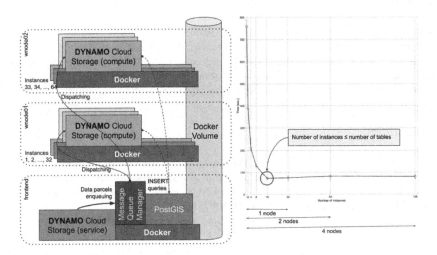

Fig. 4. The experiment setup (left) and the preliminary results (right) of the DYNAMO cloud storage.

100 times the described experiments are shown in the right side of the Fig. 4. Under the described experiment setup, the proposed methodology scales almost linearly up to 16 DYNAMO cloud storage instances as expected. This is also supported by the number of SQL tables automatically generated that is less or almost equal to the number of the compute instances. In this way, the SQL server lock is at the table level and multiple insert queries could be executed concurrently on the single SQL database server instance. As the number of instances increases, the performance gains no benefits from the concurrent inserts.

5   Conclusion and Future Directions

In this paper, we presented the latest implementation of the DYNAMO Transfer Protocol, which enforces a complete end to end encryption methodology with data signature in order to ensure data integrity, non repudiability and, above all, privacy, since the data in this context are related to people and goods and contain position and time references. The DYNAMO ecosystem could contribute to the search and rescue system: in designing the proposed framework, we considered the IoT security aspects [34] in order to avoid disruptive situations that could affect the safety at sea. All these elements lead us to believe that the DYNAMO ecosystem will gain robustness and effectiveness using the proposed data transfer approach.

Leveraging on a more sophisticated parallelization techniques [8], the final goal is building a progressively improving dataset about coastal marine environmental data [19] with a twofold utilization (i) training the next generation of deep learning models in order to carry out useful information for strategical resources management and providing assimilation data for predicting and simulating models [1,13] and workflows [23].

Acknowledgment. This research was supported by the research project "DYNAMO: Distributed leisure Yacht-carried sensor-Network for Atmosphere and Marine data crOwdsourcing applications" (DSTE373) and it is partially included in the framework of the project "MOQAP - Maritime Operation Quality Assurance Platform" and financed by Italian Ministry of Economic Development.

References

1. Ascione, I., Giunta, G., Mariani, P., Montella, R., Riccio, A.: A grid computing based virtual laboratory for environmental simulations. In: Nagel, W.E., Walter, W.V., Lehner, W. (eds.) Euro-Par 2006. LNCS, vol. 4128, pp. 1085–1094. Springer, Heidelberg (2006). https://doi.org/10.1007/11823285_114
2. Aulicino, G., et al.: Sea surface salinity and temperature in the Southern Atlantic Ocean from South African Icebreakers, 2010–2017. Earth Syst. Sci. Data **10**(3), 1227–1236 (2018)
3. Benassai, G., Di Luccio, D., Corcione, V., Nunziata, F., Migliaccio, M.: Marine spatial planning using high-resolution synthetic aperture radar measurements. IEEE J. Oceanic Eng. **43**(3), 586–594 (2018)

4. Benassai, G., Di Luccio, D., Migliaccio, M., Cordone, V., Budillon, G., Montella, R.: High resolution remote sensing data for environmental modelling: some case studies. In: 2017 IEEE 3rd International Forum on Research and Technologies for Society and Industry (RTSI), pp. 1–5. IEEE (2017)

5. Bormann, C., Castellani, A.P., Shelby, Z.: CoAP: an application protocol for billions of tiny internet nodes. IEEE Internet Comput. **16**(2), 62–67 (2012)

6. Botta, A., De Donato, W., Persico, V., Pescapé, A.: On the integration of cloud computing and Internet of Things. In: 2014 International Conference on Future Internet of Things and Cloud (FiCloud), pp. 23–30. IEEE (2014)

7. Di Paola, G., Aucelli, P.P.C., Benassai, G., Rodríguez, G.: Coastal vulnerability to wave storms of sele littoral plain (Southern Italy). Nat. Hazards **71**(3), 1795–1819 (2014)

8. D'Amore, L., Mele, V., Laccetti, G., Murli, A.: Mathematical approach to the performance evaluation of matrix multiply algorithm. In: Wyrzykowski, R., Deelman, E., Dongarra, J., Karczewski, K., Kitowski, J., Wiatr, K. (eds.) PPAM 2015. LNCS, vol. 9574, pp. 25–34. Springer, Cham (2016). https://doi.org/10.1007/978-3-319-32152-3_3

9. Fortino, G., Trunfio, P. (eds.): Internet of Things Based on Smart Objects. IT. Springer, Cham (2014). https://doi.org/10.1007/978-3-319-00491-4

10. Foster, I.: Globus online: accelerating and democratizing science through cloud-based services. IEEE Internet Comput. **15**(3), 70–73 (2011)

11. Fujisaki, E., Okamoto, T., Pointcheval, D., Stern, J.: RSA-OAEP is secure under the RSA assumption. In: Kilian, J. (ed.) CRYPTO 2001. LNCS, vol. 2139, pp. 260–274. Springer, Heidelberg (2001). https://doi.org/10.1007/3-540-44647-8_16

12. Garcia-Molina, H., Joglekar, M., Marcus, A., Parameswaran, A., Verroios, V.: Challenges in data crowdsourcing. IEEE Trans. Knowl. Data Eng. **28**(4), 901–911 (2016)

13. Giunta, G., Montella, R., Mariani, P., Riccio, A.: Modeling and computational issues for air/water quality problems: a grid computing approach. Nuovo Cimento C Geophys. Space Phys. C **28**, 215 (2005)

14. Heipke, C.: Crowdsourcing geospatial data. ISPRS J. Photogrammetry Remote Sens. **65**(6), 550–557 (2010)

15. Hunkeler, U., Truong, H.L., Stanford-Clark, A.: MQTT-S–a publish/subscribe protocol for wireless sensor networks. In: 3rd International Conference on Communication Systems Software and Middleware and Workshops, pp. 791–798. IEEE (2008)

16. Laccetti, G., Lapegna, M., Mele, V.: A loosely coordinated model for heap-based priority queues in multicore environments. Int. J. Parallel Prog. **44**(4), 901–921 (2016)

17. Mangoni, O., et al.: Phytoplankton blooms during austral summer in the Ross Sea, Antarctica: driving factors and trophic implications. PLoS ONE **12**(4), e0176033 (2017)

18. Maratea, A., Gaglione, S., Angrisano, A., Salvi, G., Nunziata, A.: Non parametric and robust statistics for indoor distance estimation through BLE. In: 2018 IEEE International Conference on Environmental Engineering (EE), pp. 1–6, March 2018. https://doi.org/10.1109/EE1.2018.8385266

19. Marcellino, L., et al.: Using GPGPU accelerated interpolation algorithms for marine bathymetry processing with on-premises and cloud based computational resources. In: Wyrzykowski, R., Dongarra, J., Deelman, E., Karczewski, K. (eds.) PPAM 2017. LNCS, vol. 10778, pp. 14–24. Springer, Cham (2018). https://doi.org/10.1007/978-3-319-78054-2_2

20. Misic, C., Harriague, A.C., Mangoni, O., Aulicino, G., Castagno, P., Cotroneo, Y.: Effects of physical constraints on the lability of POM during summer in the Ross Sea. J. Mar. Syst. **166**, 132–143 (2017)

21. Montella, R., Di Luccio, D., Kosta, S., Giunta, G., Foster, I.: Performance, resilience, and security in moving data from the Fog to the Cloud: the DYNAMO transfer framework approach. In: Xiang, Y., Sun, J., Fortino, G., Guerrieri, A., Jung, J.J. (eds.) IDCS 2018. LNCS, vol. 11226, pp. 197–208. Springer, Cham (2018). https://doi.org/10.1007/978-3-030-02738-4_17

22. Montella, R., et al.: Processing of crowd-sourced data from an internet of floating things. In: 12th Workshop on Workflows in Support of Large-Scale Science, p. 8. ACM (2017)

23. Montella, R., Di Luccio, D., Troiano, P., Riccio, A., Brizius, A., Foster, I.: WaComM: a parallel water quality community model for pollutant transport and dispersion operational predictions. In: 2016 12th International Conference on Signal-Image Technology & Internet-Based Systems (SITIS), pp. 717–724. IEEE (2016)

24. Montella, R., Kosta, S., Foster, I.: DYNAMO: distributed leisure yacht-carried sensor-network for atmosphere and marine data crowdsourcing applications. In: 2018 IEEE International Conference on Cloud Engineering (IC2E), pp. 333–339. IEEE (2018)

25. Montella, R., Ruggieri, M., Kosta, S.: A fast, secure, reliable, and resilient data transfer framework for pervasive IoT applications. In: IEEE INFOCOM 2018-IEEE Conference on Computer Communications Workshops (INFOCOM WKSHPS), pp. 710–715. IEEE (2018)

26. Paw, J.N., Thia-Eng, C.: Climate changes and sea level rise: implications on coastal area utilization and management in South-East Asia. Ocean Shorel. Manag. **15**(3), 205–232 (1991)

27. Rivaro, P., et al.: Analysis of physical and biogeochemical control mechanisms onsummertime surface carbonate system variability in the western Ross Sea (Antarctica) using in situ and satellite data. Remote Sens. **11**(3), 238 (2019)

28. Scott, K.L., Burleigh, S.: Bundle protocol specification. RFC 5050, RFC Editor, November 2007. https://tools.ietf.org/html/rfc5050

29. Sedaghat, L., Hersey, J., McGuire, M.P.: Detecting spatio-temporal outliers in crowdsourced bathymetry data. In: 2nd ACM SIGSPATIAL International Workshop on Crowdsourced and Volunteered Geographic Information, pp. 55–62. ACM (2013)

30. Small, C., Gornitz, V., Cohen, J.E.: Coastal hazards and the global distribution of human population. Environ. Geosci. **7**(1), 3–12 (2000)

31. Thangavel, D., Ma, X., Valera, A., Tan, H.X., Tan, C.K.Y.: Performance evaluation of MQTT and CoAP via a common middleware. In: IEEE Ninth International Conference on Intelligent Sensors, Sensor Networks and Information Processing (ISSNIP), pp. 1–6. IEEE (2014)

32. Turner, A.: Introduction to Neogeography. O'Reilly Media Inc., Sebastopol (2006)

33. Zhang, Q., Cheng, L., Boutaba, R.: Cloud computing: state-of-the-art and research challenges. J. Internet Serv. Appl. **1**(1), 7–18 (2010)

34. Zhou, J., Cao, Z., Dong, X., Vasilakos, A.V.: Security and privacy for cloud-based IoT: challenges. IEEE Commun. Mag. **55**(1), 26–33 (2017)

Workshop on Power and Energy Aspects of Computations (PEAC 2019)

Performance/Energy Aware Optimization of Parallel Applications on GPUs Under Power Capping

Adam Krzywaniak⬤ and Paweł Czarnul$^{(\boxtimes)}$⬤

Faculty of Electronics, Telecommunications and Informatics,
Gdansk University of Technology, Gdańsk, Poland
adam.krzywaniak@pg.edu.pl, pczarnul@eti.pg.edu.pl

Abstract. In the paper we present an approach and results from application of the modern power capping mechanism available for NVIDIA GPUs to the benchmarks such as NAS Parallel Benchmarks BT, SP and LU as well as cublasgemm-benchmark which are widely used for assessment of high performance computing systems' performance. Specifically, depending on the benchmarks, various power cap configurations are best for desired trade-off of performance and energy consumption. We present two: both energy savings and performance drops for same power caps as well as a normalized performance-energy consumption product. It is important that optimal configurations are often non-trivial i.e. are obtained for power caps smaller than default and larger than minimal allowed limits. Tests have been performed for two modern GPUs of Pascal and Turing generations i.e. NVIDIA GTX 1070 and NVIDIA RTX 2080 respectively and thus results can be useful for many applications with profiles similar to the benchmarks executed on modern GPU based systems.

Keywords: Performance/energy optimization · Power capping · GPU · NAS Parallel Benchmarks

1 Introduction

In the paper, we present results of research on performance and energy aware optimization of parallel applications run on modern GPUs under power capping. Power capping has been introduced as a feature available for modern server, desktop and mobile CPUs as well as GPUs through tools such as Intel's RAPL, AMD's APM, IBM's Energyscale for CPUs and NVIDIA's NVML/nvidia-smi for NVIDIA GPUs [6,8].

We provide a follow-up to our previous research [12,14] for CPU-based systems aimed at finding interesting performance/energy configurations obtained by setting various power caps. Within this paper, we provide results of running selected and widely considered benchmarks such as: NPB-CUDA which is an implementation of the NAS Parallel Benchmarks (NPB) for Nvidia GPUs in

© Springer Nature Switzerland AG 2020
R. Wyrzykowski et al. (Eds.): PPAM 2019, LNCS 12044, pp. 123–133, 2020.
https://doi.org/10.1007/978-3-030-43222-5_11

CUDA [7] as well as cublasgemm-benchmark [18]. We show that it is possible to obtain visible savings in energy consumption at the cost of reasonable performance drop, in some cases the performance drop being smaller than energy saving gains, percentage wise. Following work [3], we could also observe gradual and reasonably slow drop of power consumption on a GPU right after an application has finished.

2 Related Work

Our recent review [6] of energy-aware high performance computing surveys and reveals open areas that still need to be addressed in the field of energy-aware high performance computing. While there have been several works addressing performance and energy awareness of CPU-based systems, the number of papers related to finding optimal energy-aware configurations using GPUs is still relatively limited. For instance, paper [10] looks into finding an optimal GPU configuration in terms of the number of threads per block and the number of blocks. Paper [16] finds best GPU architectures in terms of performance/energy usage.

In paper [15] authors presented a power model for GPUs and showed averaged absolute error of 9.9% for NVIDIA GTX 480 card as well as 13.4% for NVIDIA Quadro FX5600, using RODINIA and ISPASS benchmarks. Furthermore, they presented that coarse-grained DVFS could achieve energy savings of 13.2% while fine-grained DVFS 14.4% at only 3% loss of performance for workloads with phase behavior, exploiting period of memory operations. Streaming multiprocessor (SM) cluster-level DVFS allowed to decrease energy consumption by 7% for the HRTWL workload – some SMs become idle at a certain point due to load imbalance.

Similarly, in [1] authors showed that by proper management of voltage and frequency levels of GPUs they were able to reduce energy by up to 28% at the cost of only 1% performance drop for the hotspot Rodinia benchmark, by a proper selection of memory, GPU and CPU clocks, with varying performance-energy trade-offs for other applications.

In [9], authors proposed a GPU power model and show that they can save on average approximately 10% of runtime energy consumption of memory bandwidth limited applications by using a smaller number of cores. Another model – MEMPower for detailed empirical GPU power memory access modeling is presented in [17].

There are also some survey type works on GPU power-aware computations. Paper [2] presents a survey of modeling and profiling methods for GPU power.

Paper [19] discusses techniques that can potentially improve energy efficiency of GPUs, including DVFS, division of workloads among CPUs and GPUs, architectural components of GPUs, exploiting workload variation and application level techniques.

Paper [20] explores trade-off between accuracy of computations and energy consumption that was reduced up to 36% for MPDATA computations using GPU clusters.

There are relatively few works on power capping in CPU-GPU environments. Selected works on load partitioning among CPUs and GPUs are discussed in [19]. Device frequencies and task mapping are used for CPU-GPU environments in [11]. In paper [21], desired frequencies for CPUs and GPUs are obtained with dynamic adjustment for a GPU at application runtime are used for controlling power consumption.

3   Proposed Approach

In our previous works [12,14] we investigated the impact on performance and energy consumption while using power capping in modern Intel CPUs. We have used an Intel Running Average Power Limit (RAPL) driver which allows for monitoring CPU energy consumption and controlling CPU power limits through Model-Specific Registers (MSR). Based on RAPL we have implemented an automatic tool called EnergyProfiler that allows for finding the energy characteristic of a device-application pair in a function of power limit. We evaluated our prototype tool on several Intel CPUs and presented that when we consider some performance impact and accept its drops power capping might result in significant (up to 35% of) energy savings. We were also able to determine such configurations of power caps that energy consumption is minimal for a particular workload and also configurations where energy consumption and execution time product is minimal. The latter metrics allow to find such power limits for which the energy savings are greater than performance loss.

In this paper we adapt our approach to Nvidia GPUs. We use power limiting and power monitoring features available in modern Nvidia graphic cards and extend EnergyProfiler to work on a new device type. Using the extended EnergyProfiler we can investigate the impact of limiting the power on performance and energy consumption on GPUs running HPC workloads.

3.1   Power Capping API

The power management API available on Nvidia GPUs is included in Nvidia Management Library (NVML). It is a C-based programmatic interface for monitoring and controlling various states within NVIDIA GPUs. Nvidia also shared a command line utility `nvidia-smi` which is a user friendly wrapper for the features available in NVML.

We based our prototype extension of EnergyProfiler on `nvidia-smi`. For controlling the power limit there is command `nvidia-smi -pl <limit>` available. The limit that we can set must fit into the range between minimal and maximal power limit which are specific for the GPU model. Monitoring and reporting the total energy consumption had to use quite a different approach than in our previous work as the NVML allows for reading the current power consumption while Intel RAPL lets the user to read the counters representing energy consumed. That caused that for evaluating the energy consumption of GPU running our testbed workload we needed to integrate the current power readings gathered while test run. Therefore, we estimated the total energy consumption as a sum of current power readings and sampling period products.

To monitor the power the `nvidia-smi dmon -s p -o T -f <filename>` was used. The parameter `-s p` specifies that we observe power and temperature, the parameter `-o T` adds the time point to each entry reported and the parameter `-f <filename>` stores the output of dmon to a file specified by `<filename>`.

The prototype extension of the EnergyProfiler tool runs a given application in parallel with the dmon, when the testbed application finishes, the tool analyses the log file and reports the energy consumption and the average power. Unfortunately, the minimal value of a sampling period in `nvidia-smi dmon` is 1 s, which means that the measurements might be inaccurate for the last sample which is taken always after the testbed application has finished.

For the testbed workloads with a really short execution time such an error would be unacceptable. In our experiments we have used workloads for which execution time varies from 20 to 200 s which means that the maximal error of energy consumption readings is less than 5% and the minimal error for the longest test runs is less than 0.5%. The execution time reported is based on precise measurements using `std::chrono::high_resolution_clock` C++11 library.

3.2   Methodology of Tests

The methodology of our research is similar to the one in [14]. We run the testbed application automatically for different values of power limit and read the execution time, the total energy consumption and the average power for the run. For each power limit we execute a series of five test runs and average the result. We observed that when Nvidia GPU temperature raises the energy consumption and average power for the same power limit settings is higher. To eliminate the impact of that phenomenon on our test runs firstly we execute a series of dummy tests which warm the GPU up. This ensures the same conditions for all test runs regardless of the position in a sequence.

In contrast to CPU, where the default settings is no power cap, on the GPU the default power limit is lower than the maximal available power limit. Another difference is that on Intel CPUs it is possible to force the power cap which is lower than the idle processor power request while on Nvidia GPU has the minimal power limit defined relatively high. Therefore, we run a series of tests starting from the maximum power limit with a 5 W step until we reach the minimal possible power limit. We refer the results of energy consumption for each result to the values obtained for the power limit set to the default value. For instance, for the Nvidia GeForce RTX 2080 the maximal power limit is 240 W, the minimal power limit is 125 W and the default power limit is 215 W.

4   Experiments

4.1   Testbed GPUs and Systems

The experiments have been performed on two testbed systems with modern Nvidia GPUs based on Pascal and Turing architecture. The first tested GPU is Nvidia GeForce GTX 1070 (Pascal architecture) with 1920 Nvidia CUDA cores with 1506 MHz base Core frequency, 8 GB of GDDR5 memory and Power Limit range between 95 W and 200 W. The other system is equipped with Nvidia GeForce RTX 2080 (Turing) with 2944 Nvidia CUDA cores with 1515 MHz base Core frequency, 8 GB of GDDR6 memory and Power Limit Range between 125 W and 240 W. Table 1 collects all details including CPU models and CUDA version used in both testbed systems.

Table 1. Testbed environments used in the experiments

System	CPU model	GPU model	RAM	CUDA version	GPU Default Power Limit	GPU Power Limit range
GTX	Intel® Core® i7-7700 (Kaby Lake)	Nvidia GeForce GTX 1070 (Pascal)	16 GB	10.0	190 W	95 W–200 W
RTX	Intel® Xeon® Gold 6130 (Skylake-X)	Nvidia GeForce RTX 2080 (Turing)	32 GB	10.1	215 W	125 W–240 W

4.2   Testbed Applications and Benchmarks

For the experiments we have selected four representative computational workloads with different computation intensity. Three of the testbed applications were selected from the well known NAS Parallel Benchmark (NPB) collection implemented in CUDA to be used on GPU [7].

The kernels that were used for the tests included: Block Tri-diagonal solver (BT), Scalar Penta-diagonal solver (SP) and Lower-Upper Gauss-Seidel solver (LU). By default kernel BT executes 200 iterations, kernel SP executes 400 iterations and kernel LU runs for 250 iterations. Similarly to the CPU version, GPU NPB may also be run for various input data sizes represented by classes A, B, C, D, E, S and W. The aforementioned classes represent the sizes of input data which are equal to 64^3, 102^3, 162^3, 408^3, 12^3 and 33^3 elements respectively. Classes A, B, W and S were not preferred in our experiments as the execution times of the kernels with such input data sizes were too short and – as it was mentioned in the previous section, we could not accept the measurement error caused by minimal sampling resolution equal to 1s. On the other hand, classes

D and E were not able to allocate enough data space due to GPU memory limitations. We decided to use all of three kernels with the same input data size – class C.

The fourth application used in the experiments was cublasgemm-benchmark [18] implementing General Matrix Multiplication (GEMM) kernel. We have modified the benchmark application to execute a series of 10 matrix multiplication (MM) operations with given square matrix size. To fulfill our requirement regarding long enough testbed application execution times we decided to use the input square matrix sizes of at least 16384 × 16384. Due to GPU memory limitations for the system with the GTX card the maximal matrix size we were able to run was only 24576 × 24576. For the RTX system we could run 32768 × 32768.

The total execution times (t) as well as total energy (E) and average power (P) consumed by each of aforementioned applications run on both testbed systems (GTX and RTX) in the default configuration of Power Limit were collected in Table 2. These values were our baseline for the relative energy savings and performance drop calculations.

Table 2. Baseline values of total execution time, total energy and average power consumption for the default Power Limit settings.

System		GTX			RTX		
App		t [s]	E [J]	P [W]	t [s]	E [J]	P [W]
BT		150,1	13493	87,6	73,9	14851	198,0
LU		44,4	3756	87,4	27,35	4972	184,2
SP		24,2	2518	109,5	17,24	3345	196,8
GEMM	16k	34,3	3030	91,8	26,3	3213	123,6
	24k	94,9	8696	89,7	69,2	8767	125,2
	32k	–	–	–	194,7	27295	135,8

4.3   Tests Results

The test results presented in the figures are organized in columns which represent the testbed workload type (GEMM for various matrix size on the left and NPB kernels on the right) and in rows which represent the observed physical magnitudes in the order as follows: relative Energy savings evaluated in percent, relative performance drop evaluated in percent, normalized energy-time product and the total average power consumption. Each figure in one column has the same horizontal axis which is the Power Limit level evaluated in Watts. All relative results are compared to the results obtained for the default Power Limit which is 190 W for GTX and 215 W for the RTX system. It is important to note that GEMM and NPB kernels present different power usage profiles. While NPB kernels begin the computation right after the application started, GEMM execution has two clear phases differing in power consumption level. The initial phase is a data preparation phase and the latter is actual MM computation phase. Both phases significantly differ in power consumption which is illustrated in Fig. 1 where sample series of test runs with different Power Limit have been presented. The sample we present was collected on the RTX system for GEMM size 16384 and NPB BT kernel.

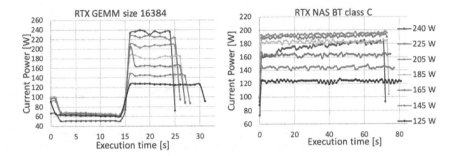

Fig. 1. Sample series of test runs with different Power Limit on RTX system.

Figure 2 presents results obtained for the first testbed system with the GTX 1070 GPU. For GEMM we observed the impact of limiting for the values below 150 W. Above 150 W the limit is neutral for the computations as the maximum spike power consumed by GEMM on GTX was around 148 W. For the lower Power Limit levels we can observe a linear falloff in power consumption. The energy consumption is also decreasing and the maximal energy saving that we could reach for that application-device pair was 15%. That value corresponded with less than 10% of performance drop and was obtained for the power limit value of 110 W. Below that value the energy savings are also observable but the benefits of limiting the power consumption are worse as the performance drops even more (up to 20%) while only 10% of energy can be saved.

While the clear energy consumption minimum seem to be found at 110 W, the other target metric suggests that a better configuration would be to set the limit between 115 W and 120 W. With such a scenario we are able to save almost 13% of energy while sacrificing only 5–6% of performance.

The results for NPB kernels obtained on the GTX system are less impressive mostly because the power consumption values while executing these testbed workloads were close or even below the minimum Power Limit level possible to be set on GTX 1070. Only for the SP kernel we can observe some interesting level of energy savings up to 17% for the lowest value of limit. What is more interesting is that the energy was saved with no cost as the performance has not dropped. This might mean that the Base Core frequency which is lowered while limiting the power has no meaningful impact on the SP execution time so the boundary in that case was memory speed.

Figure 3 presents results obtained for the testbed system with RTX 2080 GPU. With the same testbed workload RTX system present different energy

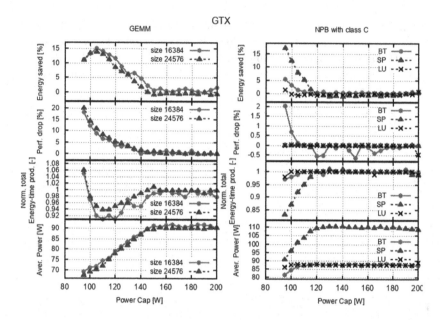

Fig. 2. Results of tests for GTX 1070 for three different problem sizes of GEMM kernel (left charts) and for three NAS Parallel Benchmarks applications run with problem size class C (right charts).

characteristics than the GTX system. Firstly what we observed is that none of the testbed workloads' average power consumption with default settings is below the minimal Power Limit available one RTX system (125 W). This implies having more abilities to adjust the power and energy consumption level.

For GEMM we observe a linear falloff in power consumption in the whole available power limit range. For the same matrix sizes as was used in the GTX system we observe similar energy characteristics with clear energy minimum (up to 15% of energy saved) for the limits in range 140 W–160 W and corresponding performance drop less than 10%. On the other hand, the other target metric which is energy-time product suggests its minimum for the Power Limit in range 160 W–170 W where we save up to 10% of energy loosing only 5% of performance. More interesting results were observed for the matrix size that was possible to run only on the RTX system which is 32786 × 32786. When the input data size was increased the energy savings are reaching 25% while the corresponding performance drop is only 10%. We see that the performance characteristics slope is less than for the smaller matrices so the energy savings are more profitable. This may suggest that for big input data when the memory is a bottleneck lowering the power using Power Limits available on the Nvidia GPUs is a really good way to lower the costs of computations.

For all three NPB kernels we observed similar energy characteristics which shows that below the 200W Power Limit we obtain stable reduction of energy consumption with the maximum 30% of energy savings. The performance drop

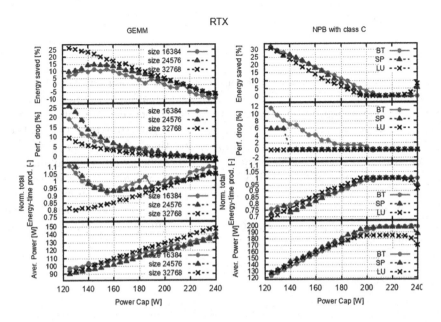

Fig. 3. Results of tests for RTX 2080 for four different problem sizes of GEMM kernel (left charts) and for three NAS Parallel Benchmarks applications run with problem size class C (right charts).

for that type of testbed workloads seem to encourage for limiting the power as its maximal value is only 12% and was observed only for the BT kernel. We see that for SP and LU kernels the performance drops are much smaller, even close to 0%. This confirms the observations from the GTX system which may suggest that the NPB kernels are more memory than computation bound.

5   Conclusions and Future Work

Our research presented in this paper showed that it is possible and even worth to lower the energy consumption by using software power caps in modern HPC systems. After our first research focused on Intel CPUs [12,14] we tested another popular in HPC manufacturer and devices: Nvidia's GPUs. Using various testbed workloads with different input data size, different computation intensity and also non trivial power consumption profile we tested two modern Nvidia's GPUs.

Our research showed that:

1. depending on the workload type it is possible to reach up to 30% of energy savings using Power Limits available on Nvidia's GPUs while corresponding performance drop evaluated in percent is usually smaller than benefits of lowering the costs,
2. power limiting in order to minimize the costs is a non-trivial task as the Power Limits for which we observed the maximum of energy savings was specific to the application-device pair and was not the minimal possible Power Limit value,

3. the second target metric we used – energy-time product, allows for finding even more profitable configurations of Power Limits as the energy savings are close to maximal possible but the performance drop is significantly lower.

In the future we plan to extend our research by exploring other types of workloads specific to GPU like Deep Learning benchmarks or trainings and compare power capping approach proposed in this paper between presented desktop GPUs and popular server GPUs. Additionally, we plan to extend the analysis of workload types beyond characterization of whether applications are memory or computation bound using the Roofline Model of energy [4]. We also aim at a hybrid power capping approach (RAPL + NVML) for hybrid applications such as parallelization of large vector similarity computations [5] run on hybrid (Intel CPU + Nvidia GPU) systems. Finally, we plan to explore the impact of power capping on scalability [13] of parallel computations run in representative modern, homogeneous and heterogeneous, HPC systems.

References

1. Abe, Y., Sasaki, H., Peres, M., Inoue, K., Murakami, K., Kato, S.: Power and performance analysis of GPU-accelerated systems. Presented as part of the 2012 Workshop on Power-Aware Computing and Systems, Hollywood, CA. USENIX (2012)
2. Bridges, R.A., Imam, N., Mintz, T.M.: Understanding GPU power: a survey of profiling, modeling, and simulation methods. ACM Comput. Surv. **49**(3), 41:1–41:27 (2016)
3. Carreño, E.D., Sarates Jr., A.S., Navaux, P.O.A.: A mechanism to reduce energy waste in the post-execution of GPU applications. J. Phys.: Conf. Ser. **649**(1), 012002 (2015)
4. Choi, J.W., Bedard, D., Fowler, R., Vuduc, R.: A roofline model of energy. In: 2013 IEEE 27th International Symposium on Parallel and Distributed Processing, pp. 661–672, May 2013
5. Czarnul, P.: Parallelization of large vector similarity computations in a hybrid CPU+GPU environment. J. Supercomput. **74**(2), 768–786 (2017). https://doi.org/10.1007/s11227-017-2159-7
6. Czarnul, P., Proficz, J., Krzywaniak, A.: Energy-aware high-performance computing: survey of state-of-the-art tools, techniques, and environments. Sci. Program. **2019**, 8348791:1–8348791:19 (2019)
7. Dümmler, J.: NPB-CUDA. Technische Universitat Chemnitz. https://www.tu-chemnitz.de/informatik/PI/sonstiges/downloads/npb-gpu/index.php.en
8. Ge, R., Vogt, R., Majumder, J., Alam, A., Burtscher, M., Zong, Z.: Effects of dynamic voltage and frequency scaling on a k20 GPU. In: 2013 42nd International Conference on Parallel Processing, pp. 826–833, October 2013
9. Hong, S., Kim, H.: An integrated GPU power and performance model. SIGARCH Comput. Archit. News **38**(3), 280–289 (2010)
10. Huang, S., Xiao, S., Feng, W.: On the energy efficiency of graphics processing units for scientific computing. In: 2009 IEEE International Symposium on Parallel Distributed Processing, pp. 1–8, May 2009

11. Komoda, T., Hayashi, S., Nakada, T., Miwa, S., Nakamura, H.: Power capping of CPU-GPU heterogeneous systems through coordinating DVFS and task mapping. In: 2013 IEEE 31st International Conference on Computer Design (ICCD), pp. 349–356, October 2013

12. Krzywaniak, A., Proficz, J., Czarnul, P.: Analyzing energy/performance trade-offs with power capping for parallel applications on modern multi and many core processors. In: 2018 Federated Conference on Computer Science and Information Systems (FedCSIS), pp. 339–346, September 2018

13. Krzywaniak, A., Czarnul, P.: Parallelization of selected algorithms on multi-core CPUs, a cluster and in a hybrid CPU+Xeon Phi environment. In: Borzemski, L., Świątek, J., Wilimowska, Z. (eds.) ISAT 2017. AISC, vol. 655, pp. 292–301. Springer, Cham (2018). https://doi.org/10.1007/978-3-319-67220-5_27

14. Krzywaniak, A., Czarnul, P., Proficz, J.: Extended investigation of performance-energy trade-offs under power capping in HPC environments. Accepted for International Conference on High Performance Computing & Simulation (HPCS 2019), Dublin, Ireland (in press)

15. Leng, J., et al.: GPUWattch: enabling energy optimizations in GPGPUs. SIGARCH Comput. Archit. News 41(3), 487–498 (2013)

16. Libuschewski, P., Marwedel, P., Siedhoff, D., Müller, H.: Multi-objective, energy-aware GPGPU design space exploration for medical or industrial applications. In: 2014 Tenth International Conference on Signal-Image Technology and Internet-Based Systems, pp. 637–644, November 2014

17. Lucas, J., Juurlink, B.: MEMPower: data-aware GPU memory power model. In: Schoeberl, M., Hochberger, C., Uhrig, S., Brehm, J., Pionteck, T. (eds.) ARCS 2019. LNCS, vol. 11479, pp. 195–207. Springer, Cham (2019). https://doi.org/10.1007/978-3-030-18656-2_15

18. He Ma. cublasgemm-benchmark. University of Guelph, Canada. https://github.com/hma02/cublasgemm-benchmark

19. Mittal, S., Vetter, J.S.: A survey of methods for analyzing and improving GPU energy efficiency. CoRR, abs/1404.4629 (2014)

20. Rojek, K.: Machine learning method for energy reduction by utilizing dynamic mixed precision on GPU-based supercomputers. Concurr. Comput.: Pract. Exp. 31(6), e4644 (2019)

21. Tsuzuku, K., Endo, T.: Power capping of CPU-GPU heterogeneous systems using power and performance models. In: 2015 International Conference on Smart Cities and Green ICT Systems (SMARTGREENS), pp. 1–8, May 2015

Improving Energy Consumption in Iterative Problems Using Machine Learning

Alberto Cabrera$^{(\boxtimes)}$ ⓘ, Francisco Almeida ⓘ, Vicente Blanco ⓘ,
and Dagoberto Castellanos–Nieves ⓘ

HPC Group, Escuela Superior de Ingeniería y Tecnología,
Universidad de La Laguna, ULL, 38270 La Laguna, Tenerife, Spain
Alberto.Cabrera@ull.edu.es

Abstract. To reach the new milestone in High Performance Computing, energy and power constraints have to be considered. Optimal workload distributions are necessary in heterogeneous architectures to avoid inefficient usage of computational resources. Static load balancing techniques are not able to provide optimal workload distributions for problems of irregular nature. On the other hand, dynamic load balancing algorithms are coerced by energy metrics that are usually slow and difficult to obtain. We present a methodology based on Machine Learning to perform dynamic load balancing in iterative problems. Machine Learning models are trained using data acquired during previous executions. We compare this new approach to two dynamic workload balancing techniques already proven in the literature. Inference times for the Machine Learning models are fast enough to be applied in this environment. These new predictive models further improve the workload distribution, reducing the energy consumption of iterative problems.

Keywords: Machine Learning · Dynamic load balancing · Energy efficiency · Iterative algorithms

1 Introduction

In the last decades, High Performance Computing (HPC) architectures have evolved to systems highly heterogeneous. Most powerful supercomputers, currently listed in the TOP500 list, increase their performance by using co–processors units, mainly Nvidia Graphic Processing Units (GPUs). Accelerator based architectures intend to maximize de energy efficiency of high performance computers, while the trend may change in the future, scientists will use this systems in the years to come, and existing codes have to be flexible and portable to adapt to the changes in the future.

Traditionally, due to the nature of the underlying problem, many scientific applications have been implemented as an iterative method. Examples of the

© Springer Nature Switzerland AG 2020
R. Wyrzykowski et al. (Eds.): PPAM 2019, LNCS 12044, pp. 134–143, 2020.
https://doi.org/10.1007/978-3-030-43222-5_12

variety of these methods are the dynamic programming approach to the Longest Common Subsequence problem, the Jacobi method, signal processing and image processing codes. In the structure of the iterative methods, data is partitioned among the parallel processors, a series of independent calculations are performed over each partition and then, a synchronization phase takes place. Load balancing techniques were designed in order to minimize waiting times in the synchronization phase. Dynamic load balancing techniques are of special interest due to the flexibility they provide, as all the information they require is acquired during the execution of previous stages of the iterative algorithm.

In this paper, we propose to incorporate Machine Learning techniques to improve existing dynamic load balancing algorithms. We propose to extract knowledge from the dynamic load balancing implementations to train Machine Learning models in order to adapt better to heterogeneous systems. Once trained, the information provided to our models is able to predict future behaviors during the execution of different applications. This is specially beneficial for energy consumption, as metrics are difficult to obtain in our context, require specific code design for acquiring data and introduce overhead to the application. Following are the main contributions of our paper:

- We propose a methodology to incorporate Machine Learning models to improve dynamic load balancing in iterative problems.
- We perform an extensive evaluation of the methodology with a use–case where we train, test and validate three Machine Learning models with acceptable overhead, and improve the results obtained using a previously validated dynamic workload balancing library from the literature.
- We introduce the hardware and software energetic behavior as part of the Machine Learning model to avoid the difficulties associated to energy measurement during the execution of an algorithm.

Our contributions have been validated over several dynamic programming algorithms originally designed for homogeneous systems, where the energy efficiency and the execution time is non–optimal when distributing the workload equally among multiple GPUs. The problems are representative of the many procedures used in scientific computing. By incorporating Machine Learning, the improvement of the workload distribution notably reduces energy consumption in the majority of the presented cases and demonstrates the usefulness of the proposed methodology, with the minimal overhead introduced by model inference.

The rest of the paper is organized as follows. In Sect. 2 we discuss the load balancing techniques presented in the literature. In Sect. 3 we present the specifics of load balancing in iterative problems. Section 4 illustrates our Machine Learning approach for dynamic load balancing. In Section 5 we discuss our methodology and present a series of experiments that validate our proposals, and finally conclude in Sect. 6.

2 Related Work

Load balancing techniques are of special interest in distributed systems, due to the heterogeneity of each system and the requirement for a minimum quality of service. Multiple approaches of both static and dynamic load balancing techniques [17], have been developed to address different levels of reliability, complexity, performance, stability and resource utilization. In cloud computing, algorithms have been designed for highly heterogeneous environments due to the available hardware, as is the case of dynamic load balancing for cloud services [6] or energy–aware scheduling algorithms for virtual machines deployments [12].

In High Performance Computing, extensive work has been done in the field of dynamic load balancing and adaptive workloads. ALEPH [15] solves a bi–objective optimization problem for both performance and energy by doing a load unbalancing technique in manycore architectures. Specifically for iterative problems, E-ADITHE [10] was introduced to obtain the optimal process allocation and workload distribution for integrated CPU-GPUs. To improve the usage of computational resources, malleable jobs have been proposed to significantly improve resource usage in parallel systems. A job resize mechanism was also presented using Charm++ to prove the benefits in performance of having malleable processes [11]. In Flex–MPI, an energy–aware module was implemented to modify at runtime the number of MPI processes in a parallel application through monitorization and a computational prediction model [16]. While our approach is also energy–aware, the application of the proposed methodology is not specifically designed for any metric, and could be translated to minimize execution time. In previous work, we introduced a library, Ull_Calibrate_Lib [1] to redistribute workload based on the current performance of the parallel processes to reduce the waiting times that occur in the synchronization phases of iterative problems, and a Generic Resource Optimization Heuristic [4], to redistribute workload to optimize a user–defined resource, which in our case was energy consumption. Both techniques estimate the performance and the user–defined resource with data obtained during the execution of an iterative algorithm. In this paper, we use the knowledge obtained at previous iterations, to improve the decision making from the load balancing using Machine Learning techniques.

The potential of Machine Learning to improve load balancing has also been introduced previously in HPC environments. Learning everywhere was proposed as the paradigm generated by the interaction between traditional HPC and Machine Learning, and an analysis of how both fields can interact has been presented in [8]. One example of this interaction between HPC and Machine Learning is an energy–aware allocation and workload redistribution approach introduced to incorporate techniques to learn the behavior of complex infrastructures [3]. Machine Learning predictive models have also been proposed to design a static load balancing technique in climate modeling codes, specifically the sea–ice module in the Community Earth System Model [2]. Our approach is similar to this last proposal, however we outline a generic methodology to then apply it to a wide variety of problems where the load balancing problem is present, instead of proposing an ad–hoc solution. We apply our methodology as part of a dynamic load balancing technique for iterative problems, and models are trained from the data extracted in previous executions without performing an extensive study of the specific problem. As iterative algorithms appear very frequently in scientific applications, our technique could be easily applied to many other classes of problems as those appearing in image processing or in stencil codes.

3   Dynamic Load Balancing

Load balancing is a technique usually applied to improve the performance in parallel applications when heterogeneity is present, be it in the hardware or in the application itself. We address the dynamic load balancing problem for iterative algorithms in heterogeneous environments. Usually, applying an homogeneous distribution over an heterogeneous architecture yields to a bad performance for both execution time and energy consumption, as each computing element has a different computing capability. In iterative problems this effect is augmented, as these kind of problems usually have a synchronization phase where partial results are shared between processes. To gather these results, the fastest processes will have to wait for the slower ones, heavily affecting the overall performance of the application. Redistributing the workload appropriately between all the computing elements of the application minimizes inefficient resource usage and improves the overall performance and energy consumption of a given application.

Both static and dynamic load balancing techniques try to minimize the impact of the synchronization phase in the iterative problems. When heterogeneity is present in the algorithm, a different amount of effort is needed to solve each workload unit of the iterative problem. In these irregular applications, the optimal workload distribution varies as iterations are solved. A static workload distribution, by definition, is not able to optimize properly every iteration in these specific algorithms, thus, a dynamic load balancing technique is necessary to achieve the optimal solution.

As examples of iterative algorithms, we have chosen four different cases implemented using dynamic programming techniques. These algorithms are the Knapsack Problem (KP), the Resource Allocation Problem (RAP), the Cutting Stock Problem (CSP) and the Triangulation of Convex Polygons (TCP) [7]. These algorithms have been chosen to represent different iterative problem properties, depending on the memory requirements, the computational granularity and the workload regularity. The KP is a regular, memory bound problem, and each state of the problem requires little computation and most of the performance is lost in data movements. The RAP, CSP and TCP are compute intensive problems, and each of them introduce irregularity in the workload distribution. Although these are algorithms with a fixed number of iterations, the ideas and concepts managed along the paper according to load balancing can be applied without loss of generality to problems with different convergence criteria as those appearing in linear algebra.

4 Machine Learning Based Workload Distribution

Dynamic load balancing algorithms are usually designed to redistribute workload using knowledge acquired during the execution of an application. Specifically, in iterative problems, metrics obtained in previous iterations help to estimate the outcome of the subsequent operations. Statistical procedures can be applied using the data gathered during the experimentation to construct predictive models that anticipate the outcome of future events with limited information. Using these techniques, we propose a three step methodology to train machine learning models with the knowledge acquired during the execution of specific applications. In a first step, the target application has to be executed using any dynamic workload balancing technique to build the input data set for our predictive models. Each record in the data set has to contain the workload, execution time and energy consumption for each process in our parallel application. Once the input data set is gathered, it can be used to train a Machine Learning model. Finally, the dynamic workload balancing technique will use the trained model as a better mechanism to redistribute the workload of our target application, minimizing data movement and bad decision making due to the lack of knowledge.

As a use–case, we will implement our proposal to improve the workload distribution of a series of iterative problems. In the past, we presented an heuristic dynamic load balancing algorithm that was agnostic to both the specific iterative problem and the resource to optimize, which we will refer as Generic Resource Optimization Heuristic, GROH. This heuristic is implemented following the principles of skeleton programming where only the specifics of a given metric have to be defined in order to be optimized, which in our case would be execution time or energy consumption. We will apply our new proposal and will compare it to GROH, as Machine Learning models incorporate knowledge to perform the minimum data movements.

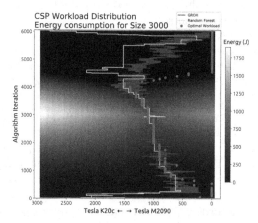

Fig. 1. Dynamic load balancing workload algorithm comparative.

Figure 1 illustrates three different workloads in an specific instance of the Cutting Stock Problem (CSP) of size 3000. The energy consumption in the CSP problem space is represented through a heatmap, where the bottom of the plot is the starting point of the problem execution and the top represents the last iterations of the algorithm. The X axis represents the workload distribution between two different GPU processes. Moving to the left of the axis would increment the workload given to process 0, executed in a *Tesla K20c*, and moving in the opposite direction would give more workload to a *Tesla M2090*.

In this Figure, we illustrate the optimal workload distribution obtained doing a brute force analysis of this CSP instance through the execution of high number enumerated workload distributions of the solution space. It is represented using a series of dots. Our Machine Learning proposal, using a Random Forest model, and GROH are plotted using two lines in the problem space to directly compare the workload distributions obtained through the execution of the iterative problem. In this problem space, the irregularity of the CSP problem is visible, specially between iterations 1000 and 4000. There is more total workload during these iterations, thus the increased energy consumption. This section is critical as bad decision making in this segment involves high losses in energy and performance. We observe how the workload distribution of the Random Forest is very close to the optimal workload, while GROH performs more data movements due to the lack of problem knowledge. The crucial difference between both cases is the objective function used to redistribute the workload between the parallel processes. GROH uses tries to minimize the energy consumption without knowledge of the problem, and data movements are calculated based on the current efficiency for each process. The data movement overhead is, however, negligible, as it takes advantage of the synchronization phase of the iterative algorithm. As workload varies at each iteration, we observe how GROH redistributes continuously the workload, and despite staying very close to the optimal solution in the critical section, it is unable to reach the optimal workload. The Random Forest however, performs better compared to the previous method, as, in practice, acts as a function that incorporates knowledge from the algorithm and the architecture. In both cases, at the beginning and at the end of the execution, both dynamic load balancing techniques differ from the optimal as iterations are really fast with negligible execution time and energy consumption.

5   Computational Experience

Supervised Machine Learning algorithms are specially designed to find functions to connect our input and our problem data, the energy consumption through the execution of the application. Using data gathered during the execution of iterative problems, we will train, test, predict and validate an alternative methodology to incorporate knowledge to improve the workload distributions of the dynamic load balancing technique. We have chosen the R software package *caret* [14] to build and train our predictive models. This package provides more than 200 statistical models from which we have evaluated only three to present an example methodology. The selected models and the reasoning behind their selection are:

Table 1. GPU cluster and Machine Learning algorithm details.

(a)Heterogeneous Cluster				(b) Machine Learning Algorithms		
Nodes	CPUs (Xeon)	GPU	Memory	Algorithm	Type	R name
Verode16	2x E5-2660	M2090	64 GB	Linear Model	Regression	lm
Verode17	2x E5-2660	K20c	64 GB	MARS	Regression	earth
Verode18	2x E5-2660	K40m	64 GB	Random Forest	Ensemble method	rf
Verode20	2x E5-2698 v3	M2090	128 GB			

GPU Type	M2090	K20c	K40m	Algorithm	Training Time	R^2
# Cores	512	2496	2880	Linear Model	1 min.	$[0.13, 0.17]$
RAM	6GB	5GB	12GB	MARS	10 min.	$[0.51, 0.55]$
Mem BW	177.6 GB/s	208 GB/s	288 GB/s	Random Forest	30 min.	$[0.83, 0.92]$
Power	225 W	225 W	235 W			

- Linear Model, *LM*, the simplest statistical regression approach. In R, it is referred as the *lm* model.
- Multivariate Adaptive Regression Splines [9], (*MARS*), a more flexible regression modeling for high dimensional data. In R, it is referred as *earth*.
- Random Forest [13], *RF*, one of the most widely used ensemble method in the Machine Learning literature. Again, it is referred as *rf* in R.

The specific model details, training times and R^2 values are shown in Table 1b. It contains the information related to the models, including the type of Machine Learning Model, the corresponding name for the model in R (lm, earth and rf), their training times and a statistical measure to represent the variance of the model respecting our data, in our case R^2. The R^2 is presented as an interval since there is one value of R^2 associated with each problem, yielding a total of 4 values per Machine Learning Model. The Linear Model R^2 also require further clarification, as the low value is not indicative of the fitness of the model. We will present later in our computational results how we obtain very good workload distributions using this model for a specific set of the presented iterative problems despite these values. Each model was trained specifically for each problem, yielding a total of 12 different Machine Learning models and the input data set used in the training phase was obtained using data from our previous experiments. Specifically, a maximum of 25 executions from each problem was used, a relatively small dataset in Machine Learning environments.

Our experimentation was performed in a heterogeneous cluster composed by 4 different nodes. Each node has a unique pair of CPU and GPU installed, detailed in Table 1a. To obtain energy measurements, we accessed the Nvidia GPUs provided data using the Nvidia Management Library (NVML) the measurement interface, EML [5]. These nodes have installed a Debian 9 with a kernel version 4.9.0-2-amd64. Our build and execution environments have GCC version 4.8.5, using $-O2$ as the only optimization flag, OpenMPI 3.0.0, CUDA 7.5.

Table 2 illustrates the average energy measurements for our target problems, to compare multiple problem sizes to the homogeneous distribution, two already implemented dynamic load balancing methods, GROH and Ull_Calibrate_Lib,

Table 2. Dynamic load balancing techniques energy consumption comparison.

	(a) Regular Algorithms					(b) Irregular Algorithms				
KP	Energy (J)				**CSP**	Energy (J)				
Size	ref	calib	GROH	RF	LM	Size	ref	calib	GROH	RF	LM
2000	137.3	**95.2**	124.5	115.2	110.0	1000	1359.5	1325.2	1360.8	1177.7	**1171.6**
3000	272.8	210.5	221.5	**187.0**	211.0	1500	4503.3	4094.2	4363.4	**3902.2**	3996.1
4000	468.1	322.8	387.9	312.8	**309.8**	2000	10660	9762.6	9446.5	**8855.3**	9186.6
5000	709.3	502.6	603.7	**442.9**	463.4	2500	21015	18620	18089	**16710**	17230

RAP	Energy (J)				**TCP**	Energy (J)				
Size	ref	calib	GROH	RF	LM	Size	ref	calib	GROH	RF	LM
1500	483.3	180.3	**119.2**	134.5	119.2	1500	1824.0	1307.0	1496.4	**1291.1**	1527.7
2000	993.6	359.6	266.7	268.4	**237.8**	2000	4315.8	2952.9	3869.0	**2915.5**	3653.1
2500	1746.1	622.3	456.0	407.9	**352.0**	2500	8142.9	5681.1	6206.2	**5650.3**	7910.9
3000	2904.8	963.6	610.4	598.2	**521.6**	3000	14139.3	9924.5	11218.7	**9899.3**	15312.1

and the best Machine Learning Models that adapted better to our needs. The standard deviation is lower than 3.4% for every presented case.

The Ull_Calibrate_Lib method does the dynamic load balancing by calculating a performance for each parallel process at a given iteration, and proportionally redistributing the workload based on this value. On the other hand, the energy consumption heuristic is the specific implementation of the GROH, that uses energy measurements to redistribute the workload in the iterative problem. The problems are grouped by the nature of the algorithm, where Table 2a contains data of the regular cases and Table 2b the data of the irregular ones. In each table, we will present the homogeneous distribution as the reference, which is labeled *ref*, Ull_Calibrate_Lib is labeled *calib* and the energy consumption heuristic labeled as *GROH*. Finally, the Machine learning models selected will be the Random Forest (*RF*) and the Linear Model (*LM*). The results for the Multivariate Adaptive Regression Splines (*MARS*) model were discarded, as the model performed better than the reference, but worse than every other case presented in these Tables. None of the Machine Learning experiments include the energy consumption of the training phase, as Table 2 contains only the measurements gathered during the execution of each algorithm.

In all the presented cases, we are able to reduce the energy consumption of the algorithms using any dynamic load balancing technique. Execution time is highly correlated for these architectures and, in all cases, the best energy consumption is achieved by the fastest execution. For the KP, in Table 2a, data indicates that for the smallest presented size the best results are achieved by Ull_Calibrate_Lib. However, for every other case, the best energy consumption is achieved by the Machine Learning models. A similar situation is observed in the RAP, also in Table 2a. The best results are obtained using the trained Linear Model. This is a very interesting result, as the linear model proposes only one workload distribution for the whole execution. While we could be criticize with our methodology stating that a dynamic workload balancing technique is worse in comparison, the focus of this work is to illustrate that a Machine Learning approach allows us to reach this conclusion with the data acquired using the dynamic workload balancing techniques. Additionally, it is expected that a static workload distribution is the best option for iterative problems where the total workload is invariable through the execution. In the irregular cases CSP and TCP, in Table 2b, we observe a different outcome. Except for size 1000 in the CSP, the best results are obtained using the Random Forest models. In these cases, dynamic workload techniques adapt better as they are able to dynamically change the total workload during the execution of the algorithm. These results would vary if different models were selected for training or if input data was different. However, the methodology presented is still justified and applicable for any iterative problem, where we improve the obtained workload distributions for the majority of the presented cases.

6   Conclusion

In this paper, we present a technique to apply traditional Machine Learning models in dynamic load balancing techniques in iterative problems. We show that these models achieve good workload distributions in both regular and irregular problems, with very little training extracted from previous executions of the algorithm. In particular, we have studied a very small set of supervised models that have been proven to improve the workload distribution obtained using two dynamic load balancing algorithms from the literature. Our proposal improves the best workload distribution in the majority of the presented cases, reducing the overall energy consumption of the executed applications. As the next logical step due to the low inference cost, we intend to study the use of deep learning techniques and consider the viability of training a neural network taking into account the cost of their training. Long short-term memory (LSTM) networks seem to be suitable for predictions based on time series data, since there can be lags of unknown duration between important events in these series related to the energy measurement difficulties and Machine Learning for Streams, through monitoring energy consumption or execution times.

Acknowledgments. This work was supported by the Spanish Ministry of Education and Science through the TIN2016-78919-R project, the Government of the Canary Islands, with the project ProID2017010130 and the grant TESIS2017010134, which is co-financed by the Ministry of Economy, Industry, Commerce and Knowledge of Canary Islands and the European Social Funds (ESF), operative program integrated of Canary Islands 2014–2020 Strategy Aim 3, Priority Topic 74 (85%); the Spanish network CAPAP-H.

References

1. Acosta, A., Blanco, V., Almeida, F.: Dynamic load balancing on heterogeneous multi-GPU systems. Comput. Electr. Eng. **39**(8), 2591–2602 (2013)
2. Balaprakash, P., Alexeev, Y., Mickelson, S.A., Leyffer, S., Jacob, R., Craig, A.: Machine-learning-based load balancing for community ice code component in CESM. In: Daydé, M., Marques, O., Nakajima, K. (eds.) VECPAR 2014. LNCS, vol. 8969, pp. 79–91. Springer, Cham (2015). https://doi.org/10.1007/978-3-319-17353-5_7
3. Berral, J.L., et al.: Towards energy-aware scheduling in data centers using machine learning. In: Proceedings of the 1st International Conference on Energy-Efficient Computing and Networking. e-Energy 2010, pp. 215–224. ACM, New York (2010). https://doi.org/10.1145/1791314.1791349
4. Cabrera, A., Acosta, A., Almeida, F., Blanco, V.: A heuristic technique to improve energy efficiency with dynamic load balancing. J. Supercomput. **75**(3), 1610–1624 (2019). https://doi.org/10.1007/s11227-018-2718-6
5. Cabrera, A., Almeida, F., Arteaga, J., Blanco, V.: Measuring energy consumption using eml (energy measurement library). Comput. Sci. Res. Dev. **30**(2), 135–143 (2015). https://doi.org/10.1007/s00450-014-0269-5
6. Chen, S.L., Chen, Y.Y., Kuo, S.H.: CLB: a novel load balancing architecture and algorithm for cloud services. Comput. Electr. Eng. **58**, 154–160 (2017)
7. Cormen, T.H., Leiserson, C.E., Rivest, R.L., Stein, C.: Introduction to Algorithms, 3rd edn. The MIT Press, Cambridge (2009)
8. Fox, G., et al.: Learning everywhere: pervasive machine learning for effective high-performance computation. arXiv preprint arXiv:1902.10810 (2019)
9. Friedman, J.H., et al.: Multivariate adaptive regression splines. Ann. Stat. **19**(1), 1–67 (1991)
10. Garzón, E.M., Moreno, J.J., Martínez, J.A.: An approach to optimise the energy efficiency of iterative computation on integrated GPU-CPU systems. J. Supercomput. **73**(1), 114–125 (2017). https://doi.org/10.1007/s11227-016-1643-9
11. Gupta, A., Acun, B., Sarood, O., Kalé, L.V.: Towards realizing the potential of malleable jobs. In: 2014 21st International Conference on High Performance Computing (HiPC), pp. 1–10. IEEE (2014)
12. Guzek, M., Kliazovich, D., Bouvry, P.: HEROS: energy-efficient load balancing for heterogeneous data centers. In: 2015 IEEE 8th International Conference on Cloud Computing, pp. 742–749. IEEE (2015)
13. Ho, T.K.: Random decision forests. In: Proceedings of 3rd International Conference on Document Analysis and Recognition, vol. 1, pp. 278–282. IEEE (1995)
14. Kuhn, M.: Building predictive models in R using the caret package. J. Stat. Softw. **28**(5), 1–26 (2008). https://doi.org/10.18637/jss.v028.i05
15. Manumachu, R.R., Lastovetsky, A.: Bi-objective optimization of data-parallel applications on homogeneous multicore clusters for performance and energy. IEEE Trans. Comput. **67**(2), 160–177 (2017)
16. Martín, G., Singh, D.E., Marinescu, M.C., Carretero, J.: Enhancing the performance of malleable MPI applications by using performance-aware dynamic reconfiguration. Parallel Comput. **46**, 60–77 (2015)
17. Soundarabai, P.B., Sahai, R., Thriveni, J., Venugopal, K., Patnaik, L.: Comparative study on load balancing techniques in distributed systems. Int. J. Inf. Technol. Knowl. Manag. **6**(1), 53–60 (2012)

Automatic Software Tuning of Parallel Programs for Energy-Aware Executions

Sébastien Varrette[1,2](✉), Frédéric Pinel[1](✉), Emmanuel Kieffer[1](✉),
Grégoire Danoy[1,2](✉), and Pascal Bouvry[1,2](✉)

[1] Department of Computer Science (DCS), University of Luxembourg,
Luxembourg City, Luxembourg
{sebastien.varrette,frederic.pinel,emmanuel.kieffer
gregoire.danoy,pascal.bouvry}@uni.lu
[2] Interdisciplinary Centre for Security Reliability and Trust (SnT),
University of Luxembourg, Luxembourg City, Luxembourg

Abstract. For large scale systems, such as data centers, energy efficiency has proven to be key for reducing capital, operational expenses and environmental impact. Power drainage of a system is closely related to the type and characteristics of workload that the device is running. For this reason, this paper presents an automatic software tuning method for parallel program generation able to adapt and exploit the hardware features available on a target computing system such as an HPC facility or a cloud system in a better way than traditional compiler infrastructures. We propose a search based approach combining both exact methods and approximated heuristics evolving programs in order to find optimized configurations relying on an ever-increasing number of tunable knobs i.e., code transformation and execution options (such as the number of OpenMP threads and/or the CPU frequency settings). The main objective is to outperform the configurations generated by traditional compiling infrastructures for selected KPIs i.e., performance, energy and power usage (for both for the CPU and DRAM), as well as the runtime. First experimental results tied to the local optimization phase of the proposed framework are encouraging, demonstrating between 8% and 41% improvement for all considered metrics on a reference benchmarking application (i.e., Linpack). This brings novel perspectives for the global optimization step currently under investigation within the presented framework, with the ambition to pave the way toward automatic tuning of energy-aware applications beyond the performance of the current state-of-the-art compiler infrastructures.

Keywords: HPC · Performance evaluation · Energy efficiency · Compiler infrastructure · Automatic tuning · MOEA · Hyper-parameter optimization

1 Introduction

With the advent of the Cloud Computing (CC) paradigm, the last decade has seen massive investments in large-scale High Performance Computing (HPC) and

© Springer Nature Switzerland AG 2020
R. Wyrzykowski et al. (Eds.): PPAM 2019, LNCS 12044, pp. 144–155, 2020.
https://doi.org/10.1007/978-3-030-43222-5_13

storage systems aiming at hosting the surging demand for processing and data-analytic capabilities. The integration of these systems in our daily life has never been so tied, with native access enabled within our laptops, mobile phones or smart Artificial Intelligence (AI) voice assistants. Outside the continuous expansion of the supporting infrastructures performed in the private sector to sustain their economic development, HPC is established as a strategic priority in the public sector for most countries and governments. For large scale systems, energy efficiency has proven to be key for reducing all kinds of costs related to capital, operational expenses and environmental impact. A brief overview of the latest Top 500 list (Nov. 2019) provides a concrete indication of the current power consumption in such large-scale systems and projections for the Exaflop machines foreseen by 2021 with a *revised* power envelop of 40 MW. Reaching this target involves combined solutions mixing hardware, middleware and software improvements, when power drainage of a system is closely related to the type and characteristics of the workload. While many computing systems remain heterogeneous with the increased availability of accelerated systems in HPC centers and the renewed global interest for AI methods, the energy efficiency challenge is rendered more complex by the fact that pure performance and resource usage optimization are also Key Performance Indicators (KPIs). In this context, this paper aims at extending HPC middleware into a framework able to transparently address the runtime adaptation of execution optimizing priority KPIs i.e., performance, energy and power usage (for both the CPU and DRAM), as well as the runtime in an attempt to solve the following question: *Can we produce energy-aware HPC workloads through source code evolution on heterogeneous HPC resources?* To that end, we propose EVOCODE, an automatic software tuning method for parallel program generation able to better exploit hardware features available on a target computing system such as an HPC facility or a cloud system.

This paper is organized as follows: Sect. 2 details the background of this work and reviews related works. Then, the EVOCODE framework is presented in Sect. 3. Implementation details of the framework, as well as the first experimental results validating the approach, are expounded in Sect. 4. Based on a reference benchmarking application (i.e., Linpack, measuring a system's floating point computing power), the initial hyper-parameter optimization phase already demonstrate concrete KPIs improvements with **8%** performance and runtime gains, up to **19%** energy and power savings and even **41%** of energy and power usage decrease at the DRAM level. Finally, Sect. 5 concludes the article and provides future directions and perspectives.

2   Context and Motivations

Recent hardware developments support energy management at various levels and allow for the dynamic scaling of the power (or frequency) for both the CPU and Memory through integrated techniques such as Dynamic Voltage and Frequency Scaling (DVFS) able also to handle idle states. These embedded sensors permit recent hardware to measure energy and performance metrics at a fine grain, aggregating instruction-level measurements to offer an accurate report of code region or process-level contributions. This can be done through *low-level* power measurement interfaces such as **Intel's Running Average Power Limit (RAPL)** interface. Introduced in 2011 as part of the SandyBridge micro-architecture, RAPL is an advanced power-capping infrastructure which allows the users (or the operating system) to specify maximum power limits on processor packages and DRAM. This allows a monitoring and control program to dynamically limit the maximum average power, such that the processor can run at the highest possible speed while automatically throttling back to stay within the expected power and cooling budget. To respect these power limits, the awareness of the current power usage is required. Direct measures being often unfeasible at the processor level, power estimates are performed within a model exploiting performance counters and temperature sensors, among others. These estimations are made available to the user via a Model Specific Register (MSR) or specific daemons which can be used when characterizing workloads. Thus RAPL energy results provide a convenient interface to collect feedback when optimizing code for a diverse range of modern computing systems. This allows for unprecedented easy access to energy information when designing and optimizing energy-aware code. Moreover, on the most recent hardware architectures and DRAMs, it was demonstrated that RAPL readings are providing accurate results with negligible performance overhead [4,7]. Finally, it is also worth to note that non-Intel processors such as the recent AMD architectures (Ryzen, Epyc) also expose RAPL interfaces which can be used via the AMD μProf performance analysis tool.

At the NVIDIA GPU accelerator level, A C-based API called **NVidia Management Library (NVML)** permits to monitor and manage various states of GPU cards. The runtime version of NVML is embedded with the NVIDIA display driver, and direct access to the queries and commands are exposed via `nvidia-smi`.

In all cases, these fine-grained interfaces (i.e., RAPL, NVML...) are used in general-purpose tools able to collect low level performance metrics. Table 1 reviews the main performance analysis tools embedding Hardware counter measurement able to report fine-grain power measurements, as well as global profiling suites that eventually build on top of these low-level hardware counter interfaces.

Optimization and Auto-Tuning of Parallel Programs. Optimizing parallel programs becomes increasingly difficult with the rising complexity of modern parallel architectures. On the one hand, parallel algorithmic improvement requires a deep understanding of performance bottleneck to tune the code application with the objective to run optimally on high-end machines. This assumes a complete workflow of performance engineering of effective scientific applications (based on standard MPI, OpenMP, an hybrid combination of both or accelerators frameworks), including instrumentation, measurement (i.e., profiling and tracing, timing and hardware counters), data storage, analysis, and visualization. Table 1 presents the main performance and profiling analysis tools sustaining this complete workflow.

Table 1. Main performance analysis tools embedding hardware counter measurement for fine-grained power monitoring.

Name	Version	Description
Low-level performance analysis tools		
Perf	4.10	Main interface in the Linux kernel and a corresponding user-space tool to measure hardware counters
PAPI	5.7.0	Performance Application Programming Interface
LikWid	5.0.1	CLI applications & API to measure hardware events
Generic performance and profiling analysis tools		
ARM Forge/Perf. Report	20.0	Profiling and Debugging for C, C++, and Fortran High Performance code
TAU	2.29	Tuning & Analysis Utilities to instrument code
Score-P	6.0	A Scalable Perf. Measurement Infra. for Parallel Codes
HPC-Toolkit	2018.09	Integrated suite of tools/performance analysis of optimized parallel programs

However, none of these tools embed automatic software tuning solutions. To that end, *Auto-tuning* [10] arose as an attempt to better exploit hardware features by automatically tuning applications to run optimally on a given high-end machine. An auto-tuner tries to find promising *configurations* for a given program executing on a given target platform to influence the non-functional behavior of this program such as runtime, energy consumption or resource usage. A configuration can be created by applying code changes to a program, also called **static** *tunable knobs* or *code transformations*. Alternatively, **runtime** tuning knobs such as the number of threads, the affinity or mapping of threads onto physical processors, or the frequency at which the cores are clocked can be adapted. The literature offers numerous studies dedicated to the optimization of the runtime knobs, much less on the code transformation exploration since this requires the use of advanced compiler infrastructures such as LLVM [9]. Furthermore, optimization is often limited to a single criteria such as runtime, while it is desirable to improve multiple objectives simultaneously which is more difficult as criteria may be conflicting. For instance, optimizing for performance may reduce energy efficiency and vice versa. In all cases, the ever-increasing number of tunable knobs (both static or runtime), coupled with the rise and complexity escalation of HPC applications, lead to an *intractable and prohibitively large search space* since the order of the code transformations applied matters. This explains why a wider adoption of auto-tuning systems to optimize real world applications is still far from reality and all modern compilers such as GCC or LLVM rely on static heuristics known to produce good results on average but may even impede performances in some cases. The huge search space induced by the quest of optimal sequences of tunable knobs for each region composing a given source code represents a severe challenge for intelligent systems. Two approaches are traditionally considered: (1) Machine Learning (ML) [1], recognized to speed up the time required to tune applications but is too dependent on rare training data and thus fail to guarantee the finding (local and global) of optimal solutions. (2) Iterative search-based methods, relying on exact or approximated (i.e., Evolutionary Algorithm (EA) inspired) heuristics. Identified as computationally expensive [6], this approach mainly targets performance or execution time optimization. Moreover, their suitability for a specific application depends on the shape of its associated search space of possible configurations. Nevertheless, search-based approaches remain the most promising and effective ones despite their identified concerns. To optimize simultaneously multiple objectives i.e., performance, runtime, energy and power usage (for both the CPU and DRAM), while minimising the time consuming evaluations of the objective vector on the target computing system, we propose EVOCODE, a search-based framework for automatic tuning of parallel programs which permits to evolve a given source code to produce optimized energy-aware versions. Combining both exact and approximated heuristics in a two-stage Multi-Objective Evolutionary Algorithm (MOEA) optimization phase relying on the LLVM Compiler Infrastructure, the proposed approach is detailed in the next section.

3   Toward Automatic Software Tuning of Parallel Programs for Energy-Aware Executions

An overview of the EVOCODE framework is proposed in Fig. 1 and is now depicted. It aims at tuning an input program denoted as $\mathcal{P}_{ref}$ for an optimized execution on a target computing system such as an HPC facility or a cloud system. "*Optimized*" in this context means the generation of semantically equivalent programs $\mathcal{P}_1, \mathcal{P}_2, ...$ demonstrating improvement for selected KPIs i.e., performance, energy and power usage (both for the CPU and DRAM), as well as runtime. In practice, we assume that $\mathcal{P}_{ref}$ is composed of multiple regions $\{R_1, ..., R_r\}$, where each region delimits a single-entry single-exit section of the source code subjected to tuning. For instance, an OpenMP section, an outermost loop within a loop nest, or a function definition associated to a function call (i.e., at least the `main()` function). The identification and analysis of these regions in $\mathcal{P}_{ref}$ (eventually to isolate code portions that can be ignored) corresponds to the **Step A and B** in EVOCODE. Note that some regions may exist as CUDA kernels for hybrid (CPU+GPU) runs. Then in **Step C**, EVOCODE will operate a two-stage MOEA optimization phase aiming at the automatic evolution of $\mathcal{P}_{ref}$ as follows: (1) a *local* optimization is achieved aiming at the best configuration selection for each region $R_{i,j}$. Typically, a categorical hyper-parameter optimization for the foreseen tunable knobs is performed for the selected KPIs leading to different versions of these regions. (2) A *global* MOEA combines the regions to measure the effect on the entire program instead of considering the effect only for individual region executions. In this way, we optimize the whole program execution instead of focusing on specific regions, to provide semantically equivalent programs $\mathcal{P}_1, \mathcal{P}_2, ...$ based on approximated Pareto-optimal solutions. In practice, new multi-objective surrogate-based approaches [8] hybridizing multi-objective meta-heuristics (e.g., NSGA-III [3]) and Machine Learning models based on Gaussian Processes are proposed to minimize the time consuming evaluations of the objective vector on the target computing system. More precisely, configurations are evaluated using surrogate versions of the objectives functions handled by an oracle. If the prediction error ε is smaller than a predefined threshold, the oracle will consider that evaluations are accurate (and thus do not need to be executed in the target system), else it will update the surrogate models with the true objectives values, obtained from the running evaluation of the selected configurations. After the Pareto set for the whole program is computed, a set of code configurations for the entire program can be selected from the Pareto set, either manually or automatically to allow for the **Step D** of EvoCode, to help for the decision making phase. For the initial developments of EvoCode, preferences rankings provided from the decision-maker (i.e., to avoid providing specific weight values between the objective functions) will be used as proposed in [2], where the pruning method restricts the considered solutions within the

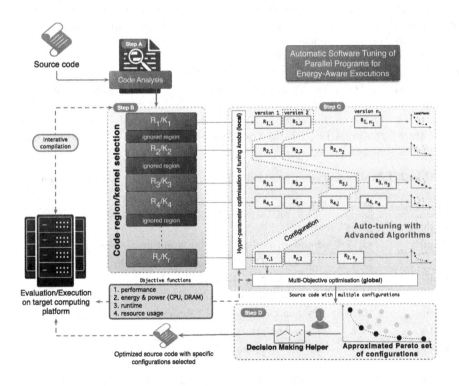

Fig. 1. Overview of the EvoCode framework.

Pareto set using threshold boundaries (e.g., afford for 10% performance penalties while reducing by at least 25% of the energy). Other heuristics relying on clustering methods i.e., grouping solutions with similar characteristics, will be then considered to improve the decision making process.

4   Validation and First Experimental Results

This section presents the first experimental results obtained by the EVOCODE framework implemented as a dedicated Python module. The technical details of the environment supporting this implementation are provided in the Table 2. The experiments detailed in the sequel were conducted on the HPC facility of the University of Luxembourg [11], more precisely on the *"regular"* computing nodes of the iris cluster, each featuring 2 Intel Skylake Xeon Gold 6132 processors (totalling 28 cores, 2.6 GHz per node). For this reason, it was important to favor libraries able to scale and exploit effectively these parallel resources. For instance, the choice of DEAP was motivated by the fact that this framework works in perfect harmony with parallelisation mechanism such as multiprocessing and SCOOP. Then the application of the static tunable knobs (the only ones considered at this early stage of developments) was done through the LLVM compiler infrastructure. In practice, EVOCODE exploits the flexibility offered by this suite to represent each program and source code from its LLVM bytecode or Internal Representation (IR) obtained using the appropriate front-end i.e., Clang for programs written in the C language family (mainly C, C++, OpenCL and CUDA). In particular, EVOCODE takes as input the reference IR representation $\mathcal{I}^{ref}$ of the program P_{ref} to optimize, and the target computing system expected to run the derived programs (for instance the Skylake nodes of the iris cluster in this section). The static tunable knobs are sequences of codes transformations i.e., LLVM transformation *passes*. 54 such passes exist

Table 2. Libraries and components details part of EVOCODE implementation.

Component	Version	Description
Python	3.7.4	n/a
NumPy	1.17.4	Fundamental package for scientific computing in Python
DEAP	1.3.0	Distributed Evolutionary Algorithms in Python
Optuna	0.19.0	Define-by-Run Hyperparameter Optimization Framework
Pandas	0.25.3	Python Data Analysis Library
plotly	4.4.1	Data Analytic Visualization Framework
LLVM/ Clang	8.0.0	LLVM Compiler Infrastructure and its front-end for the C language family (C, C++, OpenCL, CUDA...)
LikWid	5.0.0	Performance monitoring suite for RAPL hardware counter

Table 3. LikWid-based objective values reported on EVOCODE individuals evaluation.

Metric name	Counter	Event	Description
perf [MFlops]	n/a	n/a	Program result (Ex: Linpack)
runtime [s]	n/a	n/a	time: Runtime (RDTSC)
energy [J]	PWR0	PWR_PKG_ENERGY	RAPL Energy contribution
power [W]	PWR0	PWR_PKG_ENERGY/time	RAPL Power contribution
dram_energy [J]	PWR3	PWR_DRAM_ENERGY	DRAM Energy contribution
dram_power [W]	PWR3	PWR_DRAM_ENERGY/time	DRAM Power contribution

on the considered version, and examples of such transformations include dce (Dead Code Elimination), dse (Dead Store Elimination), loop-reduce (Loop Strength Reduction), loop-unroll (Unroll loops) or sroa (Scalar Replacement of Aggregates). It follows that an individual $\mathcal{I}_i$ in EVOCODE corresponds to the ordered sequence of applied transformations and the resulting LLVM byte-code obtained using the LLVM optimizer opt to apply the transforms on the reference IR code $\mathcal{I}^{ref}$ or sub-part of it i.e., the identified code regions. The **generation** of the individuals, either in the local or the global phase, consists then in aggregating the regions, compiling the LLVM bytecode into an assembly language specific to the target computing architecture using the LLVM static compiler llc, before producing the final binary from the linking phase using the LLVM front-end i.e., clang. The program $\mathcal{P}_i$ is normally *semantically* equivalent to P_{ref} since built from, and validated against, the reference $\mathcal{I}^{ref}$. Checking this equivalence is left outside the scope of EVOCODE which only validates the viability of the generated individuals from the fact that (1) the generation is successful, (2) the produced binary executes successfully on the target platform and (3) the outputs of the execution on a pre-defined set of random inputs (common to all individuals and initiated in the Step A) are equal to the ones produced upon invocation of the reference program. Then the time consuming **evaluation** of an individual consists in running and monitoring the hardware counters attached to the generated binary execution on the target platform. The energy metrics are collected from the ENERGY performance group of LikWid which supports the PWR_{PKG,PP0,PP1,DRAM,PLATFORM}_ENERGY energy counter from the RAPL interface on the Intel Skylake and Broadwell micro-architecture present on the considered computing platform. In particular, the reported fitness values are composed by a vector of the metrics presented in the Table 3, more precisely on the mean values obtained from at least 20 runs. The validation proposed in this section was performed against a set of reference benchmarking applications i.e., the C version of **Linpack** [5], STREAM (the industry standard for measuring node-level sustained memory bandwidth) or FFT. For the sake of simplicity and space savings, only the results tied to the reference Linpack benchmarks are now presented. The focus of this study was *not* to maximize the benchmark results, but to set a common input parameter set enabling fast evaluations for *all* individuals. The Linpack source code (in its C version) is structured in

13 functions, used as regions $R_1, \ldots, R_{13}$ optimized by the **local** optimization phase of EvoCode. An hyperparameter optimization is performed for each of these regions, and the complete program is for the moment rebuilt from the best configurations obtained for each region when it is planned for EvoCode to perform the **global** MOEA-based optimization phase to rebuilt the program. Thus the results presented in this paper focus on the sole local optimization phase. The Fig. 2 presents the optimization history of all trials in the EvoCode Hyper-parameter study for the `perf`,`runtime`,`energy`,`power`,`dram_energy` and `dram_power` metrics upon reconstruction of the full program from the individual region evolution. The Table 4 summarizes the best results obtained from the EvoCode auto-tuning, demonstrating improvement obtained for all criteria i.e., performance, runtime, energy, power, DRAM energy and DRAM power metrics. The improvement obtained at the DRAM level are quite astonishing (demonstrating up to 41% of energy and power savings), but the associated contribution in the energy and power dissipation is relatively small. For more classical metrics, the auto-tuning performed by EvoCode still exhibits 8% performance and runtime improvement, and up to 19% energy and power savings. This demonstrates quite significant gains, especially when considering that these results have been

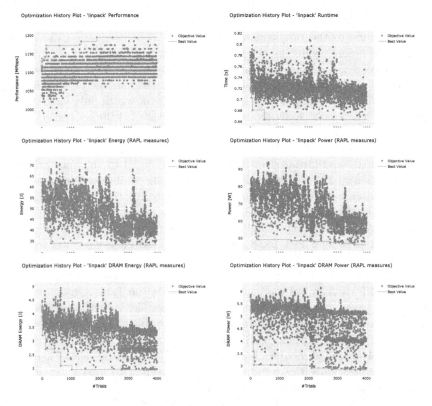

Fig. 2. Linpack benchmark Automatic tuning in EvoCode: local hyper-parameter optimization for performance, runtime, energy, power, DRAM energy & power metrics).

obtained against the program compiled with the highest level of optimization flags known by the compiler infrastructure (i.e., -O3).

The Fig. 3 reports the slice parameter relationships obtained for the sole energy optimization over the *reconstructed* program (thus considered in this case as a single region). Other similar figures were generated for the other fitness metrics, i.e., performance, runtime, power, DRAM energy and DRAM power, yet could not be presented for obvious space reasons. The objective of these analyses is to identify during the local optimization phase of EvoCode and for each optimized region the most sensitive code transformations to prune at an early stage of the global optimization unpromising configurations. Of course, it is crucial to correctly size the window for this local search strategy to avoid a premature convergence toward a local optima that may result in a non-diversity of the population. This type of evaluation is at the heart of the NSGA-III [3] heuristic currently under investigation within EvoCode.

Table 4. Best results obtained by EvoCode on the Linpack benchmark.

Metric		P_{ref} (-O3 optimized)	Best EvoCode		
perf	(MFlops)	1109.39	**1194.43**	85,04	**+8%**
runtime	(s)	0.70	**0,65**	−0,05	**−8%**
energy	(J)	40.20	**33.08**	−7,11	**−18%**
power	(W)	57.27	**46.24**	−11,03	**−19%**
dram_energy	(J)	3.28	**1.93**	−1,35	**−41%**
dram_power	(W)	4.68	**2.85**	−1,83	**−39%**

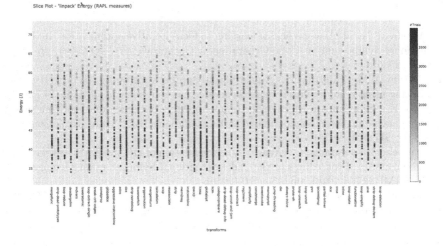

Fig. 3. Slice parameter relationships for the energy optimization tied to a single region (the full Linpack program).

5   Conclusion and Perspectives

This position paper presented the EvoCode framework aiming at the automatic software tuning of parallel programs for energy-aware executions. A reference source code is evolved using a two-stage MOEA heuristic (local and global) exploiting a compiler infrastructure (LLVM in this case) to apply **static** *tunable knobs* or *code transformations* to generate individuals[1]. The objective remains to address simultaneously multiple KPIs optimization i.e., performance, energy and power usage (for both for the CPU and DRAM) and the runtime, bringing a set of optimized binaries derived (and a priori semantically equivalent) from the reference program used as input of the EvoCode framework together with the target computing system. Our framework will also integrate a decision making process through post-Pareto-front analysis to suggest the best trade-off between the obtained solutions. EvoCode has been implemented and validated over a set of reference benchmarking applications being auto-tuned. The preliminary experimental results presented in this article (restricted to the most well-known benchmark i.e., Linpack) are quite promising. While illustrating and validating the local optimization strategy performed within EvoCode, they already demonstrate improvement for all considered metrics (ranging from 8% to 41%) when compared to the most optimized configuration set by the compiler on the reference program. The validation of the global MOEA phase within EvoCode through NSGA-III is under investigation and is bringing further improvements which will be presented in an extension of this work.

Acknowledgments. The experiments presented in this paper were carried out using the HPC facilities of the University of Luxembourg [11] – see hpc.uni.lu.

References

1. Agakov, F., et al.: Using machine learning to focus iterative optimization. In: International Symposium on Code Generation & Optimization (CGO 2006), pp. 295–305 (2006)
2. Carrillo, V.M., Taboada, H.: A post-pareto approach for multi-objective decision making using a non-uniform weight generator method, vol. 12, pp. 116–121 (2012)
3. Deb, K., Jain, H.: An evolutionary many-objective optimization algorithm using reference-point-based nondominated sorting approach. IEEE Trans. Evol. Comput. **18**(4), 577–601 (2014)
4. Desrochers, S., Paradis, C., Weaver, V.: A validation of DRAM RAPL power measurements. In: Proceedings of the 2nd International Symposium on Memory Systems (MEMSYS 2016), pp. 455–470 (2016)
5. Dongarra, J.J., Moler, C.B., Bunch, J.R., Stewart, G.W.: LINPACK Users' Guide. Society for Industrial and Applied Mathematics, Philadelphia (1979)

---

[1] Dynamic tunable knobs are also expected to be covered at a relative short term, but this will assume to rely on other compiling frameworks more suitable for such cases such as Insieme - see http://www.insieme-compiler.org.

6. Durillo, J.J., Fahringer, T.: From single- to multi-objective auto-tuning of programs: advantages and implications. Sci. Program. **22**, 285–297 (2014). https://doi.org/10.1155/2014/818579
7. Khan, K.N., Hirki, M., Niemi, T., Nurminen, J.K., Ou, Z.: RAPL in action: experiences in using RAPL for power measurements. TOMPECS **3**(2), 1–26 (2018)
8. Kieffer, E., Danoy, G., Bouvry, P., Nagih, A.: Bayesian optimization approach of general bi-level problems. In: Proceedings of the Genetic and Evolutionary Computation Conference Companion (GECCO 2017), pp. 1614–1621 (2017)
9. Lattner, C., Adve, V.: LLVM: a compilation framework for lifelong program analysis & transformation. In: Proceedings of the International Symposium on Code Generation and Optimization (CGO 2004), Palo Alto, California, March 2004
10. Naono, K., Teranishi, K., Cavazos, J., Suda, R.: Software Automatic Tuning (From Concepts to State-of-the-Art Results). Springer, New York (2010). https://doi.org/10.1007/978-1-4419-6935-4
11. Varrette, S., Bouvry, P., Cartiaux, H., Georgatos, F.: Management of an academic HPC cluster: the UL experience. In: International Conference on High Performance Computing & Simulation (HPCS 2014), Bologna, Italy, pp. 959–967. IEEE, July 2014

Special Session on Tools for Energy Efficient Computing

Overview of Application Instrumentation for Performance Analysis and Tuning

Ondrej Vysocky[1]([✉]) [ID], Lubomir Riha[1] [ID], and Andrea Bartolini[2] [ID]

[1] IT4Innovations National Supercomputing Center,
VŠB – Technical University of Ostrava, Ostrava, Czech Republic
{ondrej.vysocky,lubomir.riha}@vsb.cz
[2] University of Bologna, DEI, via Risorgimento 2, 40136 Bologna, Italy
a.bartolini@unibo.it

Abstract. Profiling and tuning of parallel applications is an essential part of HPC. Analysis and improvement of the hot spots of an application can be done using one of many available tools, that provides measurement of resources consumption for each instrumented part of the code. Since complex applications show different behavior in each part of the code, it is desired to insert instrumentation to separate these parts.

Besides manual instrumentation, some profiling libraries provide different ways of instrumentation. Out of these, the binary patching is the most universal mechanism, that highly improves user-friendliness and robustness of the tool. We provide an overview of the most often used binary patching tools and show a workflow of how to use them to implement a binary instrumentation tool for any profiler or autotuner. We have also evaluated the minimum overhead of the manual and binary instrumentation.

Keywords: Binary instrumentation · Performance analysis · Code optimization · High performance computing

1 Introduction

Developers of HPC applications are forced to optimize their applications to reach maximum possible performance and scalability. This request makes the performance analysis tools very important elements of the HPC systems, that have a goal in the identification of the hot spots of the code that provides space for improvement. Except basic, single-purpose applications every region of an application may have different requirements on the underlying hardware. In general, we may speak about several application kernels, that are bounded due to different (micro-)architectural components (e.g. compute, memory bandwidth, communication or I/O kernels) presented in [1] and evaluated in [13,29].

An application performance analysis tool provides profiling of the application - stores current time, hardware performance counters et cetera, to provide information about the program status at the given time. In general there are

© Springer Nature Switzerland AG 2020
R. Wyrzykowski et al. (Eds.): PPAM 2019, LNCS 12044, pp. 159–168, 2020.
https://doi.org/10.1007/978-3-030-43222-5_14

several ways how to connect the profiling library with the target application to select when the application's state should be captured, (1) insert the profiling library API functions into the code of the profiled application, or (2) the profiling library implements a middleware for a specific functions (e.g. memory access, I/O or MPI etc). Another option could be monitoring or simulating the application process, however these approaches may have a problem to profile exactly the application performance. The advantage of the monitoring is that it does not require to instrument the application, which means that identification of exact location in such code is ambiguous. Instrumentation can be inserted manually to the source code, by a compiler at the compilation time or the application's binary can be patched using dynamic or static instrumentation tools.

1.1 Motivation

The list of the HPC applications profiling tools is quite long, and despite many features are shared among them, every tool brings something extra to provide slightly different insight into the application's behavior. New HPC machines come with new challenges that require different ways how to optimize the code. With the upcoming HPC exascale era, there is pressure to reduce energy consumption of the system and the applications too. Several projects develop autotuning tools for energy savings based on CPU frequencies scaling or using Intel RAPL power capping [9], e.g. GEOPM [6], COUNTDOW [5], Adagio [22] or READEX [20, 23].

One of the READEX tools is MERIC [14, 29] library, that has been developed, to provide application behavior analysis and information about its energy consumption when different application or system parameters are tuned. MERIC dynamically changes the tuned parameters and searches for the configuration in which each part of the application fully utilizes the system, not to waste the resources and bring energy or time savings. This way user can detect that some parts of the target application when uses just one of two sockets is as fast as when using them both due to strong NUMA (Non-Uniform Memory Access) effect, or that the frequency of the CPU cores can be significantly reduced, due to inefficient memory access pattern. MERIC supports manual instrumentation only, which we have identified as a weak spot on a way to reach maximum possible savings. First of all process of localization where to insert the manual instrumentation to the source code is time consuming, which may lead to the situation that some parts of the code will not be sufficiently covered, and due to that the code analysis may miss identification some of the code's dynamicity and result configuration will be sub-optimal.

It is barely possible to specify a single rule for all autotuning frameworks that decides which parts of a code should be instrumented, but the most universal way is specification of a minimum region size. Under the READEX project has been specified that the minimum size of an instrumented function is 100 ms to the tuning framework be able to change the system settings of the contemporary Intel x86 processors and provide reliable energy measurement for all the instrumented regions (Intel RAPL counters [12] and HDEEM [11], 1 kHz power-sampling energy measurement systems have been used in the project).

To reach maximum possible savings, the application should contain the maximum possible amount of regions, that may show different behavior. It results in search for all regions that last more than the selected threshold and instrument them, nevertheless the threshold can be extended if the instrumentation is too heavy. In general, too detailed instrumentation can be handled also at the tuning framework side, that may ignore some of the regions, but anyway even this solution will lead to some minimal overhead depending on the framework's implementation (e.g. minimal time between regions' starts, maximum level of nesting). For purpose of identification regions with runtime longer than a specified threshold a Timeprof library [14] has been developed. The library does time measurement of the application's functions and provides a list of functions that fulfill the condition.

2 Performance Analysis Tools

List of HPC tools for application performance analysis is very long so we decided to focus on open-source tools that are selected by OpenHPC [19] project whose mission is to provide a reference collection of open-source HPC software components and best practices, lowering barriers to deployment, advancement, and use of modern HPC methods and tools. The project mentions the following tools:[1]

LIKWID [27] is one of the performance monitoring and benchmarking suite of command-line applications. Extrae [24] is a multi-platform trace-file generator to monitor the application's performance. Score-P is a library for profiling and tracing, that provides core measurement services for other libraries - Scalasca [7], TAU [25], Vampir [17] and Periscope Tuning Framework (PTF) [8]. Scalasca and TAU are very similar profiling and tracing tools that can also cooperate - e.g. Scalasca's trace-files can be visualized using TAU's profile visualizer. Vampir framework provides event tracing and focuses mainly on the visualization part of the analysis process. On the other hand, PTF is an autotuning framework, providing many plugins to tune the application from various perspectives. GEOPM is an autotuning tool focused on x86 systems, that dynamically coordinating hardware settings across all compute nodes used by an application according to the application's behavior and requests from the resource manager. The last tool from our list is the mpiP [21], which is a lightweight profiling library for MPI applications, based on middleware of the MPI functions, despite that it also has a limited list of C API functions to manually instrument the application, as well as all the mentioned tools.

Besides splitting the application into different parts of the code, some tools also provide an opportunity to instrument the most time consuming loops of the target application (e.g. in case of Score-P we speak about a Phase region, GEOPM terminology uses word Epoch, etc.). This kind of annotation is useful

---

[1] OpenHPC project list of performance analysis tools besides mentioned libraries contain tools without API (e.g. visualization libraries) and also PAPI [26].

especially in case of tools that do not only analyze the application but also provide the opportunity to tune the application performance using some kind of optimization.

3   Manual and Compiler Inserted Instrumentation

Manual instrumentation usually wraps a function, block of functions (with the similar behavior) or is inserted inside a loop body, to detect different behavior within the iterations, or in case of autotuning tools to identify optimal configuration by switching the configuration in each iteration.

Manual source code instrumentation requires access to the source code to insert the API functions and at least a basic knowledge of the application behavior, to instrument the most significant regions. The application must be recompiled for each change in the instrumentation. Due to these requirements, manual instrumentation is time-consuming and inconvenient.

Despite some of the performance analysis tools provides options how to analyze the application without doing changes in the source code, using the middleware (mpiP), compiler instrumentation (Score-P) or binary instrumentation (extrae, TAU), anyway all of the mentioned tools have their own API to let the application user/developer extend the instrumentation about specific parts of the application.

Compiler instrumentation is provided by the Score-P or by the GNU profiler gprof [10], it provides a possibility to wrap applications' functions with the instrumentation at the compilation time. In comparing to the manual instrumentation it removes the requirement to browse the source code to locate the requested functions, however, the handicap of accessing the source code persists. In default settings compiler instruments all the application's functions, without any limit on the function size, which in many cases may cause high overhead of the profiling, when measuring performance of the shortest regions too. Due to that, the compiler provides an option on how to select/filter a subset of the functions to instrument. Unfortunately, it leads to repeated compilation of the target application, which is usually slower than plain compilation (e.g. Score-P does not support parallel compilation).

4   Binary Patching

Binary patching means a modification of an application execution without recompilation of the source code. The modifications can be done dynamically during the application run or statically rewrite the binary with all the necessary changes and store the edited binary into a new file.

Dynamic Binary Instrumentation (DBI) tools [4,16,18] interrupt the analyzed application process and switch context to the tool at a certain point that should be instrumented, and execute a required action. This approach causes an overhead that is usually not acceptable for autotuning or performance analysis. On the other hand, a binary generated by a Static Binary Instrumentation

(SBI) tool should not cause any extra overhead in comparison to manually instrumented code, which confirms our measurements presented later in this section.

SBI tools not only insert functions calls at certain positions in the instrumented binary, but also add all the necessary dependencies to the shared libraries, so it is not required to recompile the application for its analysis. Also, SBI tools can access both mangled and demangled names of the functions even though the application has been compiled without debug information. SBI tools are provided by TAU (using Dyninst [3] or Pebil [15] or MAQAO [2]) and extrae (using Dyninst) and Score-P uses Dyninst to instrument the code by its compiler.

PEBIL is a binary rewriting tool allowing to patch ELF files for the x86-64 architecture. Unfortunately, PEBIL project is closed since 2017, so support for new platforms is not guaranteed. Due to that, we will focus on Dyninst and MAQAO only, from which MAQAO-2.7.0 supports the IA-64 and Xeon Phi architectures only, on the other hand, Dyninst-10.0.0 InstructionAPI implementation supports the IA-32, IA-64, AMD-64, SPARC, POWER, and PowerPC instruction sets and ARMv8 is in experimental status.

We have evaluated overhead of instrumentation when inserted manually with statically inserted instrumentation by MAQAO and Dyninst. We have used MERIC library for this measurement, that reads requested system information and store the value in memory. A single thread application (to remove the influence of an MPI/OpenMP barriers on the measurement) contained one region, that had been performed thousand times. We have not seen any difference in the overhead of manual instrumentation and SBI. Overhead of one instrumentation call on an Intel Xeon E5-2697v4 is:

- 175 μs – when reading timestamp
- 375 μs – when reading energy consumption using Intel RAPL (read four hardware counters and timestamp)

In the case of binary patching of a complex application, the time that is required to insert the instrumentation should not exceed the time needed for the application compilation. According Valensi MAQAO is able to insert 18 000 function calls in less than a minute [28].

4.1   A Binary Parsing

Dyninst as well as MAQAO holds the executable in a structure of components as the application was decomposed by a compiler. The components and their relations are illustrated in the Fig. 1. A binary base element is one or several images, which is a handle to the executable file associated with this binary. Each image contains a list of functions and global variables. A function can be also inspected for local variables and basic blocks (BBs), which is a sequence of the instructions with a single entry point and single exit point. The BBs are organized in a control-flow graph (CFG), that represents the branches of the code. From a basic block, it is also possible to access its instructions.

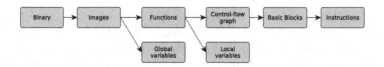

Fig. 1. Components of an application binary produced by a compiler.

When using Dyninst to browse through an application binary for its analysis or patching all the components on higher levels must be accessed first, on the other hand, MAQAO interface allows a user to access them directly. Anyway, we are primarily interested in the insertion of a function call before and after selected functions, we may stay at the level of functions.

4.2   Workflow

In this section, we will present a process of an SBI using MAQAO or Dyninst, with a goal insert a profiler function call before and after a select application function. The patching libraries provide much more functionality than presented (e.g. static binary analysis or insertion of a function call at more general locations), however for most of the profilers and autotuners wrapping a function with its instrumentation should be sufficient.

```
BPatch bpatch;
BPatch_binaryEdit *appBin = bpatch.openBinary("a.out", false);
BPatch_image *appImage = appBin->getImage();
// prepare function printf with its paramters to be inserted
std::vector<BPatch_function*> insertFunc;
appImage->findFunction("printf", insertFunc, true, true, true);
std::vector<BPatch_snippet*> args;
BPatch_snippet* param1 = new BPatch_constExpr("FUNC %s\n");
BPatch_snippet* param2 = new BPatch_constExpr("main");
args.push_back(param1);
args.push_back(param2);
// identify target location for insertion
std::vector<BPatch_function*> functions;
std::vector<BPatch_point *> *points;
appImage->findFunction("main", functions);
points = functions[0]->findPoint(BPatch_entry);
// function call insertion and store the new binary to a file
BPatch_funcCallExpr insertCall(*(insertFunc[0]), args);
appBin->insertSnippet(insertCall, *points);
appBin->writeFile ("b.out");
```

Listing 1: Dyninst code to instrument main function in a.out binary.

Both Dyninst and MAQAO open the binary and starts with its decomposition into the components as it was previously presented. We can select a function to insert from the dependent shared libraries of the application. If the application has been compiled without the profiling library, the first step should be adding all the necessary dependencies, which is a single function call.

With all the necessary dependencies, it is possible to find the function we want to insert under the application image, as well as the functions we want to wrap into the profiler instrumentation. To find the function that should be instrumented, the binary (modules in case Dyninst) must be browsed for this function. The function may have several code locations that could be instrumented, from which we are interested in its entry and exit points (addresses). With this point, it is possible to associate a function call with the requested list of arguments (be aware that there is no argument type control). This change must be committed to the binary. The edited binary is then stored and is ready to be executed to analyze its performance.

Listings present a code snapshots that insert *printf* call, that will print "FUNC main" at the beginning of execution main function of a C application a.out and stores the binary as b.out using Dyninst (Listing 1) or MAQAO (Listing 2) libraries. The examples assume that printf function is available to be added, otherwise relevant shared library dependency must be added too, also return codes are ignored to reduce size of the Listings.

```
project_t* proj = project_new("instrument_proj");
asmfile_t* asmf = project_load_file(proj, "a.out", NULL);
elfdis_t* elf = madras_load_parsed (asmf);
madras_modifs_init (elf, STACK_KEEP, 0);
fct_t* func = hashtable_lookup(asmf->ht_functions, "main");
if (func != NULL) //if main function has been found in the binary
{   //search for entry instructions of the main function
  queue_t * instructions = fct_get_entry_insns(func);
  list_t* iter = queue_iterator(instructions);
  while (iter != NULL)
  {   //insert printf function call and its parameters
    insn_t * inst = iter->data;
    modif_t* ifct = madras_fctcall_new(elf, "printf", NULL, inst->address, 0, NULL, 0);
    madras_fctcall_addparam_fromglobvar(elf,ifct,NULL,"FUNC %s\n",'a');
    madras_fctcall_addparam_fromglobvar(elf,ifct,NULL,"main",'a');
    iter = iter->next;
  }
  madras_modifs_commit(elf, "b.out"); //store the edited binary to a file
}
project_free(proj);
madras_terminate(elf);
```

Listing 2: MAQAO code to instrument main function in a.out binary.

5   Conclusion

Compiler and binary instrumentation are solution for a fully automatized application analysis and following optimized run of the application, but only in the case that such instrumentation does not lead to significantly higher overhead than in case of the manual instrumentation. Our measurements have not seen any measurable difference in manual and static binary instrumentation provided by MAQAO or Dyninst. We consider SBI as simple and the most powerful solution and based on this conclusion when writing a tool for an application behavior analysis we recommend to provide also an SBI support and present samples of code using both Dyninst and MAQAO to show how simple a basic SBI tool is.

The problem of the *ideal instrumentation* (amount and location of the probes) has a massive impact on the effectiveness of every auto-tuning framework. Autoinstrumentation tool can be written to instrument the analyzed application according to the requirements of the autotuner and its way of tuning the application. Timeprof library helps to identify the significant regions of the application to analyze their behavior. We can easily measure the runtime of all the functions of the application with Timeprof, which will provide us a selection of the regions. Afterward, the identified regions are instrumented with the selected library.

Acknowledgment. This work was supported by The Ministry of Education, Youth and Sports from the Large Infrastructures for Research, Experimental Development and Innovations project IT4Innovations National Supercomputing Center LM2015070. This work was supported by the Moravian-Silesian Region from the programme "Support of science and research in the Moravian-Silesian Region 2017" (RRC/10/2017). This work was also partially supported by the SGC grant No. SP2019/59 "Infrastructure research and development of HPC libraries and tools", VŠB - Technical University of Ostrava, Czech Republic.

References

1. Asanovic, K., et al.: The landscape of parallel computing research: a view from Berkeley. Technical report UCB/EECS-2006-183, EECS Department, University of California, Berkeley, December 2006. http://www2.eecs.berkeley.edu/Pubs/TechRpts/2006/EECS-2006-183.html

2. Barthou, D., Charif Rubial, A., Jalby, W., Koliai, S., Valensi, C.: Performance tuning of x86 OpenMP codes with MAQAO. In: Müller, M.S., Resch, M.M., Schulz, A., Nagel, W.E. (eds.) Tools for High Performance Computing 2009, pp. 95–113. Springer, Heidelberg (2010). https://doi.org/10.1007/978-3-642-11261-4_7

3. Bernat, A.R., Miller, B.P.: Anywhere, any-time binary instrumentation. In: Proceedings of the 10th ACM SIGPLAN-SIGSOFT Workshop on Program Analysis for Software Tools. PASTE 2011, pp. 9–16. ACM, New York (2011). https://doi.org/10.1145/2024569.2024572

4. Bruening, D., Zhao, Q., Amarasinghe, S.: Transparent dynamic instrumentation. In: Proceedings of the 8th ACM SIGPLAN/SIGOPS Conference on Virtual Execution Environments. VEE 2012, pp. 133–144. ACM, New York (2012). https://doi.org/10.1145/2151024.2151043

5. Cesarini, D., Bartolini, A., Bonfà, P., Cavazzoni, C., Benini, L.: COUNTDOWN - three, two, one, low power! A run-time library for energy saving in MPI communication primitives. CoRR abs/1806.07258 (2018). http://arxiv.org/abs/1806.07258

6. Eastep, J., et al.: Global extensible open power manager: a vehicle for HPC community collaboration on co-designed energy management solutions. In: Kunkel, J.M., Yokota, R., Balaji, P., Keyes, D. (eds.) ISC 2017. LNCS, vol. 10266, pp. 394–412. Springer, Cham (2017). https://doi.org/10.1007/978-3-319-58667-0_21

7. Geimer, M., Wolf, F., Wylie, B.J.N., Ábrahám, E., Becker, D., Mohr, B.: The scalasca performance toolset architecture. Concurr. Comput.: Pract. Exp. **22**(6), 702–719 (2010). https://doi.org/10.1002/cpe.v22:6

8. Gerndt, M., Cesar, E., Benkner, S.: Automatic tuning of HPC applications - the periscope tuning framework (PTF). In: Automatic Tuning of HPC Applications - The Periscope Tuning Framework (PTF). Shaker Verlag (2015). http://eprints.cs.univie.ac.at/4556/

9. Gholkar, N., Mueller, F., Rountree, B.: Power tuning HPC jobs on power-constrained systems. In: Proceedings of the 2016 International Conference on Parallel Architectures and Compilation. PACT 2016, pp. 179–191. ACM, New York (2016). https://doi.org/10.1145/2967938.2967961

10. Graham, S.L., Kessler, P.B., McKusick, M.K.: Gprof: A call graph execution profiler. SIGPLAN Not. **39**(4), 49–57 (2004). https://doi.org/10.1145/989393.989401

11. Hackenberg, D., et al.: HDEEM: high definition energy efficiency monitoring. In: 2014 Energy Efficient Supercomputing Workshop, pp. 1–10, November 2014. https://doi.org/10.1109/E2SC.2014.13

12. Hähnel, M., Döbel, B., Völp, M., Härtig, H.: Measuring energy consumption for short code paths using RAPL. SIGMETRICS Perform. Eval. Rev. **40**(3), 13–17 (2012). https://doi.org/10.1145/2425248.2425252

13. Haidar, A., Jagode, H., Vaccaro, P., YarKhan, A., Tomov, S., Dongarra, J.: Investigating power capping toward energy-efficient scientific applications. Concurr. Comput.: Pract. Exp. **31**(6), e4485 (2019). https://doi.org/10.1002/cpe.4485

14. IT4Innovations: MERIC library. https://code.it4i.cz/vys0053/meric. Accessed 21 Apr 2019

15. Laurenzano, M.A., Tikir, M.M., Carrington, L., Snavely, A.: PEBIL: efficient static binary instrumentation for Linux. In: 2010 IEEE International Symposium on Performance Analysis of Systems Software (ISPASS), pp. 175–183, March 2010. https://doi.org/10.1109/ISPASS.2010.5452024

16. Luk, C.K., et al.: Pin: building customized program analysis tools with dynamic instrumentation. In: Proceedings of the 2005 ACM SIGPLAN Conference on Programming Language Design and Implementation. PLDI 2005, pp. 190–200. ACM, New York (2005). https://doi.org/10.1145/1065010.1065034

17. Müller, M.S., et al.: Developing scalable applications with Vampir, VampirServer and VampirTrace. In: PARCO (2007)

18. Nethercote, N., Seward, J.: Valgrind: a framework for heavyweight dynamic binary instrumentation. SIGPLAN Not. **42**(6), 89–100 (2007). https://doi.org/10.1145/1273442.1250746

19. OpenHPC: Community building blocks for HPC systems. https://openhpc.community/. Accessed 21 Apr 2019

20. READEX: Horizon 2020 READEX project (2018). https://www.readex.eu

21. Roth, P.C., Meredith, J.S., Vetter, J.S.: Automated characterization of parallel application communication patterns. In: Proceedings of the 24th International Symposium on High-Performance Parallel and Distributed Computing. HPDC 2015, pp. 73–84. ACM, New York (2015). https://doi.org/10.1145/2749246.2749278

22. Rountree, B., Lowenthal, D.K., de Supinski, B.R., Schulz, M., Freeh, V.W., Bletsch, T.K.: Adagio: making DVS practical for complex HPC applications. In: Proceedings of the 23rd International Conference on Supercomputing. ICS 2009, pp. 460–469. ACM, New York (2009). https://doi.org/10.1145/1542275.1542340

23. Schuchart, J., et al.: The READEX formalism for automatic tuning for energy efficiency. Computing **99**(8), 727–745 (2017). https://doi.org/10.1007/s00607-016-0532-7

24. Servat, H., Llort, G., Huck, K., Giménez, J., Labarta, J.: Framework for a productive performance optimization. Parallel Comput. **39**(8), 336–353 (2013). https://doi.org/10.1016/j.parco.2013.05.004, http://www.sciencedirect.com/science/article/pii/S0167819113000707
25. Shende, S.S., Malony, A.D.: The TAU parallel performance system. Int. J. High Perform. Comput. Appl. **20**(2), 287–311 (2006). https://doi.org/10.1177/1094342006064482
26. Terpstra, D., Jagode, H., You, H., Dongarra, J.: Collecting performance data with PAPI-C. In: Müller, M.S., Resch, M.M., Schulz, A., Nagel, W.E. (eds.) Tools for High Performance Computing 2009, pp. 157–173. Springer, Heidelberg (2010). https://doi.org/10.1007/978-3-642-11261-4_11
27. Treibig, J., Hager, G., Wellein, G.: LIKWID: lightweight performance tools. In: Bischof, C., Hegering, H.G., Nagel, W.E., Wittum, G. (eds.) Competence in High Performance Computing 2010, pp. 165–175. Springer, Heidelberg (2012). https://doi.org/10.1007/978-3-642-24025-6_14
28. Valensi, C.: A generic approach to the definition of low-level components for multi-architecture binary analysis. Ph.D. thesis, Université de Versailles Saint-Quentin-en-Yvelines, July 2014
29. Vysocky, O., et al.: Evaluation of the HPC applications dynamic behavior in terms of energy consumption. In: Proceedings of the Fifth International Conference on Parallel, Distributed, Grid and Cloud Computing for Engineering, pp. 1–19 (2017). https://doi.org/10.4203/ccp.111.3. Paper 3, 2017

Energy-Efficiency Tuning of a Lattice Boltzmann Simulation Using MERIC

Enrico Calore[1]([✉]) [iD], Alessandro Gabbana[1,2] [iD], Sebastiano Fabio Schifano[1,2] [iD], and Raffaele Tripiccione[1,2] [iD]

[1] INFN Ferrara, Ferrara, Italy
enrico.calore@fe.infn.it
[2] Università degli Studi di Ferrara, Ferrara, Italy

Abstract. Energy-efficiency is already of paramount importance for High Performance Computing (HPC) systems operation, and tools to monitor power usage and tune relevant hardware parameters are already available and in use at major supercomputing centres. On the other hand, HPC application developers and users still usually focus just on performance, even if they will probably be soon required to look also at the energy-efficiency of their jobs. Only few software tools allow to energy-profile a generic application, and even less are able to tune energy-related hardware parameters from the application itself. In this work we use the MERIC library and the RADAR analyzer, developed within the EU READEX project, to profile and tune for efficiency the execution parameters of a real-life Lattice Boltzmann code. Profiling methodology and details are described, and results are presented and compared with the ones measured in a previous work using different methodologies and tools.

Keywords: MERIC · Optimization · Lattice Boltzmann · Energy · Efficiency

1 Introduction and Related Works

The performances of current HPC systems are increasingly bounded by their energy consumption and ownership costs for large computing facilities are increasingly shaped by the electricity bill. For several years, computing centers have considered the option to charge users not only for core/hours, but also for the energy dissipation of their compute jobs, in the hope to encourage users to optimize their applications also from the energy-efficiency point of view.

This approach could not be easily applied in the past since several tools were still missing, both from the data center side, such as tools for a fine grained energy accounting [1,3], and also from the user side, such us tools to energy-profile applications [18] and to tune hardware parameters [12,28].

There are different approaches to reduce the energy cost of a given application on a given hardware architecture. In general, the main idea is to try to match the

© Springer Nature Switzerland AG 2020
R. Wyrzykowski et al. (Eds.): PPAM 2019, LNCS 12044, pp. 169–180, 2020.
https://doi.org/10.1007/978-3-030-43222-5_15

available hardware resources with the application requirements. To accomplish this, most tools rely on *Dynamic Voltage and Frequency Scaling* (DVFS) [29], making it possible to tune the clock frequencies of the processor [13] and in some cases even of memory [11].

DVFS has an immediate impact on power drain, but a power reduction does not automatically translate to energy saving, since the average power drain of the system has to be integrated over the execution time (or *time-to-solution*, T_S) to obtain the consumed energy (or *energy-to-solution*, E_S). As can be seen in Eq. 1, in some cases an increase in T_S may actually increase E_S, in spite of a lower average power drain P_{avg}.

$$E_s = T_s \times P_{avg} \tag{1}$$

On a specific architecture, the optimal frequencies for a given application, or a given function inside an application, is defined by several factors. At a first approximation, a generic Von Neumann architecture can be seen as two different sub-systems, a compute module, characterized by a maximum instruction throughput and a memory module, characterized by a maximum bandwidth. The ratio between the two (Instructions/s and Byte/s), specific of each hardware architecture, is known as the *machine-balance* [20]. The clock frequency of the processor directly impacts the compute sub-system performance, while the clock frequency of the memory impacts memory bandwidth. Whenever one of the two sub-systems is underutilized (*e.g.* a memory-bound function which do not fully exploit the computing sub-system), there is an opportunity to lower its clock frequency to decrease power drain without impacting the execution time, and thus to increase the overall energy-efficiency.

Several attempts were made in the past in order to predict the best DVFS clock frequencies for a given application, but the large number of software and hardware parameters involved has not allowed yet to have general and widely used solutions. Despite of this, the interest towards the use of DVFS to increase the energy-efficiency of HPC applications is still vivid [15]. The European READEX project (*Run-Time Exploitation of Application Dynamism for Energy-Efficient Exascale Computing*) [21,24] is one of the latest efforts in this direction; it has developed several tools to support users in improving the energy-efficiency of their HPC applications [17]. In this work, we evaluate two tools, MERIC and RADAR [28], which allow users to profile their applications and help them identify the best performing configurations. These tools are not meant to directly model and predict the application behavior to perform auto-tuning of hardware parameters, but allow to easily explore the whole configuration space through an exhaustive search. This could also help to draw general models, but in the meanwhile it gives an opportunity to application developers to empirically find the most energy-efficient configuration.

In previous works we have developed similar tools to study the energy behaviour of our applications [14], but MERIC and RADAR have integrated these in a more general, complete and rich tool set. In [11] we have analyzed the trade-off between computing performance and energy-efficiency, for an HPC application on low-power systems using an exhaustive search, using a custom hardware and software infrastructure to monitor power consumption. In [9] we have performed a similar analysis on high-end systems, taking into account both CPUs and GPUs. Also in this case we have developed a custom software library [6] to instrument the code; it is based on PAPI APIs [14], and allows to collect power/energy related metrics from hardware registers using architecture specific interfaces, such as the *Running Average Power Limit* (RAPL) for Intel CPUs. Our custom library [6] allows to use also the *NVIDIA Management Library* (NVML) to acquire metrics from NVIDIA GPUs, which is still not possible with MERIC. However, the major drawback of our library is that, in order to change the clock frequency, one needs to modify the application code; moreover, only the clock frequencies of the processor cores could be handled. MERIC offers similar options but its use is far less invasive and allows to tune different hardware parameters, as we show in later sections.

2   The Lattice Boltzmann Application

In this work, we consider the same Lattice Boltzmann simulation code that we have analyzed in [9] using our custom library [6], for a more direct comparison with previous results. This application is a real-life example of a typical HPC workload, so it is a good candidate to elucidate the trade-off available between performance and energy-efficiency when using DVFS techniques on modern processors. Choosing the same application, with the same problem size and simulation parameters, running on the same hardware, we are able to directly compare the results obtained. This application has been used for convective turbulence studies [4,5], and it has been deeply optimized and extensively used as a benchmarking code for several programming models and HPC hardware architectures [7,8]. In this work we use an implementation specifically developed for Intel CPUs [19], highly optimized with intrinsics functions and fully exploiting the AVX2 vector instruction set available on recent Intel Haswell CPUs.

Lattice Boltzmann methods (LB) are widely used in computational fluid dynamics, to describe flows in two and three dimensions. LB methods [27] – discrete in position and momentum spaces – are based on the synthetic dynamics of *populations* sitting at the sites of a discrete lattice. At each time step, populations *propagate* from lattice-site to lattice-site and then incoming populations *collide* among each other. The collision step mixes the populations updating the physical parameters defining each population. LB models in n dimensions with p populations are labeled as $DnQp$ and in this work we consider a state-of-the-art $D2Q37$ model that correctly reproduces the thermo-hydrodynamical evolution of a fluid in two dimensions, and enforces the equation of state of a perfect gas ($p = \rho T$) [22,23].

For the purpose of this paper it is enough to remember that in a Lattice Boltzmann simulation, after an initial assignment of the populations values, the application iterates for each lattice site, and for as many time-steps as requested, two critical kernel functions:

– the *propagate* function, moves populations across lattice sites collecting at each site all populations that will interact at the next phase;
– the *collide* function, performs double precision floating-point operations local to each lattice site to compute the physics variables defining the state of the population after the collision. Input data for this phase are the populations gathered by the previous *propagate* phase.

These two functions take most of the execution time of any LB simulation and in particular, it has to be noticed that: *propagate*, is strongly memory-bound and performs a large number of sparse memory accesses; while *collide*, is strongly compute-bound (on most architectures), having an arithmetic intensity of ≈13 FLOP/Byte [9].

To fully exploit the high level of parallelism made available by modern HPC architectures, this implementation uses MPI (Message Passing Interface) to divide the lattice data domain across several processes, OpenMP to further divide each process chunk of lattice across multiple threads, and AVX2 *intrinsics* to control SIMD vector units. Whenever available, Fused Multiply-Add instructions (FMA) are used as well.

In all tests presented in this work, we analyze a simulation involving a 2-dimensional fluid described by a lattice of $L_x \times L_y$ sites. We run with one N_p MPI process binding it to one CPU socket. The only process we use handles the whole lattice which is further divided across N_t OpenMP threads, each of them handling a sub-lattice of size: $\frac{L_x/N_p}{N_t} \times L_y$.

3   The MERIC Library and the RADAR Generator

MERIC [28] is a lightweight C/C++ library developed within the READEX project [21,24], with the goal of reducing energy consumption of HPC applications. The library is originally developed for x86 systems, and tested on Intel Haswell, Broadwell and KNL Xeon Phi. Then has been added support also for IBM OpenPOWER8+ and some Arm based systems.

MERIC allows to monitor the whole application or part of it, acquiring relevant metrics from hardware counters and other available hardware sensors while running the code. In fact, it runs application codes with different parameters (*e.g.* processor clock frequencies), and automatically measures performance and energy consumption, in order to identify optimal settings for a specific metric and/or the available pareto-front in a multi-objective optimization.

Different hardware architectures and specific HPC installations, provide different ways to monitor energy consumption and to tune the corresponding hardware parameters. MERIC support several different hardware-specific monitoring systems; in our case MERIC interfaces to the RAPL energy counters that monitor Intel CPUs energy consumption, and to the *libmsr*[1] and *msr-safe*[2] libraries to change processors clock frequencies. RAPL counters are available on modern

---

[1] https://github.com/LLNL/libmsr.
[2] https://github.com/LLNL/msr-safe.

Listing 1.1. LBM application instrumentation to define the selected regions to profile. Various parts of the code are omitted to highlight just the instrumentation lines and the two main functions.

```
MPI_Init_thread(NULL, NULL, request, &provided);

READEX_INIT();

READEX_REGION_DEFINE(LBM);
READEX_REGION_DEFINE(iteration);
READEX_REGION_DEFINE(PROPAGATE);
READEX_REGION_DEFINE(COLLIDE);

READEX_REGION_START(LBM, "LBM", SCOREP_USER_REGION_TYPE_COMMON);

for ( i = 1; i <= NITER; i++ ) {
   READEX_REGION_START(iteration, "iteration", SCOREP_...);


   READEX_REGION_START(PROPAGATE, "PROPAGATE", SCOREP_...);
      propagate();
   READEX_REGION_STOP(PROPAGATE);

   READEX_REGION_START(COLLIDE, "COLLIDE", SCOREP_...);
      collide();
   READEX_REGION_STOP(COLLIDE);


   READEX_REGION_STOP(iteration);
}

READEX_REGION_STOP(LBM);
READEX_CLOSE();
MPI_Finalize();
```

Intel processors, allowing to measure consumed energy with a sampling frequency of 1 kHz, both for Package and DRAM sub-systems of each CPU Socket.

In addition to what we have done in [9], using MERIC, *libmsr* and *msr-safe*, we have been able to tune the *core frequency* and also the *uncore frequency*; the latter refers to the clock of subsystems in the physical processor package shared by all cores, such as the Last Level Cache (LLC). This allows a much wider search space in the hardware parameters.

MERIC APIs allow to easily mark code regions of a generic application. Then, the same application is executed multiple times, changing the values of several hardware parameters, specified via environment variables. In particular, we have chose as parameters the processor core and uncore frequencies, and the number of OpenMP threads. For each run, with a different set of parameters, MERIC saves a log file with relevant metrics (*e.g.*execution time, Package and DRAM energy, PAPI counters, etc.) for each instrumented code region.

Analyzing with the RADAR tool the output files produced by MERIC, a *pdf* file can be automatically generated containing a report allowing the user to identify optimal parameters for the whole application or for each instrumented code region. This enable the user to easily obtain all the required information to apply both *static tuning* (*i.e.*use a single set of parameters for the whole application execution), and *dynamic tuning* (*i.e.*a different set of parameters for each code region) while running the application.

We show in Listing 1.1 all the instrumentation lines that we have added to our application to profile separately the two functions of interest, namely *propagate* and *collide*, which are applied in sequence inside each iteration loop over the time-steps of a simulation.

4   Performed Experiments

All tests have been run on a single node of the COKA (COmputing on Kepler Architecture) Cluster at the University of Ferrara. Each node has 2 × Intel Haswell E5-2630v3 CPUs and 8 × dual GPU NVIDIA K80 Boards, but in this work we use CPUs only.

The LBM application described in Sect. 2 has been instrumented using the MERIC library, as described in Sect. 3, in order to profile the whole application and specifically *propagate* and *collide* functions. MERIC has been configured in order to read RAPL energy counters available on the Intel Haswell CPUs [14, 16], while changing the OpenMP threads between 4 and 8, and sweeping all the core and uncore frequencies available on the Intel E5-2630v3 CPU.

In our first test, we have run the same simulation described in [9], running our LBM code on a lattice of 1024 × 8192 sites, on a single socket of one COKA node. This allows us to directly compare the overall performance and energy consumption with the results obtained in our previous work, where we were able to tune only the core frequency.

In a second test, we have run the same analysis on a larger lattice, of 4096 × 8192 points, to increase the execution time. In this case, a larger number of samples of hardware counters can be collected, allowing a more accurate measure to differentiate between the two main functions of this application. In this case we cannot compare directly with the results of [9], but we can study with finer details the origin of the energy-saving, inspecting separately the two main kernels. Optimal frequencies are anyhow independent wrt the input data sizes for this kind of simulation, since for each lattice point are applied exactly the same instructions.

5   Results

In our previous work [9], we have run our LBM code on a 1024 × 8192 lattice, on a COKA CPU, and we have measured from RAPL counters that one full iteration (*i.e.*execution of *propagate* followed by *collide*), requires ≈0.67 s and

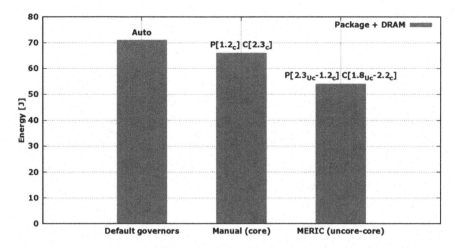

Fig. 1. Energy consumption for one iteration over a 1024×8192 lattice. The bar on the left correspond to the use of default frequency governors (*i.e.performance, conservative* or *ondemand*), the bar in the middle correspond to the manual tuning of core frequencies performed in [9], while the bar on the right show the result obtained using MERIC. Energy optimal frequencies are reported on top of the bars for *propagate* ($P[$ $]$) and *collide* ($C[$ $]$), for *core* (x_c) and *uncore* (y_{Uc}).

$\approx$71 J (CPU Package, plus DRAM contribution), using the default frequency governors with Intel Turbo Boost enabled.

On the Intel Haswell CPU adopted, the cores frequencies could be manually set in a range between 1.2 and 2.4 GHz, in 100 Mhz steps, thus we swept all of them and monitored the effects on the energy consumption and execution time. We have found that the optimal frequencies giving the lowest energy consumption were: 1.2 GHz for the *propagate* function and 2.3 GHz for the *collide*. Performing *dynamic tuning* of the CPU core frequency (*i.e.*setting the best frequency independently for each of the two functions) allowed to lower each iteration energy requirement to $\approx$66 J, giving $\approx$7% energy saving comparing to the default frequency governors, while increasing the execution time by $\approx$8%.

In this work we have performed the same test using MERIC. In this case modifications to the application code are less and easier, requiring to just instrument the original code inserting function calls as shown in Listing 1.1. Furthermore, in this case, custom data analysis scripts are not required, since the RADAR tool is provided with MERIC, in order to analyze experimental result files.

For the first experiment, we have run our application with the same simulation parameters used in [9], obtaining comparable results. Using MERIC, thanks to the wider search space including uncore frequencies (ranging between 1.2 GHz and 3.0 GHz on this CPU), we have found an energy-optimal configuration which further lower the energy consumption for a single full iteration to $\approx$54 J. This translate to $\approx$24% energy-saving with respect to the default frequency governors

Table 1. One view of the report produced by the RADAR tool. Synthesis of the overall application savings, optimizing for Package and DRAM energy consumption, using static and dynamic tuning. The application performed 10 iterations over a 4096×8192 lattice.

	Default settings	Default values	Best static config.	Static savings	Dynamic savings
RAPL_PCKG_0 [J]	8 Thr.,	2015.00 J	8 Thr.,	189.00 J (9.38%)	64.0 J of 1826.00 J (3.5%)
	3.0_{Uc} GHz,		1.8_{Uc} GHz,		
	2.4_c GHz		2.2_c GHz		
RAPL_RAM_0 [J]	8 Thr.,	499.00 J	8 Thr.,	0.00 J (0.00%)	1.0 J of 499.00 J (0.2%)
	3.0_{Uc} GHz,		3.0_{Uc} GHz		
	2.4_c GHz		2.4_c GHz		

and to $\approx 18\%$ with respect to the energy-optimal configuration in our previous work, as summarized in Fig. 1. Moreover, the energy-optimal configuration in our previous work had the drawback of an increased execution time of $\approx 8\%$ while the penalty of the energy-optimal configuration tuned with MERIC is now just $\approx 2\%$.

In the second experiment, we have run the same simulation on a 4096×8192 lattice and we report the overall application results, as presented by the RADAR tool in Table 1. In this table, the *static tuning* savings are reported with respect to the default settings which have been set to the maximum frequency values, both for core and uncore (not with respect to the default frequency governors). The *dynamic tuning* savings are reported with respect to the *static tuning* ones, and thus they have to be interpreted as a further optimization. Each of the two row of Table 1 present optimal *static tuning* and *dynamic tuning* values, when optimizing respectively for Package and DRAM energy.

In summary, we can see that adopting *static tuning* (*i.e.* 8 threads, 1.8 GHz uncore and 2.2 GHz core frequencies) for the whole application we can save more than 9% of the CPU Package energy. This can be increased by an additional 3.5% adopting *dynamic tuning* (*i.e.* switching to 2.3 GHz uncore and 1.2 GHz core while running *propagate*). Interestingly, with the energy optimal configuration, the performance is lowered by less than 3%.

As one can expect, the DRAM energy consumption is not affected by core and uncore frequency changes, while the best performance is obtained having both core and uncore frequencies close to the maximum values.

To have better insights on the energy optimal configurations for each of the two main functions of our application, we report in Table 2 another view provided by RADAR. Here we see the contribution of each of the two functions to the

Table 2. One view of the report produced by the RADAR tool. Optimal settings for each function, in terms of Package energy. The application performed 10 iterations over a 4096×8192 lattice. All the reported values concern a single iteration.

COUNTERS - RAPL: RAPL_PCKG_0 [J]						
Region	Phase %	Best static	Value	Best Dyn.	Value	Dynamic savings
COLLIDE	80.43	8 Thr., 1.8$_{Uc}$ GHz, 2.2$_c$ GHz	145.5 J	8 Thr., 1.8$_{Uc}$ GHz, 2.2$_c$ GHz	145.5 J	0.00 J (0.00%)
PROPAGATE	19.57	8 Thr., 1.8$_{Uc}$ GHz, 2.2$_c$ GHz	35.4 J	8 Thr., 2.3$_{Uc}$ GHz, 1.2$_c$ GHz	29.0 J	6.4 J (18.08%)
Total value for static tuning				$145.5 + 35.4 = 180.9$ J		
Total savings for dynamic tuning				$0.0 + 6.4 = 6.4$ J of 180.9 J (3.54 %)		

overall Package energy-saving when applying *dynamic tuning*, with respect to the static one. Since *collide* takes most of the execution time (*i.e.* 80.43%), the best overall static parameters are exactly the *collide* optimal ones. Anyhow, we also see that enabling *dynamic tuning* we achieve an impressive reduction of 18% in the energy consumption of the *propagate* function. This is given by increasing the uncore frequency, while lowering the core frequency when running *propagate*. This is clearly due to the characteristic of this kernel which is strongly memory-bound and heavily use the L3 cache [10].

6   Conclusions

In this work we have used the MERIC library and the RADAR analyzer, to energy profile a Lattice Boltzmann HPC application, running on an Intel Haswell CPU hosted in the COKA cluster. We have easily instrumented our application, and run it scanning all available core and uncore clock frequencies of the processor, with 4 and 8 threads. This analysis has identified an energy optimal set of parameters for the whole application, exploitable for *static tuning*, and also a set for each single function, exploitable for *dynamic tuning*. Interestingly, once collected optimal parameters, the same annotated code, linked to the MERIC library, can be run for production automatically performing static or dynamic tuning of the core and uncore frequencies.

Comparing with results in our previous work [9], we see that tuning uncore clock frequency allows to more than double the benefits with respect to acting only on the core frequency. In particular, acting on both we have identified a setting of clock frequencies that allow an energy saving of 24% compared to the default frequency governors, improving our previous result by a further 18%.

In a future work we plan to perform a similar analysis involving more than one CPU, and also more than one node, in order to estimate the energy-saving potential, not only in terms of a single CPU, but also at a cluster level. We also plan to test other tools which allows an automatic tuning of optimal parameters [2,25,26], not requiring a pre-characterization of the application, with the aim of comparing the obtainable energy-savings and performance.

Acknowledgements. This work was done in the framework of the COKA, and COSA projects of INFN. We thank Università degli Studi di Ferrara for access to their HPC systems. Enrico Calore was partially founded by "Contributo 5 per mille assegnato all'Università degli Studi di Ferrara - dichiarazione dei redditi dell'anno 2014".

References

1. Ahmad, W.A., et al.: Design of an energy aware petaflops class high performance cluster based on power architecture. In: 2017 IEEE International Parallel and Distributed Processing Symposium Workshops (IPDPSW), pp. 964–973 (2017). https://doi.org/10.1109/IPDPSW.2017.22
2. Alessi, F., Thoman, P., Georgakoudis, G., Fahringer, T., Nikolopoulos, D.S.: Application-level energy awareness for OpenMP. In: Terboven, C., de Supinski, B.R., Reble, P., Chapman, B.M., Müller, M.S. (eds.) IWOMP 2015. LNCS, vol. 9342, pp. 219–232. Springer, Cham (2015). https://doi.org/10.1007/978-3-319-24595-9_16
3. Beneventi, F., Bartolini, A., Cavazzoni, C., Benini, L.: Continuous learning of HPC infrastructure models using big data analytics and in-memory processing tools. In: Proceedings of the Conference on Design, Automation & Test in Europe. DATE 2017, pp. 1038–1043 (2017)
4. Biferale, L., Mantovani, F., Sbragaglia, M., Scagliarini, A., Toschi, F., Tripiccione, R.: Reactive Rayleigh-Taylor systems: front propagation and non-stationarity. EPL **94**(5), 54004 (2011). https://doi.org/10.1209/0295-5075/94/54004
5. Biferale, L., Mantovani, F., Sbragaglia, M., Scagliarini, A., Toschi, F., Tripiccione, R.: Second-order closure in stratified turbulence: simulations and modeling of bulk and entrainment regions. Phys. Rev. E **84**(1), 016305 (2011). https://doi.org/10.1103/PhysRevE.84.016305
6. Calore, E.: https://baltig.infn.it/COKA/PAPI-power-reader
7. Calore, E., Gabbana, A., Kraus, J., Pellegrini, E., Schifano, S.F., Tripiccione, R.: Massively parallel lattice-Boltzmann codes on large GPU clusters. Parallel Comput. **58**, 1–24 (2016). https://doi.org/10.1016/j.parco.2016.08.005
8. Calore, E., Gabbana, A., Kraus, J., Schifano, S.F., Tripiccione, R.: Performance and portability of accelerated lattice Boltzmann applications with OpenACC. Concurr. Computat.: Pract. Exp. **28**(12), 3485–3502 (2016). https://doi.org/10.1002/cpe.3862
9. Calore, E., Gabbana, A., Schifano, S.F., Tripiccione, R.: Evaluation of DVFS techniques on modern HPC processors and accelerators for energy-aware applications. Concurr. Comput.: Pract. Exp. **29**(12), 1–19 (2017). https://doi.org/10.1002/cpe.4143
10. Calore, E., Mantovani, F., Ruiz, D.: Advanced performance analysis of HPC workloads on Cavium ThunderX. In: 2018 International Conference on High Performance Computing Simulation (HPCS), pp. 375–382 (2018). https://doi.org/10.1109/HPCS.2018.00068

11. Calore, E., Schifano, S.F., Tripiccione, R.: Energy-performance tradeoffs for HPC applications on low power processors. In: Hunold, S., et al. (eds.) Euro-Par 2015. LNCS, vol. 9523, pp. 737–748. Springer, Cham (2015). https://doi.org/10.1007/978-3-319-27308-2_59

12. Cesarini, D., Bartolini, A., Bonfà, P., Cavazzoni, C., Benini, L.: COUNTDOWN: a run-time library for application-agnostic energy saving in MPI communication primitives. In: Proceedings of the 2nd Workshop on AutotuniNg and aDaptivity AppRoaches for Energy-efficient HPC Systems. ANDARE 2018, pp. 2:1–2:6 (2018). https://doi.org/10.1145/3295816.3295818

13. Dick, B., Vogel, A., Khabi, D., Rupp, M., Küster, U., Wittum, G.: Utilization of empirically determined energy-optimal CPU-frequencies in a numerical simulation code. Comput. Vis. Sci. **17**(2), 89–97 (2015). https://doi.org/10.1007/s00791-015-0251-1

14. Dongarra, J., London, K., Moore, S., Mucci, P., Terpstra, D.: Using PAPI for hardware performance monitoring on Linux systems. In: Conference on Linux Clusters: The HPC Revolution, vol. 5. Linux Clusters Institute (2001)

15. Etinski, M., Corbalán, J., Labarta, J., Valero, M.: Understanding the future of energy-performance trade-off via DVFS in HPC environments. J. Parallel Distrib. Comput. **72**(4), 579–590 (2012). https://doi.org/10.1016/j.jpdc.2012.01.006

16. Hackenberg, D., Schone, R., Ilsche, T., Molka, D., Schuchart, J., Geyer, R.: An energy efficiency feature survey of the Intel Haswell processor. In: 2015 IEEE International Parallel and Distributed Processing Symposium Workshop (IPDPSW), pp. 896–904 (2015). https://doi.org/10.1109/IPDPSW.2015.70

17. Kjeldsberg, P.G., et al.: Run-time exploitation of application dynamism for energy-efficient exascale computing. System-Scenario-Based Design Principles and Applications, pp. 113–126. Springer, Cham (2020). https://doi.org/10.1007/978-3-030-20343-6_6

18. Mantovani, F., Calore, E.: Performance and power analysis of HPC workloads on heterogeneous multi-node clusters. J. Low Power Electron. Appl. **8**(2) (2018). https://doi.org/10.3390/jlpea8020013

19. Mantovani, F., Pivanti, M., Schifano, S.F., Tripiccione, R.: Performance issues on many-core processors: a D2Q37 lattice Boltzmann scheme as a test-case. Comput. Fluids **88**, 743–752 (2013). https://doi.org/10.1016/j.compfluid.2013.05.014

20. McCalpin, J.D.: Memory bandwidth and machine balance in current high performance computers. IEEE Technical Committee on Computer Architecture (TCCA) Newsletter (1995)

21. Oleynik, Y., Gerndt, M., Schuchart, J., Kjeldsberg, P.G., Nagel, W.E.: Run-time exploitation of application dynamism for energy-efficient exascale computing (READEX). In: 2015 IEEE 18th International Conference on Computational Science and Engineering, pp. 347–350 (2015). https://doi.org/10.1109/CSE.2015.55

22. Sbragaglia, M., Benzi, R., Biferale, L., Chen, H., Shan, X., Succi, S.: Lattice Boltzmann method with self-consistent thermo-hydrodynamic equilibria. J. Fluid Mech. **628**, 299–309 (2009). https://doi.org/10.1017/S002211200900665X

23. Scagliarini, A., Biferale, L., Sbragaglia, M., Sugiyama, K., Toschi, F.: Lattice Boltzmann methods for thermal flows: continuum limit and applications to compressible Rayleigh-Taylor systems. Phys. Fluids (1994-present) **22**(5), 055101 (2010). https://doi.org/10.1063/1.3392774

24. Schuchart, J., et al.: The readex formalism for automatic tuning for energy efficiency. Computing **99**(8), 727–745 (2017). https://doi.org/10.1007/s00607-016-0532-7

25. Sensi, D.D., Matteis, T.D., Danelutto, M.: Simplifying self-adaptive and power-aware computing with Nornir. Future Gener. Comput. Syst. **87**, 136–151 (2018). https://doi.org/10.1016/j.future.2018.05.012

26. Shafik, R.A., Das, A., Yang, S., Merrett, G., Al-Hashimi, B.M.: Adaptive energy minimization of OpenMP parallel applications on many-core systems. In: Proceedings of the 6th Workshop on Parallel Programming and Run-Time Management Techniques for Many-core Architectures. PARMA-DITAM 2015, pp. 19–24. ACM (2015). https://doi.org/10.1145/2701310.2701311

27. Succi, S.: The Lattice-Boltzmann Equation. Oxford University Press, Oxford (2001)

28. Vysocky, O., Beseda, M., Říha, L., Zapletal, J., Lysaght, M., Kannan, V.: MERIC and RADAR generator: tools for energy evaluation and runtime tuning of HPC applications. In: Kozubek, T., et al. (eds.) HPCSE 2017. LNCS, vol. 11087, pp. 144–159. Springer, Cham (2018). https://doi.org/10.1007/978-3-319-97136-0_11

29. Wu, Q., et al.: A dynamic compilation framework for controlling microprocessor energy and performance. In: Proceedings of the 38th Annual IEEE/ACM International Symposium on Microarchitecture, pp. 271–282. IEEE Computer Society (2005)

Evaluating the Advantage of Reactive MPI-aware Power Control Policies

Daniele Cesarini[1]([✉]) [iD], Carlo Cavazzoni[2][iD], and Andrea Bartolini[1][iD]

[1] University of Bologna, viale del Risorgimento, 2, 40136 Bologna, Italy
{daniele.cesarini,a.bartolini}@unibo.it
[2] Cineca, Via Magnanelli, 6/3, 40033 Casalecchio di Reno, Italy
c.cavazzoni@cineca.it

Abstract. Power consumption is an essential factor that worsens the performance and costs of today and future supercomputer installations. In state-of-the-art works, some approaches have been proposed to reduce the energy consumption of scientific applications by reducing the operating frequency of the computational elements during MPI communication regions. State-of-the-art algorithms rely on the capability of predicting at execution time the duration of these communication regions before their execution. The COUNTDOWN approach tries to do the same by mean of a purely reactive timer based policy. In this paper, we compare the COUNTDOWN algorithm with state-of-the-art predictive-based algorithm, showing that timer based policies are more effective in extract power saving opportunities and reducing energy waste with a lower overhead. When running in a Tier1 system, COUNTDOWN achieves 5% more energy saving with lower overhead than state-of-the-art proactive policy. This suggests that reactive policies are more suited then proactive approaches for communication-aware power management algorithms.

Keywords: HPC · MPI · Power management · Reactive policy · DVFS · NPB · Energy efficiency · Parallel programming

1 Introduction

Due to the end of Dennard's scaling, in the last decade, digital electronics have faced a progressive increase of the power density at which each new processor generation operates when at its maximum performance. Today, it results that the total power consumption of high-performance computing systems and supercomputers limit the practical, achievable performance. Besides, higher power density generates more heat to be dissipated and increases cooling costs. These altogether worsen the total costs of ownership (TCO) and operational costs: limiting de facto the budget for the supercomputer computational capacity [6,8].

Low power design strategies enable computing resources to trade-off their performance for power consumption by mean of low power modes of operation. These states obtained by Dynamic and Voltage Frequency Scaling (DVFS)

© Springer Nature Switzerland AG 2020
R. Wyrzykowski et al. (Eds.): PPAM 2019, LNCS 12044, pp. 181–190, 2020.
https://doi.org/10.1007/978-3-030-43222-5_16

(P-states [1]), clock gating or throttling states (T-states), and idle states which switch off unused resources (C-states [1]). Power states transitions are controlled by hardware policies [16,22], operating system (OS) policies, and with an increasing emphasis in recent years, at user-space by the final users [2,13,15,17] and at execution time [19,24].

The first family of approaches aims to trade-off power consumption and performance to gain energy efficiency [2,14,15,17]. These approaches explore the use of HW power management knobs and application parameters to study the execution time (Time-to-Solution, TtS), average power, and energy (energy-to-solution, EtS) dependency with respect to these knobs and parameters. While these approaches can be used in combination with autotuners and resource management frameworks to explore the EtS-TtS Pareto curve, these have a limited potential in the current supercomputing scenario: slowing down applications is almost always detrimental to the total cost of ownership (TCO) due to the large contribution related to the depreciation cost of the IT equipment [8].

The second family of approaches focuses on improving application performance under a power cap [7,13,18]. These approaches target power limited systems, computing nodes, and processing elements. They rely on the runtime capability of tracking the critical task in the application; then, the power budget of the node/socket/core running the critical task is dynamically relaxed while tightening the power budget of the non-critical resources. This not only involves software approaches [7] but also HW power management solutions, like Intel ®Turbo mode, and RAPL [11]. These approaches are tailored to power capped supercomputing systems that still belong to a niche [21].

The third and last family of approaches aim at cutting the IT energy waste by reducing the performance of the processing elements when the application is in a region with communication slack available [5,9,10,20,23–25]. These approaches try to isolate at runtime regions of the application execution flow which can be executed at a reduced P-state without impacting the application performance (not in the critical task). While the hardware power management logic in today processing elements is effective in reducing the power consumption of idle resources, in large-scale MPI parallel applications that fully utilize all the assigned processing elements workload unbalance, synchronization, and communication slack can be exploited to save energy. These approaches depend on the capability of predicting critical tasks.

Rountree et al. [23] analyze the energy savings which can be achieved on MPI parallel applications by slowing down the frequencies of processors that are not in the critical path. The authors of the paper define tasks as the region of code between two MPI communication calls. The critical path is defined as the chain of the tasks which bounds the application execution time. Indeed, cores executing tasks in the critical path will be the latest ones to reach the MPI synchronization points, forcing the other cores to wait.

A later work of the same authors [24], implements an online algorithm to identify the task and the minimum frequency at which it can be executed without worsening the critical path. In addition, if after this policy some slack remain, a slack reclamation policy (namely *Fermata*) which is based on the measurement of the previous blocking time duration is used to lower the frequency to the minimum value. If the previous blocking time duration was at least twice longer than an empirical time threshold (100 ms) when the same task is executed again, a timer is set to the empirical threshold. If the MPI phase expires before the timer ends, nothing happens. Otherwise, when the timer expires, the core's frequency is set to the minimum one. This implements a last-value prediction logic to determine if there will be enough blocking time which could be exploited to save energy.

Authors of [9, 10] *COUNTDOWN* uses a timeout policy as well, but propose to apply it for each MPI phase without trying to predict its duration. This is a significant difference w.r.t to the [24] which makes it robust to miss-predictions [4]. However previous works do not compare the *COUNTDOWN* logic with the slack reclamation logic in [24].

In this work, we study the communication slack available in widely adopted applications from the NAS parallel benchmark suite [3] in a tier1 HPC production environment. We evaluate the difference between *COUNTDOWN* and *Fermata* in extracting the available communication slack. Our study demonstrates that the reactive logic in *COUNTDOWN* is more effective in isolating the communication slack than the task-based last-value prediction logic of *Fermata* with a gap which can be up to the 20% for some NAS benchmark. We then implemented *Fermata* in the *COUNTDOWN* framework, and we evaluate the overheads and power saving that the two algorithms achieve for the NAS benchmarks. In average *COUNTDOWN* achieves a lower application overhead than *Fermata* while saving an additional 5% of energy.

The paper is organized as follows. Section 2 introduces the *COUNTDOWN* framework and Sect. 3 the experimental results.

2   Framework

COUNTDOWN is a simple run-time library for profiling and fine-grain power management written in C language. COUNTDOWN implements profiler capabilities, and it can inject run-time code in the application to inspect and react to the MPI primitives. Every time an application calls an MPI primitive, COUNTDOWN profiles the call and uses a timeout strategy [4] to avoid changing the power state of the cores during too fast application and MPI context switches, where doing so may result only in an increment of the overhead without significant energy and power reduction. As we will see later in this Section, each time the MPI library asks to enter in low power mode, COUNTDOWN defers the decision for a defined amount of time. If the MPI phase terminates within this amount of time COUNTDOWN does not enter in the low power states, filtering out too short MPI phases to enter in a low power state and save energy, but costly in terms of overheads.

COUNTDOWN exposes the same interface of a standard MPI library, and it can intercept all MPI calls from the application. COUNTDOWN implements two wrappers to intercept MPI calls: (i) the first wrapper is used for C/C++ MPI libraries, (ii) the second one is used for FORTRAN MPI libraries. This is mandatory due to C/C++, and FORTRAN MPI libraries produce assembly symbols that are not application binary (ABI) compatible. The FORTRAN wrapper implements a marshaling and unmarshalling interface to bind MPI FORTRAN handlers incompatible MPI C/C++ handlers. This allows COUNTDOWN to interact with MPI libraries in FORTRAN applications. When COUNTDOWN is injected in the application, every MPI call is enclosed in a corresponding wrapper routine that implements the same signature. The wrapper routine leverage on the profiling interface of MPI called PMPI to intercept MPI primitives. The wrapper implement a *prologue* and a *epilogue* routine for each call. Both routines are used to inject profiling capabilities and power management strategies in the application. COUNTDOWN interacts with the *HW power manager* through specific events of the library. The events can also be triggered by system signals registered as callbacks for timing purposes. COUNTDOWN can be configured using environment variables where is possible to specify the verbosity of logging, the type of HW performance counters to monitor, and so on.

The library targets the instrumentation of applications through dynamic linking without user intervention. When dynamic linking is not possible COUNTDOWN has also a fall-back, a static-linking library, which can be used in the toolchain of the application to inject COUNTDOWN at compilation time. The advantage of using the dynamic linking is the possibility to instrument every MPI-based application without any modifications of the source code nor the toolchain. Linking COUNTDOWN to the application is straightforward: it is enough to configure the environment variable *LD_PRELOAD* with the path of COUNTDOWN library and start the application as usual.

COUNTDOWN is endowed with profiler capabilities which allow a detailed analysis of the application which relies on the raw HW performance counter of Intel CPU. The profiler uses the Intel Running Average Power Limit (RAPL) registers to monitor the energy/power consumed by the CPU. The energy measurements presented in the rest of this work always refer to both CPU package and DRAM consumption.

3   Experimental Results

For all the experiments we use a Tier-1 HPC system based on an IBM NeXtScale cluster which is currently ranked in the Top500 supercomputer list [12]. The compute nodes of the HPC system, are equipped with 2 Intel Broadwell E5-2697 v4 CPUs, with 18 cores at 2.3 GHz nominal clock frequency and 145W TDP and 128 GB of DDR4. Each node runs the Centos 7 OS and Linux kernel 3.10.0., nodes are interconnected with an Intel QDR (40 Gb/s) Infiniband high-performance network.

We compile all our benchmarks using the Intel *ICC/IFORT 18.0* as our toolchain, coupled with Intel *MPI Library 5.1* as the communication library. We chose the Intel software stack because it is the default production environment of our target systems as well as being supported in most HPC machines based on Intel architectures.

The default configuration for the power management in the target system is with the Linux *cpufreq* driver at the maximum P-state with turbo mode enabled. This is the baseline for our experimental results and we refer to this configuration lately as *Default*.

The NAS Parallel Benchmark suite (NPB) is a set of popular HPC benchmarks developed by the NASA Advanced Supercomputing division. The NPB consist of benchmarks and kernel widely used in different scientific areas such as spectral transform, fast Fourier transform, fluid dynamics and soon. We use the NPB version 3.3.1, and we tested different configurations, trying to balance the duration of all benchmarks at around 10 min of execution time. For CG, FT, and LU we ran on 29 nodes using 1024 cores with data set E. While for EP, MG and IS we use four nodes and 128 cores. For EP and MG we use data set E, while for IS we use data set D because it is the largest available one for that benchmark.

3.1   Fermata

We introduce *Fermata* [23,24] for comparison with *COUNTDOWN*. *Fermata* implements a simple algorithm to reduce the cores' P-state in communication regions (*Tcomm*). *Fermata* uses a prediction algorithm to decide when scaling down the P-state; the prediction is determined by the amount of time spent in communication during the previous call. If the duration is greater than or equal to twice the switching threshold, *Fermata* sets a timeout to expire at the threshold time. The threshold time is empirically set to 100 ms. Calls are identified as specific MPI primitives in the application code through the hash of the pointer that makes up the stack trace. The hash is generated when the application encounters an MPI primitive; hence, each MPI primitive in the code is uniquely identified. The information about the last call is stored in a look-up table used to choose if to set the timer in the next call.

In *COUNTDOWN* we implemented two versions of the Fermata policy, one with the original empirical switching threshold value of 100 ms [24], and one with an empirical switching threshold tuned for the target system of 500 μs [16].

3.2   Results

In this subsection, we analyze the capability of *COUNTDOWN* in comparisons with state-of-the-art *Fermata* approaches in taking advantage of the communication time to reduce energy consumption while discarding shorter communication regions.

For this purpose we analyzed with *COUNTDOWN Event Profiler* the NPB benchmarks in the default configuration (*Default*). For each benchmark, we extract information about the MPI primitives. These information recorded by *COUNTDOWN Event Profiler* during the execution of the test applications have

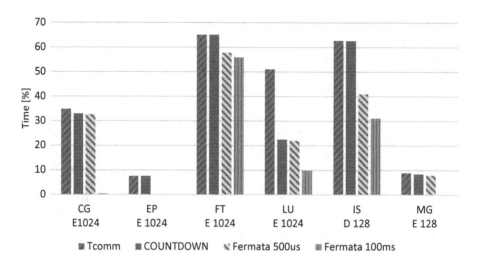

Fig. 1. Communication (MPI) time [%] which can be exploited by the power saving algorithms for the NPB.

been taken under default node power management settings (*Default* configuration). On these information, we evaluated the impact on *Fermata* and *COUNTDOWN* as described in Sects. 2, 3.1. For the *Fermata* algorithm we report both two versions with the empirical switching threshold set at 100 ms (as described in the [24]) and at 500 µs (adapted to the following the characteristics of the target HW [10,16]).

Figure 1 shows the results of this test. On the x-axis, we report the different applications, and on the y-axis, we report the percentage of MPI time. The bars report the communication (MPI) time of each benchmark (*Tcomm*) and the time that each algorithm is able to intercept to apply its power strategy. Each set of bars show the benchmark name, the dataset used, and the number of MPI processes/core involved (e.g. CG benchmark running dataset E with 1024 MPI processes/cores).

From the same Figure, we can observe that the different applications are characterized by significant communication time, which can be up to 60% of the entire application execution time. We can notice that as expected [23] 500 µs outperforms [23] 100 ms for all the benchmarks as the authors of [24] extracted the 100 ms empirical switching threshold on an older cluster machine with different power management characteristics than the one used in this study. We also highlight that in moving from 100 ms to 500 µs, the potential energy saving increases drastically for CG, LU, IS and MG.

When comparing *Fermata* 500 µs with *COUNTDOWN* we can notice that the reactive timeout policy of *COUNTDOWN* is always more effective than the proactive timeout policy of *Fermata*. While MG, LU, and CG *Fermata* achieve similarly to *COUNTDOWN*, for EP, IS and FT *COUNTDOWN* is capable of

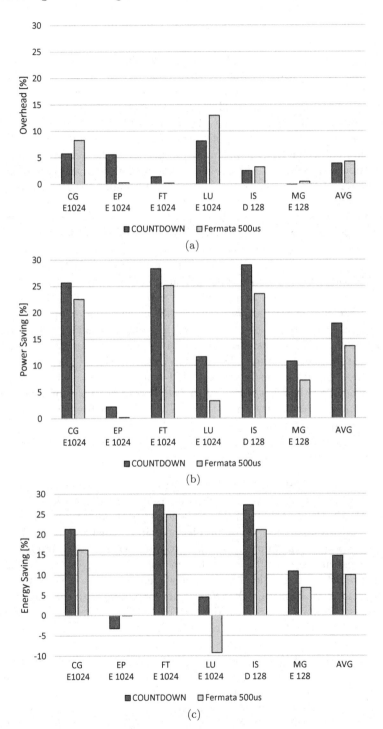

Fig. 2. *COUNTDOWN* and *Fermata* execution time overhead (a), power saving (b), and energy saving (c) for NAS benchmarks

exploiting larger communication time than *Fermata*, reaching more than the 60% of potential savings for IS and FT.

COUNTDOWN outperforms *Fermata* 500 μs differently in the benchmarks due the number of MPI primitive iterations in the application workflow. *COUNTDOWN*, which implements a reactive policy, is able to intercept and to apply its power strategy in all MPI primitives while *Fermata* need to profile at least one time each MPI primitives before to apply its power strategy.

To validate these findings we applied the *COUNTDOWN* and *Fermata* algorithm in the real Tier1 HPC nodes, and we measured the execution time overhead, the power saving, and the energy saving achieved by the two algorithms for the different NPB applications. These values are normalized with respect to the *Default* case.

Figure 2a reports the measured overheads for the two approaches. We can notice that on average *COUNTDOWN* has a lower overhead than *Fermata*. From Fig. 2b, we can see that this average lower overhead does not mean lower power savings which are on average higher for *COUNTDOWN* than *Fermata*. The combination of these two results is visible in Fig. 2c when we can see that *COUNTDOWN* achieves on average the 5% more energy saving than *Fermata*.

COUNTDOWN overcomes *Fermata* in all benchmarks except *EP E 1024*, this happens because EP is an embarrassing parallel application where communication is not significant, as shown in Fig. 1. While the computation of EP is embarrassing parallel the initial distribution of the workload and the reduction phase of the application conclusion are MPI-bound regions. But due the short communications on these regions, *COUNTDOWN* is not able to have an effective energy impact instead the short time spent in the low frequency cause up to 5% of performance overhead.

4   Conclusions

In this paper, we have studied the effectiveness of power management run times targeting the energy reduction in scientific computing applications employing hardware DVFS (P-states). Parallel scientific applications are characterized by significant synchronization and communication time, which can be exploited for reducing the power consumption of HPC systems with a lower impact on the application execution time. State-of-the-art approaches use a combination of last-value prediction and region of code knowledge to predict the duration of the upcoming communication *COUNTDOWN* and accordingly to that deciding if it is worth to enter in a low power state. A recently proposed approach, COUNTDOWN uses a reactive timeout approach applied to each communication primitive to decide if it is worth to enter a low power state. In this manuscript, we compare *COUNTDOWN* with the state-of-the-art *Fermata* approach in a Tier1 HPC system running the NAS parallel benchmarks. We demonstrated that the *COUNTDOWN* logic is capable of extracting more power saving opportunities and reducing the application overheads introduced by the power management policies. Overall, *COUNTDOWN* can save up to 25% of energy with an improvement of the 5% with respect to a proactive approach like *Fermata*.

Acknowledgment. Work supported by the EU FETHPC project ANTAREX (g.a. 671623), EU project ExaNoDe (g.a. 671578), and CINECA research grant on Energy-Efficient HPC systems.

References

1. Advanced Configuration and Power Interface (ACPI) Specification (2019). http://www.acpi.info/spec.htm. Accessed 29 Mar 2019
2. Auweter, A., et al.: A case study of energy aware scheduling on SuperMUC. In: Kunkel, J.M., Ludwig, T., Meuer, H.W. (eds.) ISC 2014. LNCS, vol. 8488, pp. 394–409. Springer, Cham (2014). https://doi.org/10.1007/978-3-319-07518-1_25
3. Bailey, D.H.: NAS parallel benchmarks. In: Padua, D. (ed.) Encyclopedia of Parallel Computing, pp. 1254–1259. Springer, Boston (2011). https://doi.org/10.1007/978-0-387-09766-4_133
4. Benini, L., Bogliolo, A., De Micheli, G.: A survey of design techniques for system-level dynamic power management. IEEE Trans. Very Large Scale Integr. (VLSI) Syst. **8**(3), 299–316 (2000). https://doi.org/10.1109/92.845896
5. Bhalachandra, S., Porterfield, A., Olivier, S.L., Prins, J.F.: An adaptive core-specific runtime for energy efficiency. In: 2017 IEEE International Parallel and Distributed Processing Symposium (IPDPS), pp. 947–956, May 2017. https://doi.org/10.1109/IPDPS.2017.114
6. Borghesi, A., Bartolini, A., Lombardi, M., Milano, M., Benini, L.: Predictive Modeling for job power consumption in HPC systems. In: Kunkel, J.M., Balaji, P., Dongarra, J. (eds.) ISC High Performance 2016. LNCS, vol. 9697, pp. 181–199. Springer, Cham (2016). https://doi.org/10.1007/978-3-319-41321-1_10
7. Borghesi, A., Bartolini, A., Lombardi, M., Milano, M., Benini, L.: Scheduling-based power capping in high performance computing systems. Sustain. Comput.: Inf. Syst. **19**, 1–13 (2018)
8. Borghesi, A., Bartolini, A., Milano, M., Benini, L.: Pricing schemes for energy-efficient HPC systems: design and exploration. Int. J. High Perform. Comput. Appl. **33**, 716–734 (2019). https://doi.org/10.1177/1094342018814593
9. Cesarini, D., Bartolini, A., Bonfà, P., Cavazzoni, C., Benini, L.: Countdown: a run-time library for application-agnostic energy saving in MPI communication primitives. In: Proceedings of the 2nd Workshop on AutotuniNg and aDaptivity AppRoaches for Energy Efficient HPC Systems, ANDARE 2018, pp. 2:1–2:6. ACM, New York (2018). https://doi.org/10.1145/3295816.3295818
10. Cesarini, D., Bartolini, A., Bonfà, P., Cavazzoni, C., Benini, L.: COUNTDOWN - three, two, one, low power! A run-time library for energy saving in MPI communication primitives. CoRR abs/1806.07258 (2018). http://arxiv.org/abs/1806.07258
11. David, H., Gorbatov, E., Hanebutte, U.R., Khanna, R., Le, C.: Rapl: memory power estimation and capping. In: Proceedings of the 16th ACM/IEEE International Symposium on Low Power Electronics and Design, ISLPED 2010, pp. 189–194. ACM, New York (2010). https://doi.org/10.1145/1840845.1840883
12. Dongarra, J.J., Meuer, H.W., Strohmaier, E., et al.: Top500 supercomputer sites (2019). https://www.top500.org/lists. Accessed 29 Mar 2019
13. Eastep, J., et al.: Global extensible open power manager: a vehicle for HPC community collaboration on co-designed energy management solutions. In: Kunkel, J.M., Yokota, R., Balaji, P., Keyes, D. (eds.) ISC 2017. LNCS, vol. 10266, pp. 394–412. Springer, Cham (2017). https://doi.org/10.1007/978-3-319-58667-0_21

14. Fraternali, F., Bartolini, A., Cavazzoni, C., Benini, L.: Quantifying the impact of variability and heterogeneity on the energy efficiency for a next-generation ultra-green supercomputer. IEEE Trans. Parallel Distrib. Syst. **29**(7), 1575–1588 (2018)
15. Fraternali, F., Bartolini, A., Cavazzoni, C., Tecchiolli, G., Benini, L.: Quantifying the impact of variability on the energy efficiency for a next-generation ultra-green supercomputer. In: International Symposium on Low Power Electronics and Design, ISLPED 2014, La Jolla, CA, USA, 11–13 August 2014, pp. 295–298 (2014). https://doi.org/10.1145/2627369.2627659
16. Hackenberg, D., Schöne, R., Ilsche, T., Molka, D., Schuchart, J., Geyer, R.: An energy efficiency feature survey of the intel Haswell processor. In: 2015 IEEE International Parallel and Distributed Processing Symposium Workshop, pp. 896–904, May 2015. https://doi.org/10.1109/IPDPSW.2015.70
17. Hsu, C., Feng, W.: A power-aware run-time system for high-performance computing. In: Proceedings of the 2005 ACM/IEEE Conference on Supercomputing, SC 2005, p. 1, November 2005. https://doi.org/10.1109/SC.2005.3
18. Kappiah, N., Freeh, V.W., Lowenthal, D.K.: Just-in-time dynamic voltage scaling: exploiting inter-node slack to save energy in MPI programs. In: Proceedings of the 2005 ACM/IEEE Conference on Supercomputing, SC 2005, p. 33, November 2005. https://doi.org/10.1109/SC.2005.39
19. Li, D., de Supinski, B.R., Schulz, M., Cameron, K., Nikolopoulos, D.S.: Hybrid MPI/OpenMP power-aware computing. In: 2010 IEEE International Symposium on Parallel Distributed Processing (IPDPS), pp. 1–12, April 2010. https://doi.org/10.1109/IPDPS.2010.5470463
20. Lim, M.Y., Freeh, V.W., Lowenthal, D.K.: Adaptive, transparent frequency and voltage scaling of communication phases in MPI programs. In: Proceedings of the 2006 ACM/IEEE Conference on Supercomputing, SC 2006, p. 14, November 2006. https://doi.org/10.1109/SC.2006.11
21. Maiterth, M., et al.: Energy and power aware job scheduling and resource management: global survey–initial analysis. In: 2018 IEEE International Parallel and Distributed Processing Symposium Workshops (IPDPSW), pp. 685–693. IEEE (2018)
22. Rosedahl, T., Broyles, M., Lefurgy, C., Christensen, B., Feng, W.: Power/Performance controlling techniques in OpenPOWER. In: Kunkel, J.M., Yokota, R., Taufer, M., Shalf, J. (eds.) ISC High Performance 2017. LNCS, vol. 10524, pp. 275–289. Springer, Cham (2017). https://doi.org/10.1007/978-3-319-67630-2_21
23. Rountree, B., Lowenthal, D.K., Funk, S., Freeh, V.W., de Supinski, B.R., Schulz, M.: Bounding energy consumption in large-scale MPI programs. In: Proceedings of the 2007 ACM/IEEE Conference on Supercomputing, SC 2007, pp. 49:1–49:9. ACM, New York (2007). https://doi.org/10.1145/1362622.1362688
24. Rountree, B., Lownenthal, D.K., de Supinski, B.R., Schulz, M., Freeh, V.W., Bletsch, T.: Adagio: making DVS practical for complex HPC applications. In: Proceedings of the 23rd International Conference on Supercomputing, ICS 2009, pp. 460–469. ACM, New York (2009). https://doi.org/10.1145/1542275.1542340
25. Venkatesh, A., et al.: A case for application-oblivious energy-efficient MPI runtime. In: Proceedings of the International Conference for High Performance Computing, Networking, Storage and Analysis, SC 2015, pp. 1–12, November 2015. https://doi.org/10.1145/2807591.2807658

Application-Aware Power Capping
Using Nornir

Daniele De Sensi[(✉)] and Marco Danelutto

Computer Science Department, University of Pisa, Pisa, Italy
{desensi,marcod}@di.unipi.it
http://pages.di.unipi.it/desensi, http://pages.di.unipi.it/marcod

Abstract. Power consumption of IT infrastructure is a major concern for data centre operators. Since data centres power supply is usually dimensioned for an average-case scenario, uncorrelated and simultaneous power spikes in multiple servers could lead to catastrophic effects such as power outages. To avoid such situations, power capping solutions are usually put in place by data centre operators, to control power consumption of individual server and to avoid the datacenter exceeding safe operational limits. However, most power capping solutions rely on Dynamic Voltage and Frequency Scaling (DVFS), which is not always able to guarantee the power cap specified by the user, especially for low power budget values. In this work, we propose a power-capping algorithm that uses a combination of DVFS and Thread Packing. We implement this algorithm in the NORNIR framework and we validate it on some real applications by comparing it to the Intel RAPL power capping algorithm and another state of the art power capping algorithm.

Keywords: Power capping · RAPL · Self-aware computing · Green computing

1 Introduction

Power consumption management is becoming a critical factor in designing applications and computing systems. In data centres, the energy cost is quickly going to overcome the cost of the physical system itself [4]. Moreover, besides economic considerations, power consumption has a considerable impact on the environment, since during 2010 the CO_2 emissions of U.S. data centres were on par with those of an entire country like Argentina or Netherlands [19].

Traditionally, to avoid possible electric surges, data centre operators have over-provisioned data centre power, considering a worst-case power consumption [15]. Albeit this ensures reliability with high confidence, it is wasteful in terms of power infrastructure utilization. To improve efficiency, researchers are investigating the possibility to over-subscribe data centre power [15,16,20].

This work has been partially supported by Univ. of Pisa PRA_2018_66 DECLware: Declarative methodologies for designing and deploying applications.

R. Wyrzykowski et al. (Eds.): PPAM 2019, LNCS 12044, pp. 191–202, 2020.
https://doi.org/10.1007/978-3-030-43222-5_17

Namely, the data centre power demand could intentionally be allowed to exceed the power supply, under the assumption that correlated spikes in servers' power consumption are infrequent. However, this exposes data centers to the risk of power outages, caused by unpredictable power spikes (e.g. due to an increase in the power consumption of more servers at the same time). Such an event would have catastrophic effects since it would lead to degradation in the final user experience or service outages. For these reasons, to achieve power safety and to avoid having under-utilized power provisioning, *power capping* techniques have been recently proposed [11,18,21,22]. These techniques monitor the data centre power consumption and, when it gets close to the available capacity, request the servers to reduce their power consumption, usually by applying Dynamic Voltage and Frequency Scaling (DVFS) [6].

One of the most commonly used techniques is Intel RAPL power capping [23]. However, it can only operate in a predefined range according to the processor specifications, and any value outside this range will be ignored. However, by extending the range of values enforceable by a power capping mechanism it would be possible to better distribute the power budget on the available servers, for example by setting low budgets for servers running non-critical applications and by letting the computing nodes running important applications run without any power cap. In this work, we address this issue by proposing a power capping algorithm which combines DVFS and Thread Packing. Thread packing [8] is a technique which forces N threads to run on a number of cores C, with $C \leq N$, thus allowing the operating system to put some cores in sleep states. Moreover, we provide a working implementation of this algorithm by adding it to the NORNIR framework, which would allow us to apply power capping to a specific application without any need to change the application code.

The main contributions of this work may be summarized as follows:

– We propose a power capping algorithm which, given a power cap, can find the most performing configuration in terms of clock frequency and the number of cores used.
– We implement this algorithm inside the NORNIR framework.
– We validate the algorithm by comparing it against Intel RAPL power capping [11] and another state of the art algorithm [13], showing improvement in the performance of the selected configuration up to 2X.

The rest of the paper is organized as follows. Section 2 describes some related works, highlighting the strengths and weaknesses of each of them. Then, in Sect. 3 we provide some background by describing the NORNIR framework. The design and implementation of the algorithm are described in Sect. 4 and it is then evaluated in Sect. 5. Eventually, Sect. 6 concludes and outlines some possible future directions for this work.

2   Related Work

Several works proposed different power capping algorithms and techniques. The most commonly used solution is Intel RAPL power capping [11], which is provided as a tool on Intel architectures. The tool dynamically scales the clock frequency and the voltage of the cores in order to enforce the power budget required by the user. However, as shown in the motivating example in Sect. 1, by only using DVFS it is not possible to decrease the power consumption below a certain threshold.

However, some works propose solutions using DVFS in conjunction with other techniques. For example, Conoci et al. [9] propose a power capping algorithm that uses DVFS and *concurrency throttling* (i.e. dynamically changing the number of threads at runtime). However, threads can be dynamically removed and added only for applications based on the thread pool model, thus limiting the applicability of the approach. On the contrary, our approach does not assume any particular application structure, since it relies on *thread packing*.

Other works also use *thread packing* [8]. However, differently from our approach, they require a training phase to be performed offline, before running the application. During the training phase, data about different applications will be collected to build a model to predict the performance and power consumption of the application in different configurations. Our algorithm relies on the opposite approach, by not requiring any training and by taking decisions only based on what is observed during application execution.

Some existing solutions do not require an offline training phase and use DVFS together with *thread packing* or *concurrency throttling* [2,13], similarly to what is done in this paper. However, such solutions either require to modify the source code of the application or are tied to some specific programming model such as OpenMP. On the contrary, our approach does not make any assumption on the application and does not require any modification to existing applications.

Eventually, some works propose techniques to coordinate power capping at the datacenter level [22]. However, since they rely on Intel RAPL for power capping, such solutions are still affected by the problem outlined in Sect. 1.

3 The Nornir Framework

NORNIR [12] is a customizable framework which can be used to add power-aware capabilities to applications. On one side, NORNIR can be used to enforce power consumption and performance requirements to applications. For example, users could ask NORNIR to dynamically change the number of resources used by a video processing application so that the application will consume no more than 60 W but, at the same time, it will process at least 20 frames per second. On the other side, it can be customized by researchers by adding new algorithms for selecting the proper amount of resources given the user constraints.

To monitor the application and to apply some decisions (e.g. dynamically remove threads from the application), NORNIR needs to be interfaced with the application. This can be done in different ways:

1. The user could implement a parallel application from scratch, by using the programming API provided by NORNIR

2. NORNIR can natively interact with some parallel runtimes (OpenMP and FastFlow [1] are currently supported). If the application uses one of these runtimes, NORNIR can interact with the application with minimal modification to the application code.

3. Otherwise (e.g. for applications using *pthreads*), the user could insert a couple of instrumentation calls in the application. These calls will monitor the performance of the application and will send these data to NORNIR, which will use it to decide how many resources to allocate to the application.

4. Eventually, if the user can not modify the application, NORNIR can still monitor it by relying on hardware performance counters (e.g. number of instructions executed per time unit). Despite this solution have been previously described [12] (denoted as *black-box*), its efficacy has never been evaluated. It is worth mentioning that in this case, because NORNIR doesn't explicitly interact with the application, performance requirements can only be expressed in terms of instructions executed per time unit.

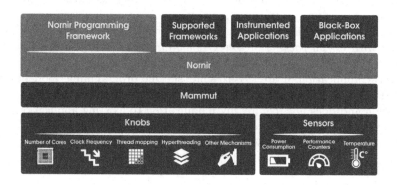

Fig. 1. Nornir architecture.

NORNIR architecture is depicted in Fig. 1. Since in this work we would like to provide a solution which could be better than Intel RAPL power capping while not being worst in terms of user effort, we will focus on the case where the user can not modify the application, forcing NORNIR to monitor the application only through hardware performance counters.

NORNIR works by following a classical *Monitor-Analyze-Plan-Execute* (MAPE) autonomic loop where, at fixed timesteps, it monitors the current performance and power consumption of the application in its current configuration. Based on this knowledge it decides if and how to change the number of resources allocated to the application, by using DVFS to scale the clock frequency and thread packing to change the number of used cores. In the *Monitor* phase, instructions per second and power consumption are collected by using MAMMUT library [14]. Among others, this library is also used by NORNIR in the *Execute* phase to apply DVFS and thread packing.

To set a specific power cap, it is sufficient to use an executable file provided by NORNIR, which takes as arguments the process identifier (*pid*) of the process we would like to control and the value of the power cap we would like to set. Note that at the moment it is only possible to control one application at a time. In the future, we will extend this approach to control multiple concurrent applications.

4 Algorithm Design

At each timestep, when NORNIR executes the *Analyze* and *Plan* phases, our algorithm is invoked. Based on the information gathered in the *Monitor* phase, we must decide which frequency f and how many cores n we would like our application to use. Since both f and n have an impact on both the performance and the power consumption, our algorithm must estimate how performance and power consumption change when f and n are increased or decreased.

Concerning the performance modeling, since we are monitoring the application by using hardware performance counters, performance in our case are represented by the number of instructions executed per time unit. We denote with $I(n, f)$ the number of instruction executed for a given number of cores and clock frequency, with $\overline{n}$ the number of cores currently used and with $\overline{f}$ the clock frequency currently used. Taking inspiration from the performance model presented in [10], we assume $I(n, f)$ to scale linearly with both n and f, i.e.

$$I(n, f) = I(\overline{n}, \overline{f}) \cdot \frac{n \cdot f}{\overline{n} \cdot \overline{f}} \tag{1}$$

Accordingly, given the current measurement $I(\overline{n}, \overline{f})$ we can estimate the performance of any other configuration by assuming that the performance will change proportionally to the changes in the number of used cores n and the clock frequency f. It is worth noting that we are assuming that the application linearly scales, i.e. by doubling the number of cores we would double the instructions executed per time unit. Of course, this is not always the case and this approximation is more severe as larger is the distance between the current configuration and the predicted one. However, as we will see in Sect. 5, this is not an issue in practice since, even if the algorithm selects a wrong configuration, another decision will be taken in the following time step. As a consequence, after a small number of steps, the algorithm will get closer to the correct configuration.

Concerning the power consumption $P(n, f)$, it is composed by a static quantity (which does not depend on n and f) and a dynamic quantity [3,7,17][1]. Since the dynamic power is also dependent on the supply voltage v, we must include it into our equation, i.e.:

$$P(n, f, v) = P_{static} + P_{dyn}(n, f, v) = P_{static} + \alpha \cdot C \cdot n \cdot v^2 \cdot f \tag{2}$$

---

[1] Actually, static power could change when changing the frequency f. However, this is a common approximation and, as we will see in Sect. 5, it does not alter the accuracy of our algorithm.

C and α represent the capacitance of the circuit and the activity factor (i.e. the fraction of the gates which are active, on average). However, the voltage v usually depends on the frequency f and the number of cores n. Accordingly, we can rewrite the equation as:

$$P(n, f) = P_{static} + P_{dyn} = P_{static} + \alpha \cdot C \cdot n \cdot V(n, f)^2 \cdot f \tag{3}$$

where $V(n, f)$ is a function which returns the voltage associated to a specific n and f. This function, in tabular form, is computed and stored by NORNIR when it is first installed on the system by using Mammut [14]. On our system, Mammut computes the voltage by accessing the PERF_STATUS [47:32] MSR register. P_{static} is constant, and it is computed and stored by NORNIR when it is installed on the system, by measuring the average idle power consumption on a one minute interval. Accordingly, we only need to estimate $P_{dynamic}$. Because n, f and $V(n, f)$ are known for all the configurations, we only need to estimate $\alpha \cdot C$. This can be done starting from Eq. 3 by considering the power consumption in the current configuration:

$$\alpha \cdot C = \frac{P(\overline{n}, \overline{f}) - P_{static}}{\overline{n} \cdot V(\overline{n}, \overline{f})^2 \cdot \overline{f}} \tag{4}$$

Because all the needed quantities are known, we can estimate the power consumption in any configuration as:

$$P(n, f) = P_{static} + (P(\overline{n}, \overline{f}) - P_{static}) \cdot \frac{n \cdot V(n, f)^2 \cdot f}{\overline{n} \cdot V(\overline{n}, \overline{f})^2 \cdot \overline{f}} \tag{5}$$

It is worth noting that this approach does not require any application characterization. NORNIR will monitor the application throughout its execution, selecting the optimal number of cores and frequency according to the predictions made by the performance and the power consumption models.

5   Experimental Evaluation

We validate our algorithm by comparing it against two algorithms: (i) Intel *RAPL* power capping, which applies DVFS and clock modulation to control the power consumption of the system; (ii) *Online Learning* [13], which uses an online learning approach where a part of the application execution is used to collect data about different configuration and to build a prediction model, which will be used to select the optimal configuration. This algorithm is one of those already provided by NORNIR.

We selected the *blackscholes*, *bodytrack* and *streamcluster* benchmarks from the PARSEC benchmark suite [5]. All the experiments have been executed on a Dual-socket NUMA machine with two Intel Xeon E5-2695 Ivy Bridge CPUs running at 2.40 GHz featuring 24 hyper-threaded cores (12 per socket). Each hyper-threaded core has 32 KB private L1, 256 KB private L2 and 30 MB of L3 shared with the cores on the same socket. The machine has 64 GB of DDR3

RAM and a Thermal Design Power (TDP) of 230 W. We did not use the hyper-threading, and the applications used at most 24 cores in our experiments. Due to hardware limitations, on this machine, it is not possible to set the frequency of each core individually. However, this is not a problem since our prediction models assume that the frequency of all the cores will be the same. The software environment consists of Linux $3.14.49 \times 86\_64$ shipped with CentOS 7.1 and gcc version 4.8.5. When using Intel RAPL power capping, we split the power budget evenly among the two packages (CPU). For example, when setting a 100 W power cap, we will set a 50 W power cap on each CPU. We express the power budget as a percentage of the TDP. For example, a power budget of 10% represents a 23 W power budget. We consider up to a 50% power budget because we did not observe any significant difference above that level. For all the approaches, we enforce the power cap on a window of one second. The static power was $\sim$37 W on the machine we used for our experiments. All the power consumption data presented include static power. According to the specifications of this processor, power capping values cannot be lower than $\sim$64 W for each package. However, we experimentally found that the actual limit which can be reached by Intel RAPL power capping is around $\sim$30 W for each package.

We evaluated each algorithm over each application for different power budgets, by analyzing the following two metrics:

Violation. Let us suppose that the application runs for s seconds, that the power cap required by the user was c Watts and that the measured power consumption at a given time t is $P(t)$. We also define with V the set of samples t such that $P(t) > c$, i.e. the set of samples where the power cap was violated. Then, this metric is defined as:

$$\frac{\sum_{t \in V}(P(t) - c)}{s}$$

This metric includes both the number and the amplitude of the power budget violations and represents the average violation of the power cap. A lower value implies a better algorithm.

Execution Time. The performance of the application, expressed as the execution time normalized to the execution time when using Intel RAPL power capping. A lower value implies a better algorithm. This metric will be shown only for experiments where RAPL correctly enforce the power cap. Indeed, when the power budget is exceeded performance would be higher than the performance achieved by solutions which properly enforce the power cap, and the comparison in such cases would not be fair.

For each of the following plots we show on the x-axis the maximum power budget we set for the application and on the y-axis the value for the metric. On the x-axis label we have on the three lines:

1. The power budget expressed as a percentage of the TDP.
2. The average number of cores set by our algorithm.
3. The clock frequency set by our algorithm.

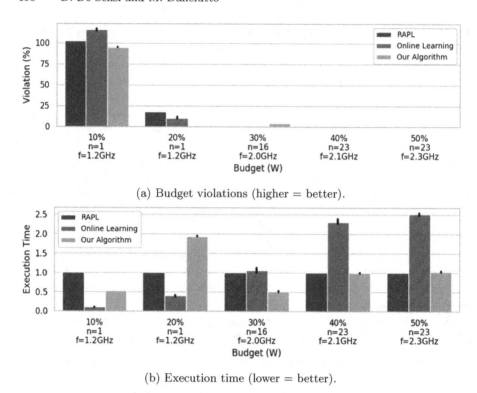

(a) Budget violations (higher = better).

(b) Execution time (lower = better).

Fig. 2. Analysis of different algorithms on the *blackscholes* application.

The black vertical bar on the top of the histograms represents the 95% confidence interval from the mean.

We report in Fig. 2 the analysis for the *blackscholes* application. By analyzing the plots, we see how *RAPL* fails in enforcing power caps with a budget lower or equal than 20%. Whereas no algorithms can reach the 10% power budget, our algorithm correctly enforces the 20% power budget, because it reduces the number of cores allocated to the application to one, while with *RAPL* all the available cores are used. By observing the performance results, our algorithm is always characterized by the best performance, except for the 10% and 20% cases, because *RAPL* and *Online Learning* violate more often the available power budget, using more resources and obtaining higher performance. For the 30% case our algorithm outperforms the *online learning* and *RAPL* algorithms, finding configurations that, while still satisfying the required power cap, are characterized by a higher performance (up to 2x). The reason why our algorithm performs better than the *online learning* one is that the latter needs more time to find a suitable configuration, due to the training phase it needs to perform to gather data about different configurations. Since during the training phase

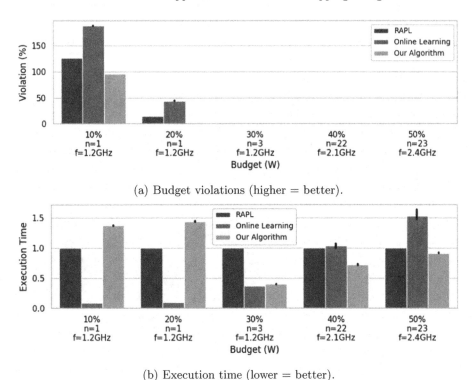

(a) Budget violations (higher = better).

(b) Execution time (lower = better).

Fig. 3. Analysis of different algorithms on the *bodytrack* application.

some low-performing configurations may be visited, this increases the execution time. Moreover, the algorithm needs to be trained again for different application phases, introducing additional overhead. For power budgets higher than 30% all the algorithm can properly enforce the power cap.

Similar results have been obtained for *bodytrack*, as shown in Fig. 3. In this case, the performance gap between our algorithm and the *online learning* one are even more evident, with a speedup higher than ∼2X when the power cap is set to 30%. Moreover, our solution can find configurations which are more performing than those selected by *RAPL*, even for higher power budgets (40% and 50%). This happens for the same reason why *RAPL* fails in enforcing low power budgets, i.e. since it only uses DVFS, the set of choices it can make are much more limited compared to our algorithm.

Eventually, Fig. 4 reports the results for the *streamcluster* application. Even in this case, the results reflect what we observed for the other two applications, with our algorithm being able to enforce 20% power caps, and providing more than ∼2X performance improvement on higher power caps compared to *RAPL*.

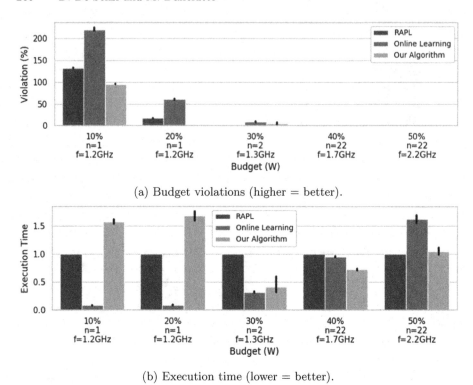

(a) Budget violations (higher = better).

(b) Execution time (lower = better).

Fig. 4. Analysis of different algorithms on the *streamcluster* application.

6 Conclusions and Future Work

In this work, we presented a power capping algorithm which used DVFS and thread packing to extend the range of reachable power caps compared to RAPL. We implemented this algorithm in the NORNIR framework, and we used its ability to control applications to test our algorithm. We then compared our algorithm with RAPL and with another state of the art approach, showing that it can satisfy the required power cap even when RAPL is not able to do so. Moreover, even when both algorithms correctly enforce the power budget required, there are cases where our algorithm can find configurations characterized by better performance, in some cases more than ∼2X more performing than those found by the other two algorithms.

In the future, we would like to extend this work for controlling multiple concurrent applications, possibly by having different power budgets for different applications according to their importance/priority. Moreover, we would like to extend the algorithm to also consider other control mechanisms such as Dynamic Clock Modulation (DCM) or DVFS for memory and *uncore* components.

References

1. Aldinucci, M., Danelutto, M., Kilpatrick, P., Torquati, M.: Fastflow: high-level and efficient streaming on multi-core. In: Programming Multi-core and Many-core Computing Systems, chap. 13 (2014)
2. Alessi, F., Thoman, P., Georgakoudis, G., Fahringer, T., Nikolopoulos, D.S.: Application-level energy awareness for OpenMP. In: Terboven, C., de Supinski, B.R., Reble, P., Chapman, B.M., Müller, M.S. (eds.) IWOMP 2015. LNCS, vol. 9342, pp. 219–232. Springer, Cham (2015). https://doi.org/10.1007/978-3-319-24595-9_16
3. Alonso, P., Dolz, M.F., Mayo, R., Quintana-Ortí, E.S.: Modeling power and energy of the task-parallel Cholesky factorization on multicore processors. Comput. Sci. - Res. Dev. **29**(2), 105–112 (2012). https://doi.org/10.1007/s00450-012-0227-z
4. Barroso, L.A., Holzle, U.: The case for energy-proportional computing. Computer **40**(12), 33–37 (2007)
5. Bienia, C., Kumar, S., Singh, J.P., Li, K.: The PARSEC benchmark suite: characterization and architectural implications. In: 17th International Conference on Parallel Architectures and Compilation Techniques (2008)
6. Burd, T., Pering, T., Stratakos, A., Brodersen, R.: A dynamic voltage scaled microprocessor system. In: ISSCC 2000, pp. 294–295 (2000)
7. Chandrakasan, A.P., Brodersen, R.W.: Minimizing power consumption in digital CMOS circuits. Proc. IEEE **83**(4), 498–523 (1995)
8. Cochran, R., Hankendi, C., Coskun, A.K., Reda, S.: Pack & Cap: adaptive DVFS and thread packing under power caps. In: Proceedings of the 44th Annual IEEE/ACM International Symposium on Microarchitecture, December 2011
9. Conoci, S., Di Sanzo, P., Ciciani, B., Quaglia, F.: Adaptive performance optimization under power constraint in multi-thread applications with diverse scalability. In: Proceedings of ICPE 2018, pp. 16–27 (2018)
10. Danelutto, M., De Sensi, D., Torquati, M.: Energy driven adaptivity in stream parallel computations. In: 2015 23rd International Conference on Parallel, Distributed and Network-Based Processing (PDP), Turku, Finland. IEEE, March 2015
11. David, H., Gorbatov, E., Hanebutte, U.R., Khanna, R., Le, C.: RAPL: memory power estimation and capping. In: 2010 ACM/IEEE International Symposium on Low-Power Electronics and Design (ISLPED), pp. 189–194 (2010)
12. De Sensi, D., De Matteis, T., Danelutto, M.: Simplifying self-adaptive and power-aware computing with Nornir. Future Gener. Comput. Syst. **87**, 136–151 (2018)
13. De Sensi, D., Torquati, M., Danelutto, M.: A reconfiguration algorithm for power-aware parallel applications. ACM TACO **13**(4), 1–25 (2016)
14. De Sensi, D., Torquati, M., Danelutto, M.: Mammut: high-level management of system knobs and sensors. SoftwareX **6**, 150–154 (2017)
15. Fan, X., Weber, W.-D., Barroso, L.A.: Power provisioning for a warehouse-sized computer. SIGARCH Comput. Archit. News **35**, 13–23 (2007)
16. Fu, X., Wang, X., Lefurgy, C.: How much power oversubscription is safe and allowed in data centers. In: Proceedings of the 8th ACM International Conference on Autonomic Computing, pp. 21–30 (2011)
17. Kim, N.S., et al.: Leakage current: Moore's law meets static power. Computer **36**(12), 68–75 (2003)
18. Lefurgy, C., Wang, X., Ware, M.: Power capping: a prelude to power shifting. Cluster Comput. **11**(2), 183–195 (2008)

19. Lucente, E.J.: The coming "c" change in data centers (2010). http://www.hpcwire.com/2010/06/15/the_coming_c_change_in_datacenters/
20. Maiterth, M.: Energy and power aware job scheduling and resource management: global survey – initial analysis. In: 2018 IEEE International Parallel and Distributed Processing Symposium Workshops (IPDPSW), pp. 685–693, May 2018
21. Wang, X., Chen, M., Lefurgy, C., Keller, T.W.: Ship: a scalable hierarchical power control architecture for large-scale data centers. IEEE Trans. Parallel Distrib. Syst. **23**(1), 168–176 (2012)
22. Wu, Q., et al.: Dynamo: facebook's data center-wide power management system. In: 2016 ACM/IEEE 43rd Annual International Symposium on Computer Architecture (ISCA), Seoul, Korea, pp. 469–480, June 2016
23. Zhang, H., Hoffmann, H.: A quantitative evaluation of the RAPL power control system (2014)

Workshop on Scheduling for Parallel Computing (SPC 2019)

A New Hardware Counters Based Thread Migration Strategy for NUMA Systems

Oscar García Lorenzo[1]([⊠]) [iD], Rubén Laso Rodríguez[1] [iD],
Tomás Fernández Pena[1] [iD], Jose Carlos Cabaleiro Domínguez[1] [iD],
Francisco Fernández Rivera[1] [iD], and Juan Ángel Lorenzo del Castillo[2] [iD]

[1] CiTIUS, Universidade de Santiago de Compostela, Santiago de Compostela, Spain
oscar.garcia@usc.es
[2] Quartz Research Lab - EISTI, Pau, France

Abstract. Multicore NUMA systems present on-board memory hierarchies and communication networks that influence performance when executing shared memory parallel codes. Characterising this influence is complex, and understanding the effect of particular hardware configurations on different codes is of paramount importance. In this paper, monitoring information extracted from hardware counters at runtime is used to characterise the behaviour of each thread in the processes running in the system. This characterisation is given in terms of number of instructions per second, operational intensity, and latency of memory access. We propose to use all this information to guide a thread migration strategy that improves execution efficiency by increasing locality and affinity. Different configurations of NAS Parallel OpenMP benchmarks running concurrently on multicore systems were used to validate the benefits of the proposed thread migration strategy. Our proposal produces up to 25% improvement over the OS for heterogeneous workloads, under different and realistic locality and affinity scenarios.

Keywords: Roofline model · Hardware counters · Performance · Thread migration

1 Introduction

Current multicores feature a diverse set of compute units and on-board memory hierarchies connected by increasingly complex communication networks and protocols. For a parallel code to be correctly and efficiently executed in a multicore system, it must be carefully programmed, and memory sharing stands out as a *sine qua non* for general purpose programming. The behaviour of the code depends also on the status of the processes currently executed in the system. A programming challenge for these systems is to partition application tasks, mapping them to one of many possible thread-to-core configuration to achieve a desired performance in terms of throughput, delay, power, and resource consumption, among others [11]. The behaviour of the system can dynamically change

© Springer Nature Switzerland AG 2020
R. Wyrzykowski et al. (Eds.): PPAM 2019, LNCS 12044, pp. 205–216, 2020.
https://doi.org/10.1007/978-3-030-43222-5_18

when multiple processes are running with several threads each. The number of mapping choices increases as the number of cores and threads do. Note that, in general purpose systems, the number of multithreaded processes can be large and change dynamically. Concerning architectural features, particularly those that determine the behaviour of memory accesses, it is critical to improve locality and affinity among threads, data, and cores. Performance issues that are impacted by this information are, among others, data locality, thread affinity, and load balancing. Therefore, addressing these issues is important to improve performance.

A number of performance models have been proposed to understand the performance of a code running on a particular system [1,4,6,17]. In particular, the roofline model (RM) [18] offers a balance between simplicity and descriptiveness based on the number of FLOPS (Floating Point Operations per Second) and the operational intensity, defined as the number of FLOPS per byte of DRAM traffic (flopsB). The original RM presented drawbacks that were taken into account by the 3DyRM model [14], which extends the RM model with an additional parameter, the memory access latency, measured in number of cycles. Also, 3DyRM shows the dynamic evolution of these parameters. This model uses the information provided by Precise Event Based Sampling (PEBS) [8,9] on Intel processors to obtain its defining parameters (flopsB, GFLOPS, and latency). These parameters identify three important factors that influence performance of parallel codes when executed in a shared memory system, and in particular, in non-uniform memory access (NUMA) systems. In a NUMA system, distance and connection to memory cells from different cores may induce variations in memory latency, and so the same code may perform differently depending on where it was scheduled, which may not be detectable in terms of the traditional RM.

Moving threads close to where their data reside can help alleviate memory related performance issues, especially in NUMA systems. Note that when threads migrate, the corresponding data usually stays in the original memory module, and they are accessed remotely by the migrated thread [3]. In this paper, we use the 3DyRM model to implement strategies for migrating threads in shared memory systems and, in particular, multicore NUMA servers, possible with multiple concurrent users. The concept is to use the defining parameters of 3DyRM as objective functions to be optimised. Thus, the problem can be defined in terms of a multiobjective optimisation problem. The proposed technique is an iterative method inspired from evolutionary optimisation algorithms. To this end, an individual utility function to represent the relative importance of the 3DyRM parameters is defined. This function uses the number of instructions executed, operational intensity, and average memory latency values, for providing a characterisation of the performance of each parallel thread in terms of locality and affinity.

2   Characterisation of the Performance of Threads

The main bottleneck in shared memory codes is often the connection between the processors and memory. 3DyRM relates processor performance to off-chip memory traffic. The Operational Intensity (OI) is the floating operations per byte of DRAM traffic (measured in flopsB). OI measures traffic between the caches and main memory rather than between the processor and caches. Thus, OI incorporates the DRAM bandwidth required by a processor in a particular computer, and the cache hierarchy, since better use of cache memories would mean less use of main memory. Note that OI is insufficient to fully characterise memory performance, particularly in NUMA systems. Extending RM with the mean latency of memory access provides a better model of performance. Thus, we employ the 3DyRM model, which provides a three dimensional representation of thread performance on a particular placement.

PEBS is an advanced sampling feature of Intel Core based processors, where the processor directly records samples from specific hardware counters into a designated memory region. The use of PEBS as a tool to monitor a program execution was already implemented in [15], providing runtime dynamic information about the behaviour of the code with low overhead [2], as well as an initial version of a thread migration tool tested with a toy examples. The migration tool presented in this work continuously gathers performance information in terms of the 3DyRM, i.e. GFLOPS, flopsB, and latency, for each core and thread. However, the information about floating point operations provided by PEBS may sometimes be inaccurate [9] or difficult to obtain. In addition, accurate information about retired instructions can be easily obtained, so giga instructions per second (GIPS) and instructions retired per byte (instB) may be used rather than GFLOPS and flopsB, respectively. For this reason, GIPS and instB are used in this work.

3   A New Thread Migration Strategy

We introduce a new strategy for guiding thread migration in NUMA systems. The proposed algorithm is executed every T milliseconds to eventually perform threads migrations. The idea is to consider the 3DyRM parameters as objective functions to be optimised, so increasing GFLOPS (or GIPS) and flopsB (or instB), and decreasing latency in each thread improve performance in the parallel code. There is a close relation between this and multiobjective optimisation (MOO) problems, which have been extensively studied [5]. The aim of many MOO solutions is to obtain the Pareto optimality numerically. However, this task is usually computationally intensive, and consequently a number of heuristic approaches have been proposed.

In our case, there are no specific functions to be optimised. Rather, we have a set of values that are continuously measured in the system. Our proposal is to apply MOO methods to address the problem using the 3DyRM parameters. Thread migration is then used to modify the state of each thread to simultaneously optimise the parameters. Therefore, we propose to characterise each thread using an aggregate objective function, P, that combines these three parameters.

Consider a system with N computational nodes or cores in which, at certain time, multiple multithreaded processes are running. Let P_{ijk} be the performance for the i-th thread of the j-th process when executed on the k-th computational node. We define the aggregate function as

$$P_{ijk} = \frac{\mathrm{GIPS}_{ijk} \cdot \mathrm{intsB}_{ijk}}{\mathrm{latency}_{ijk}}, \tag{1}$$

where GIPS_{ijk} is the GIPS of the thread, and instB_{ijk} and $\mathrm{latency}_{ijk}$ are the instB and average latency values, respectively. Note that, larger values of P_{ijk} imply better performance.

Initially, no values of P_{ijk} are available for any thread on any node. On each time interval, P_{ijk} is computed for every thread on the system according to the values read by the hardware counters. In every interval some values of P_{ijk} are updated, for those nodes k where each thread was executed, while others store the performance information of each thread when it was executed in a different node (if available). Thus, the algorithm adapts to possible behaviour changes for the threads. As threads migrate and are executed on different nodes, more values of P_{ijk} are progressively filled up.

To compare threads from different processes, each individual P_{ijk} is divided by the mean P_{ijk} of all threads of the same process, i.e. the j-th process,

$$\widehat{P}_{ijk} = \frac{P_{ijk}}{\sum_{m=1}^{n_j} P_{mjh}/n_j}, \tag{2}$$

where n_j is the number of threads of process j, and h is, for each thread m of the j-th process, the last node where it was running.

Every T milliseconds, once the new values of P_{ijk} are computed, the thread with the worst current performance, in terms of P_{ijk}, is selected to be migrated. Thus, for each process, those threads with $\widehat{P}_{ijk} < 1$ are currently performing worse than the mean of the threads in the same process, and the worst performing thread in the system is considered to be the one with the lowest $\widehat{P}_{ijk}$, i.e., the thread performing worse when compared to the other threads of its process. This is identified as the migration thread, and denoted by Θ_m.

Note that the migration can be to any core in a node other than the current node in which Θ_m resides. A weighted random process is used to choose the destination core, based on the stored performance values. In order to consider all possible migrations, all P_{ijk} values have to be taken into account. Therefore, it is important to fill as many entries of P_{ijk} as possible.

A lottery strategy is proposed in such a way that every possible destination is granted a number of tickets defined by the user, some examples are shown in Table 1, according to the likelihood of that migration improving performance. The destination with the larger likelihood has a greater chance of being chosen. Migration may take place to an empty core, where no other thread is currently being executed, or to a core occupied with other threads. If there are already threads in the core, one would have to be exchanged with Θ_m. The swap thread is denoted as Θ_g, and all threads are candidates to be Θ_g. Note that, although not all threads may be selected to be Θ_m (e.g. a process with a single thread

Table 1. List of tickets.

Ticket	Description	Default value
MEM_CELL_WORSE	Previous data show worst performance in a given node	1
MEM_CELL_NO_DATA	No previous data in a given memory node	2
MEM_CELL_BETTER	Previous data show better performance in a given node	4
FREE_CORE	It is possible to migrate a thread to a free core	2
PREF_NODE	It is possible to migrate a thread to a core located in the node in which it makes most of its memory accesses	4
THREAD_UNDER_PERF	It is possible to interchange a thread with another whose relative performance in under a determined threshold	3

would always have $\widehat{P}_{ijk} = 1$ and so never be selected), they may still be considered to be Θ_g to ensure the best performance for the whole system. When all tickets have been assigned, a final destination core is randomly selected based on the awarded tickets. If the destination core is free, a simple migration will be performed. Otherwise, an interchanging thread, Θ_g, is chosen from those currently being executed on that core. Once the threads to be migrated are selected, the migrations are actually performed.

Migrations may affect not only the involved threads, Θ_m and Θ_g, but all threads in the system due to synchronisation or other collateral relations among threads. The total performance for each iteration can be calculated as the sum of all P_{ijk} for all threads. Thus, the current total performance, $Pt_{current}$, characterises a thread configuration, independently of the processes being executed. The total performance of the previous iteration is stored as Pt_{last}. On any interval, $Pt_{current}$ may increase or decrease relatively to Pt_{last}. Depending on this variation, decisions are made regarding the next step of the algorithm.

Our algorithm dynamically adjusts the number of migrations per unit of time by changing T between a given minimum, T_{min}, and maximum, T_{max}, doubling or halving the previous value. To do that, a ratio, $0 \leq \omega \leq 1$ is defined for $Pt_{current}/Pt_{last}$, to limit an acceptable decrement in performance. So, if a thread placement has a lower total performance, more migrations should be performed to try to get a better thread placement, because they are likely to increase performance ($Pt_{current} \geq \omega Pt_{last}$). This way, T is decreased to perform migrations more often and reach optimal placement quicker. However, if current thread placement has high total performance, migrations have a greater chance of being detrimental. In this case, if $Pt_{current} < \omega Pt_{last}$, T is increased. Additionally, a rollback mechanism is implemented, to undo migrations if they result in a significant loss of performance, returning migrated threads to their former locations. Summarising, the rules guiding our algorithm are:

- If $Pt_{current} \geq \omega Pt_{last}$ then the total performance improves, so, migrations are considered productive, T is halved ($T \to T/2$), and a new migration is performed according to the rules indicated previously.
- If $Pt_{current} < \omega Pt_{last}$ then the total performance decreases more than a given threshold ω, so, migrations are considered counter-productive, T is doubled ($T \to 2 \times T$), and the last migration is rolled back.

This algorithm is named Interchange and Migration Algorithm with Performance Record and Rollback (IMAR^2). To simplify notation, IMAR^2 and its parameters are denoted as IMAR$^2[T_{min}, T_{max}; \omega]$.

4   Experimental Results

NPB-OMP benchmarks [10] were used to study the effect of the memory allocation. They are broadly used and their diverse behaviour when executed is well known. These benchmarks are well suited for multicore processors, although they do not greatly stress the memory of large servers. To study the effects of NUMA memory allocation, different memory stress situations were considered using the numactl tool [12], which allows the memory cell to store specific data, and threads to be pinned to specific cores or processors. Two servers were used to test NUMA effects. Both processors have one memory controller with four memory channels for connecting DIMM memory chips. In both systems node 0 has greater affinity with cell 0, node 1 with cell 1, and so on. Also, a NUMA aware Linux kernel was used. More specifically:

- Server A: An Ubuntu 14, with Linux kernel 3.10, NUMA server with four nodes, each has one octo-core Xeon E5-4620 (32 physical cores in total), Sandy Bridge architecture, 16 MB L3 cache, 2.2 GHz–2.6 GHz, and 512 GB of RAM. This server has memory chips connected in all four memory channels and may use all the available memory bandwidth.
- Server B: A Debian GNU/Linux 9, kernel version 5.1.15 composed by four nodes with Intel Xeon E5-4620 v4 with 10 cores each (40 in total), Broadwell-EP architecture, 25 MB L3 cache, 2.1 GHz–2.6 GHz, and 128 GB of RAM. Only one memory channel is used in this server, increasing the chances of memory congestion in remote accesses and increasing NUMA effects.

We designed experiments in which four instances of the NPB-OMP benchmarks are executed concurrently, and the placement of each can be controlled. Each benchmark instance was executed as a multi-threaded process with just enough threads to fill one node. We tested a representative set of memory and thread placements. The memory placements are:

- FREE: No specific memory placement is selected, the OS decides where to place the data of each benchmark. This is the most common case for regular users.

- DIRECT: Each benchmark have its preferred memory set to a different cell. In the case of four benchmarks, each one have one memory cell for its data, as long as its memory is large enough. This is a common option used by experienced users who know the limits of their applications [13,16].
- INTERLEAVED: Each benchmark have its memory set to interleaved, with each consecutive memory page set to a different memory cell in a round robin fashion. This is a common option used by experienced users who do not know the specific characteristics of their programs or want to use all the available bandwidth.

and the thread one's:

- OS: The OS decides where to place the threads, as well as their possible migrations. Note that the four benchmarks can not be initiated at exactly the same time, but only one at a time. This fact influences the initial thread placement. This is the most common case for regular users.
- PINNED: Each benchmark had its threads pinned to one node. When combined with the DIRECT memory placement the same node is used for one benchmark. This is a common option used by experienced users [7].
- IMAR^2: The IMAR^2 algorithm is used to place and migrate the threads.

Different combinations of these memory and thread placements were tested. Results of four class C NPB-OMP codes were selected to be shown in this paper: lu.C, sp.C, bt.C and ua.C. Benchmarks were compiled with gcc and O2 optimisation. This selection was made according to three following criteria: First, these are codes with different memory access patterns and different computing requirements. The DyRM model was used to select two benchmarks with low flopsB (lu.C and sp.C) and two with high flopsB (bt.C and ua.C). Second, since the execution times of these codes are similar, they remain in concurrent execution most of the time. This helps studying the effect of thread migrations. Third, they are representative to understand the behaviour of our proposal.

Each test was executed on the four nodes, combined as four processes of the same code that produced four combinations, named **4 lu.C, 4 sp.C, 4 bt.C**, and **4 ua.C**, and four processes of different codes, that produced one combination named (**lu.C/sp.C/bt.C/ua.C**). Tables 2 and 3 show the results for servers A and B, respectively. The times for all benchmarks of **lu.C/sp.C/bt.C/ua.C** are shown, whereas only the times of the slowest instances are shown for the four equal benchmarks. A graphical comparison is shown in Fig. 1, where times of each test are normalised to the time of a normal OS execution, the FREE memory placement with OS thread placement, with times in the first column of Tables 2 and 3 are shown as a percentage. A percentage greater that 100 means a worse execution time, while a result under 100 shows a better execution time.

Table 2. Times for four NAS benchmarks in server A. When all benchmarks are of the same kind only the time of the slowest is shown. Best time on each row is remarked in **bold**. Best time for each memory policy is shown in *italics*.

Test		Time (s)							
Benchmarks		FREE		DIRECT			INTERLEAVED		
		OS	IMAR²	OS	PINNED	IMAR²	OS	PINNED	IMAR²
lu.C/sp.C bt.C/ua.C	lu	*220.24*	245.05	344.82	**210.00**	223.33	*300.55*	428.41	310.68
	sp	**235.53**	238.39	544.63	267.89	*267.86*	*350.73*	557.39	367.57
	bt	*201.69*	214.50	321.39	**180.77**	217.15	271.34	*260.46*	270.52
	ua	*197.03*	222.02	409.35	**190.26**	212.27	307.57	316.26	*299.89*
4 lu.C		*215.84*	313.24	428.85	**212.20**	258.43	401.49	452.15	*392.84*
4 sp.C		*287.49*	324.00	1397.28	**267.71**	323.59	616.40	763.88	*610.91*
4 bt.C		*185.37*	200.70	395.95	**182.29**	207.21	241.76	246.90	*223.57*
4 ua.C		*203.54*	211.21	545.63	**190.46**	220.65	319.67	313.59	*297.92*

Table 3. Times for four NAS benchmarks in server B. When all benchmarks are of the same kind only the time of the slowest is shown. Best time on each row is remarked in **bold**. Best time for each memory policy is shown in *italics*.

Test		Time (s)							
Benchmarks		FREE		DIRECT			INTERLEAVED		
		OS	IMAR²	OS	PINNED	IMAR²	OS	PINNED	IMAR²
lu.C/sp.C bt.C/ua.C	lu	305.00	*187.00*	177.08	**176.37**	217.99	417.75	355.42	*194.01*
	sp	476.00	**354.95**	474.79	453.60	*412.59*	494.71	469.10	*402.60*
	bt	*276.75*	281.97	241.27	*229.83*	289.74	417.75	**222.39**	310.07
	ua	371.87	*326.74*	319.47	**298.33**	335.64	376.74	430.46	*363.81*
4 lu.C		*263.26*	341.69	**199.27**	259.40	326.85	*293.02*	317.60	449.19
4 sp.C		758.59	*592.73*	619.26	642.74	**569.48**	780.20	762.14	*627.22*
4 bt.C		322.58	*291.79*	**225.72**	232.30	267.85	305.67	299.00	*280.73*
4 ua.C		*316.93*	378.06	**297.95**	348.99	364.65	400.66	409.65	*358.03*

4.1 Server A

Note that the DIRECT memory placement with PINNED threads gets the best execution time (it is below 100), while INTERLEAVED memory and PINNED threads is not a good solution in this case. In Fig. 1(a) the results of using IMAR² with FREE memory placement are also shown and, in this case, the migrations do not improve, but actually decrease performance. This is due to the fact that IMAR² does not move memory, it depends on the OS for that, so it cannot reach as good results as the DIRECT memory with PINNED threads. Note that in this case the sp.C benchmark takes a longer time to execute, so it is favoured in the end by having the whole system for itself; both IMAR² and OS are able to take it into account and reach a similar end time. In Fig. 1(c), results with DIRECT memory are shown. In this case the OS does not migrate threads or memory

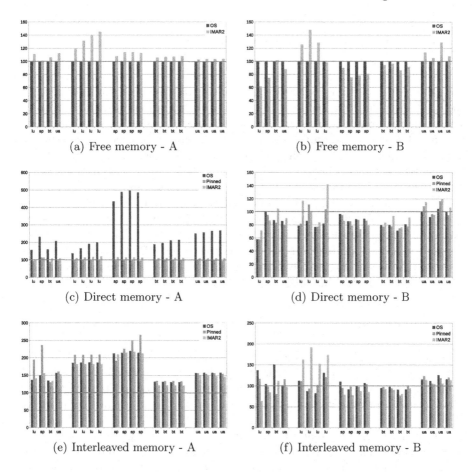

Fig. 1. Normalised execution times (in seconds, Y axis) against FREE - OS for all tests

taking into account that the benchmarks have their memory on just one node, so it results in worse performance; meanwhile the IMAR^2 migrations are able to move the threads to their correct location and performance does not suffer much. In Fig. 1(e), the results with INTERLEAVED memory are shown. Neither in this case is the OS able to fix the memory or thread placement, and results are not worse that leaving the OS alone; IMAR^2 migrations are able to improve the OS somewhat, but, since they cannot move the memory, the margin for improvement is low. In conclusion, in this system, while migrations may improve the OS in certain cases, the OS does a good work and there is little margin for improvement.

4.2   Server B

Figure 1(b) shows that global performance is greatly improved when different benchmarks run concurrently compared to the OS scheduling when memory policy is FREE. Improvements reach up to 38% the individual execution times, and up to 25% in total execution time. Note that, a migration strategy has more chances for improving performance when different processes are executed, as they may have different memory requirements. When a set of instances of the same benchmark is executed, results depend heavily on the behaviour of the code. As mentioned before, the influence of migrations is huge in **4 sp.C**, since memory latency is critical. The case for **4 bt.C** is similar, which improves too. For **4 lu.C** and **4 ua.C**, memory saturation makes almost impossible to improve the results, and even migrations cause a performance slowdown. When the memory is directly mapped to a node, see Fig. 1(d), OS outperforms the PINNED scheduling in many of the cases. Due to the work balance, OS mitigates the possible memory congestion caused when all the data is placed in a single memory node. Is must be noted that in this situation IMAR^2 improves the execution times of sp.C, the most memory intensive benchmark. Finally, when the INTERLEAVED strategy for memory is used, Fig. 1(f), IMAR^2 succeeds in achieving the best performance in the memory intensive benchmarks, thanks to a better thread placement through the cores of the server.

5   Conclusions

Thread and data allocation significantly influence the performance of modern computers, being this fact particularly true in NUMA systems. When the distribution of threads and data is far from being the optimum, the OS by itself is not able to optimise it. In this paper, a dynamic thread migration algorithm to deal with this issue is proposed. It is based on the optimisation of the operational intensity, the GIPS, and the memory latency, parameters that define the 3DyRM model. The proposed technique improves execution times when thread locality is poor and the OS is unable to improve thread placement in runtime.

In this paper, we define a product that combines the three 3DyRM parameters in a single value, which can be considered a fair representation of the whole performance of the system in terms of locality and affinity. To optimise this value, we propose a migration algorithm, named IMAR^2, based on a weighted lottery strategy. Hardware counters allow us to obtain information about the performance of each thread in the system in runtime with low overhead. IMAR^2 uses this information to quantify the 3DyRM parameters and then performs thread migration and allocation in runtime. Using benchmarks from the NPB-OMP, we evaluate IMAR^2 in a variety of scenarios. Results show that our algorithm improves execution time by up to 25% in realistic scenarios in terms of locality and affinity. Besides, only small performance losses were obtained in cases where the thread configuration was initially good. Rollbacks and changes in the time between migrations are mechanisms to adapt dynamically to the current behaviour of the system as a whole. These provide better results in cases where migrations are unnecessary, while still improving the performance in cases with low initial performance.

Several improvements might be considered as future work, like a precise measurement of FLOPS, including vector extensions, that could improve both performance estimation and migration decisions. Also, some modifications of the current objective function might be explored, like weighing its parameters or even testing different functions. Finally, other migration algorithms could be considered, maybe based on stochastic scheduling, optimisation techniques, or other state of the art approaches.

Acknowledgements. This work has received financial support from the Consellería de Cultura, Educación e Ordenación Universitaria (accreditation 2016-2019, ED431G/08 and reference competitive group 2019-2021, ED431C 2018/19) and the European Regional Development Fund (ERDF). It was also funded by the Ministerio de Economía, Industria y Competitividad within the project TIN2016-76373-P.

References

1. Adhianto, L., Banerjee, S., Fagan, M., et al.: HPCToolkit: tools for performance analysis of optimized parallel programs. Concurr. Comput.: Pract. Exp. **22**(6), 685–701 (2010). https://doi.org/10.1002/cpe.1553
2. Akiyama, S., Hirofuchi, T.: Quantitative evaluation of intel PEBS overhead for online system-noise analysis. In: Proceedings of the 7th International Workshop on Runtime and Operating Systems for Supercomputers ROSS 2017, ROSS 2017, pp. 3:1–3:8. ACM, New York (2017). https://doi.org/10.1145/3095770.3095773
3. Chasparis, G.C., Rossbory, M.: Efficient dynamic pinning of parallelized applications by distributed reinforcement learning. Int. J. Parallel Program. **47**(1), 24–38 (2017). https://doi.org/10.1007/s10766-017-0541-y
4. Cheung, A., Madden, S.: Performance profiling with EndoScope, an acquisitional software monitoring framework. Proc. VLDB Endow. **1**(1), 42–53 (2008). https://doi.org/10.14778/1453856.1453866
5. Cho, J.H., Wang, Y., Chen, R., Chan, K.S., Swami, A.: A survey on modeling and optimizing multi-objective systems. IEEE Commun. Surv. Tutor. **19**, 1867–1901 (2017). https://doi.org/10.1109/COMST.2017.2698366
6. Geimer, M., Wolf, F., Wylie, B.J.N., Ábrahám, E., Becker, D., Mohr, B.: The Scalasca performance toolset architecture. Concurr. Comput.: Pract. Exp. **22**(6), 702–719 (2010). https://doi.org/10.1002/cpe.1556
7. Goumas, G., Kourtis, K., Anastopoulos, N., Karakasis, V., Koziris, N.: Performance evaluation of the sparse matrix-vector multiplication on modern architectures. J. Supercomput. **50**(1), 36–77 (2009). https://doi.org/10.1007/s11227-008-0251-8
8. Intel Corp.: Intel 64 and IA-32 Architectures Software Developer Manuals (2017). https://software.intel.com/articles/intel-sdm. Accessed Nov 2019
9. Intel Developer Zone: Fluctuating FLOP count on Sandy Bridge (2013). http://software.intel.com/en-us/forums/topic/375320. Accessed Nov 2019
10. Jin, H., Frumkin, M., Yan, J.: The OpenMP implementation of NAS parallel benchmarks and its performance. Technical report, Technical Report NAS-99-011, NASA Ames Research Center (1999)
11. Ju, M., Jung, H., Che, H.: A performance analysis methodology for multicore, multithreaded processors. IEEE Trans. Comput. **63**(2), 276–289 (2014). https://doi.org/10.1109/TC.2012.223

12. Kleen, A.: A NUMA API for Linux. Novel Inc. (2005)
13. Lameter, C., et al.: NUMA (non-uniform memory access): an overview. ACM Queue **11**(7), 40 (2013). https://queue.acm.org/detail.cfm?id=2513149
14. Lorenzo, O.G., Pena, T.F., Cabaleiro, J.C., Pichel, J.C., Rivera, F.F.: 3DyRM: a dynamic roofline model including memory latency information. J. Supercomput. **70**(2), 696–708 (2014). https://doi.org/10.1007/s11227-014-1163-4
15. Lorenzo, O.G., Pena, T.F., Cabaleiro, J.C., Pichel, J.C., Rivera, F.F.: Multiobjective optimization technique based on monitoring information to increase the performance of thread migration on multicores. In: 2014 IEEE International Conference on Cluster Computing (CLUSTER), pp. 416–423. IEEE (2014). https://doi.org/10.1109/CLUSTER.2014.6968733
16. Rane, A., Stanzione, D.: Experiences in tuning performance of hybrid MPI/OpenMP applications on quad-core systems. In: Proceedings of 10th LCI International Conference on High-Performance Clustered Computing (2009)
17. Schulz, M., de Supinski, B.R.: PNMPI tools: a whole lot greater than the sum of their parts. In: Proceedings of the 2007 ACM/IEEE Conference on Supercomputing (2007). https://doi.org/10.1145/1362622.1362663
18. Williams, S., Waterman, A., Patterson, D.: Roofline: an insightful visual performance model for multicore architectures. Commun. ACM **52**(4), 65–76 (2009). https://doi.org/10.1145/1498765.1498785

Alea – Complex Job Scheduling Simulator

Dalibor Klusáček[1(✉)], Mehmet Soysal[2], and Frédéric Suter[3]

[1] CESNET a.l.e., Brno, Czech Republic
klusacek@cesnet.cz
[2] Steinbuch Centre for Computing, Karlsruhe Institute of Technology,
Karlsruhe, Germany
mehmet.soysal@kit.edu
[3] IN2P3 Computing Center/CNRS, Lyon-Villeurbanne, France
frederic.suter@cc.in2p3.fr

Abstract. Using large computer systems such as HPC clusters up to their full potential can be hard. Many problems and inefficiencies relate to the interactions of user workloads and system-level policies. These policies enable various setup choices of the resource management system (RMS) as well as the applied scheduling policy. While expert's assessment and well known best practices do their job when tuning the performance, there is usually plenty of room for further improvements, e.g., by considering more efficient system setups or even radically new scheduling policies. For such potentially damaging modifications it is very suitable to use some form of a simulator first, which allows for repeated evaluations of various setups in a fully controlled manner. This paper presents the latest improvements and advanced simulation capabilities of the *Alea* job scheduling simulator that has been actively developed for over 10 years now. We present both recently added advanced simulation capabilities as well as a set of real-life based case studies where Alea has been used to evaluate major modifications of real HPC and HTC systems.

Keywords: Alea · Simulation · Scheduling · HPC · HTC

1 Introduction

The actual performance of a real RMS depends on many variables that include the type (mix) of users' workloads (e.g., parallel vs. sequential jobs, short vs. long jobs), applied job scheduler and its scheduling algorithm (e.g., trivial First Come First Served (FCFS) or backfilling [14]) and also additional system configuration that typically defines job mapping to queues and their priorities and various operational limits (e.g., max. number of CPUs available to a given user). Therefore, designing a proper configuration is the most important, yet truly daunting process. Due to the complexity of the whole system even straightforward (conservative) changes in the configuration of the production system can have highly unexpected and often counterintuitive side effects that emerge from

© Springer Nature Switzerland AG 2020
R. Wyrzykowski et al. (Eds.): PPAM 2019, LNCS 12044, pp. 217–229, 2020.
https://doi.org/10.1007/978-3-030-43222-5_19

the mutual interplay of various policies and components of the RMS and scheduler [10]. Therefore, simulators that can emulate a particular production system and its configuration represent highly useful tools for both resource owners, system administrators and researchers in general.

Alea jobs scheduling simulator has been first introduced in 2007 as a basic simulator and underwent a major upgrade in 2010 [9] that mainly focused on improving the rather slow simulation speed and also introduced some visualization capabilities. In 2016, Alea was the first mainstream open source simulator to enable the use of so called *dynamically adapted workloads*, where the performance of the simulated scheduler directly influences the submission rates (arrival times) of jobs from the workload [22], providing an important step to mimic the natural user feedback to the system performance [12].

Since then, many new features have been implemented and the simulator has been successfully used for various purposes, both as a purely research tool as well as when testing new setups and new scheduling policies for production HPC and HTC systems. The main contribution of this paper is that (1) we describe recent improvements in the simulator, that allow for truly complex simulations that involve several detailed setups that correspond to typical real-life based scenarios, (2) we describe the recent speedup of the simulator that enables us to run truly large-scale simulations involving millions of jobs and thousands of nodes that complete in just a few hours, (3) we compare the performance of Alea with existing simulators, and (4) we provide several real-life based case studies where Alea has been used to develop and evaluate effects of major modifications of real HPC and HTC systems.

In Sect. 2 we provide a brief overview of existing related work. Next, Sect. 3 shows the current design of Alea and its major features and simulation capabilities. Section 4 presents several real-life examples demonstrating how Alea has been used in practice in order to improve the performance of production systems. Finally, we conclude the paper in Sect. 5.

2   Related Work

Throughout the years, there have been many grid, HPC and cloud simulators. In most cases, each such simulator falls into one of three main groups. The first group represents ad hoc simulators that are built from scratch. Those include, e.g., the recent AccaSim or Qsim. AccaSim is freely available library for Python, thus compatible with any major operating system, and executable on a wide range of computers thanks to its lightweight installation and light memory footprint [5]. Qsim is an event-driven scheduling simulator for Cobalt, which is an HPC job management suite supporting compute clusters of the IBM BlueGene series [21]. It is using exactly the same scheduling and job allocation schemes used (or proposed) for Cobalt and replays the job scheduling behavior using historic workloads analyzing how a new scheduling policy can affect system performance. Still, both simulators are somehow limited. Qsim is aiming primarily on BlueGene-like architectures, while AccaSim's capabilities (e.g., supported scheduling policies) are still rather limited as of late 2019[1].

Second group of simulators is typically using some underlying simulation toolkit, e.g., SimGrid, GridSim or CloudSim. This group is represented, e.g., by the recent Batsim, Simbatch or GSSIM [3]. Batsim is built on top of SimGrid [4]. It is made such that any event-based scheduling algorithm can be plugged to it and tested. Thus, it allows to compare various scheduling algorithms from different domains. Such schedulers must follow a text-based protocol to communicate with Batsim properly. In the paper on Batsim [4], this is demonstrated by using OAR resource manager's scheduler with the Batsim simulator [4]. This is certainly a very interesting feature adding to the realism of the simulations. Still, it is not very straightforward to use existing schedulers in this way as they are typically tightly coupled with the remaining parts of a given resource manager and cannot be easily used in a standalone fashion. Simbatch and GSSIM [3] were using SimGrid and Gridsim respectively, but their development is currently discontinued for many years.

Finally, the last group typically uses some real-life RMS executed in a simulation mode. For example, the ScSF simulator [16] emulates a real system by using Slurm Workload Manager inside its core to realistically mimic the real RMS. ScSF extends an existing Slurm Simulator [18], improving its internal synchronization to speed up its execution. Also, it adds the capability to generate synthetic workloads. Similar "simulation mode" was supported in Moab in the past[2] but has been discontinued in the recent versions. In all cases, simulators using a real RMS cannot process workload as quickly as the simulators from the first two groups. This is caused by the fact that these simulators must follow the complex timing model of a real RMS (see Sect. 3.4).

Alea simulator, which will be thoroughly described in the next section, represents the second group of simulators using an underlying simulation toolkit. The major weakness of Alea is that it cannot use an existing scheduler and/or RMS. Instead, the RMS/scheduler must be simulated using Alea and GridSim. While this fact can be somehow limiting in certain cases, Alea offers a large set of features that mimic the functionality of real schedulers (see Sect. 3). At the same time, it allows to simulate large workloads and big systems in a very competitive time (see Sect. 3.4) while remaining fully deterministic. This is not the case for simulators using real RMS that are subject to varying "system jitter" from the used RMS [18]. The aforementioned list of existing simulators is not exhaustive and more details can be found in [5,9].

3   Architecture and Major Functionality

Alea is platform-independent event-driven discrete time simulator written in Java. It is built on the top of the GridSim simulation toolkit [20]. GridSim provides the basic functionality to model various entities in a simulated computing

---

[1] https://accasim.readthedocs.io/.
[2] http://docs.adaptivecomputing.com/mwm/archive/6-0/2.5initialtesting.php.

system, as well as methods to handle the simulation events. The behavior of the simulator is driven by an event-passing protocol. For each simulated event—such as job arrival or completion—one message defining this event is created. It contains the identifier of the message recipient, the type of the event, the time when the event will occur and the message data. Alea extends this basic Grid-Sim's functionality and provides a model allowing for detailed simulation of the whole scheduling process in a typical HPC/HTC system. To do that, Alea either extends existing GridSim classes (e.g., `GridResource` or `AllocationPolicy`) or it provides new classes on its own, especially the core `Scheduler` class and classes responsible for data parsing and collection/visualization of simulation results. Figure 1 shows the overall scheme of Alea simulator, where boxes denote major functional parts and arrows express communication and/or data exchange within the simulator. The blue color denotes recently added or heavily upgraded components of the simulator.

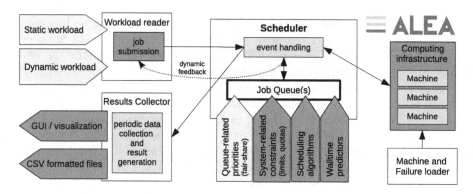

Fig. 1. Main parts of the Alea simulator (blue color denotes new functionality). (Color figure online)

The main part of the simulator is the centralized job scheduler. The scheduler manages the communication with other parts of the simulator. Also, it maintains important data structures such as job queue(s). Job scheduling decisions are performed by scheduling algorithms that can be easily added using existing simple interfaces. Furthermore, scheduling process can be further influenced by using additional system policies, e.g., the fair-sharing policy which dynamically prioritizes job queue(s). Also, various limits that further refine how various job classes are handled are supported by Alea. Additional parts simulate the actual computing infrastructure, including the emulation of machine failures/restarts. Workload readers are used to feed the simulator with input data about jobs being executed and the simulator also provides means for visualization and generation of simulation outputs. Alea is freely available at GitHub [1].

The primary goal of our job scheduling simulator is to allow for realistic evaluation of new scheduling policies or new setups of computing systems. For this purpose, it is necessary to model all important features that have significant impact on the performance of the system. Our own "hands on" experience from operating production systems have taught us that many promising "theoretical" works based on simulations are not usable in practice, due to the overly simplified nature of performed simulations. Often, researchers focus solely on particular scheduling algorithm while ignoring additional system-related constraints and policies. However, production systems use literally dozens of additional parameters, rules and policies that significantly influence the scheduler's decisions and thus the performance of the system [10,17]. Therefore, following subsections provide an overview of the advanced simulation capabilities that make Alea a very useful tool for detailed simulations of actual systems.

3.1   Detailed System Simulation Capabilities

As we have observed in practice, system performance can be largely affected by the interactions of various components and parameters of an actual RMS. While their nature or scope can be basic and limited, they can easily turn a well functioning system into a troublesome one. Therefore, the simulator should be able to mimic these features within the simulation. These features include:

– queues and their priorities, constraints and various limits
– quotas limiting user CPU usage
– mechanisms to calculate job priorities such as fair-share
– common scheduling algorithms aware of aforementioned features

Queues, Limits and Quotas. First of all, Alea allows to specify the number of job queues, their names, priorities and queue-related constraints such as the maximum number of CPUs that can be used by jobs from that queue at any given moment. Multiple queues are common in systems with heterogeneous workloads. Here, system resources are usually partitioned into several (sometimes overlapping) pools, where each pool has a corresponding queue. Users' workloads (jobs) are then mapped to these queues. Queue limits then avoid potentially dangerous situations such as saturation of the whole system—either with jobs from a single user, or with a single class of jobs [7]. For example, it would be very unwise to fill the whole system with long running jobs as this would cause huge wait times for shorter jobs. Also users and/or groups are often subject to a upper bound on the amount of resources they can use simultaneously. For this purpose, Alea now provides CPU quotas, that guarantee that a user/group will not exceed the corresponding maximum allowed share of resources [2].

Fair-Sharing. Production systems—instead of default job arrival order—often use some priority mechanism to dynamically prioritize system users. This is typically done by fair-sharing. We provide several variants of *fair-sharing mechanisms* that are used to prioritize jobs (users) within queue(s) in order to guarantee user-to-user fairness. Fair-share mechanism dynamically adjusts job/user priorities such that the use of system resources is fairly balanced among the users [7]. We support both basic fair-sharing mechanisms that only reflect CPU usage as well as more complex multi-resource implementations[3] which also reflect memory consumption.

Scheduling Algorithms. Scheduling algorithms play a critical role in RMS. Alea supports all mainstream algorithms that can be typically observed in practice, starting with trivial FCFS, Shortest Job First and Earliest Deadline First and continuing to more efficient solutions such as EASY backfilling or Conservative backfilling [14]. Alea also supports *schedule optimization methods*, that can be used to further improve initial job schedules as prepared by, e.g., the Conservative backfilling policy. Our optimization methods are based on metaheuristics and can use various objective functions to guide the metaheuristic toward improved schedule [8]. Importantly, in the recent release we provide several job walltime predictors, that can automatically refine (inaccurate) user-provided walltime estimates in order to improve the precision of scheduling decisions.

3.2   Dynamic Workloads

There is one more part which plays a significant role in job scheduling—the workload being processed by the system or the simulator. Alea supports two ways how workload can be fed into the simulator. First, it uses traditional "workload replay" mode, where jobs are submitted based on a historic workload description file (log) and their arrival times are based on the original timestamps as recorded in the log. Alternatively, Alea allows to use so called dynamic workload adaptation, where job arrival times are not fixed but can change throughout the course of the simulation, depending on the scheduler's performance. For this purpose, Alea provides a feedback loop that communicates with the workload reader and informs it upon each job completion. Using this data, the workload reader can either speed up or postpone job submissions for simulated users. This complex behavior mimics real world experience, where users react to the performance of the scheduler. In other words, real-life job arrival times are always correlated to the "user experience", thus it is unrealistic to use plain "workload replay" mode, because the results will be somehow skewed by the "embedded" influence of the original scheduler that is captured in the historic workload log, i.e., in the job arrival time pattern. Alea's implementation is based on the work of Zakay and Feitelson [22], but it also allows to write your own workload adaptation engine, having different job submission adaptation logic.

3.3   Simulation Speed

Since the start of Alea project, simulation speed was our second most important goal right after the capabilities of our simulator. During the years, Alea

---

[3] For example, we support Dominant Resource Fairness (DRF) inspired fair-share [6].

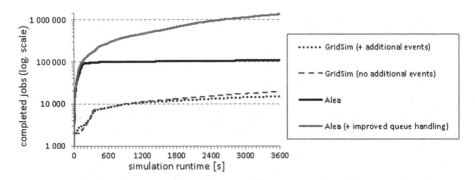

Fig. 2. Number of completed jobs (log. scale) during 1 hour-long simulation using different implementations of `SpaceShared` policy and (un)optimized queue handling.

has introduced several improvements into the GridSim's event-driven simulation model that significantly speed up the simulation. Most changes relate to the way job execution is modeled in the classes that implement job allocation policy on a modeled physical system (see, e.g., GridSim's `SpaceShared` class). As originally designed, this model was not very time-efficient. Upon each start of a job j, an internal event was generated that was scheduled to be delivered at the time $T_{compl}(j)$, which is the time when such job would complete[4]. Although this event at $T_{compl}(j)$ only corresponds to that job j, GridSim would always scan all currently executed jobs to check whether those are completed or not. Obviously, this was not very time efficient way how to proceed with a simulation. Moreover, with each such check GridSim would also generate one additional internal event to trigger a similar check (delayed by a predefined time constant) to further assure that no jobs are "forgotten" by the engine. However, this additional event generator was producing exponential-like increase of events that the GridSim core had to handle, slowing down the simulation extremely. While these inefficiencies are tolerable when dealing with small systems (hundreds of CPUs and few thousands of jobs), they became a real show-stopper for large simulations involving tens of thousands of CPUs and millions of jobs.

Therefore, in this new edition of Alea we have simplified the whole job processing model such that each job now only needs one internal event to be processed correctly. This did not change the behavior of the `SpaceShared` policy, but it introduced a huge speedup of the whole simulator. Also, we have improved the speed of scheduling algorithms. Simulations that struggle with large job queue sizes (plenty of waiting jobs) are often slowed down by the scheduling algorithm which repeatedly traverses long job queues, trying to schedule a new job. With long queues, this may slow down the simulation significantly, especially when the algorithm itself is not a trivial one. Therefore, we have introduced a more efficient *queue handling mechanism* which—based on user specified parameters—limits

---

[4] $T_{compl}(j) = T_{current} + T_{runtime}(j)$.

the number of jobs that are checked at each scheduling run. This modification brought another huge improvement.

Figure 2 shows an example of the speedup obtained by our techniques. It shows the number of completed jobs (in log. scale) that were simulated during one hour. This experiment involved large system with over 33K CPU cores and many peaks in the job queue that reached up to 5 K of waiting jobs. The results of our optimized event-processing mechanism and the queue handling mechanism are compared to the original GridSim's implementation, with the "exponential" event generator either turned on (denoted as "Gridsim + additional events") or off ("no additional events"). Clearly, there is a huge difference when the optimized event-processing mechanism is introduced (denoted as "Alea"). Even bigger improvement is reached once the more efficient queue handling mechanism is used ("Alea + improved queue handling"). This effect is amplified by the fact that this experiment often experienced very long queue of waiting jobs.

3.4   Simulation Throughput and Speedup Comparison

To give the speed of our simulator into a context, we have studied the reported speeds of different simulators and created a simple comparison of their performance. We have used the recent published data about Slurm Simulator [18], Batsim [13] and ScSF [16]. If possible, we show both the achieved *speedup* as well as the *throughput* of the simulator. The speedup is the ratio of the original makespan[5] to the wall-clock time requested by the simulator to finish the experiment. Throughput is measured as the average number of jobs simulated (completed) in one minute. Since the speedup and throughput also depends on the "size" of the experiment [18], we report the total number of CPU cores and jobs being simulated (if available). The results are shown in Table 1 and show the impressive speed and throughput of Alea. While Batsim reports a very nice speedup, it must be noted that this result was achieved on a very small problem instance (800 jobs and 32 cores) while Alea's results were achieved in a truly large setup (2,7M jobs and 33 K cores). Further comparisons (featuring Alea, AccaSim and Batsim) can be found in the AccaSim report [5].

Table 1. Throughput and speedup of various simulators.

	Jobs	Cores	Makespan (h)	Runtime (s)	Speedup	Throughput (jobs/min)
Slurm Sim	65,000	7,912	571	15,866	130	246
ScSF	N/A	322	168	43,200	14	N/A
Batsim	800	32	4	30	400	1,600
Alea	2,669,401	33,456	744	10,800	248	14,830

---

[5] Makespan denotes the time needed to process the workload in a real system.

3.5   Visualization

Alea offers `Visualizator` class that provides crucial methods to display graphical outputs during a simulation. Several metrics and outputs that are generally useful, e.g., for debugging purposes are available by default, including the visualization of created job schedule and several popular objectives. An example of such graphical output is captured in Fig. 3, which shows the average system utilization, number of waiting and running jobs, average cluster utilization and the number of used, requested and available CPUs.

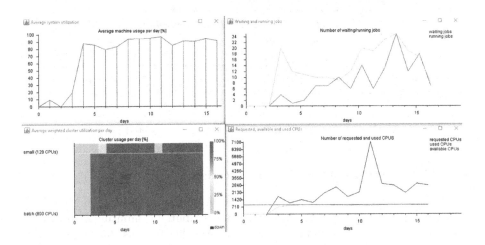

Fig. 3. Alea's visualization output showing various metrics.

Figure 4 shows the newly available visual representation of a job schedule as constructed by the scheduler. This feature is very useful especially for debugging purposes or when tuning a new algorithm. However, for larger systems the schedule cannot be reasonably displayed due to the screen resolution limitation. In this (cropped) example the vertical y-axis shows 112 CPUs of two clusters, and the x-axis denotes the planned start/completion times. The time is not to scale (linear) in order to save space. Instead, the horizontal axis uses fixed lengths between two consecutive events. An event represents either planned job start or job completion. Using this trick, the schedule can typically display rather long schedules (several days) while fitting within the limits of one screen[6].

4   Notable Usages

In this section we present four examples where Alea has been used to model an existing system and analyze the impact of new scheduling approaches. Notably, the two latter examples (Sects. 4.3 and 4.4) were achieved with the recently upgraded Alea described in this paper.

---

[6] In this case the schedule shows job-to-CPU mapping covering ∼3 days.

Fig. 4. Alea's new visualization feature showing constructed job schedule.

4.1 MetaCentrum Queue Reconfiguration

The first example is a major queue reconfiguration that took place in MetaCentrum, which is the largest Czech provider of distributed computing facilities for academic and scientific purposes. In this case, Alea has been used to evaluate the impact of new queue setup, where the goal has been to increase fairness, system utilization and wait times across different classes of jobs. Existing conservative setup with 3 major queues (short, normal and long jobs) and rather constraining limits concerning the maximum allowed number of simultaneously running long jobs has been replaced with an improved design introducing new, more fine-grained queues with more generous limits. The promising effect observed in the simulations was then also validated in practice. With the new setup being introduced in January 2014, the overall CPU utilization has increased by 43.2% while the average wait time has decreased by 17.9% (4.4 vs. 3.6 h) [11].

4.2 Plan-Based Scheduler with Metaheuristic Optimization

In July 2014, CERIT Scientific Cloud started to use a unique Torque-compatible scheduler that—instead of queuing—used planning and metaheuristics to build and optimize job schedules. This new planning-based scheduler has been first thoroughly modeled and refined in Alea and then remained in operation until 2017. It was a successful scheduler as it increased the avg. CPU utilization by 9.3% while decreasing the avg. wait time and the avg. bounded slowdown by 36.7% and 79.4%, respectively [8].

4.3 Scheduling with Advance Data Staging

The I/O subsystem is an increasing storage bottleneck on HPC Systems. The ADA-FS project [15] tries to close the bottleneck with deploying an on-demand file system and staging the data in advance to the allocated nodes. In a recent paper [19], Alea has been used to study the suitability of current mainstream scheduling algorithms such as FCFS and backfilling to accurately predict target nodes where a waiting job will be executed. Such a prediction is crucial when data is staged in advance or private file system is deployed prior to actual computation (while a job is still waiting). In this paper, we have demonstrated that current schedulers relying on inaccurate user-provided runtime estimates are unable to make reliable long-term predictions and even short-term predictions (less than 10 min ahead) are not possible for large fractions of jobs (~50% of jobs).

4.4 Improving Fairness in Large HTC System

In 2019, Alea has been used to model and then reconfigure queue and quota setup in a large HTC system. This system is shared by two different workloads— a local user workload and a grid workload that comes from LHC experiments. The motivation was to increase the fairness toward local users who often have to wait much longer than those grid-originating jobs (roughly twice as long, on average). In this work, the recently improved simulation speed of Alea was mostly important, since the HTC system is rather large (33,456 cores), processing lots of jobs each month (~2.7 millions). Using Alea, we were able to model the system and evaluate new setups for the system's queues and the per-group CPU quotas. This new setup allowed for improved fairness for local users, by better balancing their wait times with the wait times of grid-originating jobs [2].

5 Conclusion and Future Work

This paper has presented the recently upgraded complex job scheduling simulator *Alea*. We have demonstrated its capabilities and usefulness using real-life examples. Importantly, we have shown that the simulator is capable to simulate large systems and execute large workloads in an acceptable time frame. Alea can be freely obtained at GitHub [1] under the LGPL license.

Acknowledgments. We acknowledge the support and computational resources provided by the MetaCentrum under the program LM2015042, and the support provided by the project Reg. No. CZ.02.1.01/0.0/0.0/16_013/0001797 co-funded by the Ministry of Education, Youth and Sports of the Czech Republic.

References

1. Alea job scheduling simulator, April 2019. https://github.com/aleasimulator
2. Azevedo, F., Klusáček, D., Suter, F.: Improving fairness in a large scale HTC system through workload analysis and simulation. In: Yahyapour, R. (ed.) Euro-Par 2019. LNCS, vol. 11725, pp. 129–141. Springer, Cham (2019). https://doi.org/10.1007/978-3-030-29400-7_10
3. Bak, S., Krystek, M., Kurowski, K., Oleksiak, A., Piatek, W., Weglarz, J.: GSSIM - a tool for distributed computing experiments. Sci. Program. 19(4), 231–251 (2011)
4. Dutot, P.-F., Mercier, M., Poquet, M., Richard, O.: Batsim: a realistic language-independent resources and jobs management systems simulator. In: Desai, N., Cirne, W. (eds.) JSSPP 2015-2016. LNCS, vol. 10353, pp. 178–197. Springer, Cham (2017). https://doi.org/10.1007/978-3-319-61756-5_10
5. Galleguillos, C., Kiziltan, Z., Netti, A., Soto, R.: AccaSim: a customizable workload management simulator for job dispatching research in HPC systems. arXiv e-prints arXiv:1806.06728 (2018)
6. Ghodsi, A., Zaharia, M., Hindman, B., Konwinski, A., Shenker, S., Stoica, I.: Dominant resource fairness: fair allocation of multiple resource types. In: 8th USENIX Symposium on Networked Systems Design and Implementation (2011)
7. Jackson, D., Snell, Q., Clement, M.: Core algorithms of the maui scheduler. In: Feitelson, D.G., Rudolph, L. (eds.) JSSPP 2001. LNCS, vol. 2221, pp. 87–102. Springer, Heidelberg (2001). https://doi.org/10.1007/3-540-45540-X_6
8. Klusáček, D., Chlumský, V.: Planning and metaheuristic optimization in production job scheduler. In: Desai, N., Cirne, W. (eds.) JSSPP 2015-2016. LNCS, vol. 10353, pp. 198–216. Springer, Cham (2017). https://doi.org/10.1007/978-3-319-61756-5_11
9. Klusáček, D., Rudová, H.: Alea 2 - job scheduling simulator. In: 3rd International ICST Conference on Simulation Tools and Technique, ICST (2010)
10. Klusáček, D., Tóth, Š.: On interactions among scheduling policies: finding efficient queue setup using high-resolution simulations. In: Silva, F., Dutra, I., Santos Costa, V. (eds.) Euro-Par 2014. LNCS, vol. 8632, pp. 138–149. Springer, Cham (2014). https://doi.org/10.1007/978-3-319-09873-9_12
11. Klusáček, D., Tóth, Š., Podolníková, G.: Real-life experience with major reconfiguration of job scheduling system. In: Desai, N., Cirne, W. (eds.) JSSPP 2015-2016. LNCS, vol. 10353, pp. 83–101. Springer, Cham (2017). https://doi.org/10.1007/978-3-319-61756-5_5
12. Klusáček, D., Tóth, Š., Podolníková, G.: Complex job scheduling simulations with Alea 4. In: Ninth EAI International Conference on Simulation Tools and Techniques (SimuTools 2016), pp. 124–129. ACM (2016)
13. Mercier, M.: Batsim JSSPP presentation (2016). https://gitlab.inria.fr/batsim/batsim/blob/master/publications/Batsim_JSSPP_2016.pdf
14. Mu'alem, A.W., Feitelson, D.G.: Utilization, predictability, workloads, and user runtime estimates in scheduling the IBM SP2 with backfilling. IEEE Trans. Parallel Distrib. Syst. 12(6), 529–543 (2001)
15. Oeste, S., Kluge, M., Soysal, M., Streit, A., Vef, M.-A., Brinkmann, A.: Exploring opportunities for job-temporal file systems with ADA-FS. In: 1st Joint International Workshop on Parallel Data Storage and Data Intensive Scalable Computing Systems (2016)

16. Rodrigo, G.P., Elmroth, E., Östberg, P.-O., Ramakrishnan, L.: ScSF: a scheduling simulation framework. In: Klusáček, D., Cirne, W., Desai, N. (eds.) JSSPP 2017. LNCS, vol. 10773, pp. 152–173. Springer, Cham (2018). https://doi.org/10.1007/978-3-319-77398-8_9
17. Schwiegelshohn, U.: How to design a job scheduling algorithm. In: Cirne, W., Desai, N. (eds.) JSSPP 2014. LNCS, vol. 8828, pp. 147–167. Springer, Cham (2015). https://doi.org/10.1007/978-3-319-15789-4_9
18. Simakov, N.A., et al.: A slurm simulator: implementation and parametric analysis. In: Jarvis, S., Wright, S., Hammond, S. (eds.) PMBS 2017. LNCS, vol. 10724, pp. 197–217. Springer, Cham (2018). https://doi.org/10.1007/978-3-319-72971-8_10
19. Soysal, M., Berghoff, M., Klusáček, D., Streit, A.: On the quality of wall time estimates for resource allocation prediction. In: ICPP 2019 Proceedings of the 48th International Conference on Parallel Processing: Workshops. ACM (2019)
20. Sulistio, A., Cibej, U., Venugopal, S., Robic, B., Buyya, R.: A toolkit for modelling and simulating data Grids: an extension to GridSim. Concurr. Comput.: Pract. Exp. 20(13), 1591–1609 (2008)
21. Tang, W.: Qsim (2019). https://trac.mcs.anl.gov/projects/cobalt/wiki/qsim
22. Zakay, N., Feitelson, D.G.: Preserving user behavior characteristics in trace-based simulation of parallel job scheduling. In: 22nd Modeling, Analysis & Simulation of Computer and Telecommunication Systems (MASCOTS), pp. 51–60 (2014)

Makespan Minimization in Data Gathering Networks with Dataset Release Times

Joanna Berlińska[✉]

Faculty of Mathematics and Computer Science, Adam Mickiewicz University,
Uniwersytetu Poznańskiego 4, 61-614 Poznań, Poland
Joanna.Berlinska@amu.edu.pl

Abstract. In this work, we analyze scheduling in a star data gathering network. Each worker node produces a dataset of known size at a possibly different time. The datasets have to be passed to the base station for further processing. A dataset can be transferred in many separate pieces, but each sent message incurs additional time overhead. The scheduling problem is to organize the communication in the network so that the total time of data gathering and processing is as short as possible. We show that this problem is strongly NP-hard, and propose a polynomial-time 2-approximation algorithm for solving it. Computational experiments show that the algorithm delivers high quality solutions.

Keywords: Scheduling · Data gathering networks · Release times · Flow shop · Preemption penalties

1 Introduction

Gathering data is an important stage of many applications. Complex computations are often executed in distributed systems, and the scattered results have to be collected and merged. Measurement data acquired by wireless sensor networks usually cannot be processed by the sensors, due to their limited resources. Therefore, they have to be passed to a base station, which has sufficient computing power, energy etc. Thus, efficient scheduling of the data gathering process may be an important factor influencing the performance of the whole application.

Scheduling in data gathering wireless sensor networks was analyzed on the grounds of divisible load theory in [8,17]. The studied problem was to assign the amounts of measurements to the workers, and organize the data transfers, so as to minimize the total application running time. Maximizing the lifetime of a network whose worker nodes have limited memory was studied in [1]. Scheduling algorithms for networks with limited base station memory were proposed in [5]. Data gathering networks whose nodes can compress data before transferring them were analyzed in [2,15,16]. Minimizing the maximum dataset lateness in networks with dataset release times was considered in [3,4].

© Springer Nature Switzerland AG 2020
R. Wyrzykowski et al. (Eds.): PPAM 2019, LNCS 12044, pp. 230–241, 2020.
https://doi.org/10.1007/978-3-030-43222-5_20

In this paper, we analyze gathering the results of computations running in parallel in a star network. Each of the worker nodes releases a dataset containing the obtained results at a given time. The datasets have to be transferred to the base station for further processing. Communication preemptions are allowed, but starting each message requires a startup time. Thus, transferring a dataset in many separate pieces takes longer than sending it in a single message. At most one worker can communicate with the base station at a time. Therefore, the communication network can be seen as a single machine. For each dataset, two operations have to be executed: first, the dataset has to be transferred over the communication network, and then it has to be processed by the base station. Hence, the network works in a two-machine flow shop mode. Our goal is to minimize the total data gathering and processing time.

Makespan minimization in a two-machine flow shop with job release times is known to be strongly NP-hard both in the non-preemptive [12] and in the preemptive [7] scenario. Algorithms for the non-preemptive version of the problem were proposed, e.g., in [11,18,20]. However, the preemptive variant has not received much attention so far. Scheduling with preemption penalties was studied in the context of a single machine in [13,14], parallel machines in [19], and data gathering networks in [3].

The rest of this paper is organized as follows. In Sect. 2 we present the network model and formulate the scheduling problem. We analyze its complexity in Sect. 3. In Sect. 4 we propose an algorithm for solving our problem, and prove that it delivers a 2-approximation of the optimum. The average performance of the algorithm is tested by means of computational experiments in Sect. 5. The last section is dedicated to conclusions.

2   Problem Formulation

We study a data gathering network consisting of m worker nodes $P_1, \ldots, P_m$, and a single base station. Node P_j generates dataset D_j of size α_j at time r_j, where $r_1 \leq r_2 \leq \cdots \leq r_m$. Each dataset has to be sent to the base station for processing, and at most one dataset can be transferred at a time. The communication capabilities of node P_j are described by communication startup S_j and unit communication cost C_j. Thus, a message of size x from worker node P_j to the base station is sent in time $S_j + C_j x$. Hence, although communication preemptions are allowed, each additional message used for sending dataset D_j increases its total transfer time by S_j. After the whole dataset arrives at the base station, it has to be processed, which takes time $A\alpha_j$. At most one dataset can be processed at a time, but the base station can simultaneously process one dataset and receive another one. Our goal is to organize the dataset transfers so as to minimize the total time T of gathering and processing the data. To the best of our knowledge, this problem was not studied in the earlier literature.

For a fixed communication schedule, the order of processing the datasets and possible processing preemptions do not affect the makespan T, as long as no unnecessary idle times appear. Therefore, we assume without loss of generality that the datasets are processed in the order in which they were received by the base station.

3 Computational Complexity

In this section, we argue that the analyzed scheduling problem is strongly NP-hard, and indicate its special cases solvable in polynomial time.

Proposition 1. *Makespan minimization in data gathering networks with dataset release times is strongly NP-hard, even if there are no preemption penalties, i.e. $S_j = 0$ for $j = 1, \ldots, m$.*

Sketch of Proof. Let us analyze the proof of strong NP-hardness of problem $F2|pmtn, r_j|C_{max}$ given in [7]. The proof is achieved by a reduction from the 3-Partition problem. The constructed instance of problem $F2|pmtn, r_j|C_{max}$ contains exactly one job i whose second operation takes time $p_{2i} = 0$. The release time of job i is $r_i = T - p_{1i}$, where T is the required schedule length, and p_{1i} is the execution time of the first operation of job i. Thus, in order to obtain a feasible schedule, job i has to be scheduled as the last job on the first machine, in time interval $[r_i, r_i + p_{1i})$. Note that if we set p_{2i} to an arbitrary positive number, and increase the schedule length to $T' = T + p_{2i}$, then job i has to be executed in interval $[r_i, r_i + p_{1i})$ on the first machine, and in interval $[r_i + p_{1i}, T')$ on the second machine. Moreover, it has to be the last job executed on the first machine, because all other jobs have non-zero second operation execution times. Thus, all the considerations in the proof remain valid, and in consequence, problem $F2|pmtn, r_j, p_{2j} \neq 0|C_{max}$ is strongly NP-hard.

For any instance of problem $F2|pmtn, r_j, p_{2j} \neq 0|C_{max}$, we can construct an equivalent instance of our problem by taking $A = 1$, $\alpha_j = p_{2j}$, $S_j = 0$, $C_j = \frac{p_{1j}}{p_{2j}}$ for $j = 1, \ldots, m$. Hence, our scheduling problem is also strongly NP-hard. □

If $A = 0$ or if $S_j = C_j = 0$ for $j = 1, \ldots, m$, our problem reduces to a single-machine scheduling problem $1|r_j|C_{max}$. Therefore, it can be solved in $O(m \log m)$ time by sending and processing the datasets in the order of nondecreasing release times. If $r_j = r$ for $j = 1, \ldots, m$, then communication preemptions are not necessary, and our problem becomes equivalent to $F2|pmtn|C_{max}$, which can be solved in $O(m \log m)$ time by Johnson's algorithm [10]. Finally, if $S_j = 0$ and $C_j \leq A$ for $j = 1, \ldots, m$, then our problem becomes a special case of problem $F2|1\text{-}min, pmtn, r_j|C_{max}$, and can be solved in $O(m \log m)$ time by the preemptive shortest remaining transfer time rule [6].

4 Algorithm

Before proposing an algorithm for solving our problem, we will show that there exist instances such that in each optimum solution, a communication is preempted at a time when no dataset is released.

Proposition 2. *An optimum schedule for the analyzed data gathering problem may require preempting a communication at a moment when no dataset is released.*

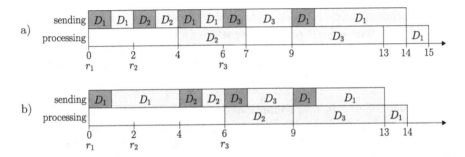

Fig. 1. Schedule structures for the proof of Proposition 2. Dark gray fields depict communication startup time, light gray fields depict data transfer and processing. (a) The shortest possible schedule under the assumption that communication preemptions take place only at dataset release times, (b) the optimum schedule.

Proof. Let $m = 3$, $A = 1$, $(\alpha_1, r_1, S_1, C_1) = (1, 0, 1, 6)$, $(\alpha_2, r_2, S_2, C_2) = (3, 2, 1, \frac{1}{3})$ and $(\alpha_3, r_3, S_3, C_3) = (4, 6, 1, \frac{1}{2})$. It is easy to check that if communication preemptions are allowed only at dataset release times, the minimum possible schedule length is 15. Still, the optimum makespan 14 can be achieved by preempting the transfer of D_1 at time 4, when no dataset is released (see Fig. 1). □

Proposition 2 shows the main difficulty in constructing an exact algorithm for our problem. Contrarily to the case of similar problems with zero startup times, we cannot limit ourselves to generating schedules with preemptions taking place only at dataset release times. Moreover, it is not known what additional time points should be taken into account. Therefore, we propose an approximation algorithm, which only preempts transfers at dataset release times.

For each $j = 1, \ldots, m$, and any $t > 0$, we define a job $K_j(t) = (p_{1j}(t), p_{2j}(t))$ corresponding to dataset D_j at time t in a given (partial) schedule. If dataset D_j is not being transferred at time t, and the part of D_j which has not yet been sent has size $\alpha'_j > 0$, then $p_{1j}(t) = S_j + C_j\alpha'_j$. If $\alpha'_j = 0$, then $p_{1j}(t) = 0$. If dataset D_j is being sent at time t, then $p_{1j}(t)$ is the time left to complete the transfer. Moreover, $p_{2j}(t) = A\alpha''_j$, where α''_j is the size of the part of dataset D_j which has not been processed by time t. Thus, $p_{1j}(t)$ and $p_{2j}(t)$ are the minimum times necessary to complete the transfer and processing of D_j, respectively.

The idea of our algorithm is based on Johnson's algorithm for problem $F2||C_{max}$ [10], which, given a set of jobs $K_j = (p_{1j}, p_{2j})$, arranges first the jobs with $p_{1j} \le p_{2j}$ in non-decreasing order of p_{1j}, and then the remaining jobs in non-increasing order of p_{2j}. Every time a dataset D_j is released during the transfer of another dataset D_i, we consider a possible preemption if $K_j(r_j)$ precedes $K_i(r_j)$ in Johnson's order. Since each preemption of D_i transfer extends the total communication time by S_i, we allow it only if this cost is small in comparison to the time until completing the transfer of D_i. The maximum acceptable cost is controlled by parameter $\gamma \in [0, \infty]$. Precisely, our algorithm, which will be called $J(\gamma)$, consists in the following scheduling rules.

1. If the communication network is idle at time t and some datasets are available (i.e. already released but not fully transferred), we start sending an available dataset D_j corresponding to job $K_j(t)$ chosen by Johnson's rule.
2. If dataset D_i is being sent at time r_j, then D_i is preempted by D_j, if $K_j(r_j)$ precedes $K_i(r_j)$ according to Johnson's rule, and $S_i < \gamma p_{1i}(r_j)$.

Note that for $\gamma = 0$, no preemptions are allowed in our algorithm. On the contrary, if $\gamma = \infty$, then our algorithm always follows the preemptive Johnson's rule. Since algorithm $J(\gamma)$ can preempt communications only at dataset release times, the maximum number of preemptions is $m - 1$, and the total number of sent messages does not exceed $2m - 1$. Hence, using a priority queue, the algorithm can be implemented to run in $O(m \log m)$ time.

In the remainder of this section, we analyze theoretical performance guarantees of our algorithm. The optimum schedule length is denoted by T^*. For a given set of jobs $\mathcal{A}$, we denote by $T_J(\mathcal{A})$ the makespan obtained by scheduling them using Johnson's algorithm, assuming that they are all available at time 0. Note that for any $t \geq 0$, and any $\mathcal{A} \subset \{K_j(t) : 1 \leq j \leq m\}$, we have

$$T_J(\mathcal{A}) \leq T^*. \tag{1}$$

Proposition 3. *For any $\gamma \in [0, \infty]$, $J(\gamma)$ is a 2-approximation algorithm for the analyzed scheduling problem.*

Proof. Let T be the makespan obtained by algorithm $J(\gamma)$. We will analyze the following two cases.

Case 1: There exists a dataset D_i whose transfer starts at time r_m. Hence, the set of jobs $\mathcal{A} = \{K_j(r_m) : D_j \text{ is not yet fully processed at time } r_m\}$ will be scheduled without preemptions, according to Johnson's rule, starting at time r_m. Thus, $T - r_m = T_J(\mathcal{A})$, and since $r_m \leq T^*$, we obtain by (1) that

$$T = r_m + T_J(\mathcal{A}) \leq 2T^*. \tag{2}$$

Case 2: No dataset transfer starts at time r_m. In other words, there exists a dataset D_i whose last part starts being sent at time $t_0 < r_m$, and arrives at the base station at time $t_i > r_m$. We divide the set of jobs $\{K_j(t_0) : j \neq i, D_j \text{ has not yet been sent by time } t_0\}$ into two subsets. Subset $\mathcal{A}$ contains the jobs which would precede job $K_i(t_0)$ according to Johnson's rule, and subset $\mathcal{B}$ contains the remaining jobs. Note that for any $K_j(t_0) \in \mathcal{A}$, we have $r_j > t_0$, as otherwise the transfer of D_j would precede sending D_i. Let $\mathcal{C} = \{K_j(t_0) : D_j \text{ has been sent but not processed by time } t_0\}$. Algorithm $J(\gamma)$ schedules jobs in sets $\mathcal{A}$ and $\mathcal{B}$ according to Johnson's rule. In consequence, the jobs from $\mathcal{A}$ precede the jobs from $\mathcal{B}$. Datasets corresponding to set $\mathcal{C}$ are processed before D_i (see Fig. 2a). As we explained in Sect. 2, the order of processing the data does not affect the makespan unless unnecessary idle times occur. Thus, without changing T, we can reorder the processing intervals as follows. First, we schedule processing datasets corresponding to jobs from $\mathcal{A}$ and $\mathcal{B}$, starting at time t_i and using Johnson's rule. Then, we schedule processing dataset D_i in consecutive idle intervals starting at time t_i. Finally, processing datasets corresponding to jobs from set $\mathcal{C}$ is scheduled in idle intervals starting at time t_0. The resulting schedule is presented in Fig. 2b. If the last dataset to be processed corresponds to a job from set $\mathcal{C}$, then data are processed in the whole interval $[t_0, T)$ without idle times. Hence,

$$T \leq t_0 + A \sum_{j=1}^{m} \alpha_j \leq r_m + T^* \leq 2T^*. \tag{3}$$

It remains to analyze the case when processing of datasets corresponding to $\mathcal{C}$ finishes before T. Without loss of generality, we can assume that $\mathcal{C} = \emptyset$ (Fig. 2c).

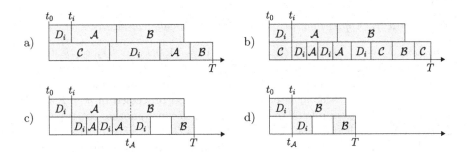

Fig. 2. Schedule modifications for the proof of Proposition 3. (a) The original schedule (from time t_0), (b) schedule after reordering processing intervals, (c) schedule under the assumption that $\mathcal{C} = \emptyset$, (d) schedule after deleting interval $[t_i, t_\mathcal{A})$.

Let $t_\mathcal{A}$ be the time when processing datasets corresponding to set $\mathcal{A}$ finishes. Since all these datasets are released after time t_0, and they are optimally scheduled by Johnson's algorithm in interval $[t_i, t_\mathcal{A})$, we have

$$t_0 + t_\mathcal{A} - t_i \leq T^*. \tag{4}$$

Let us now remove the intervals $[0, t_0)$ and $[t_i, t_\mathcal{A})$ from the schedule (see Fig. 2d). The length of the remaining schedule part is $T - t_\mathcal{A} + t_i - t_0$, and is not greater than the minimum time necessary to execute the jobs in set $\{K_i(t_0)\} \cup \mathcal{B}$. Indeed, the jobs are ordered by Johnson's rule, and moreover, some processing intervals of D_i and communication intervals of $\mathcal{B}$ may be deleted. Hence, by (1),

$$T - t_\mathcal{A} + t_i - t_0 \leq T_J(\{K_i(t_0)\} \cup \mathcal{B}) \leq T^*. \tag{5}$$

Combining (5) with (4), we obtain

$$T = (t_0 + t_\mathcal{A} - t_i) + (T - t_\mathcal{A} + t_i - t_0) \leq 2T^*. \tag{6}$$

Thus, algorithm $J(\gamma)$ delivers a 2-approximation of the optimum solution. □

Proposition 4. *Approximation ratio 2 is tight for algorithm J(0).*

Proof. Consider the following instance: $A = 1$, $m = 2$, $(\alpha_1, r_1, S_1, C_1) = (1, 0, 0, k)$, $(\alpha_2, r_2, S_2, C_2) = (k, \frac{1}{2}, 0, \frac{1}{k})$. Algorithm J(0) transfers dataset D_1 in interval $[0, k)$ and processes it in $[k, k+1)$. Dataset D_2 is sent in interval $[k, k+1)$ and processed in $[k+1, 2k+1)$. The resulting schedule length is $T = 2k+1$. In the optimum schedule, the transfer of D_1 is preempted at time $r_2 = \frac{1}{2}$. Dataset D_2 is transferred in interval $[\frac{1}{2}, \frac{3}{2})$ and processed in $[\frac{3}{2}, k+\frac{3}{2})$. Sending D_1 is completed in interval $[\frac{3}{2}, k+1)$, and this dataset is processed in interval $[k+\frac{3}{2}, k+\frac{5}{2})$. Thus, $T^* = k + \frac{5}{2}$. Hence,

$$\lim_{k \to \infty} \frac{T}{T^*} = \lim_{k \to \infty} \frac{2k+1}{k + \frac{5}{2}} = 2, \tag{7}$$

which ends the proof.   □

Proposition 5. *Approximation ratio 2 is tight for algorithm J(∞).*

Proof. Let $A = 2$, $m = 2$, $(\alpha_1, r_1, S_1, C_1) = (1, 0, k, 3)$, $(\alpha_2, r_2, S_2, C_2) = (1, k, 0, 1)$. At time $r_2 = k$, we have jobs $K_1(k) = (3, 2)$, and $K_2(k) = (1, 2)$. Hence, Johnson's rule chooses job $K_2(k)$, and a preemption is made. In consequence, dataset D_2 is sent in interval $[k, k+1)$ and processed in $[k+1, k+3)$. The transfer of D_1 is completed in $[k+1, 2k+4)$, because another startup time is required. Then, D_2 is processed in interval $[2k+4, 2k+6)$, which yields schedule of length $T = 2k+6$. Still, in the optimum schedule there is no preemption, D_1 is transferred in $[0, k+3)$ and processed in $[k+3, k+5)$, while D_2 is sent in $[k+3, k+4)$ and processed in $[k+5, k+7)$. Hence, $T^* = k+7$ and

$$\lim_{k \to \infty} \frac{T}{T^*} = \lim_{k \to \infty} \frac{2k+6}{k+7} = 2. \tag{8}$$

The claim follows.   □

To finish this section, let us note that a very simple algorithm, which schedules all jobs corresponding to datasets using Johnson's algorithm, starting at time r_m, also achieves approximation ratio 2. This can be proved using the argument given in Case 1 of the proof of Proposition 3. However, it can be expected that J(γ) will deliver much better approximations than such an algorithm for most instances. The quality of the results obtained by J(γ) for random test cases will be the subject of the next section.

5   Computational Experiments

In this section, we analyze the performance of our algorithm for different values of γ and instance parameters. The algorithm was implemented in C++ in Microsoft Visual Studio. We constructed instances with $m \in \{10, 20, \ldots, 100\}$ worker nodes. The unit processing time was $A = 1$. Dataset sizes α_j were selected randomly from interval $[1, 100]$. Preliminary experiments showed that if unit communication times C_j are not very diversified, or if all of them are significantly smaller (or greater) than A, then it is easy to obtain good schedules.

Therefore, in order to construct demanding instances, we selected basic dataset transfer times t_j randomly from $[1, 100]$, and then set $C_j = t_j/\alpha_j$. For a given value $S_{max} \in \{0, 1, \ldots, 10\}$, the startup times S_j where selected randomly from $[0, S_{max}]$. Note that if the largest release time r_m is very big, then it determines to a large degree the schedule length. Moreover, if r_m is small, then the instance is easy, because a big number of jobs which are not completed at time r_m, are afterwards scheduled optimally by Johnson's rule. Therefore, we computed the minimum possible transfer time $t_C = \sum_{j=1}^m (S_j + C_j\alpha_j)$ and generated release times r_j as follows. Dataset D_1 was released at time $r_1 = 0$, and each next release time was computed as $r_j = r_{j-1} + \delta_j \frac{t_C}{m}$, with δ_j selected randomly from $[0.5, 1.5]$. For each combination of m and S_{max}, 100 instances were constructed.

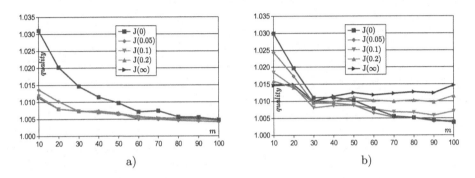

a)   b)

Fig. 3. Solution quality vs. m. (a) $S_{max} = 3$, (b) $S_{max} = 8$.

In order to assess the quality of the obtained schedules, we computed lower bounds on the optimum makespans as follows (cf. [20]). First, we take into account the lower bounds on the times necessary to transfer and process datasets released not earlier than r_k, for $k = 1, \ldots, m$. Thus, we set $LB_1 = \max_{k=1}^m \{r_k + T_J(\{K_j(r_k) : j \geq k\})\}$. Then, we consider scheduling in a network with parallel communications allowed. The resulting problem is $1|r_j|C_{max}$ with updated job release times $r'_j = r_j + S_j + C_j\alpha_j$ and processing times $p_j = A\alpha_j$. The optimum makespan LB_2 is computed by sorting the jobs according to non-decreasing release times r'_j. The last lower bound is obtained by considering parallel processing of datasets. We also assume here that the total job transfer time is $S_j + C_j\alpha_j$, and no additional preemption penalties are incurred. Thus, we obtain a preemptive single-machine problem with delivery times $A\alpha_j$, which can be solved by the extended Jackson's rule: at any time we schedule an available job with the longest delivery time [9]. The resulting optimum makespan is denoted by LB_3. Finally, we set $LB = \max\{LB_1, LB_2, LB_3\}$, and measure schedule quality by the ratio T/LB. Thus, a smaller number means better quality.

In the first set of experiments, we analyze the influence of the number of nodes m on the quality of produced schedules. We present here the results obtained for $S_{max} = 3$ and $S_{max} = 8$ (see Fig. 3). All the analyzed algorithm variants deliver very good results. The average relative errors do not exceed 3.5%. When S_{max} is small, the differences between algorithms $J(\infty)$ and $J(0.2)$ are negligible. Hence, the corresponding lines overlap in Fig. 3a. For small m, the algorithms allowing many preemptions return significantly better results than $J(0)$. However, when m increases, the instances become easier to solve, and the differences between the algorithms get smaller. For larger numbers of nodes, preempting a big fraction of communications results in a big total startup time overhead. Hence, $J(0)$ is now closer to the other algorithms, and for $m \geq 60$ the average results of algorithm $J(0.05)$ are slightly better than those obtained for $\gamma \geq 0.1$.

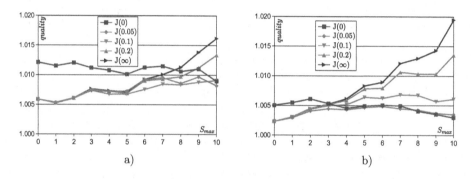

Fig. 4. Solution quality vs. S_{max}. (a) $m = 40$, (b) $m = 100$.

Figure 3b shows the quality of solutions obtained for $S_{max} = 8$. The differences between the algorithms are now larger than for $S_{max} = 3$. We can see again that the makespans obtained for small m are further from LB than for larger m. For $m = 10$, the more preemptions we allow, the better results we get. However, with growing m, the total cost of preemptions for a given $\gamma > 0$ gradually increases, and at some point it becomes larger than what we can gain by switching to sending datasets which Johnson's algorithm would prefer. Thus, for $m = 100$, a larger γ results in a longer schedule. The value of γ which gives the best solutions changes from ∞ through all the intermediate values to 0 when m increases from 10 to 100.

We also analyzed the dispersion of the results delivered by our algorithms. Let us note that a symmetric measure of dispersion may not be a good tool in our case, since the average solution quality is often close to its lower bound equal to 1. In such a situation, it can be expected that the upside dispersion is larger than the downside dispersion. Therefore, for each test setting and each algorithm, we computed the upside and downside semi-deviations of the obtained results. In most cases, the upside semi-deviation was about twice larger than the downside semi-deviation. It turned out that the dispersion of the results delivered by all algorithms decreases with growing m. The largest semi-deviations were obtained by algorithm J(0). In tests with $S_{max} = 3$ and $m = 10$, its upside semi-deviation was 0.042, and its downside semi-deviation was 0.026. However, the maximum value of the upside semi-deviation in tests with $S_{max} = 3$ and $m = 20$ (also reached by J(0)) was already 0.02, and it further decreased with growing m. Thus, we can say that the performance of our algorithm is rather stable.

Figure 4 presents the results obtained for changing S_{max} and $m \in \{40, 100\}$. When S_{max} is small, the differences between algorithm variants with $\gamma > 0$ are insignificant. All these algorithms obtain better results than the non-preemptive algorithm J(0). However, when S_{max} grows, the cost of each preemption becomes larger, and introducing too many preemptions makes the schedule long. Thus, the algorithms with $\gamma > 0$ lose quality, while J(0) performs better. The value of m determines the threshold values of S_{max} at which certain algorithms start performing better than the others. For example, J(∞) is outperformed by J(0) for $S_{max} \geq 8$ when $m = 40$, but already for $S_{max} \geq 4$ when $m = 100$. The range of solution quality changes is greater when m is bigger. It is interesting that for $m = 100$, the results returned by J(0.05) first get worse as S_{max} increases to 6, but then become better again when S_{max} grows further. Indeed, when startup times get larger in comparison to data transfer times, it is less probable that a preemption will be allowed if $\gamma < \infty$. Thus, for very big S_{max}, the performance of the algorithms with $0 < \gamma < \infty$ becomes similar to that of $J(0)$.

Table 1. Average solution quality vs. S_{max}.

S_{max}	J(0)	J(0.05)	J(0.1)	J(0.2)	J(∞)
0	1.0124	1.0053	1.0053	1.0053	1.0053
1	1.0120	1.0056	1.0056	1.0056	1.0056
2	1.0119	1.0062	1.0061	1.0061	1.0061
3	1.0118	1.0069	1.0066	1.0067	1.0067
4	1.0121	1.0081	1.0078	1.0080	1.0081
5	1.0124	1.0092	1.0085	1.0091	1.0093
6	1.0116	1.0091	1.0084	1.0094	1.0099
7	1.0107	1.0089	1.0091	1.0101	1.0111
8	1.0109	1.0095	1.0092	1.0112	1.0127
9	1.0107	1.0093	1.0087	1.0107	1.0131
10	1.0100	1.0090	1.0092	1.0125	1.0158
All	1.0115	1.0079	1.0077	1.0086	1.0094

The average results obtained for all analyzed values of S_{max} are presented in Table 1. Here we take into account instances with all values of $m \in \{10, \ldots, 100\}$. The best average results for all combinations of S_{max} and m were obtained by algorithm J(0.1). Moreover, the only values of S_{max} for which this algorithm did not have the best average over all m, were $S_{max} = 7$ and $S_{max} = 10$, when J(0.05) was slightly better. Thus, we conclude that $\gamma = 0.1$ is the best choice if we have no prior knowledge about the instance parameters.

6 Conclusions

In this paper, we analyzed makespan minimization in data gathering networks with dataset release times and communication preemption penalties. We showed that this problem is strongly NP-hard, and proposed a 2-approximation algorithm J(γ). Its average performance was tested experimentally. It turned out that for large random instances, the algorithm performs very well for all analyzed values of γ. The obtained solutions are at most several percent away from the lower bound. On average, the best results are achieved when $\gamma = 0.1$. As the schedules obtained for big m are much better than what is guaranteed by our algorithm's approximation ratio, an interesting direction for future research is to investigate its asymptotic performance guarantee. Future work can also include determining whether approximation ratio 2 is tight for algorithm J(γ) when $\gamma \in (0, \infty)$, as well as searching for polynomially solvable cases of our scheduling problem, other than the ones discussed in Sect. 3.

Acknowledgements. This research was partially supported by the National Science Centre, Poland, grant 2016/23/D/ST6/00410.

References

1. Berlińska, J.: Communication scheduling in data gathering networks with limited memory. Appl. Math. Comput. **235**, 530–537 (2014). https://doi.org/10.1016/j.amc.2014.03.024
2. Berlińska, J.: Scheduling for data gathering networks with data compression. Eur. J. Oper. Res. **246**, 744–749 (2015). https://doi.org/10.1016/j.ejor.2015.05.026
3. Berlińska, J.: Scheduling data gathering with maximum lateness objective. In: Wyrzykowski, R., Dongarra, J., Deelman, E., Karczewski, K. (eds.) PPAM 2017. LNCS, vol. 10778, pp. 135–144. Springer, Cham (2018). https://doi.org/10.1007/978-3-319-78054-2_13
4. Berlińska, J.: Scheduling in a data gathering network to minimize maximum lateness. In: Fortz, B., Labbé, M. (eds.) Operations Research Proceedings 2018. ORP, pp. 453–458. Springer, Cham (2019). https://doi.org/10.1007/978-3-030-18500-8_56
5. Berlinska, J.: Heuristics for scheduling data gathering with limited base station memory. Ann. Oper. Res. **285**, 149–159 (2020). https://doi.org/10.1007/s10479-019-03185-3
6. Cheng, J., Steiner, G., Stephenson, P.: A computational study with a new algorithm for the three-machine permutation flow-shop problem with release times. Eur. J. Oper. Res. **130**, 559–575 (2001). https://doi.org/10.1016/S0377-2217(99)00415-4
7. Cho, Y., Sahni, S.: Preemptive scheduling of independent jobs with release and due times on open, flow and job shops. Oper. Res. **29**, 511–522 (1981). https://doi.org/10.1287/opre.29.3.511

8. Choi, K., Robertazzi, T.G.: Divisible load scheduling in wireless sensor networks with information utility. In: IEEE International Performance Computing and Communications Conference 2008: IPCCC 2008, pp. 9–17 (2008). https://doi.org/10.1109/PCCC.2008.4745126

9. Horn, W.A.: Some simple scheduling algorithms. Nav. Res. Logist. Q. **21**, 177–185 (1974). https://doi.org/10.1002/nav.3800210113

10. Johnson, S.M.: Optimal two- and three-stage production schedules with setup times included. Nav. Res. Logist. Q. **1**, 61–68 (1954). https://doi.org/10.1002/nav.3800010110

11. Kalczynski, P.J., Kamburowski, J.: An empirical analysis of heuristics for solving the two-machine flow shop problem with job release times. Comput. Oper. Res. **39**, 2659–2665 (2012). https://doi.org/10.1016/j.cor.2012.01.011

12. Lenstra, J.K., Rinnooy Kan, A.H.G., Brucker, P.: Complexity of machine scheduling problems. Ann. Discret. Math. **1**, 343–362 (1977). https://doi.org/10.1016/S0167-5060(08)70743-X

13. Liu, Z., Cheng, T.C.E.: Scheduling with job release dates, delivery times and preemption penalties. Inf. Process. Lett. **82**, 107–111 (2002). https://doi.org/10.1016/S0020-0190(01)00251-4

14. Liu, Z., Cheng, T.C.E.: Minimizing total completion time subject to job release dates and preemption penalties. J. Sched. **7**, 313–327 (2004). https://doi.org/10.1023/B:JOSH.0000031424.35504.c4

15. Luo, W., Xu, Y., Gu, B., Tong, W., Goebel, R., Lin, G.: Algorithms for communication scheduling in data gathering network with data compression. Algorithmica **80**(11), 3158–3176 (2018). https://doi.org/10.1007/s00453-017-0373-6

16. Luo, W., Gu, B., Lin, G.: Communication scheduling in data gathering networks of heterogeneous sensors with data compression: algorithms and empirical experiments. Eur. J. Oper. Res. **271**, 462–473 (2018). https://doi.org/10.1016/j.ejor.2018.05.047

17. Moges, M., Robertazzi, T.G.: Wireless sensor networks: scheduling for measurement and data reporting. IEEE Trans. Aerosp. Electron. Syst. **42**, 327–340 (2006). https://doi.org/10.1109/TAES.2006.1603426

18. Potts, C.N.: Analysis of heuristics for two-machine flow-shop sequencing subject to release dates. Math. Oper. Res. **10**, 576–584 (1985). https://doi.org/10.1287/moor.10.4.576

19. Schuurman, P., Woeginger, G.J.: Preemptive scheduling with job-dependent setup times. In: Proceedings of the 10th ACM-SIAM Symposium on Discrete Algorithms, pp. 759–767 (1999)

20. Tadei, R., Gupta, J.N.D., Della Croce, F., Cortesi, M.: Minimising makespan in the two-machine flow-shop with release times. J. Oper. Res. Soc. **49**, 77–85 (1998). https://doi.org/10.1057/palgrave.jors.2600481

Workshop on Applied High Performance Numerical Algorithms for PDEs

Overlapping Schwarz Preconditioner for Fourth Order Multiscale Elliptic Problems

Leszek Marcinkowski[1]([✉]) and Talal Rahman[2]

[1] Faculty of Mathematics, Informatics, and Mechanics, University of Warsaw,
Banacha 2, 02-097 Warszawa, Poland
L.Marcinkowski@mimuw.edu.pl
[2] Department of Computing, Mathematics, and Physics,
Western Norway University of Applied Sciences,
Inndalsveien 28, 5020 Bergen, Norway
Talal.Rahman@hvl.no

Abstract. In this paper, a domain decomposition parallel preconditioner for the 4th order multiscale elliptic problem in 2D with highly heterogeneous coefficients is constructed. The problem is discretized by the conforming C^1 reduced Hsieh-Tough-Tocher (HCT) macro element. The proposed preconditioner is based on the classical overlapping Schwarz method and is constructed using an abstract framework of the Additive Schwarz Method. The coarse space consists of multiscale finite element functions associated with the edges and is enriched with functions based on solving carefully constructed generalized eigenvalue problems locally on each edge. The convergence rate of the Preconditioned Conjugate Method of our method is independent of the variations in the coefficients for a sufficient number of eigenfunctions in the coarse space.

Keywords: Fourth order problem · Finite element method · Domain Decomposition Method · Additive Schwarz Method · Abstract coarse space

1 Introduction

When modeling physical or engineering phenomena one has to numerically solve partial differential equations with highly heterogeneous contrast. The heterogeneity of the media makes many standard numerical methods to work very slowly. Domain Decomposition Methods (DDM), in particular, Schwarz methods, form a class of parallel highly efficient methods for solving a system of equations arising from the standard discretizations of elliptic partial differential equations, e.g., cf. [14] and references therein. In classical overlapping Schwarz

L. Marcinkowski—The work was partially supported by Polish Scientific Grant: National Science Center 2016/21/B/ST1/00350.

© Springer Nature Switzerland AG 2020
R. Wyrzykowski et al. (Eds.): PPAM 2019, LNCS 12044, pp. 245–255, 2020.
https://doi.org/10.1007/978-3-030-43222-5_21

method the domain is decomposed into the overlapping subdomains where the local subproblems are solved in the application of the preconditioner. We also usually add a global coarse problem obtaining proper scalability, e.g., cf. [14]. Recently, the research of DDM and in particular Schwarz method extended into to problems with highly heterogeneous coefficients, e.g., cf. [3,5,6,9–11,13]. It is common that the coarse space is built by enriching a small standard coarse space with eigenfunctions of some generalized eigenvalue problems, e.g., cf. [3–5,13]. The resulting methods are robust with respect to the heterogeneity of the coefficients, and quite often are adaptive in a sense that we can construct it automatically by adding those eigenfunctions which are associated with all respective eigenvalues below a preset threshold. The condition bounds of the obtained preconditioned problem depend only on the threshold and are independent of the coefficients.

The goal of this paper is to construct an adaptive coarse space for the standard overlapping Schwarz method with the minimal overlap for the macro finite element reduced Hsieh-Clough-Tocher (RHCT) discretization of the fourth order elliptic problem with highly heterogeneous highly varying coefficients in two dimensions. Then, the preconditioned problem is solved by the Preconditioned Conjugate Gradient Method (PCG), e.g., cf. [7]. The method is based on the abstract Schwarz framework. The coarse space is an algebraic sum of a specially constructed multiscale global space associated with the edges of the subdomains and local edge subspaces formed by eigenfunctions of generalized eigenvalue problems. This work is an extension of the recent results of [5] for the second order elliptic problem to the fourth order problem discretized by the RHCT method.

The obtained estimates are independent of the geometries of the subdomains, and the heterogeneities in the coefficients. The bounds are depended only on the parameters chosen in the eigenvalue problems, i.e. a user has to decide in a precomputational step how many eigenvectors have to be computed and included in our coarse space construction. It can be done adaptively, i.e., including the eigenfunctions for which the respective eigenvalues are below a preset threshold.

The remainder of this paper is organized as follows, in Sect. 2 we present the RHCT macro finite element discretization. In Sect. 3 our coarse space is constructed. Section 4 contains a description of the overlapping additive Schwarz preconditioner, and in Sect. 5 we briefly discuss some implementation issues.

2   Finite Element Discretization

In this section, we present our model problem and its RHCT macro element discretization.

Let Ω be a convex and polygonal domain in the plane. The differential problem is to find $u^* \in H_0^2(\Omega)$ such that

$$a(u^*, v) = \int_\Omega fv \, dx \quad \forall v \in H_0^2(\Omega), \tag{1}$$

where

$$f \in L^2(\Omega),$$

$$H_0^2(\Omega) = \{u \in H^2(\Omega) : u = \partial_n u = 0 \quad \text{on} \quad \partial\Omega\},$$

and

$$a(u,v) = \int_\Omega \beta(x)[u_{x_1x_1}v_{x_1x_1} + 2\,u_{x_1x_2}v_{x_1x_2} + u_{x_2x_2}v_{x_2x_2}]\,dx. \tag{2}$$

Here β is a strictly positive bounded function, and ∂_n is a normal unit derivative. Hence, we can always scale β by its minimal value.

We introduce a quasiuniform triangulation $T_h = T_h(\Omega)$ of Ω consisting of triangles, $h = \max_{\tau \in T_h} \operatorname{diam}(\tau)$ be the parameter of the triangulation, e.g., cf. [1]. Let $\Omega_h, \overline{\Omega}_h, \partial\Omega_h$ be the sets of vertices or the nodes of $T_h(\Omega)$, belonging to $\Omega_h, \overline{\Omega}_h, \partial\Omega_h$, respectively.

For a two-dimensional multi-index $\alpha = (\alpha_1, \alpha_2)$, where α_1, α_2 are nonnegative integers, we define

$$|\alpha| = \alpha_1 + \alpha_2, \qquad \partial^\alpha = \frac{\partial^{|\alpha|}}{\partial x_1^{\alpha_1} \partial x_2^{\alpha_2}}.$$

Further, we assume that β is piecewise constant over T_h, it may have jumps across the 1D common edges of two neighboring elements in T_h.

The reduced Hsieh-Clough-Tocher (RHCT) macro element space V_h is defined as follows, e.g., cf. Chap. 7, Sect. 46, p. 285 in [2], (also cf. Fig. 1):

$$\begin{aligned}
V_h = \{u \in C^1(\Omega) &: u \in P_3(\tau_i),\ \tau_i \in T_h(\Omega),\ \text{for triangles } \tau_i, \\
&i = 1, 2, 3,\ \text{formed by connecting the vertices of} \\
&\text{any } \tau \in T_h(\Omega_k)\ \text{to its centroid},\ \partial_n u_{|e} \in P_1(e)\ \text{for} \\
&e\ \text{an edge of } \tau\ \text{and } u = \partial_n u = 0\ \text{on } \partial\Omega\}.
\end{aligned} \tag{3}$$

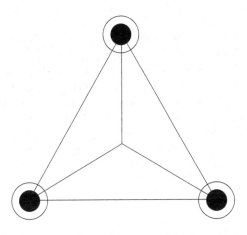

Fig. 1. The reduced Hsieh-Clough-Tocher macro element. There are three degrees of freedom at each vertex.

The degrees of freedom of the RHCT element are given by

$$\{v(p), v_{x_1}(p), v_{x_2}(p)\},$$

where p is a vertex of an element, cf. Fig. 1.

The discrete RHCT element problem will then be formulated as follows: find $u_h \in V_h$ such that

$$a(u_h, v) = \int_\Omega f v \, dx \quad \forall v \in V_h. \tag{4}$$

The problem has a unique solution by the Lax-Milgram lemma. By formulating the discrete problem in the standard RHCT nodal basis $\{\phi_i^\alpha\}_{x_i \in \Omega_h, |\alpha| \leq 1}$, we get the following system of algebraic equations

$$A_h u_h = f_h \tag{5}$$

where $A_h = (a(\phi_i^{\alpha_1}, \phi_j^{\alpha_2}))_{i,j}$, $f_h = (f_j^\alpha)_{x_j \in \Omega_h}$ with $f_j = \int_\Omega f(x) \phi_i^\alpha \, dx$, and $u_h = (u_i^\alpha)_{i,\alpha}$ with $u_i^\alpha = \partial^\alpha u_h(x_i)$. Here $u_h = \sum_{x_i \in \Omega_h} \sum_{|\alpha| \leq 1} u_i^\alpha \phi_i$. The resulting system is symmetric which is in general very ill-conditioned; any standard iterative method may perform badly due to the ill-conditioning of the system.

In this paper, we present a method for solving such systems using the preconditioned conjugate method (cf. [7]) and propose an overlapping additive Schwarz preconditioner (e.g., cf. [14]). Let assume that there exists a partition of Ω into a collection of disjoint open and connected polygonal substructures Ω_k, such that

$$\overline{\Omega} = \bigcup_{k=1}^N \overline{\Omega}_k.$$

We need another assumption, namely, let the triangulation T_h be aligned with the subdomains Ω_k, i.e. let any triangle of T_h be contained in a substructure Ω_k. Hence, each subdomain Ω_k inherits the local triangulation $T_h(\Omega_k) = \{\tau \in T_h : \tau \subset \Omega_k\}$. We make an additional assumption that the number of subdomains which share a vertex or an edge of an element of T_h is bounded by a constant. An important role plays an interface $\overline{\Gamma} = \sum_{k=1}^N (\partial \Omega_k \setminus \partial \Omega)$.

The non-empty intersection of two subdomains $\partial \Omega_i \cap \partial \Omega_j$ not on $\partial \Omega$ is either an 1D edge $\overline{\mathcal{E}}_{ij} = \partial \Omega_i \cap \partial \Omega_j$, or it is a vertex of T_h. A common vertex of substructures that is NOT on $\partial \Omega$ is called a crosspoint. The sum of closed edges of substructures, which are not on $\partial \Omega$ equals Γ the interface of this partition. We define local sets of nodal points, $\Omega_{k,h}, \mathcal{E}_{kl,h}, \overline{\Omega}_{k,h}, \overline{\mathcal{E}}_{kl,h}$ etc., as the sets of vertices of elements of T_h, which are in $\Omega_k, \mathcal{E}_{kl}, \overline{\Omega}_k, \overline{\mathcal{E}}_{kl}$ etc., respectively.

3   Coarse Space

In this section, we present an adaptive coarse space, which is a space of discrete biharmonic functions, cf. Sect. 3.1 below, consisting of two space components: the multiscale coarse space component and the generalized edge based eigenfunction space component.

3.1   Discrete Biharmonic Extensions

In this section, we define the discrete biharmonic functions. We define the local subspace $V_h(\Omega_k)$ as the space of restrictions to $\overline{\Omega}_k$, of the space V_h, i.e.,

$$V_h(\Omega_k) = \{u_{|\overline{\Omega}_k} : u \in V_h\},$$

and we let introduce its subspace of functions with zero boundary conditions (in H_0^2 sense), i.e.,

$$V_{h,0}(\Omega_k) = V_h(\Omega_k) \cap H_0^2(\Omega_k).$$

We also need a local bilinear form

$$a_k(u,v) = \int_{\Omega_k} \beta(x)[u_{x_1 x_1} v_{x_1 x_1} + 2\, u_{x_1 x_2} v_{x_1 x_2} + u_{x_2 x_2} v_{x_2 x_2}]\, dx.$$

Let $\mathcal{P}_k u \in V_{h,0}(\Omega_k)$ for any $u \in V_h(\Omega_k)$ be the a_k orthogonal projection onto $V_{h,0}(\Omega_k)$, i.e.,

$$a_k(\mathcal{P}_k u, v) = a_k(u,v) \qquad \forall v \in V_{h,0}(\Omega_k). \tag{6}$$

Then, let the local discrete biharmonic extension operator

$$\mathcal{H}_k : V_h(\Omega_k) \to V_h(\Omega_k)$$

be defined as

$$\mathcal{H}_k u = u - \mathcal{P}_k u, \tag{7}$$

or equivalently $\mathcal{H}_k u$ is the unique solution to the following local problem:

$$\begin{cases} a_k(\mathcal{H}_k u, v) = 0 & \forall v \in V_{h,0}(\Omega_k) \\ Tr\, \mathcal{H}_k u = Tr\, u & \text{on } \partial\Omega_k \end{cases}, \tag{8}$$

where $Tr\, u_k = (u_{k|\partial\Omega_k}, \nabla u_{k|\partial\Omega_k})$ for $u_k \in V_h(\Omega_k)$, e.g., cf. [8]. Since it is a discrete case, the boundary conditions are equivalently to the discrete boundary conditions: $Tr\, \mathcal{H}_k u(x) = Tr\, u(x)$ for all $x \in \partial\Omega_{k,h}$. A function $u \in V_h(\Omega_k)$ is discrete biharmonic in Ω_k if $u_{|\overline{\Omega}_k} = \mathcal{H}_k u \in V_h(\Omega_k)$. If for $u \in V_h$ all its restriction to local subdomains are discrete biharmonic, then u is piecewise discrete biharmonic in our partition. Please note, that a discrete biharmonic function in $V_h(\Omega_k)$ is uniquely defined by its values of degrees of freedom at the nodal points of $\partial\Omega_{k,h}$ and has the following minimizing property:

$$a_k(\mathcal{H}_k u, \mathcal{H}_k u) = \min\{a_k(u,u) :$$
$$u \in V_h(\Omega_k) : Tr\, u(x) = Tr\, \mathcal{H}_k u(x) \quad x \in \partial\Omega_k\}. \tag{9}$$

3.2   Multiscale Coarse Space Component

Our coarse space comprises two parts, in this section, we define the multiscale component. We need a definition of a patch around an edge $\mathcal{E}_{kl}$, namely let the

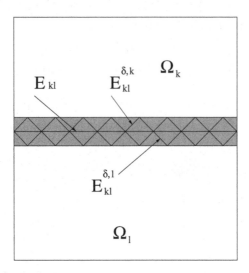

Fig. 2. The patch $\mathcal{E}_{kl}^{\delta}$ and its two subpatches related to $\mathcal{E}_{kl}$ the common edge to subdomains Ω_k and Ω_l.

patch $\mathcal{E}_{kl}^{\delta}$ an open domain which closure is the closed union of all fine triangles of $T_h(\Omega_k)$ and $T_h(\Omega_l)$ such that either one of its open edges or vertices are contained in $\mathcal{E}_{kl}$. We can naturally split the patch into two disjoint parts:

$$\overline{\mathcal{E}}_{kl}^{\delta} = \overline{\mathcal{E}}_{kl}^{\delta,k} \cup \overline{\mathcal{E}}_{kl}^{\delta,l},$$

where $\mathcal{E}_{kl}^{\delta,s} = \mathcal{E}_{kl}^{\delta} \cap \Omega_s$, $s = k, l.$, cf. Fig. 2.

The sum of all patches contained in Ω_k form a boundary layer interior to Ω defined as

$$\overline{\Omega}_k^{\delta} = \bigcup_{\Gamma_{kl} \subset \partial\Omega_k \backslash \partial\Omega} \overline{\mathcal{E}}_{kl}^{\delta,k}.$$

For simplicity of presentation we assume that all patches $\mathcal{E}_{kl}^{\delta,k}$ for a substructure Ω_k, are disjoint.

Let $V_h(\mathcal{E}_{kl}^{\delta})$ be the space of restrictions of functions from V_h to $\mathcal{E}_{kl}^{\delta}$

$$V_h(\mathcal{E}_{kl}^{\delta}) = \{u_{|\mathcal{E}_{kl}^{\delta}} : u \in V_h\}$$

and its subspace $V_{h,0}(\mathcal{E}_{kl}^{\delta})$ with zero boundary condition, i.e.,

$$V_{h,0}(\mathcal{E}_{kl}^{\delta}) = V_h(\mathcal{E}_{kl}^{\delta}) \cap H_0^2(\mathcal{E}_{kl}^{\delta}). \tag{10}$$

A function in this space is uniquely defined by the values of three degrees of freedom at all nodes in $\mathcal{E}_{kl,h}$.

Note that $Tr\, u$ onto each fine edge of $\mathcal{E}_{kl}$ can be represented as $(u, \partial_s u, \partial_n u)$, where u is a C^1 piecewise cubic function on the inherited 1D triangulation of this edge, $\partial_s u$ is its derivative, and $\partial_n u$ is a continuous piecewise linear function. We also define two bilinear forms related to an edge $\mathcal{E}_{kl}$, the first one being the restriction of the form $a(u,v)$, cf. (2), to the patch, namely,

$$a_{kl}(u,v) = \int_{\mathcal{E}_{kl}} \overline{\beta}[\partial_{ss}u \, \partial_{ss}v + \partial_{ns}u \, \partial_{ns}v \, ds] \tag{11}$$

where $\partial_{ss}u$ is the weak second order tangential derivative of the trace of u onto the edge $\mathcal{E}_{kl}$, and $\partial_{ns}u$ is the weak tangential derivative of the trace of the normal derivative on this edge. Here $\overline{\beta}$ is constant over each fine edge $e \subset \mathcal{E}_{kl}$ being the common edge of fine 2D triangles $\tau_+ \in T_h(\Omega_k)$ and $\tau_- \in T_h(\Omega_l)$. Let it be defined as $\max(\beta_{|\tau_+}, \beta_{|\tau_-})$.

The second patch bilinear form is the scaled weighted L^2 over the patch, i.e.,

$$b_{kl}(u,v) = h^{-3} \int_{\mathcal{E}_{kl}^\delta} \beta u v \, dx. \tag{12}$$

We have a simple proposition.

Proposition 1. *The both forms $a_{kl}(u,v)$ and $b_{kl}(u,v)$ are symmetric and positive definite over $V_{h,0}(\mathcal{E}_{kl}^\delta)$.*

We now introduce the multiscale coarse space.

Let a subspace $V_{ms} \subset V_h$ be formed by all discrete biharmonic functions in the sense of (8), which satisfies the following variational equality on each patch $\mathcal{E}_{kl}^\delta$:

$$a_{kl}(\hat{u}_{kl}, v) = 0 \quad \forall v \in V_{h,0}(\mathcal{E}_{kl}^\delta), \tag{13}$$

where $\hat{u}_{kl} \in V_{h,0}(\mathcal{E}_{kl}^\delta)$ satisfies $\partial^\alpha \hat{u}(x) = \partial^\alpha u(x)$ $x \in \overline{\mathcal{E}}_{kl,h}$.

We have a straightforward proposition.

Proposition 2. *A function $u \in V_{ms}$ is uniquely defined by its dofs at all crosspoints.*

Proof. Since $u \in V_{ms}$ is discrete biharmonic, it is defined by the values of its dofs at interface i.e., at crosspoints and in $\mathcal{E}_{kl,h}$ for all interfaces. Thus, it is enough to show that all dofs of u are uniquely defined on all interfaces $\mathcal{E}_{kl}$. Let define the function $\hat{u}_{kl} \in V_h(\mathcal{E}_{kl})$ such that it satisfies $\partial^\alpha \hat{u}_{kl}(x) = \partial^\alpha u(x)$ for $|\alpha| \leq 1$ and $x \in \partial \mathcal{E}_{kl}$, $\partial^\alpha \hat{u}_{kl}(x) = 0$ for all remaining fine vertices on the boundary of the patch $\mathcal{E}_{kl}$ and:

$$a_{kl}(\hat{u}_{kl}, v) = 0 \quad \forall v \in V_{h,0}(\mathcal{E}_{kl}^\delta).$$

We can represent $\hat{u}_{kl} = w_1 + w_0$ with $w_0 \in V_{h,0}(\mathcal{E}_{kl}^\delta)$. E.g., we can take w_1 with the DOFs equal to DOFs of $\hat{u}_{kl}$ on the boundary of the patch, and with zero valued remaining DOFs. Then, the last variational equation is equivalent to: find $w_0 \in V_{h,0}(\mathcal{E}_{kl}^\delta)$ such that

$$a_{kl}(w_0, v) = -a_{kl}(w_1, v) \quad \forall v \in V_{h,0}(\mathcal{E}_{kl}^\delta).$$

It follows from Proposition 1, that w_0, and thus $\hat{u}_{kl}$, is uniquely defined. It is clear that all DOFs of $\hat{u}_{kl}$ and u have the same values at $\overline{\mathcal{E}}_{kl,h}$, so u is unique.

The values of its DOFs at the nodal points of each face can be computed by solving (13), and then the values of DOFs at the nodal points of each subdomain can be computed by solving (8).

3.3   Generalized Edge Based Eigenfunction Space Component

First we define a generalized eigenproblem of the form: find $(\lambda_j^{kl}, \phi_j^{kl}) \in \mathbb{R} \times V_{h,0}(\mathcal{E}_{kl}^\delta)$ such that

$$a_{kl}(\phi_j^{kl}, v) = \lambda_j^{kl} b_{kl}(\phi, v) \qquad \forall v \in V_{h,0}(\mathcal{E}_{kl}^\delta). \tag{14}$$

Proposition 1 yields, that there are real and positive eigenvalues and their respective b_{kl} - orthonormal eigenvectors satisfying (14), such that

$$0 < \lambda_1^{kl} \leq \lambda_2^{kl} \leq \ldots \leq \lambda_M^{kl},$$

where M is the dimension of $V_{h,0}(\mathcal{E}_{kl}^\delta)$.

For any $1 \leq n = n(\mathcal{E}_{kl}) \leq M$ we can define the orthogonal projection: $\pi_n^{kl} : V_{h,0}(\mathcal{E}_{kl}^\delta) \to span\{\phi_j^{kl}\}_{j=1}^n \subset V_{h,0}(\mathcal{E}_{kl}^\delta)$ as

$$\pi_n^{kl} v = \sum_{j=1}^n b_{kl}(v, \phi_j^{kl}) \phi_j^{kl}. \tag{15}$$

Then for each eigenvector ϕ_j^{kl}, $1 \leq j \leq n(\mathcal{E}_{kl})$ we define $\Phi_j^{kl} \in V_h$ which has DOFs equal to the ones of ϕ_j^{kl} at all nodes on the edge $\mathcal{E}_{kl}$, zero DOFs on the remaining edges and at all crosspoints, and finally discrete biharmonic inside each subdomain, in the sense of (8), what defines uniquely the values of its DOFs at all interior nodes of the subdomain. Then, the edge terms of the coarse space are introduced as:

$$V_{h,n}^{kl} = span\{\Phi_j^{kl}\}_{j=1}^{n(\mathcal{E}_{kl})}, \qquad \forall \mathcal{E}_{kl} \subset \Gamma.$$

The coarse space is defined as

$$V_0 := V_{ms} + \sum_{\mathcal{E}_{kl} \subset \Gamma} V_{h,n}^{kl}. \tag{16}$$

4   Additive Schwarz Method (ASM) Preconditioner

Our preconditioner is based on the abstract framework of ASM, i.e. is based on a decomposition of the global space V_h into local subspaces and one global coarse space, equipped into respective symmetric positive definite bilinear forms, e.g., cf. [14]. Here we take only the original form $a(u, v)$, i.e. (2), as the local forms Each local subspace V_k related to Ω_k, is defined as the space formed by all functions $u \in V_h$ whose DOFs take the value zero at all nodal points that lie outside $\overline{\Omega}_k$.

The coarse space V_0 was introduced in the previous section, cf. (16). We get

$$V_h = V_0 + \sum_{k=1}^{N} V_k.$$

Then, we introduce the additive Schwarz operator $T : V_h \to V_h$ as

$$T = T_0 + \sum_{k=1}^{N} T_k,$$

where the coarse space projection operator, $T_0 : V_h \to V_0$, is defined by

$$a(T_0 u, v) = a(u, v) \qquad \forall v \in V_0,$$

and the local subspace projection operators, $T_k : V_h \to V_k$, are determined by

$$a(T_k u, v) = a(u, v) \qquad \forall v \in V_k, \qquad k = 1, \dots, N.$$

The problem (1) is then replaced as the equivalent preconditioned system,

$$T u_h = g, \tag{17}$$

where

$$g = g_0 + \sum_{k=1}^{N} g_k$$

with $g_0 = T_0 u_h^*$, $g_k = T_k u_h^*$, $k = 1, \dots, N$, and u_h^* the discrete solution, cf. (4).

Note, that the right hand side vectors, $g_k, k = 0 \cdots, N$, can be calculated without explicitly knowing the discrete solution, e.g., cf. [12,14].

4.1 An Estimate of the Condition Number

We present the main result of this paper, namely an estimate of the condition number of the preconditioned system (1).

We have the following theorem:

Theorem 1. *There exist positive constants c, C independent of h, β, and number of subdomains, such that*

$$c(\min_{kl} \lambda_{n+1}^{kl}) \, a(u, u) \le a(Tu, u) \le C \, a(u, u) \qquad \forall u \in V_h,$$

where λ_{n+1}^{kl} and $n = n(\mathcal{E}_{kl})$ are defined in Sect. 3.3.

The theorem proof uses the abstract ASM framework, e.g., cf. [14], we check the three key assumptions of this framework. The key component is to define a so called stable decomposition which is done with the help of the operators π_n^{kl}, cf. (15), and then to utilize its properties.

The number of eigenfunctions needed for the robustness of our method usually corresponds to the number of channels crossing a subdomain interface. This number can be predefined from experience or chosen adaptively by looking at the smallest eigenvalues. Note that the lower bounds in Theorem 1 is dependent on how many eigenvectors of the local face generalized eigenproblem are included in our coarse grid.

5   Implementation Issues

In this section, we briefly discuss the implementation of our ASM preconditioner. For the simplicity of presentation, we use our preconditioner with the Richardson's iteration. In practice, one uses the preconditioned conjugate gradient iteration (e.g., cf. [7]) for the system (17).

– Precomputation step. Computing the coarse grid basis.
  Constructing the coarse space requires the solution of the generalized eigenvalue problem (14) on each subdomain face (interface), the first few eigenfunctions corresponding to the smallest eigenvalues are then included in the coarse space. Prescribing a threshold λ_0, and then computing the eigenpairs with eigenvalues smaller than λ_0, we can get an automatic way to enrich the coarse space. The simplest way would be to compute a fixed number of eigenpairs, e.g. $n = 5$ or so, this may however not guarantee robustness as the number of channels crossing a face may be much larger.

– Richardson iteration.
  The Richardson iteration with the parameter τ is defined as follows: starting with any $u^{(0)}$, iterate until convergence:

$$u^{(i+1)} = u^{(i)} + \tau\,(g - T(u^{(i)})) = u^{(i)} + \tau\,T(u_h^* - u^{(i)}),$$
$$= u^{(i)} - \tau\,r^{(i)} \qquad i \geq 0.$$

Computing of $r^{(i)} = g - T(u^{(i)})$ requires solving the following problems:
  ▷ **Local subdomain problems:**

  Compute $r_k \in V_k$ $k = 1, \ldots, N$ by solving the following local problems

  $$a(r_k, v) = a(T_k(u_h^* - u^{(i)}), v) = f(v) - a(u^{(i)}, v) \qquad \forall v \in V_k.$$

  ▷ **Coarse problem:**

  Compute $r_0 \in V_0$ such that

  $$a(r_0, v) = a(T_0(u_h^* - u^{(i)}), v) = f(v) - a(u^{(i)}, v) \qquad \forall v \in V_0,$$

Then

$$r^{(i)} = r_0 + \sum_{k=1}^{N} r_k.$$

All these problems are independent and can be solved in parallel.

The local subdomain problems are solved locally on their respective subdomains. The coarse problem is global, and its dimension equals the number of the cross-points times three plus the number of local eigenfunctions.

References

1. Brenner, S.C., Scott, L.R.: The Mathematical Theory of Finite Element Methods, Texts in Applied Mathematics, vol. 15, Second edn. Springer, New York (2002). https://doi.org/10.1007/978-1-4757-3658-8
2. Ciarlet, P.G.: Basic error estimates for elliptic problems. In: Handbook of Numerical Analysis, vol. II, pp. 17–351, North-Holland, Amsterdam (1991)
3. Efendiev, Y., Galvis, J., Lazarov, R., Willems, J.: Robust domain decomposition preconditioners for abstract symmetric positive definite bilinear forms. ESAIM Math. Model. Numer. Anal. **46**(5), 1175–1199 (2012). https://doi.org/10.1051/m2an/2011073
4. Efendiev, Y., Galvis, J., Vassilevski, P.S.: Spectral element agglomerate algebraic multigrid methods for elliptic problems with high-contrast coefficients. In: Huang, Y., Kornhuber, R., Widlund, O., Xu, J. (eds.) Domain Decomposition Methods in Science and Engineering XIX. LNCSE, vol. 78, pp. 407–414. Springer, Heidelberg (2011). https://doi.org/10.1007/978-3-642-11304-8_47
5. Gander, M.J., Loneland, A., Rahman, T.: Analysis of a new harmonically enriched multiscale coarse space for domain decompostion methods. Eprint arXiv:1512.05285 (2015). https://arxiv.org/abs/1512.05285
6. Graham, I.G., Lechner, P.O., Scheichl, R.: Domain decomposition for multiscale PDEs. Numer. Math. **106**(4), 589–626 (2007). https://doi.org/10.1007/s00211-007-0074-1
7. Hackbusch, W.: Iterative Solution of Large Sparse Systems of Equations, Applied Mathematical Sciences, vol. 95. Springer, New York (1994). https://doi.org/10.1007/978-3-319-28483-5. Translated and revised from the 1991 German original
8. Le Tallec, P., Mandel, J., Vidrascu, M.: A Neumann-Neumann domain decomposition algorithm for solving plate and shell problems. SIAM J. Numer. Anal. **35**(2), 836–867 (1998)
9. Pechstein, C.: Finite and Boundary Element Tearing and Interconnecting Solvers for Multiscale Problems. LNCSE, vol. 90. Springer, Heidelberg (2013). https://doi.org/10.1007/978-3-642-23588-7
10. Pechstein, C., Scheichl, R.: Analysis of FETI methods for multiscale PDEs. Numer. Math. **111**(2), 293–333 (2008). https://doi.org/10.1007/s00211-008-0186-2
11. Scheichl, R., Vainikko, E.: Additive Schwarz with aggregation-based coarsening for elliptic problems with highly variable coefficients. Computing **80**(4), 319–343 (2007). https://doi.org/10.1007/s00607-007-0237-z
12. Smith, B.F., Bjørstad, P.E., Gropp, W.D.: Domain Decomposition: Parallel Multilevel Methods for Elliptic Partial Differential Equations. Cambridge University Press, Cambridge (1996)
13. Spillane, N., Dolean, V., Hauret, P., Nataf, F., Pechstein, C., Scheichl, R.: Abstract robust coarse spaces for systems of PDEs via generalized eigenproblems in the overlaps. Numer. Math. **126**(4), 741–770 (2014). https://doi.org/10.1007/s00211-013-0576-y
14. Toselli, A., Widlund, O.: Domain Decomposition Methods–Algorithms and Theory. SSCM, vol. 34. Springer, Berlin (2005). https://doi.org/10.1007/b137868

MATLAB Implementation of C1 Finite Elements: Bogner-Fox-Schmit Rectangle

Jan Valdman[1,2(✉)] ⓘ

[1] Institute of Mathematics, Faculty of Science, University of South Bohemia,
Branišovská 31, 37005 České Budějovice, Czech Republic
[2] Institute of Information Theory and Automation, The Czech Academy of Sciences,
Pod vodárenskou věží 4, 18208 Praha 8, Czech Republic
jan.valdman@utia.cas.cz

Abstract. Rahman and Valdman (2013) introduced a new vectorized way to assemble finite element matrices. We utilize underlying vectorization concepts and extend MATLAB codes to implementation of Bogner-Fox-Schmit C1 rectangular elements in 2D. Our focus is on the detailed construction of elements and simple computer demonstrations including energies evaluations and their visualizations.

Keywords: MATLAB vectorization · Finite elements · Energy evaluation

1 Introduction

Boundary problems with fourth order elliptic operators [3] appear in many applications including thin beams and plates and strain gradient elasticity [5]. Weak formulations and implementations of these problems require H2-conforming finite elements, leading to C^1 continuity (of functions as well as of their gradients) of approximations over elements edges. This continuity condition is generally technical to achieve and few types of finite elements are known to guarantee it. We consider probably the simplest of them, the well known Bogner-Fox-Schmit rectangle [2], i.e., a rectangular C^1 element in two space dimensions.

We are primarily interested in explaining the construction of BFS elements, their practical visualization and evalutions. Our MATLAB implementation is based on codes from [1,6,9]. The main focus of these papers were assemblies of finite element matrices and local element matrices were computed all at once by array operations and stored in multi-dimensional arrays (matrices). Here, we utilize underlying vectorization concepts without the particular interest in corresponding FEM matrices. More details on our recent implementations of C^1 models in nonlinear elastic models of solids can be found in [4,7,8]. The complementary software to this paper is available at https://www.mathworks.com/matlabcentral/fileexchange/71346 for download and testing.

The work was supported by the Czech Science Foundation (GACR) through the grant GA18-03834S.

© Springer Nature Switzerland AG 2020
R. Wyrzykowski et al. (Eds.): PPAM 2019, LNCS 12044, pp. 256–266, 2020.
https://doi.org/10.1007/978-3-030-43222-5_22

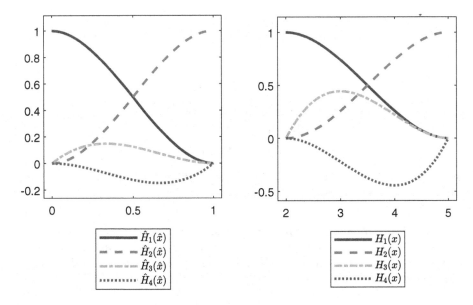

Fig. 1. Reference basis functions $\hat{H}_i, i = 1, \ldots, 4$ on $[0, 1]$ (left) and example of actual basis functions $H_i, i = 1, \ldots, 4$ on $[a, b] = [2, 5]$ (right).

2   Construction of C^1 Finite Elements

2.1   Hermite Elements in 1D

We define four cubic polynomials

$$
\begin{aligned}
\hat{H}_1(\hat{x}) &:= 2\hat{x}^3 - 3\hat{x}^2 + 1, \\
\hat{H}_2(\hat{x}) &:= -2\hat{x}^3 + 3\hat{x}^2, \\
\hat{H}_3(\hat{x}) &:= \hat{x}^3 - 2\hat{x}^2 + \hat{x}, \\
\hat{H}_4(\hat{x}) &:= \hat{x}^3 - \hat{x}^2
\end{aligned}
\tag{1}
$$

over a reference interval $\hat{I} := [0, 1]$ and can easily check the conditions:

$$
\begin{aligned}
\hat{H}_1(0) = 1, \quad & \hat{H}_1(1) = 0, \quad & \hat{H}_1'(0) = 0, \quad & \hat{H}_1'(0) = 0, \\
\hat{H}_2(0) = 0, \quad & \hat{H}_2(1) = 1, \quad & \hat{H}_2'(0) = 0, \quad & \hat{H}_2'(0) = 0, \\
\hat{H}_3(0) = 0, \quad & \hat{H}_3(1) = 0, \quad & \hat{H}_3'(0) = 1, \quad & \hat{H}_3'(0) = 0, \\
\hat{H}_4(0) = 0, \quad & \hat{H}_4(1) = 0, \quad & \hat{H}_4'(0) = 0, \quad & \hat{H}_4'(0) = 1,
\end{aligned}
\tag{2}
$$

so only one value or derivative is equal to 1 and all other three values are equal to 0. These cubic functions create a finite element basis on $\hat{I}$. More generally, we define

$$
\begin{aligned}
H_1(x) &:= \hat{H}_1(\hat{x}(x)), \\
H_2(x) &:= \hat{H}_2(\hat{x}(x)), \\
H_3(\hat{x}) &:= h\,\hat{H}_3(\hat{x}(x)), \\
H_4(\hat{x}) &:= h\,\hat{H}_4(\hat{x}(x))
\end{aligned}
\tag{3}
$$

for $x \in I := [a, b]$, where $\hat{x}(x) := (x - a)/h$ is an affine mapping from I to $\hat{I}$ and h denotes the interval I size

$$
h := b - a.
$$

These functions are also cubic polynomials and satisfy again the conditions (2) with function arguments $0, 1$ replaced by a, b. They create actual finite element basis which ensures C^1 continuity of finite element approximations. The chain rule provides higher order derivatives:

$$
\begin{aligned}
H_1'(x) &= \hat{H}_1'(\hat{x})\,/h, & H_1''(x) &= \hat{H}_1''(\hat{x})\,/h^2, \\
H_2'(x) &= \hat{H}_2'(\hat{x})\,/h, & H_2''(x) &= \hat{H}_2''(\hat{x})\,/h^2, \\
H_3'(x) &= \hat{H}_3'(\hat{x}), & H_3''(x) &= \hat{H}_3''(\hat{x})\,/h, \\
H_4'(x) &= \hat{H}_4'(\hat{x}), & H_4''(x) &= \hat{H}_4''(\hat{x})\,/h.
\end{aligned}
\tag{4}
$$

Example 1. Example of basis functions defined on reference and actual intervals are shown in Fig. 1 and pictures can be reproduced by

```
draw_C1basis_1D
```

script located in the main folder.

2.2 Bogner-Fox-Schmit Rectangular Element in 2D

Products of functions

$$
\tilde{\varphi}_{j,k}(\hat{x}, \hat{y}) := \hat{H}_j(\hat{x})\,\hat{H}_k(\hat{y}), \quad j, k = 1, \dots, 4
$$

define 16 Bogner-Fox-Schmit (BFS) basis functions on a reference rectangle $\hat{R} := [0, 1] \times [0, 1]$. For practical implementations, we reorder them as

$$
\hat{\varphi}_i(\hat{x}, \hat{y}) := \tilde{\varphi}_{j_i, k_i}(\hat{x}, \hat{y}), \quad i = 1, \dots, 16,
\tag{5}
$$

where sub-indices are ordered in a sequence

$$
(j_i, k_i)_{i=1}^{16} = \begin{aligned}
&\{(1,1),\ (2,1),\ (2,2),\ (1,2), \\
&\ (3,1),\ (4,1),\ (4,2),\ (3,2), \\
&\ (1,3),\ (2,3),\ (2,4),\ (1,4), \\
&\ (3,3),\ (4,3),\ (4,4),\ (3,4)\}.
\end{aligned}
\tag{6}
$$

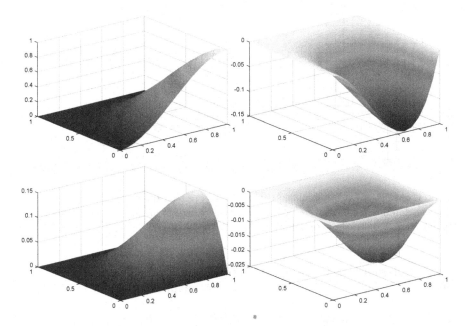

Fig. 2. Bogner-Fox-Schmit basis functions $\hat{\varphi}_i(\hat{x}, \hat{y})$ for $i = 2$ (top left), $i = 6$ (top right), $i = 8$ (bottom left), $i = 13$ (bottom right) defined over a reference rectangle $\hat{R} = [0, 1] \times [0, 1]$.

With this ordering, a finite element approximation $v \in C^1(\hat{R})$ rewrites as a linear combination

$$v(\hat{x}, \hat{y}) = \sum_{i=1}^{16} v_i \, \hat{\varphi}_i(\hat{x}, \hat{y}),$$

where:

- coefficients $v_1, \ldots, v_4$ specify values of v,
- coefficients $v_5, \ldots, v_8$ specify values of $\frac{\partial v}{\partial x}$,
- coefficients $v_9, \ldots, v_{12}$ specify values of $\frac{\partial v}{\partial y}$,
- coefficients $v_{13}, \ldots, v_{16}$ specify values of $\frac{\partial^2 v}{\partial x \partial y}$

at nodes

$$\hat{N}_1 := [0, 0], \;\; \hat{N}_2 := [1, 0], \;\; \hat{N}_3 := [1, 1], \;\; \hat{N}_4 := [0, 1].$$

Example 2. Four (out of 16) reference basis functions corresponding to the node $\hat{N}_2$ are shown in Fig. 2 and pictures can be reproduced by

```
draw_C1basis_2D
```

script located in the main folder.

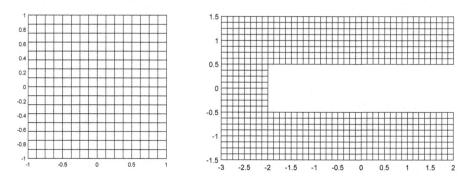

Fig. 3. Examples of a triangulation in rectangles: of a square domain (left) and of a pincers domain (right) taken from [7] and used for nonlinear elasticity simulations satisfying a non-selfpenetration condition.

More Implementation Details on Functions Evaluations. For a general rectangle $R := [a, b] \times [c, d]$, we define an affine mapping

$$(\hat{x}, \hat{y})(x, y) := ((x - a)/h_x, (y - c)/h_y),$$

from R to $\hat{R}$, where the rectangular lengths are

$$h_x := b - a, \quad h_y := d - c.$$

It enables us to define BFS basis functions on R as

$$\varphi_i(x, y) := \hat{H}_{j_i} \left(\frac{x - a}{h_x} \right) \hat{H}_{k_i} \left(\frac{y - c}{h_y} \right), \quad i = 1, \dots, 16, \tag{7}$$

Based on (4), higher order derivatives up to the second order,

$$\frac{\partial \varphi_i}{\partial x}, \ \frac{\partial \varphi_i}{\partial y}, \ \frac{\partial^2 \varphi_i}{\partial x^2}, \ \frac{\partial^2 \varphi_i}{\partial y^2}, \ \frac{\partial^2 \varphi_i}{\partial x \partial y}, \quad i = 1, \dots, 16 \tag{8}$$

can be derived as well. All basis functions (7) are evaluated by the function

    shapefun(points',etype,h)

and their derivatives (8) by the function

    shapeder(points',etype,h)

For BFS elements, we have to set `etype='C1'` and a vector of rectangular lengths `h=[hx,hy]`. The matrix `points` then contains a set of points $\hat{x} \in \hat{R}$ in a reference element at which functions are evaluated. Both functions are vectorized and their outputs are stored as vectors, matrices or three-dimensional arrays.

Example 3. The command

```
[shape]=shapefun([0.5 0.5]','C1',[1 1])
```

returns a (column) vector shape $\in \mathbb{R}^{16 \times 1}$ of all BFS basis function defined on $\hat{R} := [0,1] \times [0,1]$ and evaluated in the rectangular midpoint $[0.5, 0.5] \in \hat{R}$. The command

```
[dshape]=shapeder([0.5 0.5]','C1',[2 3])
```

returns a three-dimensional array dshape $\in \mathbb{R}^{16 \times 1 \times 5}$ of all derivatives up to the second order of all BFS basis function defined on a general rectangle with lengths $h_x = 2$ and $h_y = 3$ and evaluated in the rectangular midpoint $[0.5, 0.5] \in \hat{R}$.

For instance, if $R := [1,3] \times [2,5]$, values of all derivatives are evaluated in the rectangular midpoint $[2, 3.5] \in R$.

More generally, if points $\in \mathbb{R}^{np \times 2}$ consists of $np > 1$ points, then shapefun return a matrix of size $\mathbb{R}^{16 \times np}$ and shapeder returns a three-dimensional array of size $\mathbb{R}^{16 \times np \times 5}$.

2.3   Representation and Visualization of C^1 Functions

Let us assume a triangulation $\mathcal{T}(\Omega)$ into rectangles of a domain Ω. In correspondence with our implementation, we additionally assume that all rectangles are of the same size, i.e., with lengths $h_x, h_y > 0$. Examples of $\mathcal{T}(\Omega)$ are given in Fig. 3.

Let $\mathcal{N}$ denotes the set of all rectangular nodes and $|\mathcal{N}| := n$ the number of them. A C^1 function $v \in \mathcal{T}(\Omega)$ is represented in BSF basis by a matrix

$$VC1 = \begin{pmatrix} v(N_1), & \frac{\partial v}{\partial x}(N_1), & \frac{\partial v}{\partial y}(N_1), & \frac{\partial^2 v}{\partial x \partial y}(N_1) \\ \vdots & \vdots & \vdots & \vdots \\ v(N_n), & \frac{\partial v}{\partial x}(N_n), & \frac{\partial v}{\partial y}(N_n), & \frac{\partial^2 v}{\partial x \partial y}(N_n) \end{pmatrix}$$

containing values of $v, \frac{\partial v}{\partial x}, \frac{\partial v}{\partial y}, \frac{\partial^2 v}{\partial x \partial y}$ in all nodes of $\mathcal{T}(\Omega)$. In the spirit of the finite element method, values of v on each rectangle $T \in \mathcal{T}(\Omega)$ are obtained by an affine mapping to the reference element $\hat{R}$. Our implementations allows to evaluate and visualize continuous fields

$$v, \frac{\partial v}{\partial x}, \frac{\partial v}{\partial y}, \frac{\partial^2 v}{\partial x \partial y}$$

and also two additional (generally discontinuous) fields

$$\frac{\partial^2 v}{\partial x^2}, \frac{\partial^2 v}{\partial y^2}.$$

It is easy to evaluate a C^1 function in a particular element point for all elements (rectangles) at once. A simple matrix-matrix MATLAB multiplication

```
Vfun=VC1_elems*shape
```

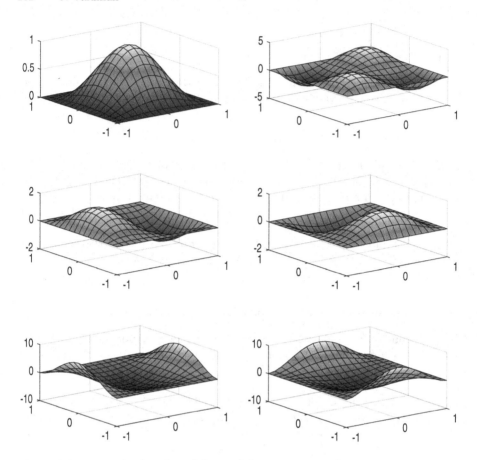

Fig. 4. A function $v(x,y) = (1-x^2)^2(1-y^2)^2$ on $\Omega = (-1,1)^2$ represented in terms of BSF elements. Separate pictures show: v (top left), $\frac{\partial^2 v}{\partial x \partial y}$ (top right), $\frac{\partial v}{\partial x}$ (middle left), $\frac{\partial v}{\partial y}$ (middle right), $\frac{\partial^2 v}{\partial x^2}$ (bottom left), $\frac{\partial^2 v}{\partial y^2}$ (bottom right).

where a matrix `VC1_elems` $\in \mathbb{R}^{ne \times 16}$ contains in each row all 16 coefficients (taken from `VC1`) corresponding to each element (ne denotes a number of elements) returns a matrix `Vfun` $\in \mathbb{R}^{ne \times np}$ containing all function values in all elements and all points. Alternate multiplications

```
V1=VC1_elems*squeeze(dshape(:,1,:));          % Dx
V2=VC1_elems*squeeze(dshape(:,2,:));          % Dy
V11=VC1_elems*squeeze(dshape(:,3,:));         % Dxx
V22=VC1_elems*squeeze(dshape(:,4,:));         % Dyy
V12=VC1_elems*squeeze(dshape(:,5,:));         % Dxy
```

return matrices `V1, V2, V11, V22, V12` $\in \mathbb{R}^{ne \times np}$ containing values of all derivatives up to the second order in all elements and all points. A modification for evaluation of function values at particular edges points is also available and essential for instance for models with energies formulated on boundary edges [8].

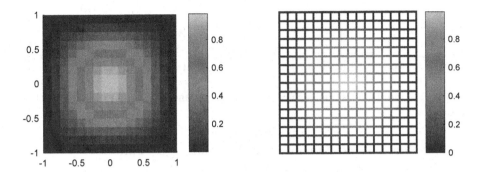

Fig. 5. A function $v(x,y) = (1 - x^2)^2(1 - y^2)^2$ on $\Omega = (-1,1)^2$ and its values in elements midpoints (left) and edges midpoints (right).

Example 4. We consider a function

$$v(x,y) = (1 - x^2)^2(1 - y^2)^2 \tag{9}$$

on the domain $\Omega = (-1,1)^2$. This function was also used in [4] as an initial vertical displacement in a time-dependent simulation of viscous von Kármán plates.

To represent v in terms of BFS elements, we additionally need to know values of

$$\frac{\partial v}{\partial x}(x,y) = -4x(1 - x^2)(1 - y^2)^2 \tag{10}$$

$$\frac{\partial v}{\partial x}(x,y) = -4y(1 - x^2)^2(1 - y^2) \tag{11}$$

$$\frac{\partial^2 v}{\partial x \partial y}(x,y) = 16xy(1 - x^2)(1 - y^2) \tag{12}$$

in nodes of a rectangular mesh $\mathcal{T}(\Omega)$. The function and its derivatives up to the second order are shown in Fig. 4 and its values in elements and edges midpoints in Fig. 5.

All pictures can be reproduced by

```
draw_C1example_2D
```

script located in the main folder.

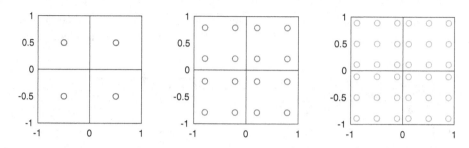

Fig. 6. 1, 4 and 9 Gauss points shown on actual rectangles of a square domain with 4 rectangles.

2.4   Evaluation and Numerical Integration of C^1 Function

Various energy formulations include evaluations of integrals of the types

$$||v||^2 := \int_\Omega |v(x,y)|^2 \, dx \, dy, \tag{13}$$

$$||\nabla v||^2 := \int_\Omega |\nabla v(x,y)|^2 \, dx \, dy, \tag{14}$$

$$||\nabla^2 v||^2 := \int_\Omega |\nabla^2 v(x,y)|^2 \, dx \, dy, \tag{15}$$

$$(f,v) := \int_\Omega f \, v \, dx \, dy, \tag{16}$$

where $v \in H^2(\Omega)$ and $f \in L^2(\Omega)$ is given. The expression

$$(||v||^2 + ||\nabla v||^2 + ||\nabla^2 v||^2)^{1/2}$$

then defines the full norm in the Sobolev space $H^2(\Omega)$. For v represented in the BFS basis we can evaluate above mentioned integrals numerically by quadrature rules. Our implementation provides three different rules with 1, 4 or 9 Gauss points. Each quadrature rule is defined by coordinates of all Gauss points and their weights.

Example 5. Gauss points are displayed in Fig. 6 and all pictures can be reproduced by

```
draw_ips
```

script located in the main folder.

Example 6. An analytical integration for the function v from (9) and $f = x^2 y^2$ reveals that

$$||v||^2 = 65536/99225 \approx 0.660478710002520,$$
$$||\nabla v||^2 = 131072/33075 \approx 3.962872260015117,$$
$$||\nabla^2 v||^2 = 65536/1225 \approx 53.498775510204084,$$
$$(f,v) = 256/11025 \approx 0.023219954648526$$

for a domain $\Omega = (-1,1)^2$. We consider a sequence of uniformly refined meshes with levels 1–10:

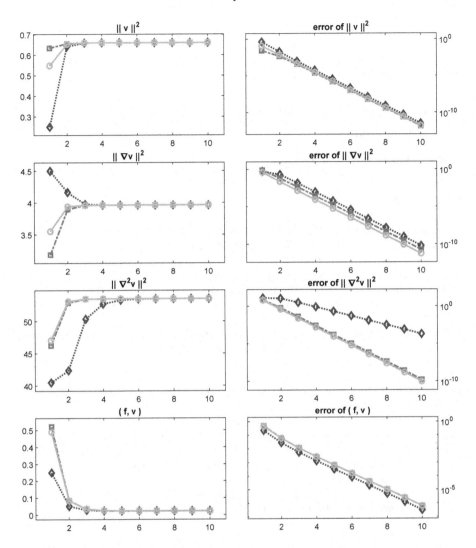

Fig. 7. Values of integrals (the left column) and their absolute error (the right column) for levels 1–10 of uniform refinements using three different quadrature rules: 1 Gauss point - blue lines with diamonds, 4 Gauss points - yellow lines with circles, 9 Gauss points - red lines with squares. (Color figure online)

- the coarsest (level = 1) mesh with 9 nodes and 4 elements is shown in Fig. 6,
- a finer (level = 4) mesh with 289 nodes and 256 elements is shown in Fig. 3 (left),
- the finest (level = 10) mesh consists of 1.050.625 nodes and 1.048.576 elements, not shown here.

Figure 7 the depicts the convergence of numerical quadratures to the exact values above. We notice that all three quadrature rules provide the same rates of convergence. The only exception is the evaluation of the second gradient integral $||\nabla^2 v||^2$, where the numerical quadrature using 1 Gauss point deteriorates the convergence.

All pictures can be reproduced by

```
start_integrate
```

script located in the main folder.

References

1. Anjam, I., Valdman, J.: Fast MATLAB assembly of FEM matrices in 2D and 3D: edge elements. Appl. Math. Comput. **267**, 252–263 (2015)
2. Bogner, F.K., Fox, R.L., Schmit, L.A.: The generation of inter-element compatible stiffness and mass matrices by the use of interpolation formulas. In: Proceedings of the Conference on Matrix Methods in Structural Mechanics, pp. 397–444 (1965)
3. Ciarlet, P.G.: The Finite Element Method for Elliptic Problems. SIAM, Philadelphia (2002)
4. Friedrich, M., Kružík, M., Valdman, J.: Numerical approximation of von Kármán viscoelastic plates. Discret. Contin. Dyn. Syst. - Ser. S (accepted)
5. Forest, S.: Micromorphic approach for gradient elasticity, viscoplasticity, and damage. J. Eng. Mech. **135**(3), 117–131 (2009)
6. Harasim, P., Valdman, J.: Verification of functional a posteriori error estimates for an obstacle problem in 2D. Kybernetika **50**(6), 978–1002 (2014)
7. Krömer, S., Valdman, J.: Global injectivity in second-gradient nonlinear elasticity and its approximation with penalty terms. Math. Mech. Solids **24**(11), 3644–3673 (2019)
8. Krömer, S., Valdman, J.: Surface penalization of self-interpenetration in second-gradient nonlinear elasticity (in preparation)
9. Rahman, T., Valdman, J.: Fast MATLAB assembly of FEM matrices in 2D and 3D: nodal elements. Appl. Math. Comput. **219**, 7151–7158 (2013)

Simple Preconditioner for a Thin Membrane Diffusion Problem

Piotr Krzyżanowski[(✉)]

University of Warsaw, Warsaw, Poland
`p.krzyzanowski@mimuw.edu.pl`

Abstract. A diffusion through a thin membrane problem discussed in [13] is discretized with a variant of the composite h-p discontinuous Galerkin method. A preconditioner based on the additive Schwarz method is introduced, and its convergence properties are investigated in numerical experiments.

Keywords: Preconditioner · Thin membrane · Additive Schwarz method · Discontinuous Galerkin

1 Introduction

In various models one has to deal with a situation when two different materials are connected through a permeable interface. One example of such situation comes from the biology of the cell, whose nucleus is surrounded by a thin membrane. Chemicals inside the cell diffuse not only inside both the nucleus and in the cytoplasm, but they also pass through the membrane as well. A mathematical model of such phenomenon has been considered in [13], leading to a system of several nonlinear, time dependent PDEs, additionally coupled by specific boundary conditions posed on the inner interface. In this paper we will investigate a simplified problem, hoping our approach can later be applied in more complicated models.

Let us denote by Ω the space occupied by the cell. It naturally decomposes into the nucleus (or nuclei), denoted here Ω_1 and the surrounding cytoplasm Ω_2, so that $\bar{\Omega} = \bar{\Omega}_1 \cup \bar{\Omega}_2$ and $\Omega_1 \cap \Omega_2 = \emptyset$ (see Fig. 1 for a bit more complicated example). The interface between the nucleus and the outer cell will be denoted $\Gamma = \partial\Omega_1 \cap \partial\Omega_2$.

The essence of the thin interface model in [13] can be expressed as a system of two PDEs of the following form, where u_1, u_2 stand for the (unknown) concentration of certain substance in Ω_1, Ω_2, respectively:

$$
\begin{aligned}
-\operatorname{div}(\varrho_1 \nabla u_1) + K_1 u_1 &= F_1 \text{ in } \Omega_1 \\
-\operatorname{div}(\varrho_2 \nabla u_2) + K_2 u_2 &= F_2 \text{ in } \Omega_2
\end{aligned}
\tag{1}
$$

This research has been partially supported by the Polish National Science Centre grant 2016/21/B/ST1/00350.

R. Wyrzykowski et al. (Eds.): PPAM 2019, LNCS 12044, pp. 267–276, 2020.
https://doi.org/10.1007/978-3-030-43222-5_23

with interface conditions

$$-\varrho_1 \nabla u_1 \cdot n_1 = G \cdot (u_1 - u_2) \text{ on } \Gamma,$$
$$-\varrho_2 \nabla u_2 \cdot n_2 = G \cdot (u_2 - u_1) \text{ on } \Gamma,$$

(2)

completed with non-permeability outer boundary condition,

$$-\varrho_2 \nabla u_2 \cdot n = 0 \text{ on } \partial\Omega.$$

(3)

Here, $\varrho_1, \varrho_2, G > 0$ and $K_1, K_2 \geq 0$ are prescribed constants, while for the source terms we assume $F_i \in L^2(\Omega_i)$, $i = 1, 2$. Let us observe that the interface condition (2) is different from standard transmission conditions,

$$u_1 = u_2, \qquad -\varrho_1 \nabla u_1 \cdot n_1 = \varrho_2 \nabla u_2 \cdot n_2,$$

which guarantee the continuity of the solution (and its flux) across Γ. Our formulation implicitly assumes that for $x \in \Gamma$ the values of $u_1(x)$ may differ from $u_2(x)$, a feature we have to address in the finite element approximation.

Positive and constant diffusion coefficients ϱ_1, ϱ_2 can differ one from another, and so the positive reaction coefficients K_1, K_2 as well. If the thickness of modeled interface is H, then the permeability constant $G \sim 1/H$; therefore for thin domains, when $H \ll 1$, one expects $G \gg 1$.

Another approach to modelling a thin interface would be to introduce it explicitly in the model, as another part of Ω, with specific material properties. Note however that such approach would usually lead to a multiscale problem, because the thickness of the interface layer would be orders of magnitude smaller than the diameter of the domain.

Choosing to model the interface by means of (2) instead introducing it as a part of the domain has even more appeal when it comes to h-p finite element approximation: one is then free to keep the mesh size h relatively large and only increase the polynomial degree p where necessary to improve the quality of the approximate solution.

In what follows we will analyze a preconditioner for a system of algebraic equations arising from a discretization of the system (1)—(3) with composite discontinuous Galerkin (cG-dG) h-p finite element method [6]. The corresponding spaces and the discrete problem are introduced below.

1.1   Variational Formulation

The interface condition (2) makes the solution discontinuous across Γ, therefore we discretize (1) by a cG–dG finite element method. Inside regions where ϱ is constant, we will use continuous h-p discretization. In order to address the boundary conditions (2), we incorporate them directly into the bilinear form.

Multiplying (1) in Ω_i by a smooth enough function ϕ and integrating over Ω_i we obtain

$$\int_{\Omega_i} -\operatorname{div}(\varrho_i \nabla u_i)\phi + K_i u_i \phi \, dx = \int_{\Omega_i} F_i \phi \, dx.$$

By integration by parts formula, we have

$$\int_{\Omega_i} -\operatorname{div}(\varrho_i \nabla u_i)\phi \, dx = \int_{\Omega_i} \varrho_i \nabla u_i \cdot \nabla \phi \, dx - \int_{\partial \Omega_i} \varrho_i \nabla u_i \cdot n_i \phi \, ds$$

where n_i denotes the outer normal unit vector to Ω_i. Taking into account boundary conditions, we have

$$-\int_{\partial \Omega_1} \varrho_1 \nabla u_1 \cdot n_1 \phi \, ds = -\int_\Gamma \varrho_1 \nabla u_1 \cdot n_1 \phi \, ds = \int_\Gamma G(u_1 - u_2)\phi \, ds$$

and analogously,

$$-\int_{\partial \Omega_2} \varrho_2 \nabla u_2 \cdot n_2 \phi \, ds = \int_\Gamma G(u_2 - u_1)\phi \, ds.$$

Thus, the weak formulation reads as follows: Find $(u_1, u_2) \in H^1(\Omega_1) \times H^1(\Omega_2)$ such that

$$\sum_{i=1,2} \int_{\Omega_i} \varrho_i \nabla u_i \cdot \nabla \phi_i + K_i u_i \phi_i \, dx + \int_\Gamma G(u_2 - u_1)(\phi_2 - \phi_1) \, ds = \sum_{i=1,2} \int_{\Omega_i} F_i \phi_i \, dx$$

for all $(\phi_1, \phi_2) \in H^1(\Omega_1) \times H^1(\Omega_2)$.

Using simplifying notation:

$$\langle f, g \rangle_\Gamma = \int_\Gamma f(x) \, g(x) \, ds \quad \text{and} \quad (f, g)_{\Omega_i} = \int_{\Omega_i} f(x) \, g(x) \, dx$$

and $[u] = u_{|\Omega_1} - u_{|\Omega_2}$ on Γ, the variational formulation of (1) reads:

Problem 1. *Find $(u_1, u_2) \in H^1(\Omega_1) \times H^1(\Omega_2)$ such that*

$$\sum_{i=1,2} \left(\varrho_i (\nabla u_i, \nabla \phi_i)_{\Omega_i} + K_i (u_i, \phi_i)_{\Omega_i} \right) + G\langle [u], [\phi] \rangle_\Gamma = \sum_{i=1,2} (F_i, \phi_i)_{\Omega_i}$$

for all $(\phi_1, \phi_2) \in H^1(\Omega_1) \times H^1(\Omega_2)$.

With $G > 0$, the bilinear form appearing in Problem 1 is symmetric and positive definite.

1.2   Finite Element Discretization

In order not to complicate the exposition of the finite element method, we will assume from now on that Ω_1 and Ω_2 are composed of polyhedrons (and, in consequence, Ω is a polyhedron as well). Of course, in biological applications, actual domains have smooth boundaries, so we may regard polyhedral Ω_1 and Ω_2 as approximations to accurate shapes.

In each Ω_i, $i = 1, 2$, we introduce an affine, shape regular, quasi-uniform and matching simplicial triangulation $\mathcal{T}_{h_i}(\Omega_i)$, where h_i is the mesh parameter,

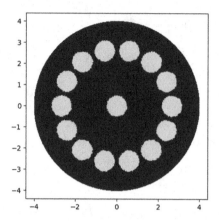

Fig. 1. Example domain and its decomposition to 15 small, circular "nuclei" (marked with light color and denoted Ω_1 in the text) surrounded with the outer cell (filled with dark color, denoted Ω_2 in the text).

i.e. $h_i = \max\{\mathrm{diam}(K) : K \in \mathcal{T}_{h_i}(\Omega_i)\}$. We will refer to $\mathcal{T}_{h_i}(\Omega_i)$ as the local triangulation of subdomain Ω_i. With $p_i \geq 1$ we define the corresponding local (continuous) finite element spaces as

$$V_{h_i}^{p_i}(\Omega_i) = \{v \in C(\Omega_i) : v_{|_K} \in \mathcal{P}^{p_i}(K) \quad \forall K \in \mathcal{T}_{h_i}(\Omega_i)\},$$

where $\mathcal{P}^{p_i}$ is the space of polynomials of degree at most p_i. This choice leads to so-called cG–dG method.

Another possibility would be to choose the space of discontinuous elementwise polynomials

$$V_{h_i}^{p_i}(\Omega_i)^{DG} = \{v \in L^2(\Omega_i) : v_{|_K} \in \mathcal{P}^{p_i}(K) \quad \forall K \in \mathcal{T}_{h_i}(\Omega_i)\},$$

(and then to replace the volume parts of the bilinear form in Problem 2 with the corresponding symmetric interior penalty form; we refer the reader to [5] for details). Such choice, however, would lead to unnecessarily larger number of unknowns as compared to the previous choice but at the same time it may be easier to deal with inside a finite element library, such as FEniCS [1].

Let $h = (h_1, h_2)$ and $p = (p_1, p_2)$ collect the parameters of local meshes and finite element spaces. We define the global fine mesh on Ω,

$$\mathcal{T}_h = \{K \in \mathcal{T}_{h_i}(\Omega_i) : i = 1, 2\}$$

and assume (for simplicity only) that this mesh is conforming. Finally we define the finite element space in which we will approximate the solution of Problem 1,

$$V_h^p = \{v \in L^2(\Omega) : v_{|_{\Omega_i}} \in V_{h_i}^{p_i}(\Omega_i), \quad i = 1, 2\}. \tag{4}$$

It consists of piecewise polynomial functions, which may be discontinuous across the interface Γ.

The finite element approximation of Problem 1 then reads:

Problem 2. *Find* $u \in V_h^p$ *such that*

$$\mathcal{A}(u, v) \equiv \sum_{i=1,2} \left(\varrho_i \left(\nabla u, \nabla \phi \right)_{\Omega_i} + K_i \left(u, \phi \right)_{\Omega_i} \right) + \sum_{e \subset \Gamma} G \cdot \langle [u], [\phi] \rangle_e = \sum_{i=1,2} (F_i, \phi)_{\Omega_i}$$

for all $\phi \in V_h^p$.

In this paper we address the practical problem of how to cope with ill conditioning of a system of algebraic equations which represent Problem 2, leaving aside theoretical questions regarding e.g. a priori error bounds of the discrete solution.

Remark 1. *Note that this form resembles the bilinear form appearing in the Symmetric Interior Penalty discontinuous Galerkin method* [6], *where on the interelement boundaries* $e = K^+ \cap K^-$ *one adds jump penalizing terms being a multiple of*

$$\frac{\max\{p^+, p^-\}^2}{\min\{h^+, h^-\}} \cdot \langle [u], [\phi] \rangle_e,$$

where $f^{\pm}$ *denotes the value of* f *attained on* $K^{\pm}$. *In our formulation, the penalizing coefficient is independent of* h *and* p_i, *but for very thin interfaces can be large.*

Our goal in this paper is to analyze a preconditioner based on the additive Schwarz method (ASM), see e.g. [14], which introduces in a natural way a coarse grain parallelism and improves the convergence rate of an iterative solver. Such problem has already been considered by Dryja in [6] for linear finite elements in 2D, where a multilevel ASM was designed and analyzed. For pure diffusion problems involving coefficient jumps discretized with now standard symmetric weighted interior penalty discontinuous Galerkin method [8], an optimal preconditioner was developed and analyzed in [3]. A less complex version, yet still robust with respect to diffusion coefficient jumps and with a high level of parallelism, was considered in [7] and [9]. Here, we introduce a preconditioner which simplifies the approach of [3] while retaining some of the ingredients of [10], and provide tests of the performance of the proposed method.

The rest of paper is organized as follows. In Sect. 2, a preconditioner based on the additive Schwarz method for solving the discrete problem is designed. We report on its performance in a series of numerical experiments in Sect. 3. We conclude with final remarks in Sect. 4.

2   An Additive Schwarz Preconditioner for Problem 2

It is well known—and easy to verify by simple numerical experiment—that the condition number of Problem 2 can be prohibitively large, affected by the degree of the polynomials used, the fine mesh size, the interface thickness G and by the magnitude of jumps in ϱ or K. Thus, for an iterative solution of Problem 2, some preconditioning is necessary. In this section we consider a preconditioner based on the nonoverlapping additive Schwarz method, first proposed, in a slightly different setting, in [2].

Let us introduce a decomposition of V_h^p:

$$V_h^p = V_0 + \sum_{i=1,2} V_i, \tag{5}$$

where for $i = 1, 2$ the local spaces are

$$V_i = \{v \in V_h^p : v_{|\Omega_j} = 0 \text{ for all } j \neq i\}, \tag{6}$$

so that V_i is a zero–extension of functions from $V_{h_i}^{p_i}(\Omega_i)$. Note that V_h^p is a direct sum of these local spaces.

The coarse space is

$$V_0 = \{v \in C(\Omega) : v_{|K} \in \mathcal{P}^q(K) \text{ for all } K \in \mathcal{T}_h\}$$

with $q = \min\{p_1, p_2\}$. Observe that functions from V_0 are continuous, also across Γ. The choice of the coarse space is inspired by Antonietti et al. [3] and notably leads to a problem whose number of unknowns is smaller than the original only by a small fraction.

Using decomposition (5) we define local operators $T_i : V_h^p \to V_i$, $i = 1, 2$, by "inexact" solvers

$$A(T_i u, v) = \mathcal{A}(u, v) \qquad \forall v \in V_i,$$

so that on each subdomain one has to solve approximately a system of linear equations for degrees of freedom restricted only to $V_{h_i}^{p_i}(\Omega_i)$; for $j \neq i$ we set $(T_i u)_{|\Omega_j} = 0$.

The coarse solve operator is $T_0 : V_h^p \to V_0$ defined analogously as

$$A_0(T_0 u, v_0) = \mathcal{A}(u, v_0) \qquad \forall v_0 \in V_0.$$

Note that on V_0, $\mathcal{A}(\cdot, \cdot)$ simplifies to a discontinuous coefficient problem in the continuous finite element space,

$$\mathcal{A}(u_0, v_0) = \sum_{i=1,2} \left(\varrho_i \left(\nabla u, \nabla \phi\right)_{\Omega_i} + K_i \left(u, \phi\right)_{\Omega_i} \right) \qquad \forall u_0, v_0 \in V_0.$$

We will assume that $A(\cdot, \cdot)$ (resp. $A_0(\cdot, \cdot)$) induces a linear operator which is spectrally equivalent to the operator induced by $\mathcal{A}(\cdot, \cdot)$ on $V_{h_i}^{p_i}(\Omega_i)$ (resp. V_0).

Finally, the preconditioned operator is

$$T = \omega_1 T_0 + \omega_2 \sum_{i=1}^{2} T_i, \tag{7}$$

where $\omega_1, \omega_2 \in \{0, 1\}$ are prescribed relaxation parameters used to "switch on/off" parts of the preconditioned operator.

2.1   The Choice of Subspace Solvers

Let us remind here that all T_i, $i = 0, \ldots, 2$, can be applied in parallel. The performance of the preconditioner is therefore affected by the specific choice of subspace solvers $A_0(\cdot, \cdot)$ and $A(\cdot, \cdot)$ and, to a smaller extent, by the choice of relaxation parameters.

For the coarse solver $A_0(\cdot, \cdot)$ we take a preconditioner for a continuous Galerkin finite element approximation of a discontinuous coefficient diffusion problem, div $\varrho \nabla u = f$. For example, a domain decomposition based, highly parallel preconditioner was investigated in [12]. Another possibility is to use an (algebraic) multigrid solver, which can also be parallelized.

For the local solvers, defined by the form $A(\cdot, \cdot)$, we choose any good preconditioner for a constant coefficient diffusion problem, or—possibly after decomposing both Ω_1 and Ω_2 into smaller subdomains if this improves load balancing and does not require a too large coarse problem—use another domain decomposition preconditioner; see [14] or [11] for a collection of such preconditioners.

3   Numerical Experiments

As the number of problem parameters is large, for the numerical tests we restrict ourselves to the case when $p_1 = p_2 =: p$ and $K_1 = K_2 =: K$. In this section we investigate, in a series of preliminary numerical experiments, how the condition number of T is affected by p, G, K and ϱ_i, $i = 1, 2$. Our implementation will use FEniCS software [1] with PETSc [4] as the linear algebra backend. As the inexact solvers $A_0(\cdot, \cdot)$ and $A(\cdot, \cdot)$ we will always choose a BoomerAMG preconditioner from the HYPRE library, with default parameters. The preconditioned system T is solved by the GMRES iterative method, with the restart parameter set to 30.

We report the number of iterations required to reduce the initial residual, measured in (unpreconditioned) ℓ^2 norm, by a factor 10^8. If the convergence does not occur in 60 iterations, we declare non-convergence, marked by a dash in the tables. The initial guess for the iteration is always equal to zero.

We consider two specific cases, depending on the choice of constants ω_1, ω_2 in (7),

$$T^{\text{Std}} = T_0 + T_1 + T_2, \qquad T^{\text{Loc}} = T_1 + T_2,$$

so that T^{Loc} uses local solvers only. For simplicity of the implementation we exclusively deal with local spaces $V_{h_i}^{p_i}(\Omega_i)^{DG}$, because composite discretizations involving continuous functions are quite difficult to work with in FEniCS. It can be expected that this choice can potentially adversely affect the performance of our preconditioners, as fully DG formulation adds another level of nonconformity to the finite element solution.

Our example domain is depicted in Fig. 1. If not specified otherwise, Ω was triangulated with unstructured, quasi-uniform mesh consisting of roughly 16,500 triangles, while the finite element functions were elementwise discontinuous polynomials of degree $p = 2$.

The default values of constants defining Problem 2 were $\varrho_1 = \varrho_2 = K = G = 1$. In subsequent experiments, we tested how the number of iterations was influenced by changing some of these values.

Table 1. Dependence of the number of iterations on p

p	1	2	4
T^{Std}	19	24	46
T^{Loc}	18	19	23

Table 2. Dependence of the number of iterations on ϱ_1, with fixed $\varrho_2 = 1$

ϱ_1	10^0	10^2	10^4	10^8
T^{Std}	19	26	27	27
T^{Loc}	18	18	18	18

It follows from Table 1 that both investigated preconditioners are robust with respect to changes in the degree of the polynomial used in h-p approximation, with T^{Loc} showing almost no dependence on p.

Table 2 confirms that the number of iterations remains independent of the jump in the diffusion coefficient, provided G is not too large. It however is quite surprising that T^{Loc} performs better that T^{Std}.

Table 3. Dependence of the number of iterations on G and ϱ_1

	G	10^0	10^2	10^4
$\varrho_1 = 10^0$	T^{Std}	24	29	—
	T^{Loc}	19	—	—
$\varrho_1 = 10^4$	T^{Std}	27	32	33
	T^{Loc}	18	—	—
$\varrho_1 = 10^8$	T^{Std}	27	43	42
	T^{Loc}	18	—	—

On the other hand, Table 3 indicates the limitations of both preconditioners when G is relatively large. As expected, T^{Std} is the only version capable of dealing with large G, because it essentially captures the most of the accurate solution (which for large G is almost continuous). Interestingly, it seems its range of applicability is limited by the condition $G \leq \varrho_1$. Another set of experiments, with ϱ_2 increasing (not provided here), seems to support this conjecture.

Table 4. Dependence of the number of iterations on G and K

	G	10^0	10^2	10^4
$K = 10^0$	T^{Std}	24	29	—
	T^{Loc}	19	—	—
$K = 10^4$	T^{Std}	12	14	—
	T^{Loc}	7	13	57
$K = 10^8$	T^{Std}	10	6	—
	T^{Loc}	6	6	41

From the point of view of discretizations of the evolutionary variant of the system (1)–(3) it is also important how the preconditioner works for large values of K. Indeed, after discretizing in time, one ends up with reaction terms $K_i \sim 1/\tau$, where $\tau \ll 1$ is the time step. It follows from Table 4 that then T^{Loc} has an edge over T^{Std}, working for a larger range of G, while being at the same time much cheaper.

4   Conclusions

An h-p finite element discretization method for a thin membrane diffusion problem has been presented, based on the composite or fully discontinuous Galerkin method. A family of relatively simple, additive Schwarz, inherently parallel preconditioners was proposed for the iterative solution of the discrete problem.

Numerical experiments for discretizations using fully discontinuous Galerkin formulation show that none of these preconditioners is perfectly robust with respect to all parameters of the discrete problem, they proved applicable in a wide range of their values, especially for very thin membranes. The case of composite discretization needs further experiments which will be presented elsewhere.

Acknowledgement. The author would like to thank two anonymous referees whose comments and remarks helped to improve the paper. This research has been partially supported by the Polish National Science Centre grant 2016/21/B/ST1/00350.

References

1. Alnæs, M.S., et al.: The FEniCS project version 1.5. Arch. Numer. Softw. **3**(100) (2015). https://doi.org/10.11588/ans.2015.100.20553
2. Antonietti, P.F., Houston, P.: A class of domain decomposition preconditioners for hp-discontinuous Galerkin finite element methods. J. Sci. Comput. **46**(1), 124–149 (2011). https://doi.org/10.1007/s10915-010-9390-1
3. Antonietti, P.F., Sarti, M., Verani, M., Zikatanov, L.T.: A uniform additive Schwarz preconditioner for high-order discontinuous Galerkin approximations of elliptic problems. J. Sci. Comput. **70**(2), 608–630 (2017). https://doi.org/10.1007/s10915-016-0259-9

4. Balay, S., et al.: PETSc users manual. Technical report ANL-95/11 - revision 3.8, Argonne National Laboratory (1995). http://www.mcs.anl.gov/petsc
5. Di Pietro, D.A., Ern, A.: Mathematical Aspects of Discontinuous Galerkin Methods, Mathématiques & Applications, vol. 69. Springer, Heidelberg (2012). https://doi.org/10.1007/978-3-642-22980-0. (Berlin) [Mathematics & Applications]
6. Dryja, M.: On discontinuous Galerkin methods for elliptic problems with discontinuous coefficients. Comput. Methods Appl. Math. **3**(1), 76–85 (2003). (electronic)
7. Dryja, M., Krzyżanowski, P.: A massively parallel nonoverlapping additive Schwarz method for discontinuous Galerkin discretization of elliptic problems. Num. Math. **132**(2), 347–367 (2015).https://doi.org/10.1007/s00211-015-0718-5
8. Ern, A., Stephansen, A.F., Zunino, P.: A discontinuous Galerkin method with weighted averages for advection-diffusion equations with locally small and anisotropic diffusivity. IMA J. Numer. Anal. **29**(2), 235–256(2009). https://doi.org/10.1093/imanum/drm050
9. Krzyżanowski, P.: On a nonoverlapping additive Schwarz method for h-p discontinuous Galerkin discretization of elliptic problems. Numer. Meth. PDEs **32**(6), 1572–1590 (2016)
10. Krzyżanowski, P.: Nonoverlapping three grid additive Schwarz for hp-DGFEM with discontinuous coefficients. In: Bjørstad, P.E., et al. (eds.) DD 2017. LNCSE, vol. 125, pp. 455–463. Springer, Cham (2018). https://doi.org/10.1007/978-3-319-93873-8_43
11. Mathew, T.: Domain Decomposition Methods for the Numerical Solution of Partial Differential Equations. LNCSE, 1st edn. Springer, Berlin (2008). https://doi.org/10.1007/978-3-540-77209-5
12. Pavarino, L.F.: Additive Schwarz methods for the p-version finite element method. Numerische Mathematik **66**(1), 493–515 (1993). https://doi.org/10.1007/BF01385709
13. Sturrock, M., Terry, A.J., Xirodimas, D.P., Thompson, A.M., Chaplain, M.A.J.: Influence of the nuclear membrane, active transport, and cell shape on the Hes1 and p53–Mdm2 pathways: insights from spatio-temporal modelling. Bull. Math. Biol. **74**(7), 1531–1579 (2012). https://doi.org/10.1007/s11538-012-9725-1
14. Toselli, A., Widlund, O.: Domain Decomposition Methods–Algorithms and Theory. SSCM, vol. 34. Springer, Berlin (2005). https://doi.org/10.1007/b137868

A Numerical Scheme
for Evacuation Dynamics

Maria Gokieli[1(✉)] and Andrzej Szczepańczyk[2]

[1] Faculty of Mathematics and Natural Sciences - School of Exact Sciences,
Cardinal Stefan Wyszyński University, Wóycickiego 1/3, 01-938 Warsaw, Poland
m.gokieli@uksw.edu.pl
[2] ICM, University of Warsaw, Tyniecka 15/17, 02-630 Warsaw, Poland
as402313@icm.edu.pl

Abstract. We give a stability condition for a semi–implicit numerical scheme and prove unconditional stability for an implicit scheme for a nonlinear advection – diffusion equation, meant as a model of crowd dynamics. Numerical stability is given for a wider class of equations and schemes.

Keywords: Finite elements method · CFL condition · Stability

1 Introduction

We consider a macroscopic description of how pedestrians exit a space, typically a room. We identify the crowd through the pedestrians' density, say $\rho = \rho(t,x)$, and assume that the crowd behavior is well described by

$$\partial_t \rho + \nabla \cdot \left(\rho \, \overrightarrow{V} \right) - \kappa \, \Delta \rho = 0 \qquad \text{in } \mathbb{R}^+ \times \Omega \,, \tag{1}$$

which is a regularization ($\kappa > 0$) of the continuity equation

$$\partial_t \rho + \text{div} \left(\rho \, \overrightarrow{V} \right) = 0 \,, \qquad \text{in } \mathbb{R}^+ \times \Omega \,, \tag{2}$$

where $\Omega \subset \mathbb{R}^2$ is the environment available to pedestrians, $\overrightarrow{V} = \overrightarrow{V}(x, \rho) \in \mathbb{R}^2$ is the velocity of the individual at x, given the presence of the density ρ. The (small) parameter $\kappa > 0$ describes the diffusion part, allowing people to spread independently of the direction they are given so as to reach the exit.

The velocity vector should be given as a function of x, possibly also of $\rho(x)$ or even some nonlocal average of ρ. Several choices for the velocity function are available in the literature, see for instance [1,4–6] for velocities depending

Supported by ICM, University of Warsaw.

R. Wyrzykowski et al. (Eds.): PPAM 2019, LNCS 12044, pp. 277–286, 2020.
https://doi.org/10.1007/978-3-030-43222-5_24

nonlocally on the density, and [7, Sect. 4.1], [2, 10] for velocities depending locally on the density. In this last case, the following assumption is usually made:

$$\vec{V} = \vec{V}(x, \rho) = v(\rho)\,\vec{w}(x), \quad \text{where } \vec{w} \text{ is a vector field in } \Omega,$$
$$\text{and } v : \mathbb{R} \to [0, v_{\max}) \text{ is a scalar function, smooth and non-increasing.} \tag{3}$$

So, $\vec{w}(x)$, often normalized, is here the direction given for an individual at x, and $v(\rho)$ its velocity value, translating the common attitude of moving faster when the density is lower. In this case, the velocity value depends on the density, but not on its direction. The correction proposed in (1) allows a direction change by spreading, and seems to be a realistic model of the crowd behaviour.

We assume that the boundary $\partial\Omega$ of Ω is a union of three disjoint parts: the walls Γ_w, the exit Γ and the corners Γ_c. The set of corners is finite; Γ_w and Γ possess a field of exterior normal vectors $\vec{n}$. Their natural functions translate into the following conditions on $\vec{V} = \vec{V}(x, \rho(x))$:

$$\partial\Omega = \Gamma \cup \Gamma_w \cup \Gamma_c; \tag{4}$$
$$\vec{V} \cdot \vec{n} = 0 \text{ on } \Gamma_w, \tag{5}$$
$$\vec{V} \cdot \vec{n} > 0 \text{ on } \Gamma. \tag{6}$$

As for the boundary conditions on ρ, there are again several choices for these. A natural one seems to be a homogeneous Dirichlet or Neumann boundary condition on the walls Γ_w. As for the condition on the exit, it can be of Dirchlet or Neummann.

The main point of interest in this particular phenomenon is the widely known now Braess paradox. It consists in the fact that what seems to improve the traffic can make it slower and, on the contrary, an obstacle to the traffic may accelerate it. In evacuation, this mean that an obstacle placed in front of the exit may shorten the evacuation time. Our aim is to check this paradox on some examples, so as to be able, later on, to compare diverse obstacles and their respective effect on evacuation.

In this paper, we propose several finite elements (FE) numerical schemes to solve our problem, in particular a semi-implicit scheme, and discuss its stability. Such a scheme has been proposed in [8] for the case inspired by hydrodynamics, where $\vec{V}$ is a function of x only, and additionally $\vec{V}$ is divergence free. The advection term is then linear. We treat here the fully nonlinear case $\vec{V} = \vec{V}(\rho, x)$, and we relax the zero divergence condition to:

$$\operatorname{div}\vec{V} = \operatorname{div}\vec{V}(x, \rho(x)) \geq 0. \tag{7}$$

We also assume the existence of a weak solution, in a sense that we define in Sect. 2. This weak formulation is formulated in more abstract terms, so as the numerical schemes that we give in Sect. 3. We prove there, in Theorem 2, stability of the semi–implicit scheme, which is the main result of this paper. The CFL condition required in Theorem 2 is given in a general abstract form, we give also its special forms in the following corollaries. In Sect. 4, we show an important example of $\overrightarrow{V}$ coming from the eikonal equation, for which the condition (7) seems to be satisfied in the numerical tests.

In the forthcoming [3] we will show the well posedeness of (1) and perform the simulations of its dynamics using the numerical schemes that we present here.

2 Weak Formulation

In what follows, we assume that our problem has a unique solution. For the well–posedeness of the problem (1), we refer to the forthcoming paper [3]. Clearly, we will be dealing with its weak solution in order to build a FE approximation. As far as weak solutions are considered, it is clear that the finite set Γ_c has no importance and we can restrict our attention to the $\Gamma_w \cup \Gamma$ part of the boundary. More regularity for the solution can be easily obtained if the boundary and the boundary conditions are regular enough.

In order to build a numerical scheme, we adapt the main ideas of [8], where such a scheme is built for the linear divergence free case (which is the case (3) with $v \equiv const$ and $\mathrm{div}\,\overrightarrow{w} = 0$). These assumptions are clearly too much restrictive for our case.

Let us define $H = L^2(\Omega)$, with $(\cdot,\cdot)$ the scalar product in H and $\|\cdot\|$ the norm in H. Let $V \subset H^1(\Omega)$ be a Hilbert space, being our working space. The choice of V shall depend on the boundary conditions imposed on our problem, e.g. it is $H^1(\Omega)$ if we impose a homogeneous Neumann boundary condition on $\partial\Omega$, and $V = \{u \in H^1(\Omega) : u = 0 \text{ on } \Gamma\}$ if a homogeneous Dirichlet condition is imposed on Γ. We call W the space containing the traces of functions from V and of their normal derivatives. This flexibility in the boundary condition will be allowed by our abstract approach. We observe the following fact relating the situation we are modeling and the boundary condition on the exit.

Lemma 1. *Consider* (1) *with the boundary assumptions* (4)–(6)*; assume additionally that a homogeneous boundary condition* $\nabla \rho \cdot \overrightarrow{n}$ *is imposed on* Γ_w*. At time* $t \geq 0$*, evacuation happens, i.e.* $\dfrac{d}{dt}\displaystyle\int_\Omega \rho(t,x)\,dx < 0$*, if and only if*

$$\int_\Gamma \left\{ \kappa \nabla \rho(t,\xi) - \rho(t,\xi)\,\overrightarrow{V}(\rho(t,\xi)) \right\} \cdot \overrightarrow{n}(\xi)\,d\xi < 0. \tag{8}$$

Proof. Integrate (1) on Ω by parts.

This lemma says that a homogeneous Neumann boundary condition on ρ on the walls as well as on the exits ensures the evacuation process.

Our aim is now to write the Eq. (1) in the abstract form as:

$$(\rho_t, \eta) + A_0(\rho)(\rho,\eta) + A_1(\rho)(\rho,\eta) - A_2(\rho)(\rho,\eta) = 0, \qquad \forall \eta \in V, \tag{9}$$

where, for $i = 0, 1, 2$, $A_i(\varphi)(\cdot,\cdot)$ are bilinear forms.

Definition 1. *Let $\alpha > 0$ be an arbitrary constant. Define the operator (nonlinear in the first, linear in the second variable),*

$$B : W \times W \to W, \quad B(\varphi)\rho = \left(\kappa \nabla\rho - \frac{1}{2}\rho \overrightarrow{V}(\varphi) \right) \cdot \overrightarrow{n}$$

and the following functionals (nonlinear in the first, linear in the other variables)

$$A_0 : V \times V^2 \to \mathbb{R}, \quad A_1, A_2 : W \times W^2 \to \mathbb{R},$$

$$A_0(\varphi)(\rho, \eta) = \int_\Omega \left\{ -\rho \overrightarrow{V}(\varphi) \cdot \nabla\eta + 2\kappa \nabla\rho \cdot \nabla\eta + \kappa \rho\Delta\eta \right\},$$

$$A_1(\varphi)(\rho, \eta) \equiv A_1^\alpha(\varphi)(\rho, \eta) = \frac{1}{2\alpha} \int_{\partial\Omega} [B(\varphi)\rho - \alpha\rho] [B(\varphi)\eta - \alpha\eta],$$

$$A_2(\varphi)(\rho, \eta) \equiv A_2^\alpha(\varphi)(\rho, \eta) = \frac{1}{2\alpha} \int_{\partial\Omega} [B(\varphi)\rho + \alpha\rho] [B(\varphi)\eta + \alpha\eta].$$

We will in general omit the dependence on α in the functionals A_1, A_2. Note that A_1, A_2 are by definition positive and symmetric and that

$$A_2(\varphi)(\rho, \eta) - A_1(\varphi)(\rho, \eta) = \int_{\partial\Omega} \left\{ \rho B(\varphi)\eta + \eta B(\varphi)\rho \right\} \tag{10}$$

Lemma 2. *With the forms A_0, A_1, A_2 given in Definition 1, the weak form of (1) can be written as (9).*

Proof. Multiply (1) by η, integrate on Ω and apply integration by parts (twice on the diffusive term), use the definition of A_0 and B and (10).

Lemma 3. *Assume (5), (6), (7). The forms A_0, A_1, A_2 given in Definition 1 satisfy, for any $\varphi, \rho \in V$*

$$A_2(\varphi)(\rho, \rho) - A_1(\varphi)(\rho, \rho) \leq 2A_0(\varphi)(\rho, \rho). \tag{11}$$

Assume also a homogenous (Neumann or Dirichlet) boundary condition for ρ on Γ. Then for any $\varphi, \rho \in V$

$$A_2(\varphi)(\rho, \rho) - A_1(\varphi)(\rho, \rho) \leq 0. \tag{12}$$

Proof. By integration by parts and noting that

$$2\int_\Omega \rho\nabla\rho \cdot \overrightarrow{V} = -\int_\Omega \rho^2 \operatorname{div}\overrightarrow{V} + \int_{\partial\Omega} \rho^2 \overrightarrow{V} \cdot \overrightarrow{n},$$

the inequality (11) writes, by virtue of (10), (5),

$$\frac{1}{2} \int_\Omega \rho^2 \operatorname{div}\overrightarrow{V}(\varphi) + \kappa \int_\Omega |\nabla\rho|^2 \geq 0,$$

which is satisfied by (7). This proves (11). Now, by virtue of (10),

$$A_2(\varphi)(\rho, \rho) - A_1(\varphi)(\rho, \rho) = 2\int_{\partial\Omega} \rho B(\varphi)\rho = 2\kappa \int_\Gamma \rho\nabla\rho \cdot \overrightarrow{n} - \int_\Gamma \rho^2 \overrightarrow{V}(\varphi) \cdot \overrightarrow{n}.$$

This is clearly negative by (6) and the homogeneous boundary condition on Γ.

Remark 1. We may also assume here $\nabla \rho \cdot \vec{n} \leq 0$ on Γ.

Corollary 1. *As $A_1(\varphi)$, $A_2(\varphi)$ are symmetric and positive, (12) implies*

$$A_2(\varphi)(u,v) \leq \frac{1}{2}\left[A_1(\varphi)(u,u) + A_2(\varphi)(v,v)\right]. \tag{13}$$

Proof. By positivity, symmetry and bilinearity of A_2 we have

$$0 \leq A_2(\varphi)(u-v, u-v) = A_2(\varphi)(u,u) - 2A_2(\varphi)(u,v) + A_2(\varphi)(v,v)$$

which together with (12) gives the desired conclusion.

Remark 2. Corollary 1 implies that conditions (11)–(12), together with the additional assumption that the bilinear form $A_2(\varphi)$ is positive, imply a condition of the form of (2.2) of [8] for $\mathcal{A}_i = A_i(\varphi)$. Take $\mathcal{A}_3 = A_2(\varphi)$.

3   Numerical Scheme and Its Stability

Lemma 4. *Let $A_i : V \times V^2 \to \mathbb{R}$ for $i = 0, 1, 2$ be such that $A_i(\varphi)(\cdot, \cdot)$ are bilinear symmetric forms satisfying (11), (12) for any $\varphi, \rho \in V$. Let ρ be the solution to (9). Then $t \mapsto (\rho, \rho)(t)$ decreases, i.e. the solution ρ is L^2 stable.*

Proof. Note that (11)–(12) imply

$$A_0(\rho)(\rho, \rho) + A_1(\rho)(\rho, \rho) - A_2(\rho)(\rho, \rho) \geq 0. \tag{14}$$

From (9) one gets directly the desired conclusion.

We will now consider a family of finite element spaces $V_h \subset V$, where H is, as usual, the mesh parameter and consider approximated solutions in these spaces.

Preserving the important solution's property given by Lemma 4 is not obvious when dealing with a numerical approximation, especially with schemes that are not fully implicit. Any disturbance of this property, however, may cause a geometrically growing error. Schemes preserving it will be called stable.

Definition 2. *Let $V_h \subset V$ be a finite-dimensional vectorial subspace of V. Denote the unknown at time step n by $\rho_h^n \in V_h$. Denote the test function as $\eta_h \in V_h$. The fully implicit first order scheme for (9) reads: find a sequence $\rho_h^n \in V_h$, $n = 0, 1, \ldots$ such that for any test function $\eta_h \in V_h$*

$$\left(\frac{\rho_h^{n+1} - \rho_h^n}{\Delta t}, \eta_h\right) + A_0\left(\rho_h^{n+1}\right)\left(\rho_h^{n+1}, \eta_h\right) + A_1\left(\rho_h^{n+1}\right)\left(\rho_h^{n+1}, \eta_h\right)$$
$$-A_2\left(\rho_h^{n+1}\right)\left(\rho_h^{n+1}, \eta_h\right) = 0. \tag{15}$$

The semi–implicit first order scheme for (9) reads: find a sequence $\rho_h^n \in V_h$, $n = 0, 1, \ldots$ such that for any test function $\eta_h \in V_h$

$$\left(\frac{\rho_h^{n+1} - \rho_h^n}{\Delta t}, \eta_h\right) + A_0\left(\rho_h^n\right)\left(\rho_h^{n+1}, \eta_h\right) + A_1\left(\rho_h^n\right)\left(\rho_h^n, \eta_h\right) - A_2\left(\rho_h^n\right)\left(\rho_h^n, \eta_h\right) = 0. \tag{16}$$

The explicit *first order scheme for* (9) *reads: find a sequence* $\rho_h^n \in V_h$, $n = 0, 1, \ldots$ *such that for any test function* $\eta_h \in V_h$

$$\left(\frac{\rho_h^{n+1} - \rho_h^n}{\Delta t}, \eta_h\right) + A_0\left(\rho_h^n\right)\left(\rho_h^n, \eta_h\right) + A_1\left(\rho_h^n\right)\left(\rho_h^n, \eta_h\right) - A_2\left(\rho_h^n\right)\left(\rho_h^n, \eta_h\right) = 0.$$

(17)

We say that a scheme is stable *if* $(\rho_h^{n+1}, \rho_h^{n+1}) \leq (\rho_h^n, \rho_h^n)$ *for any* n.

Theorem 1 (Unconditional stability). *Assume that* $A_i(\varphi)(\cdot, \cdot)$, $i = 0, 1, 2$, *are bilinear forms satisfying* (11)–(12). *Then the fully implicit scheme* (15) *is unconditionally stable. If* (12) *is replaced by a stronger condition*

$$A_2(\varphi)(\rho, \eta) - A_1(\varphi)(\rho, \eta) \leq 0 \quad \forall \rho, \eta \in V,$$

(18)

the semi–implicit scheme (16) *is also unconditionally stable. If, additionally,* (11) *is replaced by a stronger condition*

$$A_2(\varphi)(\rho, \eta) - A_1(\varphi)(\rho, \eta) \leq 2A_0(\varphi)(\rho, \eta) \quad \forall \rho, \eta \in V,$$

(19)

the explicit scheme (17) *is also unconditionally stable.*

Proof. Take $\eta_h = \rho_h^{n+1}$. For the implicit scheme, use (14) and the Schwarz inequality. For the semi–implicit scheme,

$$\left(\rho_h^{n+1}, \rho_h^{n+1}\right) \leq \left(\rho_h^n, \rho_h^{n+1}\right) + \frac{\Delta t}{2}A_1\left(\rho_h^n\right)\left(\rho_h^{n+1}, \rho_h^{n+1}\right) - \frac{\Delta t}{2}A_2\left(\rho_h^n\right)\left(\rho_h^{n+1}, \rho_h^{n+1}\right)$$
$$-\Delta t\, A_1\left(\rho_h^n\right)\left(\rho_h^n, \rho_h^{n+1}\right) + \Delta t\, A_2\left(\rho_h^n\right)\left(\rho_h^n, \rho_h^{n+1}\right) \quad \text{(by (11))}$$
$$\leq \left(\rho_h^n, \rho_h^{n+1}\right) - \frac{\Delta t}{2}A_1\left(\rho_h^n\right)\left(\rho_h^n - \rho_h^{n+1}, \rho_h^{n+1}\right) + \frac{\Delta t}{2}A_2\left(\rho_h^n\right)\left(\rho_h^n - \rho_h^{n+1}, \rho_h^{n+1}\right)$$
$$-\frac{\Delta t}{2}A_1\left(\rho_h^n\right)\left(\rho_h^n, \rho_h^{n+1}\right) + \frac{\Delta t}{2}A_2\left(\rho_h^n\right)\left(\rho_h^n, \rho_h^{n+1}\right) \quad \text{(by linearity)}$$
$$\leq \left(\rho_h^n, \rho_h^{n+1}\right) \quad \text{(by (18))}.$$

We conclude by the Schwarz inequality. For the explicit scheme, we have

$$\left(\rho_h^{n+1}, \rho_h^{n+1}\right) \leq \left(\rho_h^n, \rho_h^{n+1}\right) + \frac{\Delta t}{2}A_1\left(\rho_h^n\right)\left(\rho_h^n, \rho_h^{n+1}\right) - \frac{\Delta t}{2}A_2\left(\rho_h^n\right)\left(\rho_h^n, \rho_h^{n+1}\right)$$
$$-\Delta t\, A_1\left(\rho_h^n\right)\left(\rho_h^n, \rho_h^{n+1}\right) + \Delta t A_2\left(\rho_h^n\right)\left(\rho_h^n, \rho_h^{n+1}\right) \quad \text{(by (19))}$$
$$\leq \left(\rho_h^n, \rho_h^{n+1}\right) \quad \text{(by (18))}.$$

The conclusion comes again by the Schwarz inequality.

Remark 3. As in Theorem 1 the fully implicit scheme does not require symmetry of the forms A_1, A_2, it can also be applied for a stronger variational formulation, where the diffusive term is integrated by part just once.

Theorem 2 (CFL condition for stability). *Assume that $A_i(\varphi)$, for $i \in \{0,1,2\}$, are bilinear forms such that $A_1(\varphi)$, $A_2(\varphi)$ are symmetric and positive, and the conditions* (11), (12) *are satisfied. The semi–implicit scheme* (16) *is stable under the abstract CFL condition*

$$\Delta t\, A_1(\rho_h)(u_h, u_h) \leq (u_h, u_h) \qquad \forall \rho_h, u_h \in V_h. \tag{20}$$

The explicit scheme is stable under the above condition and (18)–(19).

Proof. Take again $\eta_h = \rho_h^{n+1}$. We have

$$
\begin{aligned}
\left(\rho_h^{n+1}, \rho_h^{n+1}\right) &\leq \left(\rho_h^n, \rho_h^{n+1}\right) - \Delta t\, A_0\left(\rho_h^n\right)\left(\rho_h^{n+1}, \rho_h^{n+1}\right) \\
&\quad - \Delta t\, A_1\left(\rho_h^n\right)\left(\rho_h^n, \rho_h^{n+1}\right) + \Delta t A_2\left(\rho_h^n\right)\left(\rho_h^n, \rho_h^{n+1}\right) \\
&\leq \left(\rho_h^n, \rho_h^{n+1}\right) - \Delta t\, A_0\left(\rho_h^n\right)\left(\rho_h^{n+1}, \rho_h^{n+1}\right) - \Delta t\, A_1\left(\rho_h^n\right)\left(\rho_h^n, \rho_h^{n+1}\right) \\
&\quad + \frac{\Delta t}{2} A_1\left(\rho_h^n\right)\left(\rho_h^n, \rho_h^n\right) + \frac{\Delta t}{2} A_2\left(\rho_h^n\right)\left(\rho_h^{n+1}, \rho_h^{n+1}\right) \quad \text{(by (13))} \\
&\leq \left(\rho_h^n, \rho_h^{n+1}\right) - \Delta t\, A_0\left(\rho_h^n\right)\left(\rho_h^{n+1}, \rho_h^{n+1}\right) \\
&\quad + \frac{\Delta t}{2} A_1\left(\rho_h^n\right)\left(\rho_h^{n+1} - \rho_h^n, \rho_h^{n+1} - \rho_h^n,\right) \\
&\quad - \frac{\Delta t}{2} A_1\left(\rho_h^n\right)\left(\rho_h^{n+1}, \rho_h^{n+1}\right) + \frac{\Delta t}{2} A_2\left(\rho_h^n\right)\left(\rho_h^{n+1}, \rho_h^{n+1}\right) \\
&\leq \left(\rho_h^n, \rho_h^{n+1}\right) + \frac{\Delta t}{2} A_1\left(\rho_h^n\right)\left(\rho_h^{n+1} - \rho_h^n, \rho_h^{n+1} - \rho_h^n\right) \quad \text{(by (11))},
\end{aligned}
$$

which gives $\left(\rho_h^{n+1}, \rho_h^{n+1}\right) \leq \left(\rho_h^n, \rho_h^n\right)$ by (20). For the explicit scheme,

$$
\begin{aligned}
\left(\rho_h^{n+1}, \rho_h^{n+1}\right) &\leq \left(\rho_h^n, \rho_h^{n+1}\right) - \Delta t\, A_0\left(\rho_h^n\right)\left(\rho_h^n, \rho_h^{n+1}\right) - \Delta t\, A_1\left(\rho_h^n\right)\left(\rho_h^n, \rho_h^{n+1}\right) \\
&\quad + \frac{\Delta t}{2} A_1\left(\rho_h^n\right)\left(\rho_h^n, \rho_h^n\right) + \frac{\Delta t}{2} A_2\left(\rho_h^n\right)\left(\rho_h^{n+1}, \rho_h^{n+1}\right) \quad \text{(by (13))} \\
&\leq \left(\rho_h^n, \rho_h^{n+1}\right) - \frac{\Delta t}{2} A_1\left(\rho_h^n\right)\left(\rho_h^n, \rho_h^{n+1}\right) - \frac{\Delta t}{2} A_2\left(\rho_h^n\right)\left(\rho_h^n, \rho_h^{n+1}\right) \\
&\quad + \frac{\Delta t}{2} A_1\left(\rho_h^n\right)\left(\rho_h^n, \rho_h^n\right) + \frac{\Delta t}{2} A_2\left(\rho_h^n\right)\left(\rho_h^{n+1}, \rho_h^{n+1}\right) \quad \text{(by (19))} \\
&\leq \left(\rho_h^n, \rho_h^{n+1}\right) - \frac{\Delta t}{2} A_1\left(\rho_h^n\right)\left(\rho_h^n, \rho_h^{n+1} - \rho_h^n\right) + \frac{\Delta t}{2} A_2(\rho_h^n)\left(\rho_h^{n+1}, \rho_h^{n+1} - \rho_h^n\right) \\
&\quad \leq \left(\rho_h^n, \rho_h^{n+1}\right) + \frac{\Delta t}{2} A_1\left(\rho_h^n\right)\left(\rho_h^{n+1} - \rho_h^n, \rho_h^{n+1} - \rho_h^n\right) \\
&\quad - \frac{\Delta t}{2} A_1\left(\rho_h^n\right)\left(\rho_h^{n+1}, \rho_h^{n+1} - \rho_h^n\right) + \frac{\Delta t}{2} A_2(\rho_h^n)\left(\rho_h^{n+1}, \rho_h^{n+1} - \rho_h^n\right) \\
&\leq \left(\rho_h^n, \rho_h^{n+1}\right) + \frac{\Delta t}{2} A_1\left(\rho_h^n\right)\left(\rho_h^{n+1} - \rho_h^n, \rho_h^{n+1} - \rho_h^n\right) \quad \text{(by (18))}.
\end{aligned}
$$

This gives again stability by (20).

Theorem 2 and Lemma 3 imply immediately the following.

Corollary 2. *Assume* (5), (6), (7) *and a homogenous Neumann boundary condition* $\nabla \rho \cdot \overrightarrow{n} = 0$ *on* $\Gamma_w \cup \Gamma$. *If* α, h *and* Δt *are positive constants such that for any* $\rho_h, u_h \in V_h$ *the CFL condition*

$$\frac{\Delta t \int_\Gamma u_h^2 \left(\overrightarrow{V}(\rho_h) \cdot \overrightarrow{n} - 2\alpha \right)^2}{8\alpha \int_{\Omega_h} u_h^2} \leq 1 \tag{21}$$

is satisfied, the semi–implicit scheme is stable.

Corollary 3. *Assume* (5), (6), (7), *a homogenous Neumann boundary condition* $\nabla \rho \cdot \overrightarrow{n} = 0$ *on* Γ_w *and a homogenous Dirichlet boundary condition* $\rho = 0$ *on* Γ. *If* α, h *and* Δt *are positive constants such that for any* $u_h \in V_h$ *the CFL condition*

$$\frac{\Delta t \int_\Gamma \left(\kappa \nabla u_h \cdot \overrightarrow{n} - \alpha u_h \right)^2}{2\alpha \int_{\Omega_h} u_h^2} \leq 1 \tag{22}$$

is satisfied, the semi–implicit scheme is stable.

Remark 4. The CFL condition giving stability of the semi–implicit scheme can also be written for the case when $\nabla \rho \cdot \overrightarrow{n} \leq 0$ on Γ.

4   Example

An interesting example that seems, by a numerical evidence, to satisfy (7), is $\overrightarrow{V} = -\nabla \Phi$ where Φ is a solution to the regularized eikonal equation

$$\begin{cases} \|\nabla \Phi\|^2 - \delta \, \Delta \Phi = 1 & x \in \Omega \\ \nabla \Phi(\xi) \cdot \overrightarrow{n}(\xi) = 0 & \xi \in \Gamma_w \\ \Phi(\xi) = 0 & \xi \in \Gamma, \end{cases}$$

If $\delta = 0$, this equation is known as the eikonal equation it comes from optics. $-\nabla \Phi$ gives the shortest path to Γ followed by a light ray. With the regularization ($\delta > 0$), the vector field $-\nabla \Phi$ can be seen as giving an approximation to the shortest way to Γ, having the advantage that different paths do not cross; see Fig. 1. We show in [2] that $-\nabla \Phi$ satisfies conditions (5), (6).

By integration by parts, it clearly follows that

$$\int_\Omega -\Delta \Phi = \int_\Omega \mathrm{div}(-\nabla \Phi) = \int_\Omega \mathrm{div} \overrightarrow{V} > 0.$$

Although we have not got a formal proof yet, the numerical simulations suggest that the same inequality is satisfied pointwise: $\|\nabla \Phi\| \leq 1$, which would mean $-\Delta \Phi = \mathrm{div} \overrightarrow{V} \geq 0$.

The density–dependent case $\overrightarrow{V} = -v(\rho)\nabla \Phi$ is to be considered. More simulations are given in [9].

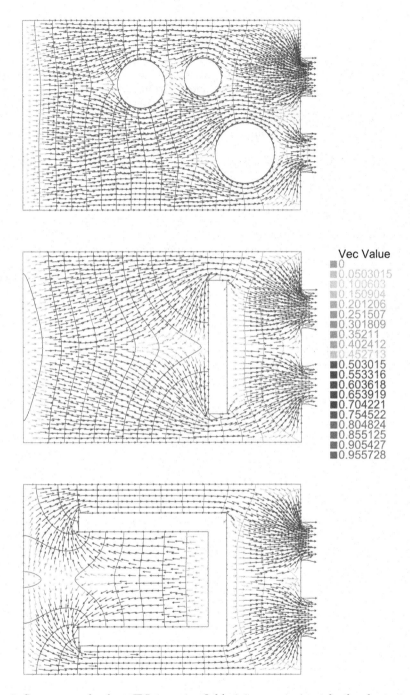

Fig. 1. Some examples for $-\nabla\Phi$, a vector field giving approximately the shortest path to the exit. The legend gives the values of $\|\nabla\Phi\|$.

References

1. Borsche, R., Colombo, R.M., Garavello, M., Meurer, A.: Differential equations modeling crowd interactions. J. Nonlinear Sci. **25**, 827–859 (2015)
2. Colombo, R.M., Gokieli, M., Rosini, M.D.: Modeling crowd dynamics through hyperbolic - elliptic equations. In: Non-Linear Partial Differential Equations, Mathematical Physics, and Stochastic Analysis – The Helge Holden Anniversary Volume, pp. 111–128. EMS Series of Congress Reports, May 2018
3. Gokieli, M.: An advection-diffusion equation as model for crowd evacuation (to appear)
4. Hughes, R.L.: A continuum theory for the flow of pedestrians. Transp. Res. Part B: Methodol. **36**(6), 507–535 (2002)
5. Hughes, R.L.: The flow of human crowds. Annu. Rev. Fluid Mech. **35**(1), 169–182 (2003)
6. Jiang, Y., Zhou, S., Tian, F.-B.: Macroscopic pedestrian flow model with degrading spatial information. J. Comput. Sci. **10**, 36–44 (2015)
7. Kachroo, P.: Pedestrian Dynamics: Mathematical Theory and Evacuation Control. CRC Press, Boca Raton (2009)
8. Kamga, J.-B.A., Després, B.: CFL condition and boundary conditions for DGM approximation of convection-diffusion. SIAM J. Numer. Anal. **44**(6), 2245–2269 (2006)
9. Szczepańczyk, A.: Master's thesis. University of Warsaw, Interdisciplinary Centre for Mathematical and Computational Modelling (ICM) (to appear)
10. Twarogowska, M., Goatin, P., Duvigneau, R.: Macroscopic modeling and simulations of room evacuation. Appl. Math. Model. **38**(24), 5781–5795 (2014)

Additive Average Schwarz with Adaptive Coarse Space for Morley FE

Salah Alrabeei[1]($\boxtimes$), Mahmood Jokar[1], and Leszek Marcinkowski[2]

[1] Department of Computing, Mathematics, and Physics,
Western Norway University of Applied Sciences,
Inndalsveien 28, 5020 Bergen, Norway
{Salah.Alrabeei,Mahmood.Jokar}@hvl.no
[2] Faculty of Mathematics, Informatics, and Mechanics, University of Warsaw,
Banacha 2, 02-097 Warszawa, Poland
L.Marcinkowski@mimuw.edu.pl

Abstract. We propose an additive average Schwarz preconditioner with two adaptively enriched coarse space for the nonconforming Morley finite element method for fourth order biharmonic equation with highly varying and discontinuous coefficients. In this paper, we extend the work of [9,10]: (additive average Schwarz with adaptive coarse spaces: scalable algorithms for multiscale problems). Our analysis shows that the condition number of the preconditioned problem is bounded independent of the jump of the coefficient, and it depends only on the ratio of the coarse to the fine mesh.

Keywords: Additive average Schwarz · Nonconforming finite element · Domain decomposition methods · Fourth order problems with highly varying coefficients

1 Introduction

Most of the physical problems are mathematically described in two or three dimensions. Solving these problems leads to solving large linear systems of equations that require a highly expensive computational cost. Several attempts have been made to find efficient methods to solve these linear systems. One of the most efficient techniques is the Additive Schwarz methods, which itself is considered the most effective preconditioning method, cf. [12].

In this paper, we consider variable coefficient fourth-order elliptic problems with Dirichlet boundary conditions. The coefficient of the problem is highly heterogeneous.

Schwarz type solvers were used for different finite elements discretazations in 2nd order elliptic problem with highly varying coefficients [1,6] and [10]. For

The work of the last author was partially supported by Polish Scientific Grant: National Science Center 2016/21/B/ST1/00350.

R. Wyrzykowski et al. (Eds.): PPAM 2019, LNCS 12044, pp. 287–297, 2020.
https://doi.org/10.1007/978-3-030-43222-5_25

a fourth order problems, only with constan coefficient problems were developed and analyzed, e.g. cf. [3,4,7,14], and references therein.

In this paper, we develop an additive Schwarz method, cf. [2], for solving fourth order ellipitc problems with highly varying coefficients. The differential problem is discretized by the nonconforming Morley element method, cf. e.g. [5,11]. We propose a new coarse space which is constructed by adding local adaptive eigenspaces defined over subdomains, cf. [10]. The outline of the paper is given as follows: in Sect. 2, the continuous problem and its approximation using Morley finite element are presented. Additive average Schwarz methods and the equivalent decomposed problem of our problems are given in Sect. 3. In Sect. 4, we show condition number and theoretically prove its bounds. In Sect. 5, we give some numerical experiments to verify our theory as well as to show the scalability and efficiency of our algorithm.

2   Discrete Problem

We consider a thin plate occupying a polygonal domain $\Omega \in \mathbb{R}^2$ with a clamped boundary $\partial\Omega$ under a distributed load $f \in L^2(\Omega)$. The deflection u is governed by the following biharmonic equation,

$$\Delta(\alpha(x)\Delta u) = f, \quad \text{in} \quad \Omega, \tag{1}$$

$$u = \frac{\partial u}{\partial \nu} = 0, \quad \text{on} \quad \partial\Omega, \tag{2}$$

where $\alpha \in L^\infty(\Omega)$ is a positive elementwise constant coefficient function. We assume that there exists an $\alpha_0 > 0$ such that $\alpha(x) \geq \alpha_0$ in Ω. (cf. [8]). The operator Δ is defined as

$$\Delta = \frac{\partial^2}{\partial x_1^2} + 2\frac{\partial^2}{\partial x_1 \partial x_2} + \frac{\partial^2}{\partial x_2^2} \tag{3}$$

To derive the weak form of the plate bending problem (1) and (2), we first define the Hilbert space

$$H_0^2(\Omega) = \{v \in H^2(\Omega) : v|_{\partial\Omega} = n \cdot \nabla v|_{\partial\Omega} = 0\}. \tag{4}$$

We are seeking $u \in H_0^2(\Omega)$ such that

$$a(u, v) = \int_\Omega fv \, dx, \quad v \in H_0^2(\Omega), \tag{5}$$

where

$$a(u, v) = \int_\Omega \alpha(x)\left(\frac{\partial^2 u}{\partial x_1^2}\frac{\partial^2 v}{\partial x_1^2} + \frac{\partial^2 u}{\partial x_2^2}\frac{\partial^2 v}{\partial x_2^2} + 2\frac{\partial^2 u}{\partial x_1 \partial x_2}\frac{\partial^2 v}{\partial x_1 \partial x_2}\right) dx \tag{6}$$

For the sake of simplicity in the practical computations, we choose the non-conforming Morley element that has much smaller number of degrees for freedom than conforming C^1 plate elements. Therefore, We define the Morely finite element space V_h as

$$
\begin{aligned}
V_h = \{v \in L^2 : v|_K \in P^2(K) \quad &\forall K \in \mathcal{T}_h, \\
v \quad &\text{continuous at the vertices,} \\
\tfrac{\partial v}{\partial \nu} \quad &\text{continuous at the edge mid-points} \\
v(x) = \tfrac{\partial v}{\partial \nu}(m) = 0, \; &x, m \in \partial\Omega; \; \text{x vertex, m midpoint}\}.
\end{aligned}
\tag{7}
$$

where $\mathcal{T}_h$ is the quasi-uniform triangulation of Ω with the mesh size h. It is worth mentioning that the Morley element are neither C^1 nor C^0 and the space $V_h \not\subset C^0(\overline{\Omega})$.

Hence, the finite element approximation of the weak form (5) is given as follows; find $u_h \in V_h$ such that

$$
a(u_h, v) = F(v), \qquad \forall v \in V_h,
\tag{8}
$$

where

$$
a(u_h, v) = \sum_{K \in \mathcal{T}_h(\Omega)} \int_K \alpha(x) \left(\frac{\partial^2 u}{\partial x_1^2} \frac{\partial^2 v}{\partial x_1^2} + \frac{\partial^2 u}{\partial x_2^2} \frac{\partial^2 v}{\partial x_2^2} + 2 \frac{\partial^2 u}{\partial x_1 \partial x_2} \frac{\partial^2 v}{\partial x_1 \partial x_2} \right) dx
\tag{9}
$$

and

$$
F(v) = \sum_{K \in \mathcal{T}_h} \int_K f v \, dx.
\tag{10}
$$

Lemma 1. *There exists two positive constants c_1 and c_2 such that*

$$
c_1 \|v\|_{h,2,\Omega}^2 \le a(v,v) \le c_2 \|v\|_{h,2,\Omega}^2, \qquad \forall v \in V_h,
\tag{11}
$$

where the norm $\|v\|_{h,2,\Omega}$ on V_h is defined as

$$
\|v\|_{h,2,\Omega} = \left(\sum_{K \in \mathcal{T}_h, K \subset \Omega} |v|_{H^2(K)}^2 \right)^{\frac{1}{2}},
\tag{12}
$$

The proof of this lemma is straightforward in [5]. The requirements of Lax-Milgram theorem are satisfied which yields us the existence and uniqueness of the weak solution (cf. e.g., [5,8]).

For any $u \in V_h$, it can be written as a linear combination of the nodal basis for the space V_h. Thus, the bilinear form (8) can be written as a system of algebraic equations,

$$
\mathbf{A}\mathbf{u} = \mathbf{b},
\tag{13}
$$

where $\mathbf{A}$ is the stiffness matrix whose components are $A_{i,j} = a(\psi_i, \psi_j)$, where ψ_i and ψ_j are the Morley basis functions. The vector of unknowns is denoted by $\mathbf{u}$, which contains the values of the solution at the vertices and the values of the normal derivatives at the midpoints, respectively. The vector $\mathbf{b}$ is the load vector containing $F(\psi_i)$. The system (13) is generally very large, which makes the computational cost using the direct solver so expensive. An alternative way to solve this kind of large systems is to use iterative methods. The condition number of the matrix $\mathbf{A}$ in Eq. (13) determines whether the system is well or ill-conditioned. Therefore, choosing a particular iterative method to solve the linear system depends on the condition number. It is well known that the condition number of of the matrix of coefficient $\mathbf{A}$ is of order $O(\frac{\max \alpha}{\min \alpha} h^{-4})$. Therefore, if the mesh size is small, or α is strongly varying, then the linear system becomes ill-conditioned and thus the usual iterative methods such as the Conjugate Gradient (CG) method or the Generalized Minimal Residual (GMRS) method may not be a good choice, cf. [7,12]. Thus, we propose an additive Schwarz method as a way of constructing a parallel preconditioner for our system and then solved the resulting preconditioned system by, e.g., a Preconditioned Conjugate Gradient method (PCG).

3   Additive Average Schwarz Method

Let the domain Ω be divided into N nonoverlapping subdomains triangles of $\{\Omega_i\}_i$ i.e $\overline{\Omega} = \bigcup_i^N \overline{\Omega}_i$ and the intersection between two subdomains $\Omega_i \cap \Omega_j$ is empty for all $i \neq j$. Let Ω_{ih} and $\partial\Omega_{ih}$ denote the set of interior and in the boundary nodes of the domain Ω_i respectively. Similarly, Ω_{ih}^* and $\partial\Omega_{ihi}^*$ denote the set of interior and in the boundary edge midpoints of Ω_i, respectively.

We define the subspace V_{hi} corresponding to the subdomain Ω_i, for all $i = 1, 2, \ldots, N$, such that

$$V_{hi} = \{v \in V_h : v(x) = 0 \qquad \forall x \in \partial\Omega_{ih}$$
$$\frac{\partial v(x)}{\partial \nu} = 0. \qquad \forall x \in \partial\Omega_{ih}^* \qquad (14)$$
$$v = 0, \text{ outside } \overline{\Omega}_i\}.$$

We assume that each subdomain has its own triangulation $\mathcal{T}_h(\Omega_i)$ inherited from $\mathcal{T}_h$. We then define the local maximums and minimums values of coefficients over a subdomain, as following

$$\underline{\alpha}_i := \min_{x \in \Omega_i} \alpha(x), \qquad \overline{\alpha}_i := \max_{x \in \Omega_i} \alpha(x). \qquad (15)$$

We also need to define projections from the the finite element space to the subspaces, i.e define $\mathcal{P}_i : V_h \to V_{hi}$, and $\mathcal{P}_0^{\text{enriching}} : V_h \to V_0^{\text{enriched}}$ corresponding to the spaces V_{hi} and V_0^{enriched} respectively as

$$a(\mathcal{P}_i u, v) = a(u, v), \qquad v \in V_{hi}, \qquad i = 1, \ldots, N, \qquad (16)$$

$$a(\mathcal{P}_0^{\text{enriching}} u, v) = a(u, v), \qquad v \in V_0^{\text{enriched}}, \qquad (17)$$

where V_0^{enriched} is the coarse space which will be defined later. Now the additive Schwarz operator $\mathcal{P}^{\text{enriching}} : V^h \to V^h$ considering the Schwarz scheme can be written as

$$\mathcal{P}^{\text{enriching}} u = \mathcal{P}_0^{\text{enriching}} u + \sum_{i=1}^{N} \mathcal{P}_i u. \qquad (18)$$

Thus the original problem (5) can be replaced with the following problem:

$$\mathcal{P}^{\text{enriching}} u_h = g^{\text{enriching}}, \tag{19}$$

where $g^{\text{enriching}} = g_0^{\text{enriching}} + \sum_i g_i$ with $g_0^{\text{enriching}} = \mathcal{P}_0^{\text{enriching}} u_h$ and $g_i = \mathcal{P}_i u_h$.

The Standard Additive Average Coarse Space: Following [2,7,10], we define the standard additive average coarse space as the image of the interpolation like operator $I_a : V_h \to V_h$, where I_a is defined as

$$I_a v = \begin{cases} v(x) & x \in \partial\Omega_{ih}, \\ \overline{v}_i & x \in \Omega_{ih}, \end{cases} \tag{20}$$

and

$$\frac{\partial I_a v}{\partial \nu} = \begin{cases} \dfrac{\partial v(x)}{\partial \nu} & x \in \partial\Omega_{ih}^*, \\[2mm] \dfrac{\partial \overline{v}(x)}{\partial \nu} & x \in \Omega_{ih}^*, \end{cases} \tag{21}$$

where $\overline{v}_i = \frac{1}{n_i} \sum_{x \in \partial\Omega_{ih}} v(x)$, and n_i is the total number of nodal points in $\partial\Omega_{ih}$. For the sake of clarity, the sum is over all the nodes on $\partial\Omega_{ih}$ and i corresponds to the subdomain index not to the index of the summation.

Enriched Additive Average Coarse Space: We enrich the standard coarse space using functions that are adaptively selected from the following generalized eigenvalue problem:

Remark 1. We were supposed to define two different enriched coarse spaces. However, we could not analytically prove the convergence of the condition number for case 2 i.e. (index = 2) in spite of the successful computational proof. Therefore, in this paper, we only consider one coarse space denoted by V_0^{enriched} and defined as in (23).

Find all eigen pairs: $(\lambda_j^i, \phi_j^i) \in \mathbb{R} \times V_{h0}(\Omega_i)$, such that

$$\begin{aligned} a_i(\phi_j^i, v) &= \lambda_j^i b_i(\phi_j^i, v), \quad v \in V_{h0}(\Omega_i), \\ b_i(\phi_j^i, \phi_j^i) &= 1, \end{aligned} \tag{22}$$

where the bilinear forms are defined as

$$a_i(u, v) := \sum_{K \in \mathcal{T}_h(\Omega_i)} \int_K \alpha(x) \left(\frac{\partial^2 u}{\partial x_1^2} \frac{\partial^2 v}{\partial x_1^2} + \frac{\partial^2 u}{\partial x_2^2} \frac{\partial^2 v}{\partial x_2^2} + 2 \frac{\partial^2 u}{\partial x_1 \partial x_2} \frac{\partial^2 v}{\partial x_1 \partial x_2} \right) dx,$$

$$b_i(u, v) := \sum_{K \in \mathcal{T}_h(\Omega_i)} \int_K \alpha_i \left(\frac{\partial^2 u}{\partial x_1^2} \frac{\partial^2 v}{\partial x_1^2} + \frac{\partial^2 u}{\partial x_2^2} \frac{\partial^2 v}{\partial x_2^2} + 2 \frac{\partial^2 u}{\partial x_1 \partial x_2} \frac{\partial^2 v}{\partial x_1 \partial x_2} \right) dx,$$

and the local space $V_h 0(\Omega_i)$ is defined as follows

$$V_{h0}(\Omega_i) = \{ u_{|\Omega_i} : u \in V_h, \quad u(x) = \partial_\nu u(m) = 0 \quad \forall x \in \partial\Omega_{ih}, \quad \forall m \in \partial\Omega_{ih}^* \}.$$

It should be noted that the corresponding eigenspaces and eigenfunctions of two different eigenvalues are both orthogonal to each other. If the multiplicity of an eigenvalue is larger than one, we consider all its eigenfunctions as one. We need to order the eigenvalues in the decreasing form as $\lambda_1^i \geq \lambda_2^i \geq \ldots, \lambda_{N_i}^i > 0$, where N_i is the dimension of $V_{h0}(\Omega_i)$. We observe that all the eigenvalues are bounded by above the ratio of the maximum to the minimum of the element-wise function α, i.e $1 \leq \lambda_j^i \leq \frac{\overline{\alpha}_i}{\underline{\alpha}_i}$. Therefore, the eigenvalues of the eigenvalue problem (22) are all ones in case if the function α is constant in each subdomain, Ω_i.

Since the solution-space of the eigenvalue problem (22) is defined locally, we extend it by zero to the rest of the domain Ω and keep the same symbol to the extended function. Now, we define the enriched coarse spaces as follows

$$V_0^{\text{enriched}} = I_a V_h + \sum_{i=1}^{N} W_i, \tag{23}$$

where

$$W_i := \text{Span}(\phi_j^i)_{j=1}^{M_i}, \tag{24}$$

when $0 \leq M_i < N_i$ is a number either preset by the user or chosen adaptively. We assume that if an eigenvalue which has been selected to be included has multiplicity larger than one, then all its eigenfunctions will be included in the W_i. Consequently, $\lambda_{M_i+1} < \lambda_{M_i}$. Thus $M_i = 0$ means enrichment is not required in the subdomain Ω_i.

For the sake of the analysis, we need to define the following two operators. Let

$$\Pi_i^{\text{enriching}} : V_{h0}(\Omega_k) \to V_{h0}(\Omega_k),$$

be b_i-orthogonal projection defined as

$$\Pi_i^{\text{enriching}} u_h = \sum_{j=1}^{M_i} b_i(u_h, \psi_j^i)\psi_j^i, \tag{25}$$

where $(\psi_j^i)_j$ is the $b_i(\cdot, \cdot)$-orthonormal eigenbasis of $V_{h0}(\Omega_i)$.

Since $\Pi_i^{\text{enriching}}(u_h - I_a u_h)_{\partial \Omega_{ih}} = 0$, the projection can be extended to the rest of domain to obtain a function in W_i, and the same symbol will denote it.

The second operator is denoted by $I_a^{\text{enriching}} : V_h \to V_0^{\text{enriching}}$ which is defined as

$$I_a^{\text{enriching}} u_h = I_a u_h + \sum_{i=1}^{N} \Pi_i^{\text{enriching}}(u_h - I_a u_h). \tag{26}$$

4   Condition Number Bound

Our main theoretical result is given in this section and is mainly the condition number bound (see Theorem (1)). In order to prove this theorem, we need first to show some estimates. From now on, for the sake for clarity we give $||u||_{a_i} = a_i(u, u)$ and $||u||_{b_i} = b_i(u, u)$.

Lemma 2. *For $u_h \in V_h$, the following inequality holds*

$$\|u_h - \Pi_i^{enriching} u_h\|_{a_i} \leq \lambda_{M_i+1} \|u_h\|_{b_i},$$

Proof. The proof follows the lines of the proof given in [10]. □

We still need to estimate the coarse interpolation operator I_a; then we will be ready to prove the condition number bounds.

Lemma 3. *For $u_h \in V_h$, the following inequality holds*

$$\|u_h - I_a^{enriching} u_h\|_a \leq \max_i \lambda_{M_i+1} \left(\frac{H}{h}\right)^3 \|u_h\|_a, \tag{27}$$

where $H = \max_{i=1,...,N} \operatorname{diam}(\Omega_i)$.

Proof. Following [10] we define $w = u_h - I_a u_h$ Clearly, $w = 0$ on the interface Γ. Note that

$$u_h - I_a u_h = \sum_i (I - \Pi_i^{enriching}) w, \tag{28}$$

which is also equal to zero on the interface Γ. Then

$$\|u_h - I_a^{enriching} u_h\|_a = \sum_i \|(I - \Pi_i^{enriching}) w\|_{a_i},$$

$$\leq \sum_i \lambda_{M_i+1} \|w\|_{b_i}, \tag{29}$$

The last inequality is obtained from Lemma 2.
We still need to bound the bilinear $\|w\|_{b_i}$ for each i as follows:

$$\|w\|_{b_i} = b_i(w, w) = \underline{\alpha}_i a_{i_C}(w, w) = \underline{\alpha}_i \|w\|_{a_{i_C}}, \tag{30}$$

where $\|w\|_{a_{i_C}} = \|\frac{1}{\sqrt{\alpha}} w\|_{a_i}$, i.e it is the same bilinear form defined in (23) except the function $\alpha = 1$ over all the triangles. This bilinear form is exactly the bilinear form in defined in [7] where it was proved the following
Following [7] and using Lemma (1)

$$\|w\|_{a_{i_K}} = \|u_h - I_a u_h\|_{a_{i_K}} \leq \left(\frac{H}{h}\right)^3 \|u_h\|_{a_{i_C}}, \tag{31}$$

Finally, from Eqs. (29)–(31), and summing over all subdomains, the proof ends. □

We have just estimated all the required operators throughout all the above lemmas. So, we are ready to prove the condition number in the following theorem.

Theorem 1. *Let $\mathcal{P}^{enriching}$ be the additive Schwarz operator, Then for all $u_h \in V_h$*

$$\left(\min_i \frac{1}{\lambda_{M_i+1}}\right) \left(\frac{h}{H}\right)^3 a(u_h, u_h) \leq a(\mathcal{P}^{enriching} u_h, u_h) \leq a(u_h, u_h),$$

where $H = \max_{i=1,...,N} \operatorname{diam}(\Omega_i)$ and λ_{M_i+1} is the $(M_i + 1)$-th eigenvalue of (22) (cf. also (24)).

Proof. We use the framework of general Schwarz (see chapter 5, [13]) to prove this theorem. Based on the framework, we have to define a decomposition of each

$$u_h = u_0 + \sum_{i=1}^{N} u_i$$

where $u_h \in V_h$, $u_i \in V_{hi}$ and $u_0 = I_a^{\text{enriching}} u_h \in V_0^{\text{enriched}}$.

We need to satisfy three assumption. First, since we defined local bilinears by the exact bilinear form $a(u_h, v)$, the stability constant $\omega = 1$ assumption is satisfied. For assumption two, since the local subspaces are orthogonal to each other, the spectral radius of the matrix of constants of the strengthened Cauchy-Schwarz inequalities $\rho(\mathcal{E}) = 1$ which satisfy the Cauchy-Schwarz relationship between the local subspaces' assumption. What remains to prove is the first assumption, i.e.,

$$\sum_{i=0}^{N} ||u_i||_{a_i} \leq \max_i \lambda_{M_i+1} \left(\frac{H}{h}\right)^3 ||u_h||_a, \tag{32}$$

where $u_h \in V_h$ and $u_i \in V_{hi}$. It can easily be proved that $u_h = u_0 + \sum_{i=1}^{N} u_i$ where $u_i \in V_{hi}$ and $u_0 = I_a^{\text{enriching}} u_h \in V_0^{\text{enriched}}$. Using the triangle inequality, we get

$$||u_0||_a \leq ||u_h||_a + ||u_h - I_a^{\text{enriching}} u_h||_a, \tag{33}$$

also

$$||u_i||_a = ||u_i||_{a_i} = ||u_h - I_a^{\text{enriching}} u_h||_{a_i}, \qquad i = 1, 2, ... N, \tag{34}$$

Summing over all i = 1, 2, ..., N, we get

$$\sum_{i=1}^{N} ||u_i||_a = \sum_{i=1}^{N} ||u_h - I_a^{\text{enriching}} u_h||_{a_i} = ||u_h - I_a^{\text{enriching}} u_h||_a. \tag{35}$$

Combining Eqs. (33) and (35) and using Lemma (3), we conclude the proof.

5   Numerical Experiments

In this section, we shall present some results using the proposed method to validate our theory. All experiments are testing our fourth-order elliptic problem (1) using the Morley finite element for discretization. In our experiment, we consider the domain as the unite square $[0, 1] \times [0, 1]$, and the right hand side in the problem (1) is set so that the exact solution $u = x_1^2 x_2^2 (x_1 - 1)^2 (x_2 - 1)^2$. The overall domain is divided into nonoverlapping rectangular subdomains, each of which contains an equal number of blocks. We use the conjugate gradient method to solve the linear system in (19) and stop iterating when the ℓ-norm of the residual is reduced by the factor $5e - 6$.

We define the distribution of the coefficient α over the whole domain as follows

$$\alpha = \begin{cases} 2 + \sin(100\pi x_1)\sin(100\pi x_2) & x \in D_b, \\ \alpha_j\big(2 + \sin(100\pi x_1)\sin(100\pi x_2)\big) & x \in D_j, \end{cases} \tag{36}$$

where α_j is a parameter describing the contrast of discontinuities in the distribution of the coefficient, D_j is the region of the channel colored in red and D_b is the background region (see Fig. 1).

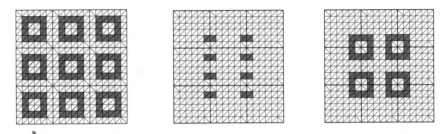

Fig. 1. Geometry with 18×18 fine mesh and 3×3 coarse mesh showing three different distributions of α for the three examples. The distributions from left to right are considered for the 1_{st}, the 2_{nd} and the 3_{rd} examples respectively. (Color figure online)

In order to numerically test our theory, we divide our experiments into two parts. In the first part, we show that the efficiency of the preconditioner is independent of both distributions of the coefficient α as well as its contrast. In order to do that, we test the algorithm by three examples (cf. Fig. 1). In the first example, we consider the distribution of α consisting of channels in the interior of the subdomains, i.e., in these channels, the contrast of α is highly jumping (cf. Fig. 1 (left)). In the second example, the jumps of α occur along subdomain interfaces (cf. Fig. 1 (middle). In example 3, we consider the case where α has jumps over subdomain layers as well as the interior of the subdomain (cf. Fig. 1 (right)). For each of the three examples, we divide the overall domain Ω into 3×3 subdomains and test our preconditioner for different value of α_j (cf. Table 1). The rows in Table 1 show that the condition number is independent of the distributions of α. Whereas, the columns show no dependency also on the contrast of α.

The second part of this section is devoted to the show how the condition number estimates depend on the parameters h and H separately. Furthermore, we present the exact minimum number of eigenfunctions (corresponding to the bad eigenvalues) required in the enrichment for the method to be robust with respect to the contrast. In the experiments in this part, we consider distribution of α as in Example 3. (cf. Fig. 1), and its contrast $\alpha_j = 1e4$.

For the first experiment in this part, we set $H = \frac{1}{3}$ and $h = \frac{1}{36}$ and run our algorithm. We run several tests and fix the number of added eigenfunctions (non-adaptive) for each subdomain in each test. See Table 2.

Table 1. Number of iterations and condition numbers (in parentheses) required until the convergence for the solution of (1). For the enrichment, we include only those eigenvalues that are greater than a given threshold $Tri = 100$.

α_j	Example 1:iter	Example 2:iter	Example 3: iter
1e1	23(1.21e1)	96(4.78e2)	55(4.16e2)
1e3	21(1.1e1)	40(3.51e1)	27(1.87e1)
1e4	21(1.11e1)	40(3.51e1)	27(1.87e1)
1e5	21(1.11e1)	40(3.51e1)	38(3.38e1)

Table 2. Number of iterations and a condition number estimates (in parentheses) for fixed number of eigenfunctions for enrichment. Here $H = 1/3$, $h = 1/36$ and $\alpha = 1e4$.

Numb. eigenvalue	0	5	10	15	16
Iter(cond. numb.)	354(1.91e+6)	303(1.15e+6)	304(9.29e+5)	66(2.20e+2)	66 (2.20e+2)

In Table 2, we can see from the condition number is getting smaller and smaller as the number of eigenfunctions is added more and more. However, at a certain number of the added eigenfunctions, the condition number stops decreasing even if we added more eigenfunctions beyond that specific number. So the minimum number of eigenfunctions, in this case, is fifteen. Even if we use adaptive enrichment, only fifteen eigenfunctions will be added.

Our final experiment is to show how the condition number depends on $\frac{H}{h}$. We run several tests to show the dependency of h and H separately, and the result has been shown in Table 3.

Table 3. Number of iterations and a condition number estimates (in parentheses) for varying H and h.

Subdomains	Blocks	Iteration	Condition number
3×3	18×18	25	1.79e1
	36×36	49	1.17e2
	72×72	108	9.82e2
3×3	48×48	95	5.61e2
6×6		61	6.74e1
12×12		14	4.5e0

The results in Table 3 support our theory. i.e, as we move downwards along the table, the condition number estimates grow by a factor of about 8, i.e, proportional to the factor $1/h^3$. In the last three rows, the condition number estimates also depends on H^3. Our numerical results thus support the theory in this paper.

Acknowledgments. The authors are deeply thankful for Prof. Talal Rahman for his invaluable comments, discussions, and suggestions in this work.

References

1. Bjørstad, P.E., Dryja, M., Rahman, T.: Additive Schwarz methods for elliptic mortar finite element problems. Numer. Math. **95**(3), 427–457 (2003)
2. Bjørstad, P.E., Dryja, M., Vainikko, E.: Additive Schwarz methods without subdomain overlap and with new coarse spaces. In: 1995 Domain Decomposition Methods in Sciences and Engineering, Beijing, pp. 141–157 (1997)
3. Brenner, S.C.: Two-level additive Schwarz preconditioners for nonconforming finite element methods. Math. Comp. **65**(215), 897–921 (1996)
4. Brenner, S.C., Sung, L.Y.: Balancing domain decomposition for nonconforming plate elements. Numer. Math. **83**(1), 25–52 (1999)
5. Ciarlet, P.G.: Basic error estimates for elliptic problems. In: Handbook of Numerical Analysis, vol. II, pp. 17–351. North-Holland, Amsterdam (1991)
6. Dryja, M., Sarkis, M.: Additive average Schwarz methods for discretization of elliptic problems with highly discontinuous coefficients. Comput. Methods Appl. Math. **10**(2), 164–176 (2010). https://doi.org/10.2478/cmam-2010-0009
7. Feng, X., Rahman, T.: An additive average Schwarz method for the plate bending problem. J. Numer. Math. **10**(2), 109–125 (2002)
8. Larson, M.G., Bengzon, F.: The Finite Element Method: Theory, Implementation, and Applications, vol. 10. Springer, Heidelberg (2013). https://doi.org/10.1007/978-3-642-33287-6
9. Marcinkowski, L., Rahman, T.: Two new enriched multiscale coarse spaces for the additive average Schwarz method. In: Lee, C.-O., et al. (eds.) Domain Decomposition Methods in Science and Engineering XXIII. LNCSE, vol. 116, pp. 389–396. Springer, Cham (2017). https://doi.org/10.1007/978-3-319-52389-7_40
10. Marcinkowski, L., Rahman, T.: Additive average Schwarz with adaptive coarse spaces: scalable algorithms for multiscale problems. Electron. Trans. Numer. Anal. **49**, 28–40 (2018). https://doi.org/10.1553/etna_vol49s28
11. Morley, L.S.D.: The triangular equilibrium problem in the solution of plate bending problems. Aero. Quart. **23**(19), 149–169 (1968)
12. Rahman, T., Xu, X., Hoppe, R.: Additive Schwarz methods for the Crouzeix-Raviart mortar finite element for elliptic problems with discontinuous coefficients. Numer. Math. **101**(3), 551–572 (2005)
13. Smith, B., Bjorstad, P., Gropp, W.: Domain Decomposition: Parallel Multilevel Methods for Elliptic Partial Differential Equations. Cambridge University Press, Cambridge (2004)
14. Xu, X., Lui, S., Rahman, T.: A two-level additive Schwarz method for the morley nonconforming element approximation of a nonlinear biharmonic equation. IMA J. Numer. Anal. **24**(1), 97–122 (2004)

Minisymposium on HPC Applications in Physical Sciences

Application of Multiscale Computational Techniques to the Study of Magnetic Nanoparticle Systems

Marianna Vasilakaki[ID], Nikolaos Ntallis[ID], and Kalliopi N. Trohidou$^{(\boxtimes)}$[ID]

Institute of Nanoscience and Nanotechnology, NCSR "Demokritos",
153 10 Aghia Paraskevi, Attiki, Greece
k.trohidou@inn.demokritos.gr

Abstract. We have employed a multiscale modeling approach that combines ab-initio electronic structure calculations with atomic and mesoscopic scale modeling to describe the magnetic behavior of assemblies of magnetic nano-particles (MNPs) with core/surface morphology. Our modeling is based on the calculated atomistic parameters and we rescale them after the reduction of the simulated number of the NPs atomic spins to the minimum necessary to represent their magnetic structure in the assemblies. Monte Carlo simulations are them performed to study their macroscopic magnetic behavior. We apply our model to (a) $CoFe_2O_4$ NPs coated with two different surfactants and (b) bovine serum albumin-coated $MnFe_2O_4$ MNPs' clusters. Our approach overcomes current computational limitations. The numerical results produced are in excellent agreement with the experimental findings illustrating the potentials of our strategy to simulate the magnetic behavior of complex magnetic nanoparticle systems and to optimize their magnetic properties for advanced energy and biotechnology nanomaterial applications.

Keywords: Multiscale modeling · Magnetic nanoparticles · DFT calculations · Monte Carlo simulations

1 Introduction

Magnetic nanoparticles (MNPs) have received large attention because of their remarkable physical properties which are different from those of the bulk materials [1]. Their intriguing properties have motivated nanomaterials engineering (e.g. cluster-like morphology, surface engineering) and innovative applications ranging from nanotechnology to biomedicine [2,3]. One of the main challenges in MNPs research field is to develop nanomaterials consisted of MNPs covered with organic ligands to achieve fluid stability and without toxicity [4].

From a theoretical point of view there are many challenges in the modeling and the calculation of the magnetic properties of these types of nanomaterials. It is a very complicated issue to study simultaneously the intra-particle interactions in the core and at the coated nanoparticle surface and the long range

© Springer Nature Switzerland AG 2020
R. Wyrzykowski et al. (Eds.): PPAM 2019, LNCS 12044, pp. 301–311, 2020.
https://doi.org/10.1007/978-3-030-43222-5_26

inter-particle interactions for the MNPs assemblies. Notably, the nano-assemblies involve different length scales extending from the atomic dimensions to nanoparticles and aggregates dimensions. The particle size effects, the morphology and the surface-coating of these MNPs in addition to the inter-particle interactions play a key role to their magnetic behavior [5]. On the other hand, extensive first-principles studies for nanoparticles of few to a few tens nanometers in sizes produced in most experiments, are beyond computational capability in most researchers' laboratories. Hence, atomic scale techniques are not enough to fully describe the magnetic properties of nano-assemblies, and multiscale modeling is fundamental to link the atomistic length-scale with the macroscopic properties of real materials given the computational time limitations.

In this study we present our multiscale strategy to study the effect of the surfactant on the magnetic behavior of assemblies of nanoparticle ferrites. We focus on spinel ferrite nanoparticles coated with a surfactant since they are essentially complicated atomic systems and gather a lot of scientific attention for energy and biotechnology applications [3]. We apply our model to different complex magnetic particle systems: (a) $CoFe_2O_4$ nanoparticles ($\sim$5 nm) coated with diethylene glycol (DEG) and oleic acid (OA) surfactants [6], which are common covering materials for several NPs applications and modify the cationic distributions of the nanoparticles, and (b) $MnFe_2O_4$ uncoated nanoparticles ($\sim$2 nm) (MFO) [7] and MnFe2O4 nanoparticle clusters coated with bovine serum albumin (MFO_ BSA) [8]. Covering procedure of albumin induces some further particle aggregation leading to the formation of clusters of nanoparticles. In this case the nanoparticles interact not only via dipolar interactions but also with exchange interactions when they are in contact. We study the magnetic properties of these NPs via Density Functional Theory electronic structure calculations related to their structure and surface properties and their magnetic behavior via Monte Carlo simulations in a mesoscopic scale calculating the field and temperature dependence of their magnetization. We compare our results with the experimental results to elucidate the role of the surfactant on their magnetic behavior.

2   Multiscale Modeling of Magnetic Nanoparticles

2.1   Electronic Structure Calculation of the Single Magnetic Nanoparticle Parameters

$CoFe_2O_4$ Nanoparticles - Effect of OA and DEG Surfactant. We performed first principles calculations, based on spin-polarized density functional theory using MPI parallelization in the Vienna Ab Initio calculations package [9,10] on ARIS high performance system for a cluster of atoms (nanoparticle $\sim$2 nm) of Co ferrite structure, with the two coatings (DEG and OA). The electronic charge density and the local potential were expressed in plane wave basis sets. Geometries are fully optimized (electronic relaxation: 10^{-4} eV; ionic relaxation: 10^{-3} eV). The exchange correlation functional chosen is the one proposed by Perdew-Burke-Ernzerhof. The interactions between the electrons and ions were described using the projector-augmented-wave method. A cutoff energy of 550eV was used.

First we performed the bulk $CoFe_2O_4$ structure calculations in order to tune the on-site Coulomb strength U and exchange coupling J parameters. In the Duradev's [11] scheme the effective parameter $U_{eff} = U - J$ is taken as input. For the effective parameter U_{eff} a value of 4.5 eV for Fe and 4eV for Co atoms reproduces the cell dimension of the bulk unit cell of $CoFe_2O_4$, i.e. $a = b = c = 8.35$ Å. Fe atoms have magnetic moment of $4.0 \mu_B$ in A sites and $4.2 \mu_B$ in B sites of the inverse spinel structure. The Co atoms have magnetic moments $2.6 \mu_B$ in B sites. The lowest energy was found when the O atoms are placed in the (x, x, x) crystal coordinates, where $x = 0.386$). As these U_{eff} parameters properly define bulk values they were used as input data to the finite system calculations.

Then, our calculations on ~ 2 nm spherical nanoparticles were performed with ionic distributions of the ferrite structure $(Fe_{0.78})[Co_{1.00}Fe_{1.22}]O_4$ in DEG case and $(Co_{0.14}Fe_{0.86})[Co_{0.86}Fe_{1.14}]O_4$ in OA case, obtained from the Mössbauer spectra of the coated MNPs [6]. Starting from these cationic distributions electronic structure calculations have been performed for the two surfactants. To avoid interaction between the periodic images, we have taken 1.5 nm of empty space along all the directions.

Figure 1 shows the relaxed structures produced by DFT (VASP) electronic calculations of the spherical particles [6]. In both samples tetrahedral Fe possesses a 10% smaller magnetic moment than octahedral Fe. The mean magnetic moment per Fe Ion is found $4.02 \mu_B$ for the DEG coated sample and $3.98 \mu_B$ for the OA coated sample and per Co Ion $2.65 \mu_B$ and $2.15 \mu_B$ respectively. Some Co atoms initially placed in pseudo octahedral sites show a reduced moment. This is attributed to the fact that due to the spatial distortion of the nanoparticles atomic configurations in comparison with the bulk structure, the O atoms are found at different distances here leading to a reduction of the d orbital moment. This affects a larger number of magnetic ions in the Oleic acid coated sample reducing the mean magnetic moment per Co ion. After performing the ionic relaxation, we calculated the magnetic anisotropy energy (MAE) taking into account the Spin-Orbit Coupling, non-self-consistently, for several spin orientations, by rotating all spins along different directions. We calculated the energy variation as a function of the squared cosine of the polar angle that represents the magnetic moment rotation. We obtained a linear dependence that indicates a uniaxial response in both cases. From the slope of the curve we extracted the MAE energy that equals to KV, where K is the anisotropy constant and V is the volume of the cell. Our calculations show that the DEG sample has a net magnetic moment $163.1 \mu_B$ approximately 1.3 times larger than that of the OA sample, whereas the OA sample has magnetic anisotropy energy $MAE = 9.68$ meV approximately 1.5 times larger than that of the DEG sample ($MAE = 6.31$ meV).

We have also calculated the exchange coupling parameters for the two nanoparticles mapping different magnetic configurations on a Heisenberg model. The mapping was performed in such a way that we had the interaction in each sublattice and between sublattices. In the DEG case J_{AA}, J_{BB}, J_{AB} are 0.8, 1.3,

a)   b)

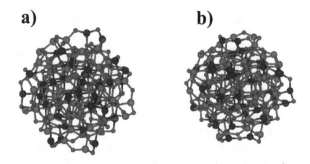

Fig. 1. Calculated relaxed structures for the (a) DEG, (b) OA coated $CoFe_2O_4$ particle (Co: blue circles, Fe: yellow circles, O: red circles) [6,12]. (Color figure online)

$-1.4\,\mathrm{meV}/\mu_B^2$ respectively while for OA case they are 0.9, 1.2, $-1.51\,\mathrm{meV}/\mu_B^2$. We observe that the exchange coupling parameters have no big variation between the two samples but they give the proper signs for the sublattices in agreement with the literature for bulk Co ferrites [13].

$MnFe_2O_4$ Nanoparticles - Effect of Albumin Surfactant. In this case starting from the bulk $MnFe_2O_4$ structure calculations - as in the previous case - to tune the on-site Coulomb strength U and exchange coupling J parameters, the effective parameter U_{eff} was taken as 4.5 eV for Fe atoms and 3.7 eV for Mn atoms. These values properly predict the dimensions of $a = b = c = 8.511$ Å of a bulk unit cell. The magnetic moments were $4.2\,\mu_B$ for Fe atoms and $4.1\,\mu_B$ for Mn atoms.

In our model, the particle was covered partially by sections of the albumin protein at the top and bottom face as shown in Fig. 2. The modeled particle consisted of four atomic layers of $MnFe_2O_4$ (inverse spinel), covered by the BSA protein. The relaxed structures produced by DFT (VASP) electronic calculations are shown in Fig. 2 [8]. The size of the obtained relaxed structure was 1.6 nm × 0.7 nm × 0.6 nm. Mn ions were found to create bonds only with O atoms, whereas Fe ions created bonds with O, C and N atoms. The mean value of the magnetic moment of the bonded atoms is increased by 1.8% with respect to the uncoated case (see Table 1).

Table 1. Mean magnetic moment per atom type and the total magnetic moment for the coated and uncoated Mn Ferrite particle.

	Mean moment per Fe (μ_B)	Mean moment per Mn (μ_B)	Saturated mag. moment (μ_B)	MAE energy (meV)
Uncoated	4.03	4.27	197.555	2.1
Coated	3.99	4.02	192.475	1.94

Due to the general expansion and distortion of the cell, neighboring magnetic atoms are found at larger distances and displaced positions with respect to the uncoated case. The general distortion affects also the environment of inner atoms leading to a reduction of the mean moment close to $0.17\,\mu_B$ with respect to the uncoated case. In Table 1, we observe that the coated particle has smaller magnetic moment per atom for each type of atoms resulting in 3% reduced total magnetic moment. This reduction is further enhanced with increasing number of bonded atoms at the NP surface according to our calculations. DFT results also indicate that in the uncoated $MnFe_2O_4$ the coordination symmetry is greatly reduced for the metal cations at the surface due to missing of some coordination oxygen atoms. In the coated with albumin NPs, the adsorbed ligands take the positions of the missing oxygen atoms. Even though the surface is spatially distorted in terms of number of neighbors, the surrounding environment tends to recover the bulk phase. This makes the crystal field of the surface metal ion to resemble closer that of the bulk material. Importantly, these changes account for the 7.6% reduction in the anisotropy of the coated sample.

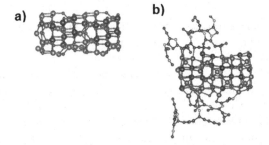

Fig. 2. Calculated relaxed structures for the (a) uncoated and (b) albumin coated $MnFe_2O_4$ particle (Mn: purple circles, Fe: yellow circles, O: red circles, C: brown circles, N: cyan circles) [8,12]. (Color figure online)

2.2 Mesoscopic Modeling of Magnetic Behavior of the Nanoparticles Assemblies

Assemblies of $CoFe_2O_4$ Nanoparticles Coated with OA and DEG Ligands. We have developed a simple mesoscopic model of 3-spins to simulate the magnetic properties of assemblies of $CoFe_2O_4$ nanoparticles with core/surface morphology, since the surface contribution of small sized MNPs is very important to their magnetic behavior. Our mesoscopic model [14] was based on the reduction of the amount of simulated spins to the minimum number necessary to describe the magnetic structure of the core/surface particles and on the introduction of the adequate exchange and anisotropy parameters between the different spin regions inside the nanoparticle. The latter were calculated starting from our DFT results, properly rescaled, to take into account the number of the spins in the core and at the surface regions based on an atomic scale model of a spinel ferrite particle of 5 nm. In this way we estimated the core and surface anisotropies, the intra-particle exchange coupling constants and the magnetic moments. In all the calculations, we took explicitly into account the coating thickness.

An assembly of N spherical ferrimagnetic nanoparticles of diameter d was considered with core/surface morphology, located randomly on the nodes of a cubic lattice inside a box of $10\alpha \times 10\alpha \times 10\alpha$ where α is the smallest inter-particle distance.

The total energy of the system for the N nanoparticles is:

$$
\begin{aligned}
E = & -\frac{1}{2}\sum_{i=1}^{N}\left[J_{c1}\left(s_{1i}\cdot s_{2i}\right) + J_{c2}\left(s_{1i}\cdot s_{3i}\right) + J_{srf}\left(s_{2i}\cdot s_{3i}\right)\right] - \sum_{i=1}^{N}K_C V_1 (s_{1i}\cdot\hat{e}_{1i})^2 \\
& -\sum_{i=1}^{N}K_{srf}V_2(s_{2i}\cdot\hat{e}_{2i})^2 - \sum_{i=1}^{N}K_{srf}V_3(s_{3i}\cdot\hat{e}_{3i})^2 - \sum_{i=1}^{N}\sum_{n=1}^{3}\mu_0 H m_{ni}(s_{ni}\cdot\hat{e}_h) \\
& -\frac{1}{2}\frac{\mu_0(M_S V)^2}{4\pi d^3}\sum_{i,j=1}^{N}\left\{\begin{array}{c}\frac{(m_{1i}s_{1i}+m_{2i}s_{2i}+m_{3i}s_{3i})\cdot(m_{1j}s_{1j}+m_{2j}s_{2j}+m_{3j}s_{3j})}{|r_{ij}|^3}\\ -3\frac{[(m_{1i}s_{1i}+m_{2i}s_{2i}+m_{3i}s_{3i})\cdot r_{ij}][(m_{1j}s_{1j}+m_{2j}s_{2j}+m_{3j}s_{3j})\cdot r_{ij}]}{|r_{ij}|^5}\end{array}\right\}
\end{aligned}
$$

(1)

Each nanoparticle is located at a lattice site (x, y, z) and it is described by a set of three classical spin vectors, one for the core s_{1i} and two s_{2i} s_{3i} for the surface where $i = 1,\ldots,N$ (total number of particles) with magnetic moment $m_n = M_n V_n / M_S V$, $n = 1$ stands for the core and $n = 2, 3$ for the "up" and "down" surface sublattices of the nanoparticle, respectively. V is the particle volume (equivalent to the number of the spins) and M_S its saturation magnetization. The first, second and third energy term in the square brackets describe the Heisenberg exchange interaction between the core spin and the two surface spins (interface coupling J_{c1} and J_{c2}), and the exchange interaction between the surface spins (surface coupling J_{srf}), respectively. The fourth, the fifth and the sixth terms give the anisotropy energy for the core and the surface spins assumed uniaxial with $\hat{e}_{1i}$, $\hat{e}_{2i}$, $\hat{e}_{3i}$ being the anisotropy random easy-axes direction. The next term is the Zeeman energy ($\hat{e}_h$ is the direction of the magnetic field). The last term gives the interparticle dipolar interactions including all the spins in the nanoparticles where the magnetic moments of the three macrospins of each particle are defined as $m_1 = M_1 V_1 / M_S V$, $m_2 = M_2 V_2 / M_S V$, and $m_3 = M_3 V_3 / M_S V$. The vector $r_{ij} = r_i - r_j$ corresponds to the particle position in a simulated box $L \times L \times L$ centered at $(0, 0, 0)$.

Starting from the bulk anisotropy values and taking into account our DFT calculations on surface anisotropies we estimated K_C and K_{srf}. The normalized volume of the three spin domains to the total volume was calculated, using an atomic scale model of an inverse spinel structured sphere where $V_1 = 0.3$, $V_2 = 0.21$, $V_3 = 0.49$, assuming the surface thickness 0.835 nm. In the normalization, we also took into account the fact that the surface layer volume of the DEG sample is smaller than that of the OA sample due to the thinner ($\sim$40%) DEG surface layer [6]. The magnetic moments for the three macrospins have been extracted from our DFT calculations taking into account also the volume in the

OA case. We normalized the energy parameters of Eq. 1 by the factor KV that is the core volume anisotropy of the nanoparticle, so they are dimensionless (see Table 2). The dipolar strength is $g = \mu_0(M_S V)^2/(4\pi d^3 K_C V_1)$. The effective exchange coupling constants were estimated by taking into account our DFT calculations and the difference in the magnetic moments of the two sublattices at the surface due to the number of the uncompensated spins [14]. The external magnetic field is denoted as H and the thermal energy as $k_B T$ (temperature T).

Table 2. Energy parameters for OA and DEG coated $CoFe_2O_4$ NPs.

OA coated	$j_{c1} = 1.3$, $j_{c2} = 1.2$, $j_{srf} = -1.2$	$k_c = 1$, $k_{srf} = 3.0$	$g = 1$	$m_1 = 0.198$, $m_2 = 0.76$, $m_3 = 0.23$
DEG coated	$j_{c1} = 1.3$, $j_{c2} = 1.2$, $j_{srf} = -1.2$	$k_c = 1$, $k_{srf} = 2.0$	$g = 1$	$m_1 = 0.198$, $m_2 = 1.07$, $m_3 = 0.26$

We use the Monte Carlo (MC) simulation technique with the implementation of the Metropolis algorithm [15]. The MC simulations results for a given temperature and applied field were averaged over 80 samples with various spin configurations, easy-axes distribution and spatial configurations for the nanoparticles. For every field and temperature value, the first 500 steps per spin are used for equilibration, and the subsequent 5000 MC steps are used to obtain thermal averages. The number of 5000MC steps is the optimum value to have the minimum statistical error in the calculation of the magnetization value, and to obtain consistency with the experimental results. For the calculation of the hysteresis loops at 5K and ZFC curves the standard experimental protocol is followed [6].

Cluster Assemblies of $MnFe_2O_4$ Nanoparticles Coated with Albumin Protein. We model the assembly of $MnFe_2O_4$ nanoparticles coated with albumin as a group of well separated small clusters of nanoparticles. The clusters are surrounded by the long albumin molecules so they are not touching each other while the nanoparticles in the clusters can be in physical contact. We employ our 3-spin mesoscopic model. The clusters of nanoparticles are created by dividing the box into eight regions with size $5\alpha \times 5\alpha \times 5\alpha$ each and variable particle concentration for each cluster, but under the constraint that the total concentration is $p = 50\%$ the same as in the case of the simulated uncoated assembly as described in ref. [7] (see Fig. 3). The total energy of the system is given by Eq. (1) adding the term of the interparticle exchange interactions between touching nanoparticles in the cluster: $E_{inter} = -\frac{1}{2}J_{inter}\Sigma_{\langle i,j \rangle}[(s_{2i} \cdot s_{3j}) + (s_{3i} \cdot s_{2j})]$ [6]. The energy parameters have been normalized by the factor $20 \times K_C V_1$ where V_1 is the core volume of the nanoparticle, so they are dimensionless.

In Table 3 the energy parameters of the model are presented. The J_{srf} for the albumin coated clusters of nanoparticles is decreased, because our DFT calculations showed that the bonding with the albumin causes expansion of the neighboring magnetic atoms at the surface of the MNPs and consequently reduction of the effective exchange coupling strength. There is not exact microscopic model

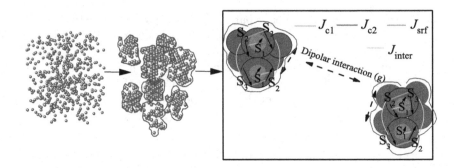

Fig. 3. Modeling of the uncoated (left) and the albumin-coated clusters (middle) ultra-small MnFe$_2$O$_4$ nanoparticles, (right) Enlarged schematic representation of two selected dipolarly interacting pairs of nanoparticles core (s$_1$)/surface (s$_2$, s$_3$) that belong to two neighboring clusters in each pair the exchange intraparticle (J_{c1}, J_{c2}, J_{srf}) and the exchange interparticle interactions (J_{inter}) between macrospins are depicted [8].

for the calculation of the J_{inter} so it is treated it as free parameters. We set $J_{inter} = -0.80$ between the nanoparticles in contact in the cluster larger than that of the uncoated case, since the presence of albumin brings closer the trapped nanoparticles in the clusters and enhances the fraction of the shell that comes into contact with the neighboring shells of these nanoparticles. The effective surface anisotropy constant in albumin case is taken 20% reduced compared to the uncoated case based on DFT results [7]. The saturation magnetization ratios have been extracted from atomic scale calculations for the spinel structure of a 2 nm diameter MnFe$_2$O$_4$ nanoparticle [7] and based on our DFT calculations where the surface magnetic moments are 10% reduced due to the existence of albumin.

3   Results and Discussion

We first examine and compare the effect of the two different ligands: the diethylene glycol (DEG) and the Oleic Acid (OA), bonded at the surface of 5 nm in size CoFe$_2$O$_4$ nanoparticles, on their magnetic behavior. In Fig. 4a,b the Monte Carlo simulated Field dependence magnetization (hysteresis loops at 5K) and temperature dependence of magnetization (Zero Field Cooled/Field Cooled Magnetization) curves of an assembly of CoFe$_2$O$_4$ nanoparticles coated with DEG (circles) and OA (squares) surfactant are presented.

Table 3. Energy parameters for uncoated and albumin coated MnFe$_2$O$_4$ NPs.

Uncoated	$j_{c1} = 0.5$, $j_{c2} = 0.45$, $j_{srf} = -1.0$, $j_{inter} = -0.50$	$k_C = 0.05$, $k_{srf} = 1.0$	$g = 3$	$m_1 = 0.1$, $m_2 = 0.5$, $m_3 = 0.4$
Coated	$j_{c1} = 0.5$, $j_{c2} = 0.45$, $j_{srf} = -0.8$, $j_{inter} = -0.80$	$k_C = 0.05$, $k_{srf} = 0.8$	$g = 3$	$m_1 = 0.1$, $m_2 = 0.45$, $m_3 = 0.4$

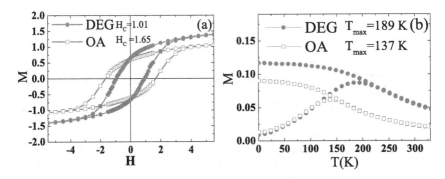

Fig. 4. Monte Carlo simulation results for the hysteresis loops at 5K (a) and the ZFC/FC curves (b) for an assembly of interacting $CoFe_2O_4$ NPs coated with DEG (circles) and OA (squares) surfactants [6].

We observe that the DFT calculated increase of the magnetic moments of the surface sublattices and the decrease of the surface anisotropy as an effect of the DEG coating, results to an increase of the saturation magnetization and a decrease of the coercivity of the system in comparison with the saturation magnetization and the coercivity values of the OA coated nanoparticles. In addition, the ZFC curves exhibit a maximum at a temperature (T_{max}) and the magnetic behavior becomes irreversible below a given temperature (T_{irr}) that we attribute to the blocking of the biggest particles. Notably, FC flattens out below T_{max}, which is typical feature of strong interparticle interactions inducing a collective state with high anisotropy. T_{irr} and T_{max} are higher for DEG comparing to OA samples. The increase in the saturation magnetization in the case of DEG sample results to the increase in the dipolar strength and consequently the increase of T_{max}. MC simulation results are in very good agreement with the experimental findings [6].

Figure 5 shows the Monte Carlo simulation results of the hysteresis loops (a) and ZFC/FC magnetization curves of the MFO sample (b) of ref. [7] together with those of the MFO_BSA sample (c) [8]. Our calculations show that the existence of ex-change coupled nanoparticles in the albumin coated clusters and the reduction of the surface anisotropy in the nanoparticles, due to the presence of albumin, causes the reduction of the coercive field H_C compared to the uncoated sample. In addition, the MC calculated ZFC magnetization curves (Fig. 5b,c) show maxima at almost the same temperature value $T_{max} \sim 0.6$ in agreement with the experimental results signaling the collective freezing of particles below T_{max} and that the albumin coating has a minor effect with respect to the dipolar interactions. On the other hand, the MFO_BSA system shows irreversible temperature ~ 2 times higher than that of the uncoated sample. This temperature difference indicates the existence of a large cluster size distribution, unlike in the uncoated system, providing also a qualitative measure of the width of the cluster size distribution. The simulated hysteresis curves and ZFC/FC curves reproduce very well the experimental results of ref. [8].

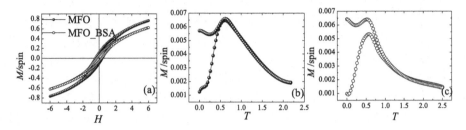

Fig. 5. Monte Carlo results of the hysteresis loops at $T = 0.01$ (a) and ZFC/FC magnetization curves for the uncoated nanoparticles (MFO sample) (b) ZFC/FC magnetization for the albumin-coated particles (MFO_BSA sample) (c) curves at $H_{\mathrm{app}} = 0.03$ [8].

4 Conclusion and Outlook

We have developed a multiscale numerical approach to study the magnetic behavior of MNP assemblies in all length scales starting from electronic to atomic and mesoscopic scale interactions on ARIS HPC system. The good agreement of our results with the experimental findings on the study of organic coated $CoFe_2O_4$ and $MnFe_2O_4$ nanoparticles demonstrates that our multiscale strategy is efficient to describe phenomena at the nanoscale and contributes to the interpretation and optimization of the magnetic properties of new high-performance magnetic nanomaterials for energy and biotechnological applications.

Acknowledgments. This work was supported by the European Union's Horizon 2020 Research and Innovation Programme: under grant agreement No. 731976 (MAGENTA). The authors acknowledge the computational time granted from the Greek Research & Technology Network (GRNET) in the Greek National HPC facility ARIS (http://hpc.grnet.gr) under project MNBIE (pr005030).

References

1. Wu, L., Mendoza-Garcia, A., Li, Q., Sun, S.: Organic phase syntheses of magnetic nanoparticles and their applications. Chem. Rev. **116**, 10473–10512 (2016). https://doi.org/10.1021/acs.chemrev.5b00687
2. Noh, S.H., Moon, S.H., Shin, T.H., Lim, Y., Cheon, J.: Recent advances of magneto-thermal capabilities of nanoparticles: from design principles to biomedical applications. Nano Today **13**, 61–76 (2017). https://doi.org/10.1016/j.nantod.2017.02.006
3. Hazra, S., Ghosh, N.N.: Preparation of nanoferrites and their applications. J. Nanosci. Nanotechnol. **14**, 1983–2000 (2014). https://doi.org/10.1166/jnn.2014.8745
4. Reddy, L.H., Arias, J.L., Nicolas, J., Couvreur, P.: Magnetic nanoparticles: design and characerization, toxicity and biocompatibility, pharmaceutical and biomedical applications. Chem. Rev. **112**, 5818–5878 (2012). https://doi.org/10.1021/cr300068p

5. Scherer, C., Neto, A.M.F.: Ferrofluids?: Properties and applications **35**, 718–727 (2005). https://doi.org/10.1590/S0103-97332005000400018
6. Vasilakaki, M., Ntallis, N., Yaacoub, N., Muscas, G., Peddis, D., Trohidou, K.N.: Optimising the magnetic performance of Co ferrite nanoparticles via organic ligand capping. Nanoscale **10**, 21244–21253 (2018). https://doi.org/10.1039/c8nr04566f
7. Vasilakaki, M., et al.: Monte Carlo study of the superspin glass behavior of interacting ultrasmall ferrimagnetic nanoparticles. Phys. Rev. B. **97**, 094413 (2018). https://doi.org/10.1103/PhysRevB.97.094413
8. Vasilakaki, M., et al.: Effect of Bovine serum albumin mediating clustering on the magnetic behavior of MnFe2O4 nanoparticles. Nanotechnology **31**, 25707 (2020). https://doi.org/10.1088/1361-6528/ab4764
9. Kresse, G., Furthmüller, J.: Efficiency of ab-initio total energy calculations for metals and semiconductors using a plane-wave basis set. Comput. Mater. Sci. **6**, 15–50 (1996). https://doi.org/10.1016/0927-0256(96)00008-0
10. Kresse, G., Furthmüller, J.: Efficient iterative schemes for ab initio total-energy calculations using a plane-wave basis set. Phys. Rev. B. **54**, 11169–11186 (1996). https://doi.org/10.1103/PhysRevB.54.11169
11. Malerba, L., et al.: Ab initio calculations and interatomic potentials for iron and iron alloys: achievements within the perfect project. J. Nuclear Mater. **406**, 7–18 (2010). https://doi.org/10.1016/j.jnucmat.2010.05.016
12. Momma, K., Izumi, F.: VESTA 3 for three-dimensional visualization of crystal, volumetric and morphology data. J. Appl. Crystallogr. **44**, 1272–1276 (2011). https://doi.org/10.1107/S0021889811038970
13. Bercoff, P.G., Bertorello, H.R.: Exchange constants and transfer integrals of spinel ferrites. J. Magn. Magn. Mater. **169**, 314–322 (1997). https://doi.org/10.1016/S0304-8853(96)00748-2
14. Margaris, G., Trohidou, K.N., Nogués, J.: Mesoscopic model for the simulation of large arrays of Bi-magnetic core/shell nanoparticles. Adv. Mater. **24**, 4331–4336 (2012). https://doi.org/10.1002/adma.201200615
15. Binder, K.: Applications of the Monte-Carlo Method in Statistical Physics. Springer, Heidelberg (1987). https://doi.org/10.1007/978-3-642-96788-7

clique: A Parallel Tool for the Molecular Nanomagnets Simulation and Modelling

Michał Antkowiak[✉], Łukasz Kucharski, and Monika Haglauer

Faculty of Physics, Adam Mickiewicz University,
ul. Uniwersytetu Poznańskiego 2, 61-614 Poznań, Poland
antekm@amu.edu.pl

Abstract. A powerful program for modelling the molecular nanomagnets is presented. The exact diagonalization method is used, which gives numerically accurate results. Its main bottleneck is the diagonalization time of large matrices, however it is removed by exploiting the symmetry of the compounds and implementing the method in the parallel computing environment. The diagonalization scheduling algorithm is implemented to increase the balance of the parallel processes workload. The running times of two different diagonalization procedures are compared.

Keywords: Molecular nanomagnets · Exact diagonalization · High performance computing

1 Introduction

Molecular nanomagnets based on transition metal ions have been very intensively investigated [14]. Their popularity is mostly due to the fact that quantum phenomena characteristic for a single molecule (like, e.g., quantum tunnelling or step like field dependence of magnetisation) can be observed in bulk samples. It is possible because nanomolecules are magnetically shielded from each other by organic ligands and the dominant interactions are those within the molecule. There are also expectations that this kind of materials may find application in quantum computing [6,7,15,23,24,29,35] and information storage [26]. A large family of molecular nanomagnets comprises ring-shaped molecules. Most of them contain even number of antiferromagnetically interacting ions. Only recently the first odd membered antiferromagnetic molecules have been reported [1,8,9,13,17,31]. They are especially interesting because of magnetic frustration which is expected to appear in this kind of materials.

Precise determination of the energy structure of molecular nanomagnets is necessary to allow the calculations of the state dependent properties such as local magnetisations or correlations [3,4,12,19,21]. An ideal tool for fulfilling this task is the exact diagonalization (ED) of Hamiltonian matrix [5,22]. In this paper we present the *clique*, a powerful new program for the simulation and modelling of the molecular nanomagnets.

© Springer Nature Switzerland AG 2020
R. Wyrzykowski et al. (Eds.): PPAM 2019, LNCS 12044, pp. 312–322, 2020.
https://doi.org/10.1007/978-3-030-43222-5_27

2   Exact Diagonalization Technique

The general form of the Hamiltonian which can be used for the molecular nano-magnets modelling is as follows:

$$\mathcal{H} = -\sum_{\langle ij \rangle} J_{ij} \boldsymbol{S}_i \cdot \boldsymbol{S}_j + \mu_{\mathrm{B}} \sum_{i=1}^{n} \boldsymbol{B} \cdot g_i \cdot \boldsymbol{S}_i + \sum_{i=1}^{n} D_i (S_i^z)^2, \tag{1}$$

where i, j denotes the positions of magnetic ions within a system, J_{ij} are the exchange integrals between sites i and j, $\boldsymbol{S}_i$ is the spin operator of the spin S of site i, D_i is the single-ion anisotropy of site i, $\boldsymbol{B}$ is the external magnetic field, g_i is the corresponding Landé factor and μ_{B} stands for the Bohr magneton.

The exact diagonalization technique allows to obtain the values of energy levels with accuracy only limited by machine arithmetic, thanks to which it is possible to obtain interesting thermodynamic quantities in a simple way. Unfortunately, the size of the Hamiltonian matrix for larger spin systems causes that numerical diagonalization becomes impossible in a realistic time, and is also difficult due to the limitation of operating memory. However, using the symmetry of the system relative to the reflection operation, the Hamiltonian matrix (1) can be divided into two smaller submatrices, and in the case of magnetic field application only in the direction of z—to $2\sum_{i=1}^{n} s_i + 1$ sub-matrices, where s_i is the spin value on i—th node, and n—the number of nodes. After considering both properties, the Hamiltonian matrix gains the form of $4\sum_{i=1}^{n} s_i$ diagonally placed submatrices in the basis created from possible projection of the total spin $S^z = \sum_{i=1}^{n} s_i^z$ on z axis. Each submatrix defined by the quantum number $M = \sum_{i=1}^{n} s_i^z$ and symmetry with respect to the reflection can be diagonalized separately.

Because the Hamiltonian matrix for the more complex systems is too large to first create it in a simple vector basis, and then to convert it to a quasi-diagonal form, we decided to develop an algorithm that allows direct creation of independent submatrices. For this purpose, we create a simple vectors basis, transform it into a symmetrized one and sort the elements according to the value of M and symmetry. In this way we get a basis divided into segments with the same value of M and symmetry, which allows us to independently create individual submatrices of Hamiltonian.

In order to show the operation of the algorithm we present its course on the example of a simple system of three spins $s = \frac{1}{2}$. The number of states of the system under consideration is $2^3 = 8$. The spin Hamiltonian (1) including the exchange anisotropy and the field along the z axis is then:

$$\mathcal{H} = \sum_{i=1}^{3} \left(\frac{1}{2} J_{\perp} \left(s_i^+ s_{i+1}^- + s_i^- s_{i+1}^+ \right) + J_{\parallel} s_i^z s_{i+1}^z + g\mu_B B_z s_i^z \right). \tag{2}$$

The simple vectors basis is as follows:

$$
\begin{aligned}
|A\rangle &= \left|\tfrac{1}{2};\tfrac{1}{2};\tfrac{1}{2}\right\rangle = \left|M=\tfrac{3}{2}\right\rangle , \\
|B\rangle &= \left|\tfrac{1}{2};\tfrac{1}{2};-\tfrac{1}{2}\right\rangle = \left|M=\tfrac{1}{2}\right\rangle , \\
|C\rangle &= \left|\tfrac{1}{2};-\tfrac{1}{2};\tfrac{1}{2}\right\rangle = \left|M=\tfrac{1}{2}\right\rangle , \\
|D\rangle &= \left|\tfrac{1}{2};-\tfrac{1}{2};-\tfrac{1}{2}\right\rangle = \left|M=-\tfrac{1}{2}\right\rangle , \\
|E\rangle &= \left|-\tfrac{1}{2};\tfrac{1}{2};\tfrac{1}{2}\right\rangle = \left|M=\tfrac{1}{2}\right\rangle , \\
|F\rangle &= \left|-\tfrac{1}{2};\tfrac{1}{2};-\tfrac{1}{2}\right\rangle = \left|M=-\tfrac{1}{2}\right\rangle , \\
|G\rangle &= \left|-\tfrac{1}{2};-\tfrac{1}{2};\tfrac{1}{2}\right\rangle = \left|M=-\tfrac{1}{2}\right\rangle , \\
|H\rangle &= \left|-\tfrac{1}{2};-\tfrac{1}{2};-\tfrac{1}{2}\right\rangle = \left|M=-\tfrac{3}{2}\right\rangle .
\end{aligned}
\tag{3}
$$

Then we transform the basis into symmetrized one and sort its elements by M and symmetry (related to the exchange of the equivalent ions 1 and 3):

$$
\begin{aligned}
|1\rangle = |A\rangle = &\quad \left|\tfrac{1}{2};\tfrac{1}{2};\tfrac{1}{2}\right\rangle &&= \left|M=\tfrac{3}{2}\right\rangle_s , \\
|2\rangle = |C\rangle = &\quad \left|\tfrac{1}{2};-\tfrac{1}{2};\tfrac{1}{2}\right\rangle &&= \left|M=\tfrac{1}{2}\right\rangle_s , \\
|3\rangle = \tfrac{1}{\sqrt{2}}(|B\rangle+|E\rangle) = &\tfrac{1}{\sqrt{2}}(\left|\tfrac{1}{2};\tfrac{1}{2};-\tfrac{1}{2}\right\rangle+\left|-\tfrac{1}{2};\tfrac{1}{2};\tfrac{1}{2}\right\rangle) &&= \left|M=\tfrac{1}{2}\right\rangle_s , \\
|4\rangle = \tfrac{1}{\sqrt{2}}(|B\rangle-|E\rangle) = &\tfrac{1}{\sqrt{2}}(\left|\tfrac{1}{2};\tfrac{1}{2};-\tfrac{1}{2}\right\rangle-\left|-\tfrac{1}{2};\tfrac{1}{2};\tfrac{1}{2}\right\rangle) &&= \left|M=\tfrac{1}{2}\right\rangle_a , \\
|5\rangle = |F\rangle = &\quad \left|-\tfrac{1}{2};\tfrac{1}{2};-\tfrac{1}{2}\right\rangle &&= \left|M=-\tfrac{1}{2}\right\rangle_s , \\
|6\rangle = \tfrac{1}{\sqrt{2}}(|D\rangle+|G\rangle) = &\tfrac{1}{\sqrt{2}}(\left|\tfrac{1}{2};-\tfrac{1}{2};-\tfrac{1}{2}\right\rangle+\left|-\tfrac{1}{2};-\tfrac{1}{2};\tfrac{1}{2}\right\rangle) = \left|M=-\tfrac{1}{2}\right\rangle_s , \\
|7\rangle = \tfrac{1}{\sqrt{2}}(|D\rangle-|G\rangle) = &\tfrac{1}{\sqrt{2}}(\left|\tfrac{1}{2};-\tfrac{1}{2};-\tfrac{1}{2}\right\rangle-\left|-\tfrac{1}{2};-\tfrac{1}{2};\tfrac{1}{2}\right\rangle) = \left|M=-\tfrac{1}{2}\right\rangle_a , \\
|8\rangle = |H\rangle = &\quad \left|-\tfrac{1}{2};-\tfrac{1}{2};-\tfrac{1}{2}\right\rangle &&= \left|M=-\tfrac{3}{2}\right\rangle_s .
\end{aligned}
\tag{4}
$$

The modified basis of vectors allows the calculation of non-zero Hamiltonian matrix elements:

$$
\begin{aligned}
\mathcal{H}_{1,1} &= \langle 1|\,\mathcal{H}\,|1\rangle = \tfrac{3}{4}J_\parallel + \tfrac{3}{2}g\mu_B B_z , \\
\mathcal{H}_{2,2} &= \langle 2|\,\mathcal{H}\,|2\rangle = -\tfrac{1}{4}J_\parallel + \tfrac{1}{2}g\mu_B B_z , \\
\mathcal{H}_{3,2} = \mathcal{H}_{2,3} &= \langle 2|\,\mathcal{H}\,|3\rangle = \langle 3|\,\mathcal{H}\,|2\rangle = \tfrac{1}{\sqrt{2}}J_\perp , \\
\mathcal{H}_{3,3} &= \langle 3|\,\mathcal{H}\,|3\rangle = \tfrac{1}{2}J_\perp - \tfrac{1}{4}J_\parallel + \tfrac{1}{2}g\mu_B B_z , \\
\mathcal{H}_{4,4} &= \langle 4|\,\mathcal{H}\,|4\rangle = -\tfrac{1}{2}J_\perp - \tfrac{1}{4}J_\parallel + \tfrac{1}{2}g\mu_B B_z , \\
\mathcal{H}_{5,5} &= \langle 5|\,\mathcal{H}\,|5\rangle = -\tfrac{1}{4}J_\parallel - \tfrac{1}{2}g\mu_B B_z , \\
\mathcal{H}_{6,5} = \mathcal{H}_{5,6} &= \langle 5|\,\mathcal{H}\,|6\rangle = \langle 6|\,\mathcal{H}\,|5\rangle = \tfrac{1}{\sqrt{2}}J_\perp , \\
\mathcal{H}_{6,6} &= \langle 6|\,\mathcal{H}\,|6\rangle = \tfrac{1}{2}J_\perp - \tfrac{1}{4}J_\parallel + \tfrac{1}{2}g\mu_B B_z , \\
\mathcal{H}_{7,7} &= \langle 7|\,\mathcal{H}\,|7\rangle = -\tfrac{1}{2}J_\perp - \tfrac{1}{4}J_\parallel - \tfrac{1}{2}g\mu_B B_z , \\
\mathcal{H}_{8,8} &= \langle 8|\,\mathcal{H}\,|8\rangle = \tfrac{3}{4}J_\parallel - \tfrac{3}{2}g\mu_B B_z .
\end{aligned}
\tag{5}
$$

The sizes of the Hamiltonian matrix, number of submatrices which can be formed using above approach, size and memory use of the largest amongst those submatrices are presented in Table 1 for chosen systems.

3   Implementation of the ED Technique in the Parallel Environment

The first system for which we have used the ED method is the Cr_8 molecule [20] containing eight spins $s = \frac{3}{2}$. The number of states in this case is $4^8 = 65536$, and the Hamiltonian matrix in the simple vectors basis has a size ($4^8 \times 4^8$). This gives 4^{16} matrix elements that will take up 32 GB of memory in double precision. However, after applying the method described in the previous section, we obtain a matrix divided into 48 blocks, the sizes of which range from 4068×4068 to 1×1.

The first version of *clique* was written in C++. Diagonalization procedures were taken from Numerical Recipes [27]. After creating a given submatrix, the program first reduces it to a tridiagonal form using the *tred2* procedure, and then calculates its eigenvalues (and eigenvectors if needed) using the *tqli* procedure. The main part of the calculation in terms of time is carried out by the first of the mentioned procedures. It is worth noting that in the case of calculating the eigenvectors, the calculation time increases by almost an order of magnitude.

Because the diagonalization of individual submatrices proceeds independently, we decided to use the MPI [34] library to parallelize diagonalization processes and prepare an application operating in a parallel environment. The main process of the program (*master*) plays only a management role, assigning tasks for diagonalization to the remaining processes (*slave*) and collecting the eigenvalues calculated by them. Due to differences in the diagonalization time of individual blocks, it was important to properly separate the tasks into individual processes. In order to make the most efficient use of the time of all processes, we used the *Longest Processing Time* (LPT) algorithm [16], which involves sorting tasks according to the size of submatrices and assigning the largest ones in the first place. The process that is the first to finish its task, receives the next largest from the list of the remaining tasks.

The *clique* allows the calculation of any molecular compound. In the input file you can specify the number of nodes n, the value of spin s_j for each node, the integrals of exchange between all spins, symmetry, anisotropy, value and angle of the magnetic field in the plane $x - z$. Using this data, the application creates a simple vectors basis, symmetrizes it and divides it into $n_b = 4 \sum_{j=1}^{n} s_j$ blocks according to the values of the quantum number M and the symmetry of the states a. The number a is 1 for symmetric states and 0 for antisymmetrical states. The number of blocks and sizes of the largest blocks for the exemplary systems is shown in Table 1.

The *master* process distributes the model and simulation parameters to all *slave* processes using the MPI_Bcast function and allocates the first p blocks, one for each of the *slave* processes, sending the appropriate fragment of the vectors basis using the MPI_Send function. Then it performs n_b loop iterations, in which it receives the results and passes subsequent tasks for the calculation. When receiving results using MPI_Recv function in the first call in the given iteration, the MPI_ANY_SOURCE parameter is used, so that it always receives the calculation results from the process that completed the calculation first.

Table 1. Space complexity for systems of n spins s. Consecutive rows for each n comprise: size of the Hamiltonian matrix, number of submatrices, size of the largest submatrix, memory use of the largest submatrix in double precision.

n \ s	$\frac{1}{2}$	1	$\frac{3}{2}$	2	$\frac{5}{2}$	3	$\frac{7}{2}$	4
2	4	9	16	25	36	49	64	81
	4	8	12	16	20	24	28	32
	1	2	2	3	3	4	4	5
	8 B	32 B	32 B	72 B	72 B	128 B	128 B	200 B
3	8	27	64	125	216	343	512	729
	6	12	18	24	30	36	42	48
	2	4	7	11	15	20	26	33
	32 B	128 B	392 B	968 B	2 kB	3 kB	5 kB	9 kB
4	16	81	256	625	1296	2401	4096	6561
	8	16	24	32	40	48	56	64
	4	11	24	45	76	119	176	249
	128 B	968 B	4 kB	16 kB	45 kB	111 kB	242 kB	484 kB
5	32	243	1024	3125	7776	16807	32768	59049
	10	20	30	40	50	60	70	80
	6	27	81	197	398	735	1244	1995
	288 B	6 kB	51 kB	303 kB	1 MB	4 MB	12 MB	30 MB
6	64	729	4096	15625	46656	117649	262144	531441
	12	24	36	48	60	72	84	96
	10	74	290	885	2166	4684	9076	16361
	800 B	43 kB	657 kB	6 MB	36 MB	167 MB	628 MB	2 GB
7	128	2187	16384	78125	279936	823543	2097152	4782969
	14	28	42	56	70	84	98	112
	19	200	1076	4095	12048	30400	67350	136711
	3 kB	312 kB	9 MB	128 MB	1 GB	7 GB	34 GB	139 GB
8	256	6561	65536	390625	1679616	5764801	16777216	43046721
	16	32	48	64	80	96	112	128
	38	563	4068	19125	68050	199399	506504	1153257
	11 kB	2 MB	126 MB	3 GB	35 GB	296 GB	2 TB	10 TB
9	512	19683	262144	1953125	10077696	40353607	134217728	387420489
	18	36	54	72	90	108	126	144
	66	1579	15180	90285	383910	1318471	3818659	9806297
	34 kB	19 MB	2 GB	61 GB	1 TB	13 TB	106 TB	700 TB
10	1024	59049	1048576	9765625	60466176	282475249	1073741824	
	20	40	60	80	100	120	140	
	126	4502	58152	428663	2197728	8769804	29099604	
	124 kB	155 MB	25 GB	1 TB	35 TB	560 TB	6 PB	
11	2048	177147	4194304	48828125	362797056	1977326743		
	22	44	66	88	110	132		
	236	12852	220397	2046303	12546213	58614304		
	435 kB	1 GB	362 GB	30 TB	1 PB	24 PB		

Each *slave* process creates a given Hamiltonian matrix block, diagonalizes it, and sends its eigenvalues to the *master* process. When the entire task queue is emptied, the *master* process sends the MPI_BOTTOM signal to the processes that have already completed the calculation. Thus, the communication is characterized by the asymmetric star topology—the processes responsible for diagonalization of the largest systems communicate the least with the *master* process. The *master* process only has the role of managing the remaining processes, which implement over 99% of calculations.

To determine the scalability of the program, we calculate the speedup S defined as the ratio of the clock run time of the sequential t_{seq} and parallel t_{par} versions of the program.

$$S = \frac{t_{seq}}{t_{par}}, \tag{6}$$

and the efficiency of parallelization E parallelism resulting from dividing the speedup of S by the number of processes p [30]:

$$E = \frac{S}{p} = \frac{t_{seq}}{t_{par}p}. \tag{7}$$

Statistics for each p value were obtained from several program starts for different values of the number of computing cores used per node.

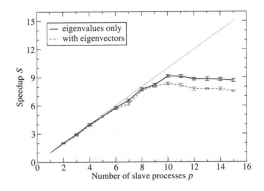

Fig. 1. The dependence of the speedup S on the number of parallel processes p for Cr$_8$ compound ($n = 8$, $s = \frac{3}{2}$). The ideal speedup is shown with a dotted line, the error bars mark standard deviation.

The graph of speedup shown in Fig. 1 indicate good scalability of our problem only to some limit $p = p_{max}$, which for the system Cr$_8$ is 10 if only eigenvalues are calculated or 9 if also eigenvectors are needed. The efficiency limitation is due to the fact that only a small number of Hamiltonian matrix blocks has a relatively large size, therefore for larger values of p program execution time is equal to the time of diagonalization of the largest submatrix, while other processes end up counting their share of tasks in a much shorter time. Therefore, for the case of $p > p_{max}$, you can observe stabilization of speedup (see Fig. 1).

The value of p_{max} increases slightly for systems larger than Cr_8, however, the diagonalization time increases with the third power of the matrix size. In the case of the Cr_9 system, for which the number of states is $4^9 = 262144$, the diagonalization time (calculation of eigenvalues only) of the largest submatrix size 15180×15180 was about seven hours and scalability increased slightly to $p_{max} = 12$. It was technically possible to carry out a simulation of this system, but it absorbed quite a lot of resources. Therefore, it was necessary to change the diagonalization algorithm to an even more efficient one.

In the next version of *clique* we have replaced the procedures from the Numerical Recipes with the *dsyev* procedure from the LAPACK [32] library. This allowed for several times faster calculations and a reduction in the use of resources. However, the use of the ScaLAPACK [33] library turned out to be crucial, thanks to which the diagonalization of the single submatrix could additionally be effectively parallelized within computing cores located in one node with shared operational memory.

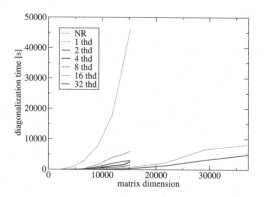

Fig. 2. The dependence of the diagonalization time on the matrix size. Yellow line shows the time of diagonalization obtained using Numerical Recipes procedures while the others—using ScaLAPACK library for different number of threads. (Color figure online)

Tests of a more advanced *clique* program were carried out on the *VIP* supercomputer located in the computing center in Garching. Each node of this computer has 32 cores built in. In Fig. 2, we compare the diagonalization times of matrices with different sizes and using different number of threads for the *dsyev* procedure from the ScaLAPACK library and procedures developed in Numerical Recipes. By thread, we mean a part of the program executed concurrently within one process; there may be multiple threads in one process. In order to test the program in a wider range of matrix sizes, we launched it with parameters of Cr_9 and additionally Ni_{12} molecules (only for 16 and 32 threads), which consists of 12 spins $s = 1$. Its Hilbert space contains a number of states equal to $3^{12} = 511441$, and the largest Hamiltonian submatrix reaches a size of 36965×36965.

The calculation time (Fig. 2) performed using procedures with Numerical Recipes significantly deviates from time for the ScaLAPACK library, even if only one thread is used, i.e. without additional parallelization. The use of many threads does not give the ideal speedup, but the efficiency of 63% with 32 threads allows for a significant acceleration of calculations and diagonalization of much larger matrices.

In order to further increase the scalability of the *clique* program in the next version, we enabled simultaneous calculations for many different systems and their parameters. The program accepts data from any number of input files and creates vectors for each of them. The calculations are carried out in accordance with the LPT algorithm, however, the scalability increases with the increase in the number of blocks coming from all those given for the simulation of the systems. The program allows simultaneous simulation of systems of different sizes, however, in a typical production run, there are usually the same systems differing only in parameters. Thanks to this, the set of blocks contains a correspondingly larger number of submatrices of the same size.

We have also eliminated the limitation of the main process to the role of managing other processes. In the latest version of the program, all processes take part in the calculation, and the results are collected at the very end of the program. Each process creates a vector basis itself, and the information about the last block taken for diagonalization is saved in a shared file, to which simultaneous access is secured by the *flock* function. This allows for even more efficient use of computing resources.

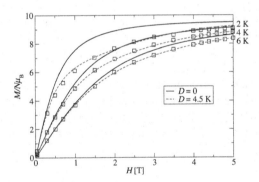

Fig. 3. The magnetization profiles of Ni_6Cr calculated using isotropic model and with added anisotropy.

4   Conclusions

Using the *clique* we were able to perform the modelling of different compounds by fitting the thermodynamic quantities obtained in the experiment. This resulted in a very good fits of the $Co_3^{II}La_3^{III}$ [25], $Mn_3^{II}Mn^{III}$ and $Mn_2^{II}Mn_2^{III}$ [28] molecules.

Our program was also used to calculate the energy spectra of the wide range of theoretical systems including frustrated antiferromagnetic rings with odd number of local spins characterized by a single bond defect or by arbitrary uniform couplings to an additional spin located at the center [2,10,11,18]. Fitting the magnetization of the Ni_6Cr we were able to justify the use of anisotropic modelling of the compound (see Fig. 3).

Precise determination of the energy structure of the simulated compound is crucial to allow the calculations of thermodynamic quantities for the whole temperature range available in the experiment. An ideal tool for fulfilling this task is the exact diagonalization of Hamiltonian matrix. The results obtained by this method are numerically accurate, but a major constraint and challenge is the exponential increase of the size of the matrix defined by $(2S+1)^n$, where n stands for the size of the system. It is very helpful to exploit fully symmetry of a given compound. If the magnetic field is directed along the z axis, the Hamiltonian takes a quasi-diagonal form in the basis formed by eigenvectors of the total spin projection S^z and can be divided into a number of submatrices labeled by quantum number M and the symmetry of the eigenstates.

Implementation of the exact diagonalization method in the parallel computing environment made the simulations possible for more complex molecules in the reasonable time. Running our optimized software on largest European supercomputers we were able to calculate the exact energy structure even for such big molecules like Ni_{12}, which has 531441 states. With *clique* we can perform the calculation for a molecular compounds with any number of ions, value of the spin of each ion, exchange interaction between all the spins, anisotropy, value and angle of magnetic field in the x-z plane. We used the MPI [34] library to parallelize the processes of the diagonalization of separate submatrices. For the most efficient use of computing time of all processes we implemented the *Longest Processing Time* algorithm [16]. In the final version of our code we applied ScaLAPACK library [33] which not only accelerates the diagonalization process, but also allows to parallelize the diagonalization of a single submatrix over all computational cores localized at a single node with shared memory. The simulations may be performed for a number of different parameter sets simultaneously, resulting in the increased scalability and better balance of the use of time by processes.

References

1. Adelnia, F., et al.: Low temperature magnetic properties and spin dynamics in single crystals of Cr8Zn antiferromagnetic molecular rings. J. Chem. Phys. **143**(24), 244321 (2015). https://doi.org/10.1063/1.4938086
2. Antkowiak, M., Florek, W., Kamieniarz, G.: Universal sequence of the ground states and energy level ordering in frustrated antiferromagnetic rings with a single bond defect. Acta Phys. Pol. A **131**, 890 (2017)
3. Antkowiak, M., Kozłowski, P., Kamieniarz, G.: Zero temperature magnetic frustration in nona-membered s=3/2 spin rings with bond defect. Acta Phys. Pol. A **121**, 1102–1104 (2012)

4. Antkowiak, M., Kozłowski, P., Kamieniarz, G., Timco, G., Tuna, F., Winpenny, R.: Detection of ground states in frustrated molecular rings by in-field local magnetization profiles. Phys. Rev. B **87**, 184430 (2013)
5. Antkowiak, M., Kucharski, L., Kamieniarz, G.: Genetic algorithm and exact diagonalization approach for molecular nanomagnets modelling. In: Wyrzykowski, R., Deelman, E., Dongarra, J., Karczewski, K., Kitowski, J., Wiatr, K. (eds.) PPAM 2015. LNCS, vol. 9574, pp. 312–320. Springer, Cham (2016). https://doi.org/10.1007/978-3-319-32152-3_29
6. Ardavan, A., et al.: Will spin-relaxation times in molecular magnets permit quantum information processing? Phys. Rev. Lett. **98**, 057201 (2007)
7. Atzori, M., et al.: Quantum coherence times enhancement in vanadium(IV)-based potential molecular qubits: the key role of the vanadyl moiety. J. Am. Chem. Soc. **138**(35), 11234–11244 (2016). https://doi.org/10.1021/jacs.6b05574. pMID:27517709
8. Baker, M., et al.: A classification of spin frustration in molecular magnets from a physical study of large odd-numbered-metal, odd electron rings. Proc. Natl. Acad. Sci. USA **109**(47), 19113–19118 (2012)
9. Cador, O., Gatteschi, D., Sessoli, R., Barra, A.L., Timco, G., Winpenny, R.: Spin frustration effects in an oddmembered antiferromagnetic ring and the magnetic Möbius strip. J. Magn. Magn. Mater. **290–291**, 55 (2005)
10. Florek, W., Antkowiak, M., Kamieniarz, G., Jaśniewicz-Pacer, K.: Highly degenerated ground states in some rings modeled by the ising spins with competing interactions. Acta Phys. Pol. A **133**, 411 (2018)
11. Florek, W., Antkowiak, M., Kamieniarz, G.: Sequences of ground states and classification of frustration in odd-numbered antiferromagnetic rings. Phys. Rev. B **94**, 224421 (2016). https://doi.org/10.1103/PhysRevB.94.224421
12. Florek, W., Kaliszan, L.A., Jaśniewicz-Pacer, K., Antkowiak, M.: Numerical analysis of magnetic states mixing in the Heisenberg model with the dihedral symmetry. In: EPJ Web of Conferences, vol. 40, p. 14003 (2013)
13. Furukawa, Y., et al.: Evidence of spin singlet ground state in the frustrated antiferromagnetic ring Cr_8Ni. Phys. Rev. B **79**, 134416 (2009)
14. Gatteschi, D., Sessoli, R., Villain, J.: Molecular Nanomagnets. Oxford University Press, Oxford (2006)
15. Georgeot, B., Mila, F.: Chirality of triangular antiferromagnetic clusters as qubit. Phys. Rev. Lett. **104**, 200502 (2010)
16. Graham, R.: Bounds of multiprocessing timing anomalies. SIAM J. Appl. Math. **17**, 416–429 (1969)
17. Hoshino, N., Nakano, M., Nojiri, H., Wernsdorfer, W., Oshio, H.: Templating odd numbered magnetic rings: oxovanadium heptagons sandwiched by β-cyclodextrins. J. Am. Chem. Soc. **131**, 15100 (2009)
18. Kamieniarz, G., Florek, W., Antkowiak, M.: Universal sequence of ground states validating the classification of frustration in antiferromagnetic rings with a single bond defect. Phys. Rev. B **92**, 140411(R) (2015)
19. Kamieniarz, G., et al.: Anisotropy, geometric structure and frustration effects in molecule-based nanomagnets. Acta Phys. Pol. A **121**, 992–998 (2012)
20. Kamieniarz, G., et al.: Phenomenological modeling of molecular-based rings beyond the strong exchange limit: bond alternation and single-ion anisotropy effects. Inorg. Chim. Acta **361**, 3690–3696 (2008). https://doi.org/10.1016/j.ica.2008.03.106
21. Kozłowski, P., Antkowiak, M., Kamieniarz, G.: Frustration signatures in the anisotropic model of a nine-spin $s = 3/2$ ring with bond defect. J. Nanopart. Res. (2011). https://doi.org/10.1007/s11051-011-0337-8

22. Kozłowski, P., Musiał, G., Antkowiak, M., Gatteschi, D.: Effective parallelization of quantum simulations: nanomagnetic molecular rings. In: Wyrzykowski, R., Dongarra, J., Karczewski, K., Waśniewski, J. (eds.) PPAM 2013. LNCS, vol. 8385, pp. 418–427. Springer, Heidelberg (2014). https://doi.org/10.1007/978-3-642-55195-6_39

23. Lehmann, J., Gaita-Ariño, A., Coronado, E., Loss, D.: Spin qubits with electrically gated polyoxometalate molecules. Nature Nanotech. **2**, 312 (2007)

24. Luis, F., et al.: Molecular prototypes for spin-based CNOT and SWAP quantum gates. Phys. Rev. Lett. **107**, 117203 (2011). https://doi.org/10.1103/PhysRevLett.107.117203

25. Majee, M.C., et al.: Synthesis and magneto-structural studies on a new family of carbonato bridged 3d–4f complexes featuring a [CoII3LnIII3(CO3)] (Ln = La, Gd, Tb, Dy and Ho) core: slow magnetic relaxation displayed by the cobalt(II)-dysprosium(III) analogue. Dalton Trans. **47**, 3425–3439 (2018). https://doi.org/10.1039/C7DT04389A

26. Mannini, M., et al.: Magnetic memory of a single-molecule quantum magnet wired to a gold surface. Nature Mat. **8**, 194 (2009)

27. Press, W., Teukolsky, S., Vetterling, W., Flannery, B.: Numerical Recipes in C: The Art of Scintific Computing. Cambridge University Press, Cambridge (1992)

28. Sobocińska, M., Antkowiak, M., Wojciechowski, M., Kamieniarz, G., Utko, J., Lis, T.: New tetranuclear manganese clusters with [MnII3MnIII] and[MnII2MnIII2] metallic cores exhibiting low and high spin ground state. Dalton Trans. **45**, 7303–7311 (2016). https://doi.org/10.1039/C5DT04869A

29. Timco, G., et al.: Engineering the coupling between molecular spin qubits by coordination chemistry. Nature Nanotech. **4**, 173–178 (2009)

30. de Velde, E.V.: Concurrent Scientific Computing. Springer, New York (1994). https://doi.org/10.1007/978-1-4612-0849-5

31. Yao, H., et al.: An iron(III) phosphonate cluster containing a nonanuclear ring. Chem. Commun. **16**, 1745–1747 (2006)

32. LAPACK - Linear Algebra PACKage. http://www.netlib.org/lapack/

33. ScaLAPACK – Scalable Linear Algebra PACKage. http://www.netlib.org/scalapack/

34. The Message Passing Interface (MPI) Standard. http://www.mcs.anl.gov/research/projects/mpi/

35. Zadrozny, J.M., Niklas, J., Poluektov, O.G., Freedman, D.E.: Millisecond coherence time in a tunable molecular electronic spin qubit. ACS Cent. Sci. **1**(9), 488–492 (2015). https://doi.org/10.1021/acscentsci.5b00338. pMID: 27163013

Modelling of Limitations of BHJ Architecture in Organic Solar Cells

Jacek Wojtkiewicz(iD) and Marek Pilch(✉)(iD)

Faculty of Physics, Warsaw University,
ul. Pasteura 5, 02-093 Warsaw, Poland
{wjacek,Marek.Pilch}@fuw.edu.pl

Abstract. Polymer solar cells are considered as very promising candidates for the development of photovoltaics of the future. They are cheap and easy to fabricate, however, up to now, they possess a fundamental drawback: low effectiveness. One ask the question how fundamental this limitation is. We propose the simple model which examines the limitations of efficiency by analysis of geometrical aspects of the bulk heterojunction (BHJ) architecture. We calculate the effective area of the donor-acceptor border in the random mixture of the donor and the acceptor nanocrystals and further compare it with an ideal "brush architecture". It turns out that in the BHJ architecture, this effective areas are very close to the value obtained in the "brush" one. Implications of this fact are discussed: we consider some other factors, which could limit the efficiency of the BHJ architecture, try to estimate its scale and speculate on possibilities of realization of another architectures and materials in the construction of solar cells.

Keywords: Photovoltaics · Organic solar cells · Bulk heterojunction architecture · Photovoltaic efficiency

1 Introduction

Organic photovoltaics is considered as one of the most perspective investigational trends in the entire topic of new types of solar cells design. Main advantages of the organic photovoltaic cells are: low cost, flexibility and small weight. Regrettably, the price we have to pay so far is low effectiveness: for few years the efficiency record has been fixed on level of 12% [1].

To gain an isight into various aspects of the problem, let us first remind the basic mechanism of action of solar cell. The conversion of light into electric current in organic cell is a complex, multistage process. One can recognize the following main stages of it [2,3].

The basic elements of the active layer of a cell are: the electron donor and the acceptor. In most cases, the donor is an organic polymer or oligomer. For the second component, i.e. the acceptor, fullerenes or their chemical derivatives are used in most cases. In the first stage of photovoltaic action, the donor absorbs

© Springer Nature Switzerland AG 2020
R. Wyrzykowski et al. (Eds.): PPAM 2019, LNCS 12044, pp. 323–332, 2020.
https://doi.org/10.1007/978-3-030-43222-5_28

photons of solar light. After absorption, an exciton is formed (i.e. a bound state of an excited electron and a hole). It diffuses to the border between the donor and the acceptor. On the border, the exciton dissociates into an electron and a hole. The hole remains confined in the donor, whereas the electron moves to the acceptor. In the last stage, the carriers of electric charge wander to the electrodes, where they accumulate. As a result, we observe a voltage between the electrodes.

An opportunity, which must be taken into account in the course of solar cells designing is a *short diffusion length* of an exciton. In most cases, it is on the order of few nanometers, rarely exceeding this value to about 20–30 nm [4]. Currently the most popular is the architecture called the BHJ (Bulk HeteroJunction) [5–8] (Fig. 1a). In a typical case, the active layer is composed of the grains of the donor and the acceptor. The characteristic grain size is on the order of tens of nanometers. This makes the area of D-A contact large and the generated exciton can get the border with the acceptor with high probability. It's a great opportunity of the BHJ architecture. Another opportunity is it's simplicity: to prepare an active BHJ blend, it suffices to mix the donor and the acceptor solutions and after evaporating the solvent, the blend is ready.

a)   b)

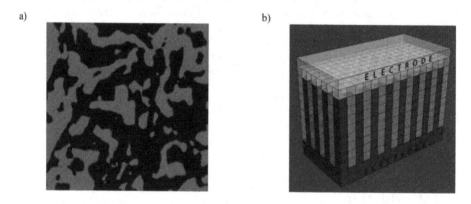

Fig. 1. (a) Very schematic view of the BHJ architecture; (b)—the brush architecture

However, the BHJ architecture has also certain drawbacks. One of them is the creation of the "islands" of the donor and the acceptor i.e. the attendance of the grains, which have no connection with the electrodes. In such a situation, even if the charge is generated on an "island", it can't go to the proper electrode. This means a loss of the cell's effectiveness. An analogical loss is caused by the attendance of the "bad peninsulas", i.e. the donor's grains against the cathode and the acceptor's grains against the anode. These negative factors were recognized very long ago, but surprisingly we could not find the estimates of the scale of these effects in the literature.

The architecture which is well-fitted to the exciton's features is so-called "brush architecture" ("comb" in the two-dimensional version) (Fig. 1b) [9].

The size of the donor/acceptor "teeths" should fit the exciton's diffusion length, i.e. their characteristic width should be on the order of 10–20 nm. A comparison of effectivenesses of the optimal "brush" architecture with that of the BHJ architecture seems very interesting. To express the effectiveness in a more quantitative manner: we compute the "geometric factor" Q, being the quotient of the average of the area of an active contact between the donor and the acceptor A_{BHJ} and analogous area in the "brush" architecture B:

$$Q = A_{BHJ}/B \tag{1}$$

The general setup of the model is as follows. We treat the donor's and the acceptor's nanograins as cubes in a simple cubic lattice. We consider also the second version of the model where the grains are the hexagonal prisms occupying the cells of a three-dimensional hexagonal lattice. In both versions, we assume that the donor's and the acceptor's grains are distributed randomly. For every distribution of the grains, we have computed the area of the border between the regions occupied by the donor and the acceptor. We allow the "parasitic" effects attendance, i.e. the fact that the "islands" and the "bad peninsulas" borders contribute nothing to the production of electricity. This way, the so-called "effective" area of the D-A border was calculated. In the next step, we compared it with the area of the border in the brush architecture. And last, we averaged over the random configurations of the donor and the acceptor grains.

Remark. The model we consider can be viewed as belonging to the broad class of models describing *percolation*. There are many kinds of this phenomenon; there exists large literature devoted to this, see for instance the textbook [10] for basic background in the subject. However, we couldn't find the information needed by us (i.e. the concrete values of the probability of the "plaquette" percolation in the systems of specific geometry) and this motivated us to undertake our own simulations.

2   The Model and the Computational Algorithm

2.1   Kinds of Blends Which We Simulate

Experimental probing of the BHJ blend structure was performed in numerous papers; an exhaustive review is [13]. Numerous methods, for instance TEM (Transmission Electron Microscopy) suggest that regions occupied by the donor and the acceptor could form irregular shapes, and borders between them could be sharp or fuzzy. Our modeling refers to such structures, where *regions occupied by the donor and the acceptor have similar size and shape, and the boundary between them is not fuzzy*. Examples of such structures are presented for instance in [12]. They are encountered in the case where both the donor and the acceptor are relatively small molecules (containing up to a few hundred atoms). On the other hand, polymers—as a rule—have a very small tendency to crystallization, and our model is less adequate for them.

2.2   Basic Technical Assumptions

- We consider two variants, defined on lattices: the cubic one (version 1) and the three-dimensional hexagonal one (version 2). We assume that in version 1 the model of the BHJ layer consists of randomly colored white (donor) and red (acceptor) cubes (Fig. 2a). In version 2, the model of the BHJ layer consists of randomly colored hexagon prisms.
- We considered subsets of lattices with sizes ranging between $10 \times 30 \times 30$ and $20 \times 200 \times 200$ lattice units. The length of the shorter side corresponds to the thickness of the BHJ layer, i.e. the distance between the electrodes. We adopted it as 10, 15 and 20 grains; this way cells with the thickness of the BHJ layer of 100–200 nm were simulated. The longer side of the BHJ layer corresponds to the characteristic size of an electrode; in a real cell it is way bigger than the thickness of the BHJ layer.
- In a lattice we can have various ratios of the white and the red cells. We adopted the ratios: 1:1, 2:3, 1:2 and 1:4.
- The border between the donor and the acceptor may be active, i.e. such that the charge created after exciton's decay can reach the proper electrode by flowing a sequence of connected donor/acceptor grains. A border can be also disactive when charges don't have such a possibility (i.e. they are trapped in an island without contact with an electrode). Figure 2 illustrates examples of active borders (marked in green) and an example of a disactive border (blue) in a blend section.
- The area of an active border is measured and compared with the area of the border in the "brush architecture".

2.3   Basic Technical Assumptions

Algorithm for the Cube Lattice. The parameters of the program are: the vertical size V (the thickness of the layer); the length, H_x and the width, H_y, of the layer, respectively; the probability P of filling a given cell by a donor grain. The P value corresponds to the ratio of the donor and the acceptor.

We start from declaring a three-dimensional array indexed with non-negative integers (i, j, k). Every individual cell stores a variable that can assume two values corresponding to the cell being occupied either by the donor or the acceptor. For illustrative purposes, every individual cell of the declared array is coloured in one of two colours: white means that a given cell is filled by a donor grain and red—by an acceptor grain. The main steps of the procedure are as follows.

Step 1. Sampling of configurations. For each cell, a random number from the interval $[0, 1]$ is generated. If this number is less than P, then the cell is marked red. Otherwise it is marked white. After step 1, the occupation of the lattice cells by the donor and the acceptor are determined.

Step 2. Local connectedness between cells. Examining the charge flow between the sequence of cubes, touching each other and containing the same content (i.e. the donor or acceptor), we are facing with the following problem.

a)

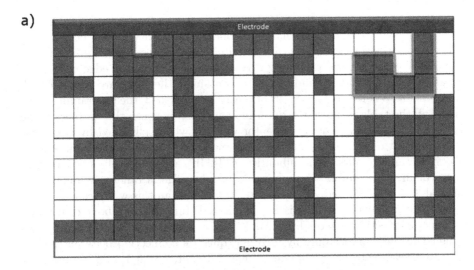

b)

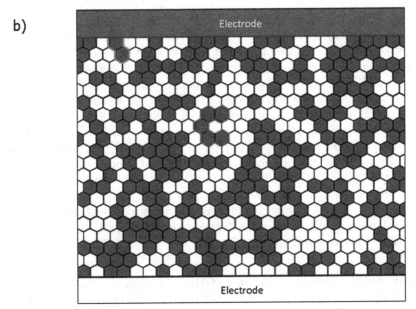

Fig. 2. A horizontal section of the model of the BHJ layer; (a) the BHJ blend is formed as a random mixture of cubic grains of the donor and the acceptor; (b) the BHJ blend is formed as a random mixture of hexagonal prisms. The blue line indicates the non-active borders, whereas the green line concerns the active borders. (Color figure online)

When two touching cubes possess the common wall, it is clear that the charge can flow from one cube to their neighbor. The situation is however different in the case where two touching cubes possess common edge only, or vertex only.

In such situations, we have to settle the query: are two touching cubes connected or not? We could assume that charge just has no possibility to flow between the cells via edges or vertexes – this case we will take into consideration in future. In this paper we adjudicate randomly if two cubes with common edge containing the acceptor are connected and two other containing the donor are disconnected, or vice versa. When coloring of cells of the array is finished, the programme again checks all the table to settle the quest of connection between the "grains" of the donor and the "grains" of the acceptor in cases when cells filled by this same colour contact only by edges or only by corners; for each cell programme checks it's colour and colours of neighbouring cells. In case when cells filled by this same colour, for example white, contact only by edge, a random number from interval $[0, 1]$ is generated. If it is greater than 0.5 than this white cells are connected and a pair of red cells is not connected. Information about connection or less of connection is scored up. Analogically, in case when cells filled by this same colour are contact only by a corner, quest of connection between them is arbitrated by generating a random number from interval $[0, 1]$ and comparing it with probability 0.25.

Step 3. When the quest of edge and corner connections between cells is resolved, the programme enters into searching cells filled with red colour, which are connected with the "red" electrode (i.e. with an edge of the BHJ layer which – as assumed – collects the electrons) by a coherent path composed with the red cells. Analogically are searched the coherent paths composed by the white cells reaching a "white" electrode (i.e. an edge of the BHJ layer which collects holes).

Step 4. In this stage, there are determined these regions from which the charges can flow into an adequate electrode. In step 4. the area of a border between these regions is calculated. After this last step, the total area of the active border is known and an algorithm is finishing one simulation. Every four-step simulation is repeated N times (we took $N = 100$). After that, standard statistical analysis of active border length is performed: We calculate the maximum, minimum, average, variance and standard deviation. Next, a quotient of the average and the length of border between the donor and the acceptor in 'brush architecture' is appointed (in percent).

Algorithm for the Hexagon Prism Lattice. The algorithm for the hexagon prism lattice (Fig. 2b) is very similar to this for the cube lattice, so we only stress the differences: donor and the acceptor grains can touch only by walls and edges but not by vertices, what facilitates the step 2. – we must to arbitrate only edges connections quest.

Remark. We began our simulations from the simpler two-dimensional situation [14]. We treated the two-dimensional model as a "warm-up" towards the three-dimensional one, which is certainly much closer to the reality. For this reason, we will neither discuss nor compare results coming from two and three dimensions.

2.4   Results

Results for Cubic Lattice. We performed simulations of blends for proportions of the donor and the acceptor: 1:1, 2:3, 1:2 and 1:4. In experiments, an arbitrary proportion of the donor and the acceptor can be taken. We took values close to the most popular values taken in the literature devoted to the experiments. A sample of obtained values of Q is presented on Fig. 3. It is seen that the largest value of Q are significantly higher than 100%, but in proportion 1:1 case it is less than 90%. High stability of quotient Q with growing horizontal sizes is observed too. The largest values of Q were observed for smallest values of V.

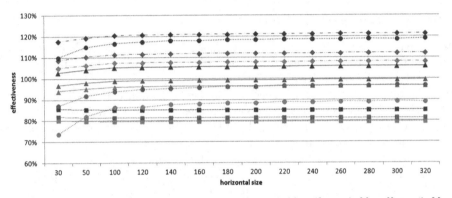

Fig. 3. Q values for the cube lattice. Results for various horizontal and vertical sizes as well as various proportions of the donor and the acceptor are presented.

For the D/A proportion being 1:1, the largest value of Q was 86%; for D/A proportion equal to 2:3, the largest value of Q was 106%; for D/A proportion 1:2, the largest value of Q was 121% and for D/A proportion 1:4 the largest value of Q was 119%.

Results for the Honeycomb Lattice. Here simulations are under development. Preliminary results (to be confirmed) give Q values much higher than in cube lattice case: the $Q(H)$ functions tend to certain limit value (depending of D/A proportion and vertical size) which ranges from 140% to 200%. Again, quotient Q appeared smallest for proportion 1:1. For instance, for proportion 1:1 and width 10, the maximal value of Q was 143% (86% for squares); for proportion 2:3 and width 10, the maximal value of Q was 175% (106% for squares); for proportion 1:2 and width 10, maximal value of Q was 196% (121% for squares) and for proportion 1:4 and width 10, maximal value of Q was 200% (119% for squares). For other widths we observe similar interrelations.

Summary of Simulations. The ultimate goal of our simulations was to find an answer to the question: Which is the area of the border between the donor and the Acceptor in the BHJ architecture A_{BHJ}, compared with the area of the border in the "brush" architecture A_{brush}. The answer we have obtained is that *the border area in the BHJ architecture is approximately the same or even bigger than in the "brush" architecture.* In a more quantitative manner, the quotient $Q = A_{\mathrm{BHJ}}/A_{\mathrm{brush}}$ took the value between 0.74 and 2.0. The value of Q depend of the thickness of the active layer, proportion of the Donor and the acceptor and the shape of their grains. An immediate consequence of this opportunity is that the efficiency of the photovoltaic device in the BHJ architecture should be similar or even significantly higher than in the "brush" architecture.

2.5   Summary and Conclusions

In the paper, we have examined the "geometric" factor which can influence the efficiency of photovoltaic cells built in the BHJ architecture. More precisely, we have calculated the effective area of the donor-acceptor border in the random mixture of the donor and the acceptor nanocrystals and compared it with an ideal "brush" architecture. As a result, it turned out that the areas in these two kinds architectures are approximately the same. In the other words, influence of the geometrical factor is of small importance.

We have also looked for other factors which make the efficiency of organic cells is far less than predicted by certain theoretical considerations, according to which the efficiency of the organic cells—even in one-junction simplest version—can achieve 20–24% [3,15]. In our opinion, lower efficiency can be caused by the presence of the regions occupied by the mixture of the donor and the acceptor, mixed in a molecular scale (Fig. 4). It is so because the blend emerges by evaporation of the solution, in which the donor and the acceptor are present. The faster evaporation is, the more fuzzy the boundary between the grains of the donor and the acceptor forms. On the other hand, the evaporation cannot be too slow, as the donor and the acceptor crystals would be too large.

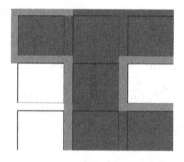

Fig. 4. Fuzzy boundary between grains of the donor and the acceptor

If the thickness of the fuzzy region is 5% of a cell's edge, then area of the sharp donor-acceptor boundary decays to approximately 80% and if 10%—to 64%. This simple calculus leads to supposition that if in region occupied by the mixture of the donor and the acceptor occurs some disadvantageous phenomena (e.g. the excitons have problem with dissociation into the electrons and the holes), this could have large impact on effectiveness.

Another potential direction of research aimed to enlarge the efficiency of photovoltaic devices could be to return to *layer* architecture. Devices constructed in a layer architecture exhibit lower (however not drastically) efficiency compared with those in BHJ architecture [11]. To improve efficiency of "layer" devices, one has to solve the main problem: *To find the substance(s), where the exciton diffusion length is comparable to the optical penetration length.* In more quantitative manner, typical value of optical penetration length is on the order of 100 nm [11], so one should find the substance where the exciton diffusion length is on the order of 100 nm. It is very difficult task, as in the most of donors or acceptors used in photovoltaic devices the exciton diffusion length is on the order of 10 nm [11]. But it seems that it is not hopeless, as there are known certain compounds (for instance, the anthracene) where the exciton diffusion length is about 100 nm! [11]. Unfortunately, the anthracene does not absorb the light in the visible range. To find the compound(s) absorbing the light in visible range and possessing the large exciton length is great challenge for the material research.

References

1. http://commons.wikimedia.org/wiki/File:PVeff(rev181217)v2.pdf. Accessed 4 Oct 2019
2. Heeger, A.J.: Semiconducting and metallic polymers: the fourth generation of polymeric materials (Nobel lecture). Angew. Chem. Int. Ed. **40**, 2591–2611 (2001)
3. Janssen, R.A.J., Nelson, J.: Factors limiting device efficiency in organic photovoltaics. Adv. Mater. **25**, 1847–1858 (2013)
4. Rand, B.P., Richter, H. (eds.): Organic Solar Cells: Fundamentals, Devices, and Upscaling, 1st edn. CRC Press, Taylor & Francis Group, Boca Raton (2014)
5. Yu, G., Heeger, A.J.: Charge separation and photovoltaic conversion in polymer composites with internal donor/acceptor heterojunctions. J. Appl. Phys. **78**, 4510–4515 (1995)
6. Halls, J.J.M., et al.: Efficient photodiodes from interpenetrating polymer networks. Nature **376**(6540), 498–500 (1995). https://doi.org/10.1038/376498a0
7. Yu, G., Gao, J., Hummelen, J.C., Wudl, F., Heeger, A.J.: Polymer photovoltaic cells: enhanced efficiencies via a network of internal donor-acceptor heterojunctions. Science **270**, 1789–1791 (1995)
8. Yang, C.Y., Heeger, A.J.: Morphology of composites of semiconducting polymers mixed with C60. Synth. Met. **83**, 85–88 (1996)
9. Gunes, S., Neugebauer, H., Sariciftci, N.S.: Conjugated polymer-based organic solar cells. Chem. Rev. **107**, 1324–1338 (2007)
10. Stauffer, D., Aharony, A.: Introduction to Percolation Theory. CRC Press, Taylor & Francis Group, Boca Raton (1994)

11. Lunt, R.R., Holmes, R.J.: Small molecule and vapor-deposited organic photo-voltaics. In: Rand, B.P., Richter, H. (eds.) Organic Solar Cells: Fundamentals, Devices, and Upscaling, 1st edn. CRC Press, Taylor & Francis Group, Boca Raton (2014)
12. Zhang, Q., et al.: Small-molecule solar cells with efficiency over 9%. Nat. Photonics **9**, 35–41 (2015)
13. Chen, W., et al.: Nanophase separation in organic solar cells. In: Qiao, Q. (ed.) Organic Solar Cells: Materials, Devices, Interfaces and Modeling. CRC Press, Taylor & Francis Group, Boca Raton (2015)
14. Wojtkiewicz, J., Pilch, M.: Modelling of limitations of bulk heterojunction architecture in organic solar cells. ArXiv:1709.01048
15. Shockley, W., Queisser, H.J.: Detailed balance limit of efficiency of p-n junction solar cells. J. Appl. Phys. **32**, 510–519 (1961)

Monte Carlo Study of Spherical and Cylindrical Micelles in Multiblock Copolymer Solutions

Krzysztof Lewandowski[1], Karolina Gębicka[1], Anna Kotlarska[1],
Agata Krzywicka[1], Aneta Łasoń[1], and Michał Banaszak[1,2]([envelope]) [ORCID]

[1] Faculty of Physics, A. Mickiewicz University,
ul. Uniwersytetu Poznańskiego 2, 61-614 Poznań, Poland
`michal.banaszak@amu.edu.pl`
[2] NanoBioMedical Centre, A. Mickiewicz University,
ul. Wszechnicy Piastowskiej 3, 61-614 Poznań, Poland

Abstract. Solutions of multiblock copolymer chains in implicit selective solvents are studied by Monte Carlo off-lattice method which employs a parallel tempering algorithm. The aggregation of block copolymers into micellar structures of spherical and cylindrical shapes is observed. The parallel tempering Monte Carlo method with feedback-optimized parallel tempering method is used.

Keywords: Block copolymer · Vesicles · Feedback-optimized parallel tempering · Monte Carlo method · Micelles

1 Introduction

Block copolymers have the property of aggregating into various structures [1–5]. When placed in a selective solvent, they can form micelles. In particular, they can be placed in water. At low concentrations, the amphiphiles are present in the solution as monomers, whereas in high concentrations they aggregate and self-assemble into micelles. The transition between low and high concentration regimes occurs at a concentration, which is referred to as critical micelle concentration (CMC). When CMC is surpassed, monomers in the solution are propelled by their hydrophobic "tails", which aggregate into micellar cores mostly via van der Waals interactions. The hydrophobic polymer blocks have low solubility in polar solvents and therefore they form the micellar core. Its corona, on the other hand, also known as the micellar outer shell, is built from hydrophilic blocks. Micelles can be formed in many different shapes, such as spheres, rods, tubes, lamellae or vesicles. The shape depends chiefly on the solvent quality, the block lengths and temperature.

Amphiphilic block copolymer micellar carriers are a popular choice for drug delivery vehicles. Their usefulness stems from their unique nanostructure.

© Springer Nature Switzerland AG 2020
R. Wyrzykowski et al. (Eds.): PPAM 2019, LNCS 12044, pp. 333–340, 2020.
https://doi.org/10.1007/978-3-030-43222-5_29

Hydrophobic drug molecules can be inserted into the micelle core and transported to the target with better solubility than normally in the bloodstream, due to the hydrophilic properties of the shell. The process is easiest to execute when the micelle is vesicle-shaped [6,7].

Vesicles are microscopic pockets enclosed by a thin molecular membrane. The membranes are usually aggregates of amphiphilic molecules. Vesicles formed by biological amphiphilic substances are essential for proper cell functioning. They are mostly lipids of molecular mass lower than 1 kDa. Block copolymers, which imitate lipids, can also self-assemble into vesicles in a diluted solution. Their molecular mass is often an order of magnitude larger than the mass of lipids, and their structural properties, as well as properties pertaining to stability, fluidity and intermembrane dynamics, are largely dependent on the properties of the polymers. This type of structure is engineered to trap substances within the space enclosed by the micellar core [8].

In this study we show the aggregation of block copolymers into micelles using off-lattice Monte Carlo simulations. The observed nanostructures are spherical and cylindrical. Since vesicles can be formed from these shapes, we expect that the cylindrical structures will evolve into vesicles given enough time, volume and proper conditions in the solution.

The aim is to use the parallel tempering Monte Carlo method with feedback-optimized parallel tempering method. This technique is known to improve the sampling of the phase space by reducing the average circulation time of replicas diffusing in temperature space. Because of the nature of parallel algorithm we use extensively the HPC environment (we employ a PC Linux-based cluster with 192 threads using the MPI protocol).

2   Methods

2.1   Simulation Methods

We use the Metropolis [9] acceptance criteria for the Monte Carlo (MC) moves. The MC moves are chain rotations, translations, crankshaft rotations, and slithering snake moves. A Monte Carlo step (MCS) is defined as an attempt to move once each monomer of the chain.

Moreover, we use parallel tempering (replica exchange) Monte Carlo [10,11] (PT) with feedback-optimized parallel tempering method [12,13] (FOPT). In the PT method M replicas of system are simulated in parallel, each at a different temperature T_i^*, with i ranging from 1 to M. After a number of MCS (in this work it is 100 MCS) we try to exchange replicas with neighboring T_i^* in random order with probability:

$$p(T_i^* \leftrightarrow T_{i+1}^*) = \min[1, \exp(-(\beta_i - \beta_{i+1})(U_{i+1} - U_i))], \qquad (1)$$

where $\beta_i = 1/k_B T_i^*$ and U_i is potential energy of replica at T_i^*.

Correctly adjusted PT method allows a better probing of the phase space of the system and prevents trapping in energy minima at low temperatures.

Thus it allows us to obtain better statistics in simulation and after a single simulation we obtain results for the selected range of temperatures. We use $M = 24, 32$, and 40 replicas.

2.2 Models

Simulation Box and Environment. Simulation is performed in a cubic box and the usual periodic boundary conditions are imposed. The simulation box size is sufficiently large for a chain to fit in, and not to interact with itself across boundary conditions. We simulate a single polymer chain and polymer-solvent interactions are included in an implicit manner in polymer-polymer interaction potential [14]. This can be considered a dilute polymer solution.

Polymer Model. We use a coarse-grained model for the polymer chain with monomers of diameter σ, taken also as the length unit [15]. In this work, by "monomer" we mean the basic building unit of the coarse-grained chain. Monomers are of two types: A and B. Neighboring monomers along the chain are connected via the bond potential:

$$U_{\text{Bonded}}(r) = \begin{cases} \infty, & \text{for } r < \sigma, \\ 0, & \text{for } \sigma \leq r \leq \sigma + \eta, \\ \infty, & \text{for } r > \sigma + \eta, \end{cases} \tag{2}$$

where $\sigma + \eta$ is the maximum bond length, and $\sigma + \frac{1}{2}\eta$ is considered to be the average bond length.

Monomers that are not adjacent along the chain (nonbonded monomers), interact via the following square well potential:

$$U_{\text{Non-bonded}}(r) = \begin{cases} \infty, & \text{for } r < \sigma, \\ \epsilon_{ij}, & \text{for } \sigma \leq r \leq \sigma + \mu, \\ 0, & \text{for } r > \sigma + \mu, \end{cases} \tag{3}$$

where $\sigma + \mu$ is range of the interaction potential, ϵ_{ij} is interaction energy between monomers of types i and j. We assume that $\mu = \frac{1}{4}\sigma$, and $\eta = \frac{1}{4}\sigma$ [16].

Chain bonds are not allowed to be broken, however they are allowed to be stretched. Interaction parameter ϵ_{ij} is defined for the selective solvent as:

$$\begin{aligned} \epsilon_{AA} &= -\epsilon, \\ \epsilon_{BB} &= \epsilon_{AB} = 0. \end{aligned} \tag{4}$$

The ϵ parameter, which is positive, serves as an energy unit to define the reduced energy per monomer E^*/N, and the reduced temperature T^* as:

$$\begin{aligned} E^*/N &= \left(\tfrac{E}{\epsilon}\right)/N, \\ T^* &= k_B T/\epsilon, \end{aligned} \tag{5}$$

where N is the number of chain monomers, and k_B is Boltzmann constant. Negative ϵ_{ij}'s indicate that there is an attraction between monomers, and the presence of the solvent is taken into account in an implicit manner [14]. By controlling the relative strength of this attraction, via T^*, we effectively vary solvent quality, from good to bad, which causes a collapse of the polymer chain, from a swollen state to a globular state [16–18]. The swollen and collapsed states are separated by the Θ solvent state, where the chain is Gaussian. This state is characterized by a temperature T_Θ^*, which is of the order of unity $T_\Theta^* \sim 1$ for this model. Since we are interested in intra-globular structures, we mostly concentrate on temperatures below the Θ-temperature.

3   Results

The behavior of A-B diblock copolymers in a solvent is here reported. At high temperatures, the chains remain unfolded. We anticipate the A-type segments to collapse as upon cooling, which should lead to the forming of micelles. The A-type blocks will aggregate, forming micellar cores. The B-type blocks will remain outside of the core, forming the corona. Repulsive interactions between individual B-type blocks will cause them to spread out, forming 180° angles between each copolymer's A and B blocks.

The systems considered were comprised of a constant number $n = 100$ of symmetric A_m-B_m type copolymer chains, with varying block length $m = 2, 3, \ldots, 10$. By symmetric copolymer we mean the chain with the same number of A and B monomers. Micellization can be reached either by changing concentration or temperature. We fixed the concentration and vary T_i^* which is more efficient than varying concentration.

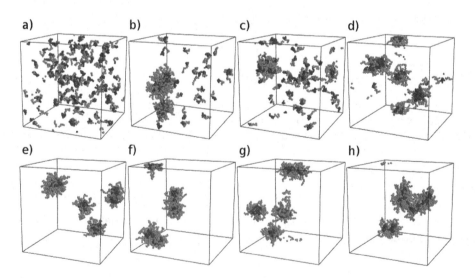

Fig. 1. Simulation snapshots of a system with $n = 100$ A_{10}–B_{10} copolymers at temperatures: (a) $T^* = 0.84$, (b) $T^* = 0.78$, (c) $T^* = 0.78$, (d) $T^* = 0.71$, (e) $T^* = 0.6$, (f) $T^* = 0.6$, (g) $T^* = 0.3$, (h) $T^* = 0.3$.

Starting with the longest blocks, $m = 10$, snapshots of the simulation box have been taken for different temperatures and are shown in Fig. 1. At $T^* = 0.84$, the polymers formed a disordered phase which means that no aggregation occurs. As temperature is lowered, the average distance between A-blocks decreases, whereas B-blocks stay at the coronas of the micelles, as expected. At $T^* = 0.78$, micelle-like aggregates begin to form. At $T^* = 0.71$ we observe free chains and four distinct micelles. At lower temperatures, $T^* = 0.6$ and $T^* = 0.3$, no free chains are observed - all copolymers have aggregated into 2, 3 or 4 distinct

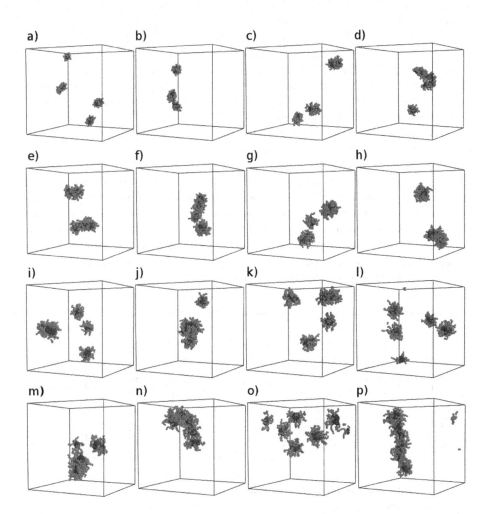

Fig. 2. Snapshots taken during a simulation of a system of A-B symmetric copolymers at temperatures $T^* = 0.2$ (a, c, e, g, i, k, m, o) and $T^* = 0.3$ (b, d, f, k, j, l, n, p) for chain lengths: (a) A_2-B_2, (b) A_2-B_2, (c) A_3-B_3, (d) A_3-B_3, (e) A_4-B_4, (f) A_4-B_4, (g) A_5-B_5, (h) A_5-B_5, (i) A_6-B_6, (j) A_6-B_6, (k) A_7-B_7, (l) A_7-B_7, (m) A_8-B_8, (n) A_8-B_8, (o) A_9-B_9, (p) A_9-B_9.

larger structures. It has to be stressed that both spherical and cylindrical micelles can be observed here. This process can be seen in the lower row of Fig. 1, as well as in another set of simulation snapshots, shown in Fig. 2, performed this time for smaller block lengths and exclusively at low temperatures. It is evident that the longer the copolymer chains, the easier it is to observe cylindrical structures in simulations. However, as seen in Fig. 2, the cylindrical micelles are formed in chains consisting of shorter blocks as well.

The chain's total size is shown in Fig. 3 as a function of temperature, for all considered block sizes m. As temperature is lowered, the chain size also decreases. At critical temperature a sudden increase of R_g^2, much higher than the individual increase of A- and B-type blocks, is observed. This is an effect of relative orientation of various blocks. The average size of the A-type blocks decreases similarly to that of the homopolymer chain, while the size of the B-type blocks remains roughly the same, and slightly increases below the cloud point temperature T_{CP}, as they spread out to accommodate more copolymer chains within the micellar structure.

Various numbers (from 1 to 5) of both spherical and cylindrical micellar structures can be found in Fig. 2 (however, at some conditions even more structures as shown in Fig. 2o). We think that the cylindrical micelles are stable because the crowns of spherical micelles are sufficiently thin for the two micelle cores to come close enough for the Van der Waals interaction to pull them together into a larger, cylindrical structure. It is well established that as the volume of sample is increased (at constant concentration), cylindrical micelles can aggregate into vesicular forms. Therefore, we also conjecture that for larger systems we can obtain vesicles in the same computational setup but this is beyond the scope of this paper. We intend to show this in a future study.

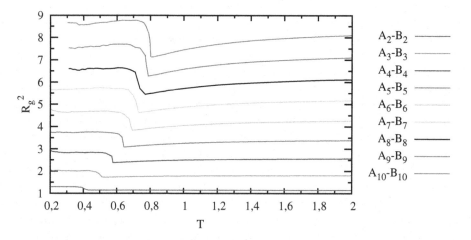

Fig. 3. Mean square of the radius of gyration R_g^2 of the whole chain for $n = 100$ symmetric A–B copolymers in temperature range $T^* \in (0.2; 2)$.

4   Conclusions

By simulating polymeric structures in a range of temperatures, we found that at fairly high temperatures the micellization cannot be observed; the polymers are in a disordered state. As temperature is lowered, around the micellization temperature $T^* = 0.78$, the micelle-like aggregates begin to form, with some fraction of free chains. Upon lowering the temperature further we observe more spherical micelles. Moreover, we also can see cylindrical nanostructures, consisting of a larger number of block chains.

The average number of clusters (spatially separate aggregates) in systems of A_{10}–B_{10} copolymers is about 3. For larger systems and at the same copolymer chain concentration, this number will be appropriately higher. Some of these aggregates will be spherical micelles; others - short cylinders. As for the radius of gyration, we found that as T^* is lowered, the mean chain length also decreases. At micellization temperature, a rapid increase of R_g^2 can be observed. It is then much larger than the average size of individual blocks. We conclude that the longer the copolymer chain length, the higher the micellization temperature is. This also means that dissolving micelles requires higher temperatures for longer chains.

One of the reasons that cylindrical micelles occur is the fact that the crowns of spherical micelles are thin enough that two micelle cores can come close enough to be pulled together into a larger, cylindrical structure. The next possible step, which is beyond the scope of this study, is the aggregation of micelles into vesicular forms.

Acknowledgments. We gratefully acknowledge the Polish National Science Centre (NCN) Grant No. UMO-2017/25/B/ST5/01970 and the computational grant from the Poznan Supercomputing and Networking Center (PCSS).

References

1. Hamersky, M.W., Smith, S.D., Gozen, A.O., Spontak, R.J.: Phase behavior of triblock copolymers varying in molecular asymmetry. Phys. Rev. Lett. **95**(16), 168306 (2005). https://doi.org/10.1103/PhysRevLett.95.168306
2. Bates, F.S., Fredrickson, G.H.: Block copolymers-designer soft materials. Phys. Today **52**(2), 32–38 (1999). https://doi.org/10.1063/1.882522
3. Lazzari, M., Liu, G., Lecommandoux, S.: Block Copolymers in Nanoscience, pp. 1–428. Wiley-VCH, Weinheim (2008)
4. Abetz, V., Simon, P.F.W.: Phase behaviour and morphologies of block copolymers. In: Abetz, V. (ed.) Block Copolymers I. Advances in Polymer Science, vol. 189, pp. 125–212. Springer, Heidelberg (2005). https://doi.org/10.1007/12_004
5. Semenov, A.N.: Contribution to the theory of microphase layering in block-copolymer melts. Sov. Phys. JETP **61**, 733–742 (1985)
6. Discher, D., Eisenberg, A.: Polymer vesicles. Science **297**, 967–973 (2002). https://doi.org/10.1126/science.1074972
7. Lewandowski, K.: Polimerowe struktury globularne i micelarne badane matodami symulacji komputerowych, p. 85. Ph.D. thesis (2012)

8. Hanafy, N., El-Kemary, M., Leporatti, S.: Micelles structure development as a strategy to improve smart cancer therapy. Cancers **10**, 238 (2018). https://doi.org/10.3390/cancers10070238

9. Metropolis, N., Rosenbluth, A., Rosenbluth, M., Teller, A., Teller, E.: Equation of state calculations by fast computing machines. J. Chem. Phys. **21**, 1087–1092 (1953). https://doi.org/10.1063/1.1699114

10. Earl, D.J., Deem, M.W.: Parallel tempering: theory, applications, and new perspectives. Phys. Chem. Chem. Phys. **7**, 3910–3916 (2005). https://doi.org/10.1039/B509983H

11. Sikorski, A.: Properties of star-branched polymer chains, Application of the replica exchange Monte Carlo method. Macromolecules **35**, 7132–7137 (2002). https://doi.org/10.1021/ma020013s

12. Katzgraber, H., Trebst, S., Huse, D., Troyer, M.: Feedback-optimized parallel tempering Monte Carlo. J. Stat. Mech.: Theory Exp. **3**, P03018 (2006). https://doi.org/10.1088/1742-5468/2006/03/P03018

13. Lewandowski, K., Knychała, P., Banaszak, M.: Parallel-tempering Monte-Carlo simulation with feedback-optimized algorithm applied to a coil-to-globule transition of a lattice homopolymer. CMST **16**, 29–35 (2010). https://doi.org/10.12921/cmst.2010.16.01.29-35

14. Zhou, Y., Hall, C., Karplus, M.: First-order disorder-to-order transition in an isolated homopolymer model. Phys. Rev. Lett. **77**, 2822 (1996). https://doi.org/10.1088/1742-5468/2006/03/P03018

15. Lewandowski, K., Banaszak, M.: Intraglobular structures in multiblock copolymer chains from a monte carlo simulation. Phys. Rev. E **84**, 011806 (2011). https://doi.org/10.1103/PhysRevE.84.011806

16. Lewandowski, K., Knychała, P., Banaszak, M.: Protein-like behavior of multiblock copolymer chains in a selective solvent by a variety of lattice and off-lattice monte carlo simulations. Phys. Status Solidi B **245**(11), 2524–2532 (2008). https://doi.org/10.1002/pssb.200880252

17. Wołoszczuk, S., Banaszak, M., Knychała, P., Lewandowski, K., Radosz, M.: Alternating multiblock copolymers exhibiting protein-like transitions in selective solvents: a Monte Carlo study. J. Non-Cryst. Solids **354**(35–39), 4138–4142 (2008). https://doi.org/10.1016/j.jnoncrysol.2008.06.022

18. Lewandowski, K., Banaszak, M.: Collapse-driven self-assembly of multiblock chains: a monte carlo off-latice study. J. Non-Cryst. Solids **355**, 1289–1294 (2009). https://doi.org/10.1016/j.jnoncrysol.2009.05.037

Electronic and Optical Properties of Carbon Nanotubes Directed to Their Applications in Solar Cells

Jacek Wojtkiewicz[1]([✉])[ID], Bartosz Brzostowski[2][ID], and Marek Pilch[1][ID]

[1] Faculty of Physics, University of Warsaw,
Pasteura 5, 02-093 Warsaw, Poland
{wjacek,Marek.Pilch}@fuw.edu.pl
[2] Institute of Physics, University of Zielona Góra,
ul. Prof. Szafrana 4a, 65-516 Zielona Góra, Poland
B.Brzostowski@if.uz.zgora.pl

Abstract. We calculate electronic and optical properties of a series of finite carbon nanotubes. Where available, our calculations exhibit good consistency with experimental data. Our study is directed towards potential application of carbon nanotubes in solar cells, constructed in a layer architecture.

Keywords: Carbon nanotubes · Photovoltaics · DFT calculations

1 Introduction

Organic photovoltaics is considered as one of the most perspective investigational trends in entire topic of new types of solar cells design. Main advantages of the organic photovoltaic cells are: low cost, flexibility and small weight. Regrettably, the price we have to pay so far is low effectiveness: for a few years the efficiency record has been fixed on the level of 12% [1]. One can ask what are perspectives to improve this effectiveness in order to achieve the performance on the level of commercial silicon-based cells (18–24% [1]). This problem has been raised in numerous papers, see for instance [2,3].

In order to recognize various aspects of the problem, let us first remind the basic mechanism of the action of solar cell. The conversion of light into electric current in organic cell is a complex, multistage process. One can recognize the following main stages of it [2,4].

Basic elements of active layer of a cell are: the electron donor and the acceptor. In most cases, the donor is an organic polymer or oligomer. For the second component, i.e. an acceptor, fullerenes or their chemical derivatives are used in most cases. In the first stage of photovoltaic action, the donor absorbs photons of solar light. After absorption, an exciton is formed (i.e. a bound state of excited electron and a hole). It diffuses to the border between donor and acceptor. On the border, the exciton dissociates onto an electron and a hole. The hole

© Springer Nature Switzerland AG 2020
R. Wyrzykowski et al. (Eds.): PPAM 2019, LNCS 12044, pp. 341–349, 2020.
https://doi.org/10.1007/978-3-030-43222-5_30

remains confined in the donor, whereas the electron moves to the acceptor. In the last stage, the carriers of electric charge diffuse to the electrodes, where they accumulate. As a result, we observe the voltage between the electrodes.

An opportunity, which must be taken into account in the course of solar cells designing is a *short diffusion length* of an exciton. In most cases, it is of the order of 10 nm, rarely exceeding this value [5]. Historically, the first organic solar cells have been built in a simple layer architecture [6]. Thickness of the layers should be comparable to the optical penetration depth, so they were of the order of 50–100 nm [6–9]. It turns out that solar cells built in the layer architecture exhibit rather limited efficiency (up to 5%). The most important factor which limits their efficiency is that majority of the excitons decays without dissociation into an electron and a hole before they achieve the donor – acceptor border.

Partial solution of this problem is given by the most popular now architecture called BHJ (Bulk HeteroJunction) [10–14]. They overcome the layer cells, however, they suffer from another (not fully recognized) limitations. For more than 5 years the top efficiency has not exceed 12% [1].

One of possible directions of research aimed to enlarge the efficiency of photovoltaic devices is return to layer architecture. But to improve efficiency of 'layer' devices, one has to solve the main problem: *Find the substance(s), where the exciton diffusion length is comparable to the optical penetration length.* In a more quantitative manner, typical value of optical penetration length is of the order 100 nm [5], so one should find the substance where the exciton diffusion length is of the same order. Unfortunately, in most of donors or acceptors used in photovoltaic devices the exciton diffusion length is of the order 10 nm [5]. Among rare exceptions, there is a remarkable one: for *carbon nanotubes* (CNTs), reported exciton diffusion length exceeds 200 nm [15,16].

The idea of using CNTs in solar cells is not new and motivations of such applications are quite diverse [17–25]. However, we haven't seen the scheme which we propose now:

The cell is constructed in the layer architecture. Nanotubes serve as a donor, and fullerenes as an acceptor.

To achieve good efficiency, the donor layer should absorb the light in the substantial part of visible and infrared regions. It is known, however, that – as a rule – absorption lines of CNTs are sharp and cover very narrow region of wavelength [18]. But there are many kinds of CNTs. They are classified by two natural numbers (n, m) (the n number is related to CNT circumference, and $m \leq n$ is a *chirality parameter* [18]. One can hope that using sufficiently many different CNTs, it will be possible to cover substantial region of solar light spectrum. Some rather crude and partial informations, based on the simple calculations of energy gaps in CNTs, show that could be in fact possible [18,19]. We are aware of some more precise calculations, but they concern only few CNTs [26,27]. To make our idea convincing, it is necessary to perform more systematic calculations.

We formulate our goal as follows:

Calculate: energies and amplitudes of optical transitions as well as HOMO and LUMO energies in a series of CNTs with growing n and m. Examine whether whole visible and IR regions of solar light could be covered by energies of optical transition of CNTs.

To realize this programme, it is necessary to compute energies and amplitudes of optical transitions for a few tens (perhaps about hundred) CNTs. We estimate that this is a large programme. Our present paper is aimed as a first step towards this direction: we present some results of calculations for a few CNTs and compare it with existing experimental data. It turns out that results of our calculations agree well with experiment. So we claim that our programme is worth further continuation.

2 Methodology of Computations

We performed our calculations mainly for finite systems. Such approach was motivated by the following opportunities.

Most of calculations concerning CNT's is performed within framework of computational scheme for periodic structures. It is reasonable point of view as the CNTs are very long in most cases (i.e. the ratio of their length and circumference is large). Such computations have been performed within quite a few approaches: non-interacting particles [18], DFT formalism with the use of various functionals, Green functions [26]. However, in most cases, calculations were performed without optimalization of geometry, assuming some given data, usually coming from experiment. But it is well known that geometry can have substantial influence on value of energies. In our computations for finite systems, we always used the geometry optimization.

The second aspect is a choice of computational scheme. The most of contemporary material calculations is performed within DFT formalism. However, there is a problem of choice: There are numerous DFT functionals, differing by quality of theoretical justification, computational complexity, universality etc. Here we decided to employ the Gaussian 09 package [28], widely used in quantum chemistry, and the functional B3LYP, which proved its effectiveness in a large class od chemical molecules [29].

We realize that our approach, based on considerations of CNTs of finite lengths, has both positive and negative aspects. The positive aspects are: good quality of computations (use of hybrid functionals, better developed in quantum chemistry packages than in the case of solid state ones) and possibility of calculations of transition amplitudes using TDDFT formalism. Calculations for finite systems constitute also the first step into the finite-size study and extrapolation. The price paid is however that at present time we are able to calculate relatively small systems – up to about 200 atoms, which means that calculated CNTs are not too long. We believe that calculated quantities saturate quickly with increasing length (we have checked this for selected systems). More profound calculations: for larger systems and using extrapolation to infinite length we plan to realize in the future.

3   Results

We performed the calculations primarily for these CNTs for which we could find experimental results. They are: (9,1), (8,3) and (7,5) CNTs. We have also calculated CNTs (8,0) and (6,0). (See Figs. 1, 2 and 3). The calculated CNTs were finite systems; so there appeared a problem how to 'end' it, in order to avoid artefacts coming from unpaired bonds. We decided to make the simplest solution, i.e. to saturate the unpaired bonds by the hydrogen atoms. We have calculated as large systems as we could with the use of computational capabilities in our disposal. The largest system calculated by us was $C_{144}H_{12}$.

Fig. 1. The finite CNT (9,1)

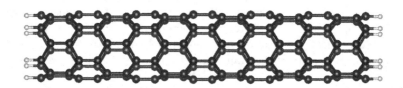

Fig. 2. The finite CNT (6,0)

Calculations were performed with the use of Gaussian 09 package [28]. We have optimized the geometry of every molecule. After that, we calculated one-electron energies and among them, the HOMO and LUMO energies. The most important quantities we calculated were energies and amplitudes of infrared and optical transitions; we calculated them with the use of TDDFT.

We encountered the problem of choice of the DFT functional. There exist 50 (or more) DFT functionals. We decided to use some as universal as possible in our preliminary study, and we choose the B3LYP functional, which has proved to give reliable results in many kinds of chemical molecules. The second aspect

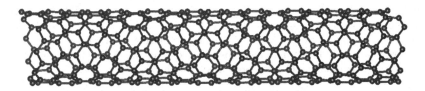

Fig. 3. Fragment of the infinite CNT (9,1)

is that the quality of calculations depend of the size of basis [29]. Again, we have chosen such a basis which guarantee reasonable accuracy and is not too time-consuming. We used the 6-31G(d) basis.

We have detected many optical transitions. But most of them possess negligible amplitude. For this reason, we present only the first (in the order of increasing transition energy) most intense one. All remaining ones (we calculated 30 transitions) possess the amplitude being 10 (or more) times lower.

For the sake of comparison, we have also calculated infinite variants of CNTs using SIESTA [30] package. At present, we were able to calculate only one-electron energies. It is known that, for CNTs, they give unsatisfacory results with respect to reproduction energies of optical transitions [26]. In general, one cannot expect that difference of HOMO and LUMO energies will precisely correspond to energy of lowest optical transition. Our results (coming for SIESTA and Gaussian) well illustrate this opportunity. In the future, we plan to extend studies of infinite systems by departing from one-particle approach.

The results of calculations for finite CNTs are collected in the Table 1, and for infinite CNTs in the Table 2. Moreover, plots of densities of states for CNTs are plotted on Fig. 4

One of absorption energies detected by us takes place within visible region (the second transition in (8,0) CNT) and another are in the infrared region. Due to limited computational capabilities, we couldn't calculate transitions of larger energy (within visible region), but we think that there are also such transitions. Which was confirmed in the (8,0) CNT [27].

It is also apparent that our results agree well with experimental ones [18] for CNTs: (7,5), (8,3) and (9,1). However, there is a point here: Our results concern an absorption, whereas data in [18] refer to the emission. Differences between these two values can be significant, so we couldn'd definitely settle the consistency of these two sets of data. We are aware also about results for CNT (8,0) [27] where values 1.60 eV (experimental) and 1.55 (theoretical) are reported. Our value is 1.64 eV – very close to these results. We also detected additional transition at 0.828 eV, absent in [27]. We don't know the reason for this discrepancy. Perhaps there is a finite-size effect, absent for long CNTs, considered in [27].

There is also another important aspect of construction of organic solar cells. Namely, the energies of HOMO and LUMO of the donor have to be higher than their corresponding values for the acceptor. The fullerene is a most natural choice for the acceptor. However, in such a case, not all CNTs give the

Table 1. Results of DFT and TDDFT calculations for selected finite CNTs with the use of Gaussian 09 package. Results for the first intense transitions are presented; for the CNTs (8,0) and (6,0), also the second transition is presented. Available experimental results are also placed for comparison. For the sake of comparison, energies of HOMO and LUMO of the fullerene C_{60} are presented. All energies are in eV.

Compound	Formula	E^{HOMO}	E^{LUMO}	ΔE_{DFT}	E^{opt}_{TDDFT}	E^{opt}_{exp}	f
CNT (7,5)	$C_{120}H_{24}$	−4.22	−2.83	1.39	1.188	1.212 [18]	0.140
CNT (8,3)	$C_{110}H_{22}$	−4.06	−2.98	1.08	1.256	1.298 [18]	0.417
CNT (9,1)	$C_{114}H_{20}$	−3.71	−3.46	0.25	1.266	1.355 [18]	0.494
CNT (8,0)	$C_{128}H_{16}$	−3.70	−3.48	0.22	0.828		0.381
CNT (8,0)	$C_{128}H_{16}$	−3.70	−3.48	0.22	1.64	1.60 [27]	0.205
CNT (6,0)	$C_{144}H_{12}$	−3.84	−3.29	0.55	1.09		0.069
CNT (6,0)	$C_{144}H_{12}$	−3.84	−3.29	0.55	1.14		0.269
Fullerene	C_{60}	−5.99	−3.23				

Table 2. One-electron energies for infinite CNTs with the use of SIESTA package. All energies are in eV.

CNT	E^{HOMO}	E^{LUMO}	ΔE_{DFT}
6-0	−4.24	−3.91	0.33
7-5	−3.91	−3.01	0.90
8-0	−4.39	−3.82	0.57
8-3	−3.86	−2.96	0.90
9-1	−3.81	−2.94	0.87

correct ordering of energy levels, see Table 1. But one can hope that by chemical modification ('functionalization') of CNTs or fullerenes it will be possible to manipulate locations of HOMO and LUMO orbitals, and to achieve the correct ordering of energy levels. We plan to examine modified CNTs and fullerenes in further studies.

4   Summary, Perspectives of Future Research

We have presented results for optical absorption for selected CNTs. For these cases where we could find a comparison, we observed good consistency of our research with GW-BSE approach as well as experimental results [18,26,27].

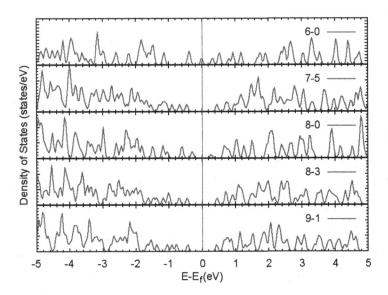

Fig. 4. Densities of states for infinite CNTs. The Fermi energy is at 0.

Some of our further goals are as follows:

(i) Continue systematic calculations for another CNTs. An access to high-performance machines will be crucial in this aspect.
(ii) Perform computations for longer CNTs and perform systematic study of finite-size effects.
(iii) Study infinite-length CNTs using formalism developed for periodic structures and compare it with results for finite systems.
 And – we estimate it as more demanding –
(iv) Develop theoretical and computational tools towards prediction of exciton diffusion length in CNTs.

Our present study is meant as first steps towards the realization of the programme sketched in the Introduction: *Examine whether most of visible and IR regions of solar light could be covered by energies of optical transitions of CNTs.* One can hope that the answer is positive, due to the following simple experimental fact: the person who saw the standard CNT sample, knows that they are black and absorb all visible light. Our calculations shows that absorption lines of specific CNT are sharp and hold at some concrete wavelengths. However, in a standard sample, very diverse kinds of nanotubes are present (differing by length, circumference, chirality, etc). Such a mixture gives the absorption spectrum being an average of absorption lines of every individual nanotube. (Example of such situation is in the book by Fox [18], Fig. 8.26). One of our goals is to identify which CNT absorb in a given wavelength, and how averaging of absorption lines of many different kinds of nanotubes holds.

In our opinion, the results obtained are quite satisfactory and encouraging, and we would like to continue this programme. Our calculations might serve as a road-map towards construction of light and cheap photovoltaic cells with high efficiency.

Acknowledgment. We are grateful to anonymous referee for questions, remarks and constructive criticism.

References

1. https://commons.wikimedia.org/wiki/File:Best_Research-Cell_Efficiencies.png
2. Janssen, R.A.J., Nelson, J.: Factors limiting device efficiency in organic photovoltaics. Adv. Mater. **25**, 1847–1858 (2013). https://doi.org/10.1002/adma.201202873
3. Scharber, M.C., Sariciftci, N.S.: Efficiency of bulk-heterojunction solar cells. Prog. Polym. Sci. **38**, 1929–1940 (2013). https://doi.org/10.1016/j.progpolymsci.2013.05.001
4. Heeger, A.J.: Semiconducting and metallic polymers: the fourth generation of polymeric materials (Nobel lecture). Angew. Chem. Int. Ed. **40**, 2591–2611 (2001). https://doi.org/10.1002/1521-3773(20010716)40:14%3C2591::AID-ANIE2591%3E3.0.CO;2-0
5. Lunt, R.R., Holmes, R.J.: Small-molecule and vapor-deposited organic photovoltaics. In: Rand, B.P., Richter, H. (eds.) Organic Solar Cells: Fundamentals, Devices, and Upscaling. CRC Press, Taylor & Francis Group, Boca Raton (2014). https://doi.org/10.4032/9789814463669
6. Tang, C.W.: Two-layer organic photovoltaic cell. Appl. Phys. Lett. **48**, 183–185 (1986). https://doi.org/10.1063/1.96937
7. Chu, C.W., Shao, Y., Shrotriya, V., Yang, Y.: Efficient photovoltaic energy conversion in tetracene-C60 based heterojunctions. Appl. Phys. Lett. **86**, 243506 (2005). https://doi.org/10.1063/1.1946184
8. Terao, Y., Sasabe, H., Adachi, C.: Correlation of hole mobility, exciton diffusion length, and solar cell characteristics in phthalocyanine/fullerene organic solar cells. Appl. Phys. Lett. **90**, 103515 (2007). https://doi.org/10.1063/1.2711525
9. Xue, J., Uchida, S., Rand, B.P., Forrest, S.R.: Asymmetric tandem organic photovoltaic cells with hybrid planar-mixed molecular heterojunctions. Appl. Phys. Lett. **85**, 5757–5759 (2004). https://doi.org/10.1063/1.1829776
10. Yu, G., Heeger, A.J.: Charge separation and photovoltaic conversion in polymer composites with internal donor/acceptor heterojunctions. J. Appl. Phys. **78**, 4510–4515 (1995). https://doi.org/10.1063/1.359792
11. Halls, J.J.M., et al.: Efficient photodiodes from interpenetrating polymer networks. Nature **376**, 498–500 (1995). https://doi.org/10.1038/376498a0
12. Yu, G., Gao, J., Hummelen, J.C., Wudl, F., Heeger, A.J.: Polymer photovoltaic cells - enhanced efficiencies via a network of internal donor-acceptor heterojunctions. Science **270**, 1789–1791 (1995). https://doi.org/10.1126/science.270.5243.1789
13. Yang, C.Y., Heeger, A.J.: Morphology of composites of semiconducting polymers mixed with C60. Synth. Met. **83**, 85–88 (1996). https://doi.org/10.1016/S0379-6779(97)80058-6
14. Collins, S.D., Ran, N.A., Heiber, M.C., Nguyen, T.-Q.: Small is powerful: recent progress in solution-processed small molecule solar cells. Adv. Energy Mater. **7**(10), 1602242 (2017). https://doi.org/10.1002/aenm.201602242

15. Moritsubo, S., et al.: Exciton diffusion in air-suspended single-walled CNTs. Phys. Rev. Lett. **104**, 247402 (2010). https://doi.org/10.1103/PhysRevLett.104.247402
16. Yoshikawa, K., Matsuda, K., Kanemitsu, Y.: Exciton transport in suspended single carbon nanotubes studied by photoluminescence imaging spectroscopy. J. Phys. Chem. C **114**, 4353–4356 (2010). https://doi.org/10.1021/jp911518h
17. Sgobba, V., Guldi, D.M.: Carbon nanotubes as integrative materials for organic photovoltaic devices. J. Mater. Chem. **18**, 153–157 (2008). https://doi.org/10.1039/B713798M
18. Fox, M.: Optical Properties of Solids. Clarendon Press, Oxford (2010)
19. Kataura, H., et al.: Optical properties of single-wall carbon nanotubes. Synth. Met. **103**, 2555–2558 (1999). https://doi.org/10.1016/S0379-6779(98)00278-1
20. Baker, B.A., Zhang, H., Cha, T.-G., Choi, J.H.: Carbon nanotubes sollar cells. In: Yamashita, S., Saito, Y., Choi, J.H. (eds.) Carbon Nanotubes and Graphene for Photonic Applications. Woodhead Publishing Series in Electronic and Optical Materials, pp. 241–269. Woodhead Publishing, Cambridge (2013)
21. Cataldo, S., Menna, E., Salice, P., Pignataro, B.: Carbon nanotubes and organic solar cells. Energy Environ. Sci. **5**(3), 5919–5940 (2012). https://doi.org/10.1039/C1EE02276H
22. Kymakis, E., Amaratunga, G.A.J.: Single-wall carbon nanotube/conjugated polymer photovoltaic devices. Appl. Phys. Lett. **80**, 112–114 (2002). https://doi.org/10.1063/1.1428416
23. Lee, U.J.: Photovoltaic effect in ideal carbon nanotube diodes. Appl. Phys. Lett. **87**, 073101 (2005). https://doi.org/10.1063/1.2010598
24. Pradhan, B., Batabyal, S.K., Pal, A.J.: Functionalized carbon nanotubes in donor/acceptor-type photovoltaic devices. Appl. Phys. Lett. **88**, 093106 (2006). https://doi.org/10.1063/1.2179372
25. Kymakis, E., Amaratunga, G.A.J.: Carbon nanotubes as electron acceptors in polymeric photovoltaics. Rev. Adv. Mater. Sci. **10**, 300–305 (2005)
26. Spataru, C.D., Ismail-Beigi, S., Capaz, R.B., Louie, S.G.: Quasiparticle and excitonic effects in the optical response of nanotubes and nanoribbons. In: Jorio, A., Dresselhaus, G., Dresselhaus, M.S. (eds.) Carbon Nanotubes. TAP, vol. 111, pp. 195–228. Springer, Heidelberg (2007). https://doi.org/10.1007/978-3-540-72865-8_6
27. Spataru, C.D., Ismail-Beigi, S., Benedict, L.X., Louie, S.G.: Quasiparticle energies, excitonic effects and optical absorption spectra of small-diameter single-walled carbon nanotubes. Appl. Phys. A **78**, 1129–1136 (2004). https://doi.org/10.1007/s00339-003-2464-2
28. Frisch, M.J., et al.: Gaussian 09, Revision A.02. Gaussian Inc., Wallingford (2009)
29. Piela, L.: Ideas of Quantum Chemistry. Elsevier, Amsterdam (2019)
30. Soler, J.M., et al.: The SIESTA method for ab initio order-N materials simulation. J. Phys.: Condens. Matter **14**, 2745–2779 (2002). https://doi.org/10.1088/0953-8984/14/11/302

Minisymposium on High Performance Computing Interval Methods

The MPFI Library: Towards IEEE 1788–2015 Compliance

(*In Memoriam Dmitry Nadezhin*)

Nathalie Revol[✉]

University of Lyon - Inria, LIP - ENS de Lyon, 46 allée d'Italie, 69007 Lyon, France
Nathalie.Revol@inria.fr
http://perso.ens-lyon.fr/nathalie.revol/

Abstract. The IEEE 1788–2015 has standardized interval arithmetic. However, few libraries for interval arithmetic are compliant with this standard. In the first part of this paper, the main features of the IEEE 1788–2015 standard are detailed, namely the structure into 4 levels, the possibility to accomodate a new mathematical theory of interval arithmetic through the notion of *flavor*, and the mechanism of *decoration* for handling exceptions. These features were not present in the libraries developed prior to the elaboration of the standard. MPFI is such a library: it is a C library, based on MPFR, for arbitrary precision interval arithmetic. MPFI is not (yet) compliant with the IEEE 1788–2015 standard for interval arithmetic: the planned modifications are presented. Some considerations about performance and HPC on interval computations based on this standard, or on MPFI, conclude the paper.

Keywords: Interval arithmetic · IEEE 1788–2015 standard · MPFI library · Compliance

1 Introduction and Context

Interval arithmetic has been defined even before the 1960s [15,27] and has continuously evolved and improved since then, with the development of algorithms to solve larger classes of problems through the 1970s and 1980s [1,16,20], then with a focus on the implementation [26] and more recently with its introduction in master courses [17,29]. However, in 2008, it was noticed and strongly resented that there were no definitions common to all authors and that it made it difficult to compare results. Under the auspices of IEEE where the standardization of floating-point arithmetic took place, leading to IEEE 754–1985 [8] and IEEE 754–2008 [9], a standardization effort was launched. It led to the IEEE 1788–2015 standard for interval arithmetic [10]. Its development phases and its main features are explained in [14,21,25].

Nevertheless, only few libraries of interval arithmetic are compliant with the IEEE 1788–2015 standard. Most libraries were developed before the standard. Regarding the libraries developed since then and compliant with the standard,

© Springer Nature Switzerland AG 2020
R. Wyrzykowski et al. (Eds.): PPAM 2019, LNCS 12044, pp. 353–363, 2020.
https://doi.org/10.1007/978-3-030-43222-5_31

let us mention `libieee1788` [19], which was developed along with the standard as a proof-of-concept. However, its author has left academia and this C++ library is no longer maintained: it does not compile any more with recent versions of g++. `JInterval` [18] is a Java library that was more used to test the compliance of interval arithmetic libraries with the standard, than used as an interval arithmetic library *per se*. Unfortunately, this library has also untimely lost its main developer, D. Nadezhin. The `interval` package for Octave [7] is the only library that is still maintained and for public use, even if its author has also left academia. `MPDI` [5] is another library for interval arithmetic that is compliant with the standard, but it is not (yet?) distributed.

This lack of libraries compliant with the IEEE 1788–2015 standard led us to consider the adaptation of our MPFI library for arbitrary precision interval arithmetic into a compliant library. We will detail the required modifications in Sect. 4, but we first detail the main features of the IEEE 1788–2015 standard in Sect. 2, and we introduce MPFI and explain how far it is from being IEEE 1788–2015 compliant in Sect. 3. We conclude with some personal considerations about the relevance of interval arithmetic computations in HPC.

2   IEEE 1788–2015 Standard for Interval Arithmetic

2.1   Structure in Four Levels

The IEEE 1788–2015 standard for interval arithmetic is structured in 4 levels, similarly to the IEEE 754–1985 standard for floating-point arithmetic. This structure clearly separates the mathematical notions from implementation issues, all the way to bit encoding.

The first level is the mathematical level: this level is about intervals of real numbers, such as $[0, \pi]$. Reasoning about intervals of real numbers, establishing and proving mathematical theorems about such intervals, are done at Level 1. No representation issue interferes with this level.

The second level addresses discretization issues: it deals with the fact that an implementation on a computer will have a discrete, finite set of numbers at its disposal for the representation of intervals, in particular for the representation of the endpoints. It specifies the existence of directed rounding modes, because it is required that every interval at Level 2 encloses the mathematical, Level 1, interval it represents. For instance, the interval $[0, \pi]$ is represented by an interval of the form $[\mathrm{rd}(0), \mathrm{ru}(\pi)]$ where rd stands for *rounding downwards* and ru stands for *rounding upwards*. This second level remains quite abstract and does not specify the set of numbers, it – only but completely – explicitly specifies how to go from the real numbers at mathematical Level 1 to a finite and discrete set of numbers at Level 2.

At Level 3, this finite and discrete set of numbers is specified: for instance it can be the set of `binary64` floating-point numbers given by the IEEE 754–2008 standard. Level 4 gives the binary encoding of the representation. Actually, the bulk of the work has been done at the floating-point (or any other numbers representation) level and the standard specifies only decorations, see Sect. 2.4.

2.2 Definitions: Intervals and Operations

Notation: following [13], intervals are denoted using boldface symbols, as in **x**.

Now that the structure of the standard is clear, let us detail the definitions adopted in the standard. Regarding intervals: everybody agreed that $[0, \pi]$ and $[1, 3]$ are intervals. Discussions were hot regarding whether $\emptyset$, $[5, \infty)$ or $[3, 1]$ should be considered as legal intervals. Thus, at Level 1, the definition for which there was a consensus, a **common** agreement, is that an interval is a non-empty bounded closed connected subset of $\mathbb{R}$: $\mathbf{x} = [\underline{x}, \bar{x}]$ with $\underline{x} \in \mathbb{R}$, $\bar{x} \in \mathbb{R}$ and $\underline{x} \leq \bar{x}$.

At Level 1, operations are defined in such a way that the FTIA holds.

Theorem 1 (FTIA: Fundamental Theorem of Interval Arithmetic). *Any operation φ evaluated on interval arguments $\mathbf{x}_1, \mathbf{x}_2, \ldots, \mathbf{x}_k$ within its domain returns its range on these arguments $\varphi(\mathbf{x}_1, \mathbf{x}_2, \ldots, \mathbf{x}_k)$.*

Implementation issues relax the FTIA to the requirement that the result encloses the range of φ on $\mathbf{x}_1, \mathbf{x}_2, \ldots, \mathbf{x}_k$. The application of this principle yields the well-known formulas for arithmetic operations such as $+$, $-$, $*$ or $\sqrt{}$:

$$[\underline{x}, \bar{x}] + [\underline{y}, \bar{y}] = [\underline{x} + \underline{y}, \bar{x} + \bar{y}],$$
$$[\underline{x}, \bar{x}] - [\underline{y}, \bar{y}] = [\underline{x} - \bar{y}, \bar{x} - \underline{y}],$$
$$[\underline{x}, \bar{x}] * [\underline{y}, \bar{y}] = [\min(\underline{x} * \underline{y}, \underline{x} * \bar{y}, \bar{x} * \underline{y}, \bar{x} * \bar{y}), \max(\underline{x} * \underline{y}, \underline{x} * \bar{y}, \bar{x} * \underline{y}, \bar{x} * \bar{y})],$$
$$\sqrt{[\underline{x}, \bar{x}]} = [\sqrt{\underline{x}}, \sqrt{\bar{x}}] \text{ if } \underline{x} \geq 0,$$

and is used to evaluate mathematical functions, e.g. $\sin([3, 5]) \subset [-1, +0.14113]$.

Other operations are also specified by the IEEE 1788–2015 standard. Some operations are specific to sets, such as the intersection or the convex hull of the union, for instance $[2, 4] \cap [3, 7] = [3, 4]$ and $[-2, -1] \cup [3, 7] = [-2, 7]$. In the latter example, the closed convex hull of the result of the union must be returned, otherwise the result has a "gap" and is not an interval. Some operations are specific to intervals, such as the endpoints (infimum and supremum): $\inf([-1, 3]) = -1$, $\sup([-1, 3]) = 3$; the width and the radius: $\operatorname{wid}([-1, 3]) = 4$, $\operatorname{rad}([-1, 3]) = 2$; the magnitude and the magnitude[1]: $\operatorname{mag}([-1, 3]) = 3$, $\operatorname{mig}([-1, 3]) = 0$.

Some operations have been added to ease constraint solving: it is known that the addition and subtraction defined above are not the reciprocal of each other. The standard specifies two operations that are respectively the reciprocal of addition, namely `cancelMinus`, and of subtraction, namely `cancelPlus`. The formulas for `cancelMinus` and `cancelPlus` are as follows

$$
\begin{aligned}
&\texttt{cancelMinus}(\mathbf{x}, \mathbf{y}) &&= \mathbf{z} \text{ such that } \mathbf{y} + \mathbf{z} = \mathbf{x} \\
\Rightarrow\ &\texttt{cancelMinus}([\underline{x}, \bar{x}], [\underline{y}, \bar{y}]) = [\underline{x} - \underline{y}, \bar{x} - \bar{y}], && \text{if } \operatorname{wid}(\mathbf{x}) \geq \operatorname{wid}(\mathbf{y}), \\
&\texttt{cancelPlus}(\mathbf{x}, \mathbf{y}) &&= \texttt{cancelMinus}(x, -y) = \mathbf{z} \text{ such that } \mathbf{z} - \mathbf{y} = \mathbf{x}, \\
\Rightarrow\ &\texttt{cancelPlus}([\underline{x}, \bar{x}], [\underline{y}, \bar{y}]) = [\underline{x} + \bar{y}, \bar{x} + \underline{y}], && \text{if } \operatorname{wid}(\mathbf{x}) \geq \operatorname{wid}(\mathbf{y}).
\end{aligned}
$$

For example, `cancelMinus([2, 5], [1, 3]) = [1, 2]` and `cancelPlus([2, 5], [1, 3]) =` $[5, 6]$. Such reciprocal operations are called *reverse operations*.

---

[1] The definition of the mignitude is $\operatorname{mig}([a, b]) = \min(|x| : x \in [a, b]) = \min(|a|, |b|)$ if $0 \notin [a, b]$ and 0 otherwise.

2.3   Flavors

This definition of an interval and the specification of these operations are the common ground of the IEEE 1788–2015 standard for interval arithmetic. However, this common ground was felt as too restrictive by many users of interval arithmetic, who are accustomed to manipulate a larger set of intervals in their daily practice. Still, it was impossible to extend the definition of an interval to simultaneously encompass all varieties of intervals and still keep a consistent theory. For instance, both $\emptyset$ and $[5, +\infty)$ are meaningful within the set-based approach of interval arithmetic, but not $[3, 1]$. Conversely, $[3, 1]$ is a valid interval in Kaucher [12] or modal arithmetic, but neither $\emptyset$ nor unbounded intervals.

The standard is thus designed to accomodate "variants" of interval arithmetic, called *flavors* in IEEE 1788–2015. After many discussions, including a clear definition of modal arithmetic [3,4], the partisans of modal arithmetic did not pursue their standardization effort. Currently, only the *set-based flavor*, derived from set theory, is specified by the IEEE 1788–2015 standard.

Let us highlight the set-based flavor. First, the set-based flavor removes some limitations on the allowed intervals: the empty set as well as unbounded intervals are legal intervals for this flavor. An interval is a closed connected subset of $\mathbb{R}$. As the empty set is a valid interval, the definition of operations and functions can be extended outside their domain, and $\sqrt{[-1, 2]}$ now has a meaning. More generally, the meaning of $\varphi(\mathbf{x}_1, \mathbf{x}_2, \ldots, \mathbf{x}_k)$ is

$$\varphi(\mathbf{x}_1, \ldots, \mathbf{x}_k) = \text{Hull} \left\{ \varphi(x_1, \ldots, x_k) \; : \; (x_1, \ldots, x_k) \in (\mathbf{x}_1, \ldots, \mathbf{x}_k) \cap \text{Dom}(\varphi) \right\}.$$

Let us go back to the example given above: $\sqrt{[-1, 2]} = \sqrt{[-1, 2] \cap Dom_{\sqrt{}}} = \sqrt{[0, 2]} = [0, \sqrt{2}]$. Similarly, $[2, 3]/[-1, 2]$ is permitted and $[2, 3]/[-1, 2] = \mathbb{R}$, whereas $[2, 3]/[0, 2] = \text{Hull}([2, 3]/(0, 2]) = [1, +\infty)$.

Another extension defined by the set-based flavor is the set of available operations, in particular of reverse operations. For instance, the reverse operation of the square operation is sqrRev, examplified here:

$$\text{sqrRev}([1, 4]) = \text{Hull} \left\{ x \; : \; x^2 \in [1, 4] \right\} = \text{Hull} \left([-2, -1] \cup [1, 2] \right) = [-2, 2].$$

Another important reverse operation is mulRevToPair, that corresponds to the extended division defined in [22]. This operation is rather peculiar, as it returns 0, 1 or 2 interval(s), as in mulRevToPair$([2, 3], [-1, 2]) = ((-\infty, -2], [1, +\infty))$. It does not return the convex hull of the result, instead it preserves the gap. This is particularly useful in Newton's method for the determination of the zeroes of a function: as this gap corresponds to a region that does not contain any zero and that can be eliminated for further exploration, it also separates zeros. Newton's method can subsequently be applied successfully to each of the two results.

2.4 Decorations

Let us have a closer look at Newton's method. A particularly useful side result is the proof of existence, and sometimes uniqueness, of a zero in the computed interval. This proof is obtained by applying Brouwer's theorem.

Theorem 2 (Brouwer's Theorem). *If the image of a compact set K by a continuous function f is enclosed in K, then f has a fixed point in K: if $f(K) \subset K$, then $\exists x_0 \in K$ such that $f(x_0) = x_0$.*

Another way of stating this result is to say that the function $g : x \mapsto x - f(x)$ has a zero in K.

In particular, if K is a non-empty bounded interval, and if the result of the evaluation of f on K returns an interval $K' \subset K$, then the existence of a fixed-point of f on K is established.

Let us consider the following example: the function

$$f : x \mapsto \sqrt{x} - 2,$$

has no real fixed point. We leave it to the reader, hint: $x - \sqrt{x} + 2$ has no real zero, or equivalently the polynomial $y^2 - y + 2$ has no real root. The evaluation of f on $[-4, 9]$ using the set-based flavor of interval arithmetic yields:

$$\sqrt{[-4, 9]} - 2 = \sqrt{[0, 9]} - 2 = [0, 3] - 2 = [-2, 1] \subset [-4, 9],$$

and a hasty application of Brouwer's theorem falsely establishes that f has a fixed point in $[-4, 9]$. The mistake here is to omit checking whether f is continuous over $[-4, 9]$. Actually f is not even defined everywhere on $[-4, 9]$. As the assumption of Brouwer's theorem is not satisfied, no conclusion can be derived.

The IEEE 1788–2015 standard must offer a mechanism to handle exceptions and to prohibit such erroneous conclusions from being drawn. After hot and long debates, the chosen mechanism is called *decoration*; it consists in a piece of information, a tag attached to or "decorating" each interval. Decorations have been deemed as the best way (or, should we say, "the least worse") to deal with the abovementioned problem:

- they avoid the inappropriate application of Brouwer's theorem: Brouwer's theorem can be used only when the tag indicates that it is valid to do so;
- they avoid the storage of any global information for exceptions, contrary to the global flags defined in the IEEE 754–1985 standard for floating-point arithmetic: such global flags are difficult to implement in a parallel context (that is, SIMD, multithreaded, or distributed).

The meaning of a decoration, in the set-based flavor, is a piece of information about the history of the computations that led to the considered interval. In particular, it indicates whether every operation involved a defined and continuous function applied to arguments within its domain or not. The user must thus consult this decoration before applying Brouwer's theorem for instance.

For the set-based flavor, the chosen decorations are listed below:
- com for common,
- dac   for   defined   and   – trv for trivial (no information),
  continuous,   – ill for ill-formed (nowhere defined).
- def for defined,

As a decoration results from the computation of the interval it is attached to, this computation must also incorporate the determination of the decoration. The set-based flavor specifies the propagation rules for decorations.

Last, every flavor must provide a FTDIA (Fundamental Theorem of Decorated Interval Arithmetic), that accounts for decorations.

Theorem 3 (FTDIA for the Set-Based Flavor). *Let* f *be an arithmetic expression denoting a real function* f. *Let* f *be evaluated, possibly in finite precision, on a validly initialized decorated box* $\mathbf{X} = \mathbf{x}_{dx}$, *to give result* $\mathbf{Y} = \mathbf{y}_{dy}$. *If some component of* $\mathbf{X}$ *is decorated* ill, *then the decoration* $dy = ill$. *If no component of* $\mathbf{X}$ *is decorated* ill, *and none of the operations* φ *of* f *is an everywhere undefined function, then* $dy \neq ill$ *and* $\mathbf{y} \supset \mathrm{Range} f(\mathbf{x})$ *and the decoration* dy *of* $\mathbf{y}$ *correctly (i.e., pessimistically) accounts for the properties of* f *over* $\mathbf{x}$.

By pessimistically, it is expected that a decoration never raises false hopes. For instance, a function can be defined and continuous while the computed decoration only states def, but the converse cannot happen.

2.5   Exact Dot Product

As the IEEE 1788–2015 standard addresses the quality of numerical computations, it also incorporates a recommendation regarding the accuracy of specific floating-point computations. Namely, it recommends that for each supported IEEE 754–1985 floating-point type, an implementation shall provide a correctly rounded version of the four reduction operations sum, dot, sumSquare and sumAbs, that take a variable number of arguments.

3   The MPFI Library

After this introduction to the IEEE 1788–2015 standard for interval arithmetic, let us now concentrate on its implementation. As already stated, the libraries that are compliant with the standard are rather rare. This section focuses on the MPFI library, developed since 2000 and thus prior to the standard by large, and on its transformation into an IEEE 1788–2015 compliant library.

MPFI stands for *Multiple Precision Floating-point reliable Interval* library. It is a library written in C that implements arbitrary (rather than multiple) precision interval arithmetic. More precisely, intervals are represented by their endpoints and these endpoints are floating-point numbers of arbitrary precision: for each endpoint, the significand can be arbitrarily precise, the only limit being the memory of the computer. The MPFI library is based on MPFR [2] for arbitrary precision floating-point arithmetic. Its development started in 2000 with

Revol and Rouillier [24], it has evolved and improved since then thanks to the contributions of S. Chevillard, C. Lauter, H. D. Nguyen and Ph. Théveny. The library is freely available at https://gforge.inria.fr/projects/mpfi/.

Before digging in the functionalities and specificities of MPFI, let us recall some justification for its development, as given by Kahan in [11]. In "How Futile are Mindless Assessments of Roundoff in Floating-Point Computation?", Kahan lists tools for assessing the numerical quality of a computed result, in the presence of roundoff errors. He exhibits examples that defeat these tools, when applied mindlessly. A typical example of mindless use of a tool such as interval arithmetic is the replacement of every floating-point datatype in the code by an interval datatype that is not more precise, before running the code again, on data of interval type(s). It is well known that, most of the time, such a mindless use of interval arithmetic produces results with widths too large to be of any help. However, if running time is not an issue, using interval arithmetic with arbitrary precision, and increasing the precision as needed, is a mindless (as opposed to artful, or expert) but cheap (in development time) and effective way to assess the numerical quality of a code. As Kahan puts it [11], *"For that price* (slow execution compared to the execution of the purely floating-point version) *we may be served better by almost foolproof extendable-precision Interval Arithmetic."*. MPFI offers the required arbitrary precision interval arithmetic.

Let us go back to MPFI and detail the definitions it uses and the functionalities it offers. MPFI is based on MPFR and thus on GMP, for accuracy, efficiency and portability. As GMP and MPFR, MPFI is developed in the C language. MPFR provides arbitrary precision floating-point arithmetic, that is compliant with the IEEE 754–1985 philosophy. In particular, for every function, MPFR guarantees that the returned result is equal to the exact result (that is, as if it were computed with infinite precision), rounded using the required rounding mode. This correct rounding is guaranteed not only for basic arithmetic operations but for every function of the mathematical library. In MPFI, an interval is any closed connected subset of $\mathbb{R}$ whose endpoints are numbers representable using MPFR. Thus the empty set and unbounded intervals are valid intervals, However, this definition corresponds to Level 2 of the IEEE 1788–2015 standard.

Regarding the functionalities offered by MPFI, they correspond to most of the requirements of IEEE 1788–2015, with some exceptions. On the one hand, MPFI offers a richer set of mathematical constants (π, Euler constant, etc.) and functions. On the other hand, there is (yet) no implementation of the reverse functions, except `mulRevToPair`. MPFI offers most of the lengthy list of conversions mandated by the standard: to and from integer, double precision floating-point numbers, exact integers and rationals (through GMP), arbitrary precision floating-point numbers (through MPFR) and text strings. MPFI also accomodates intervals with any floating-point endpoints, including infinities, signed zeroes and NaNs: again, MPFI has been designed at Level 2 of IEEE 1788–2015.

However, MPFI accounts for neither flavors nor decorations. Thus, operations are not defined according to any flavor and do not propagate decorations. Still, MPFI has a mechanism for handling exceptions, which is a "Level 2" mechanism in the sense that it is based on the floating-point, IEEE 754–1985-like, mechanism for handling exceptions. Let us illustrate this mechanism through an example: when MPFI is given $\sqrt{[-1, 2]}$, as $[-1, 2]$ contains -1 and as $\sqrt{-1}$ is an invalid result denoted as NaN in floating-point arithmetic, MPFI considers this computation as an invalid one and returns NaI: Not an Interval. In IEEE 1788–2015, the only NaIs are produced by meaningless inputs such as ["bla", 1].

To sum up, MPFI has to be reworked in several directions to be compliant with the IEEE 1788–2015 standard.

4   Towards Compliance of MPFI with IEEE 1788–2015

In order to incorporate the new concepts present in the standard, the *data structure* of a MPFI interval must be modified. First, a field `flavor` will be added to each interval, even if this was not the original intent of IEEE 1788–2015: the principle of flavors was that either a whole computation, or at least significant portions of it, would be performed using a single flavor; thus a flavor would be attached to a computation rather than to a data.

Second, parameterized by the flavor, a field `decoration` will be added and its possible values will be the ones defined by the corresponding flavor. The technicalities of "bare" intervals and "compressed" intervals will be handled in an ad hoc way (by adding a boolean indicating whether the interval is bare or not) or not implemented (in the case of compressed intervals). As these notions were not detailed in Sect. 2, they will not be discussed further here.

Then, the *code for each existing operation* needs to be updated. When entering the code of an operation, a preliminary test on the flavors of the arguments and on their compatibility will be performed, and the computation will then be branched to the corresponding part of the code. Before quitting the code, a postprocessing will be performed to determine and set the decoration of the result. Code for the missing reverse operations must be developed.

Another issue is *backward compatibility* for users of MPFI who want to preserve the existing behavior of their MPFI computations. This will be achieved by adding a new "flavor" – even if it is not really one: no clear specification at Level 1 – called `MPFIoriginal`, so that every computation behaves the same old way. When this flavor is encountered, each operation will branch to the existing and unmodified code to perform it.

5   Concluding Remarks Regarding Performance and HPC

The previous section is written in future tense, because most of the modifications are still waiting to be implemented. Indeed, a major update of MPFI is ongoing, but still not finished. This update consists not only in turning MPFI into a IEEE 1788–2015 compliant library, but also in incorporating all mathematical functions provided by MPFR, such as erf or Bernoulli. Another direction of future developments concerns the ease of use of MPFI, through a Julia interface.

Let us now conclude with a few remarks regarding performance and HPC. The author worked on the parallelization [23] of Hansen's algorithm for global optimization using interval arithmetic [6]. This algorithm is of branch-and-bound type and the original idea to parallelize it was to explore simultaneously several branches of the tree corresponding to the branch-and-bound exploration. However, it was rapidly obvious that brute force (that is, bisection of the candidate box and evaluation of the objective function over each sub-box) was not the best way to obtain speed-ups. A smarter, sequential processing of the candidate box was more efficient, either to reduce it or to prune it. The simplest solution was, as mentioned in [11], to use larger or arbitrary precision interval arithmetic. This led to the development of MPFI.

Let us go back to parallel computations, with "parallel" covering a large spectrum of possibilities, all the way from SIMD to multithreaded to multicore to distributed to heterogeneous computations. The IEEE 1788–2015 standard has tried to avoid some pitfalls, such as the use of global flags for handling exceptions. However, the mechanism of decorations has also been heavily criticized. On the one hand, adding this piece of information to each interval destroys padding efforts and other memory optimizations. On the other hand, the computation and propagation of decorations does not integrate gracefully with pipelined or SIMD operations such as AVX, SSE or SSE2. Similarly, MPFI computations do no seem suited for parallel execution. The MPFI library cannot benefit from hardware accelerators. It is also not well suited to cache optimizations strategies, as its data have irregular sizes, as opposed to fixed and constant sizes such as `binary32` or `binary64` floating-point datatypes. Furthermore, each operation in MPFI takes a large computing time, compared to the time of the same operation (such as addition, multiplication or exponential) applied to `binary64` operands. In practice, a slowdown larger than 50, for one operation, has often been observed.

However, IEEE 1788–2015 and MPFI computations are not comparable with `binary32` or `binary64` computations. First, the results they provide are guaranteed, in the sense that they contain the sought results, even in the presence of roundoff errors. Second, they are well suited for multithreaded or distributed computations: for such computations, it is well known that the communication time needed to bring the data to the computational device is much larger, by typically 3 orders of magnitude, than the computational time, that is, the time required to perform the arithmetic operations on these data. It means that there is plenty of time to apply numerical computations to the data. With interval computations and, in particular, with arbitrary precision interval computations, the computational time is much larger and becomes closer to the communication time. In other words, with interval computations, the numeric intensity is increased, as already observed in [28, Chapter 8] for the product of interval matrices with `binary64` coefficients. HPC computations leave time for interval computations and high-precision interval computations: the execution time is better balanced between communication time and computation time, with a better final accuracy and a guarantee on the results.

References

1. Alefeld, G., Herzberger, J.: Introduction to Interval Analysis. Academic Press, Cambridge (1983)
2. Fousse, L., Hanrot, G., Lefèvre, V., Pélissier, P., Zimmermann, P.: MPFR: a multiple-precision binary floating-point library with correct rounding. ACM Trans. Math. Softw. **33**(2), 13-es (2007)
3. Goldsztejn, A.: Modal intervals revisited, part 1: a generalized interval natural extension. Reliable Comput. **16**, 130–183 (2012)
4. Goldsztejn, A.: Modal intervals revisited, part 2: a generalized interval mean value extension. Reliable Comput. **16**, 184–209 (2012)
5. Graillat, S., Jeangoudoux, C., Lauter, C.: MPDI: a decimal multiple-precision interval arithmetic library. Reliable Comput. **25**, 38–52 (2017)
6. Hansen, E.R.: Global Optimization Using Interval Analysis. Marcel Dekker, New York (1992)
7. Heimlich, O.: Interval arithmetic in GNU Octave. In: SWIM 2016: Summer Workshop on Interval Methods, France (2016)
8. IEEE: Institute of Electrical and Electronic Engineers: 754–1985 - IEEE Standard for Binary Floating-Point Arithmetic. IEEE Computer Society (1985)
9. IEEE: Institute of Electrical and Electronic Engineers: 754–2008 - IEEE Standard for Floating-Point Arithmetic. IEEE Computer Society (2008)
10. IEEE: Institute of Electrical and Electronic Engineers: 1788–2015 - IEEE Standard for Interval Arithmetic. IEEE Computer Society (2015)
11. Kahan, W.: How Futile are Mindless Assessments of Roundoff in Floating-Point Computation? (2006). https://people.eecs.berkeley.edu/~wkahan/Mindless.pdf
12. Kaucher, E.: Interval analysis in the extended interval space IR. Comput. Supplementa **2**(1), 33–49 (1980). https://doi.org/10.1007/978-3-7091-8577-3_3
13. Kearfott, R.B., Nakao, M.T., Neumaier, A., Rump, S.M., Shary, S.P., van Hentenryck, P.: Standardized notation in interval analysis. Comput. Technol. **15**(1), 7–13 (2010)
14. Kearfott, R.B.: An overview of the upcoming IEEE P-1788 working group document: standard for interval arithmetic. In: IFSA/NAFIPS, pp. 460–465. IEEE, Canada (2013)
15. Moore, R.E.: Interval Analysis. Prentice Hall, Englewood Cliffs (1966)
16. Moore, R.E.: Methods and Applications of Interval Analysis. SIAM Studies in Applied Mathematics, Philadelphia (1979)
17. Moore, R.E., Kearfott, R.B., Cloud, M.J.: Introduction to Interval Analysis. SIAM, Philadelphia (2009)
18. Nadezhin, D.Y., Zhilin, S.I.: Jinterval library: principles, development, and perspectives. Reliable Comput. **19**(3), 229–247 (2014)
19. Nehmeier, M.: libieeep1788: A C++ Implementation of the IEEE interval standard P1788. In: Norbert Wiener in the 21st Century, pp. 1–6, IEEE, Australia (2014)
20. Neumaier, A.: Interval Methods for Systems of Equations. Cambridge University Press, Cambridge (1990)
21. Pryce, J.: The forthcoming IEEE standard 1788 for interval arithmetic. In: Nehmeier, M., Wolff von Gudenberg, J., Tucker, W. (eds.) SCAN 2015. LNCS, vol. 9553, pp. 23–39. Springer, Cham (2016). https://doi.org/10.1007/978-3-319-31769-4_3
22. Ratz, D.: Inclusion Isotone Extended Interval Arithmetic. Report 5 (96 pages). Institut für Angewandte Mathematik, Universität Karlsruhe (1996)

23. Revol, N., Denneulin, Y., Méhaut, J.-F., Planquelle, B.: Parallelization of continuous verified global optimization. In: 19th IFIP TC7 Conference on System Modelling and Optimization, Cambridge, United Kingdom (1999)
24. Revol, N., Rouillier, F.: Motivations for an arbitrary precision interval arithmetic and the MPFI library. Reliable Comput. **11**(4), 275–290 (2005)
25. Revol, N.: Introduction to the IEEE 1788-2015 standard for interval arithmetic. In: Abate, A., Boldo, S. (eds.) NSV 2017. LNCS, vol. 10381, pp. 14–21. Springer, Cham (2017). https://doi.org/10.1007/978-3-319-63501-9_2
26. Rump, S.M.: Verification methods: rigorous results using floating-point arithmetic. Acta Numerica **19**, 287–449 (2010)
27. Sunaga, T.: Geometry of Numerals. Master thesis, U. Tokyo, Japan (1956)
28. Théveny, P.: Numerical Quality and High Performance In Interval Linear Algebra on Multi-Core Processors. PhD thesis, ENS Lyon, France (2014)
29. Tucker, W.: Validated Numerics - A Short Introduction to Rigorous Computations. Princeton University Press, Princeton (2011)

Softmax and McFadden's Discrete Choice Under Interval (and Other) Uncertainty

Bartlomiej Jacek Kubica[1], Laxman Bokati[2], Olga Kosheleva[2] (iD),
and Vladik Kreinovich[2](✉) (iD)

[1] University of Life Sciences, Warsaw, Poland
bartlomiej.jacek.kubica@gmail.com
[2] University of Texas at El Paso, El Paso, TX 79968, USA
lbokati@miners.utep.edu, {olgak,vladik}@utep.edu

Abstract. One of the important parts of deep learning is the use of the softmax formula, that enables us to select one of the alternatives with a probability depending on its expected gain. A similar formula describes human decision making: somewhat surprisingly, when presented with several choices with different expected equivalent monetary gain, we do not just select the alternative with the largest gain; instead, we make a random choice, with probability decreasing with the gain – so that it is possible that we will select second highest and even third highest value. Both formulas assume that we know the exact value of the expected gain for each alternative. In practice, we usually know this gain only with some certainty. For example, often, we only know the lower bound $\underline{f}$ and the upper bound $\overline{f}$ on the expected gain, i.e., we only know that the actual gain f is somewhere in the interval $\left[\underline{f}, \overline{f}\right]$. In this paper, we show how to extend softmax and discrete choice formulas to interval uncertainty.

Keywords: Deep learning · Softmax · Discrete choice · Interval uncertainty

1   Formulation of the Problem

Deep Learning: A Brief Reminder. At present, the most efficient machine learning technique is *deep learning* (see, e.g., [2,7]), in particular, *reinforcement deep learning* [12], where, in addition to processing available information, the system also (if needed) automatically decides which additional information to request – and if an experimental setup is automated, to produce.

For selecting the appropriate piece of information, the system estimates, for each possible alternative, how much information this particular alternative will bring.

This work was supported in part by the National Science Foundation grants 1623190 (A Model of Change for Preparing a New Generation for Professional Practice in Computer Science) and HRD-1242122 (Cyber-ShARE Center of Excellence). The authors are thankful to the anonymous referees for valuable suggestions.

© Springer Nature Switzerland AG 2020
R. Wyrzykowski et al. (Eds.): PPAM 2019, LNCS 12044, pp. 364–373, 2020.
https://doi.org/10.1007/978-3-030-43222-5_32

It is Important to Add Randomness. And here comes an interesting part. A reader who is not familiar with details of deep learning algorithms may expect that the system selects the alternative with the largest estimate of expected information gain. This idea was indeed tried – but it did not work well: instead of finding the model that best fits the training data, the algorithm would sometimes get stuck in a local minimum of the corresponding objective function.

In numerical analysis, a usual way to get out of a local minimum is to perform some random change. This is, e.g., the main idea behind simulated annealing. Crudely speaking, it means that we do not always follow the smallest – or the largest – value of the corresponding objective function, we can follow the next smallest (largest), next next smallest, etc. – with some probability.

Softmax: How Randomness is Currently Added. Of course, the actual maximum should be selected with the highest probability, the next value with lower probability, etc. In other words, if we want to maximize some objective function $f(a)$, and we have alternatives $a_1, \ldots, a_n$ for which this function has values $f_1 \overset{\text{def}}{=} f(a_1), \ldots, f_n \overset{\text{def}}{=} f(a_n)$, then the probability p_i of selecting the i-th alternative should be increasing with f_i, i.e., we should have $p_i \sim F(f_i)$ for some increasing function $F(z)$, i.e., $p_i = c \cdot F(f_i)$, for some constant c.

We should always select one of the alternatives, so these probabilities should add up to 1: $\sum_{j=1}^{n} p_j = 1$. From this condition, we conclude that $c \cdot \sum_{j=1}^{n} F(f_j) = 1$.

Thus, $c = 1 \left/ \left(\sum_{j=1}^{n} F(f_j) \right) \right.$ and so,

$$p_i = \frac{F(f_i)}{\sum_{j=1}^{n} F(f_j)}. \tag{1}$$

Which function $F(z)$ should we choose? In deep learning – a technique that requires so many computations that it cannot exist without high performance computing – computation speed is a must. Thus, the function $F(z)$ should be fast to compute – which means, in practice, that it should be one of the basic functions for which we have already gained an experience of how to compute it fast. There are a few such functions: arithmetic functions, the power function, trigonometric functions, logarithm, exponential function, etc.

The selected function should be increasing, and it should return non-negative results for all real values z (positive or negative) – otherwise, we will end up with meaningless negative probability. Among basic functions, only one function has this property – the exponential function $F(z) = \exp(k \cdot z)$ for some $k > 0$. For this function, the probability p_i takes the form

$$p_i = \frac{\exp(k \cdot f_i)}{\sum_{j=1}^{n} \exp(k \cdot f_j)}. \tag{2}$$

This expression is known as the *softmax* formula.

It is Desirable to Further Improve Deep Learning. Deep learning has lead to many interesting results, but it is not a panacea. There are still many challenging problems where the existing deep learning algorithms has not yet led to fully successful learning. It is therefore desirable to look at all the stages of deep leaning and see if we can modify them so as to improve the overall learning performance.

Need to Generalize Softmax to the Case of Interval Uncertainty. One of the aspects of deep learning computations in which there is a potential for improvement is the use of the softmax formulas. Indeed, when we apply the softmax formula, we only take into account the corresponding estimates $f_1, \ldots, f_n$. However, in practice, we do not just have these estimates, we often have some idea of how accurate is each estimate. Some estimates may be more accurate, some may be less accurate. It is desirable to take this information about uncertainty into account.

For example, we may know the upper bound Δ_i on the absolute value

$$|f_i - f_i^{\mathrm{act}}| \tag{3}$$

of the difference between the estimate f_i and the (unknown) actual value f_i^{act} of the objective function. In this case, the only information that we have about the actual value f_i^{act} is that this value is located in the interval $[f_i - \Delta_i, f_i + \Delta_i]$.

How to take this interval information into account when computing the corresponding probabilities p_i? This is the problem that we study in this paper – and for which we provide a reasonable solution.

Another Important Case Where a Softmax-Type Formula is Used. There is another application area where a similar formula is used: the analysis of human choice. If a person needs to select between several alternatives $a_1, \ldots, a_n$, and this person knows the exact monetary values $f_1, \ldots, f_n$ associated with each alternative, then we expect this person to always select the alternative with the largest possible monetary value – actual or equivalent. We also expect that if we present the person with the exact same set of alternatives several times in a row, this person will always make the same decision – of selecting the best alternative.

Interestingly, this is *not* how most people make decisions. It turns out that we make decisions probabilistically: instead of always selecting the best alternative, we select each alternative a_i with probability p_i described exactly by the softmax-like formula (2), for some $k > 0$.

In other words, in most cases, we usually indeed select the alternative with the higher monetary value, but with some probability, we will also select the next highest, with some smaller probability, the next next highest, etc.

This fact was discovered by an economist D. McFadden – who received a Nobel Prize in Economics for this discovery; see, e.g., [10,11,13].

But Why? At first glance, such a probabilistic behavior sounds irrational – why not select the alternative with the largest possible monetary value?

A probabilistic choice would indeed be irrational if this was a stand-alone choice. In reality, however, no choice is stand-alone, it is a part of a sequence of choices, some of which involve conflict – and it is known that in conflict situations, a probabilistic choice makes sense; see, e.g., [9].

In Practice, We Usually only Know Gain with Some Certainty. McFadden's formula describes people's behavior in an idealized situation when the decision maker knows the exact monetary consequences f_i of each alternative a_i. In practice, this is rarely the case. At best, we know a lower bound $\underline{f}_i$ and an upper bound $\overline{f}_i$ of the actual (unknown) value f_i. In such situations, all we know is that the unknown value f_i is somewhere within the interval $\left[\underline{f}_i, \overline{f}_i\right]$. It is therefore desirable to extend McFadden's formula to the case of interval uncertainty.

2   Formulating the Problem in Precise Terms

Discussion. Let $\mathcal{A}$ denote the class of all possible alternatives. We would like, given any finite set of alternatives $A \subseteq \mathcal{A}$ and a specific alternative $a \in A$, to describe the probability $p(a\,|\,A)$ that out of all the alternatives from the set A, the alternative a will be selected.

Once we know these probabilities, we can then compute, for each set $B \subseteq A$, the probability $p(B\,|\,A)$ that one of the alternatives from the set B will be selected as $p(B\,|\,A) = \sum_{b \in B} p(b\,|\,A)$. In particular, we have $p(a\,|\,A) = p(\{a\}\,|\,A)$.

A natural requirement related to these conditional probabilities is that if we have $A \subseteq B \subseteq C$, then we can view the selection of A from C as either a direct selection, or as first selecting B, and then selecting A out of B. The resulting probability should be the same, so we must have $p(A\,|\,C) = p(A\,|\,B) \cdot p(B\,|\,C)$. Thus, we arrive at the following definition.

Definition 1. *Let $\mathcal{A}$ be a set. Its elements will be called* alternatives. *By a* choice function, *we mean a function $p(a\,|\,A)$ that assigns to each pair $\langle A, a\rangle$ of a finite set $A \subseteq \mathcal{A}$ and an element $a \in A$ a number from the interval $(0,1]$ in such a way that the following two conditions are satisfied:*

– for every set A, we have $\sum_{a \in A} p(a\,|\,A) = 1$, and
– whenever $A \subseteq B \subseteq C$, we have $p(A\,|\,C) = p(A\,|\,B) \cdot p(B\,|\,C)$, where

$$p(B\,|\,A) \stackrel{\text{def}}{=} \sum_{b \in B} p(b\,|\,A). \tag{4}$$

Proposition 1. *For each set $\mathcal{A}$, the following two conditions are equivalent to each other:*

– the function $p(a\,|\,A)$ is a choice function, and

– there exists a function $v : \mathcal{A} \to \mathbb{R}^+$ that assigns a positive number to each alternative $a \in \mathcal{A}$ such that

$$p(a \mid A) = \frac{v(a)}{\sum\limits_{b \in A} v(b)}. \tag{5}$$

Proof. It is easy to check that for every function v, the expression (5) indeed defines a choice function. So, to complete the proof, it is sufficient to prove that every choice function has the form (5).

Indeed, let $p(a \mid A)$ be a choice function. Let us pick any $a_0 \in \mathcal{A}$, and let us define a function v as

$$v(a) \stackrel{\text{def}}{=} \frac{p(a \mid \{a, a_0\})}{p(a_0 \mid \{a, a_0\})}. \tag{6}$$

In particular, for $a = a_0$, both probabilities $p(a \mid \{a, a_0\})$ and $p(a_0 \mid \{a, a_0\})$ are equal to 1, so the ratio $v(a_0)$ is also equal to 1. Let us show that the choice function has the form (5) for this function v.

By definition of $v(a)$, for each a, we have $p(a \mid \{a, a_0\}) = v(a) \cdot p(a_0 \mid \{a, a_0\})$.

By definition of a choice function, for each set A containing a_0, we have $p(a \mid A) = p(a \mid \{a, a_0\}) \cdot p(\{a, a_0\} \mid A)$ and $p(a_0 \mid A) = p(a_0 \mid \{a, a_0\}) \cdot p(\{a, a_0\} \mid A)$. Dividing the first equality by the second one, we get

$$\frac{p(a \mid A)}{p(a_0 \mid A)} = \frac{p(a \mid \{a, a_0\})}{p(a_0 \mid \{a, a_0\})}. \tag{7}$$

By definition of $v(a)$, this means that

$$\frac{p(a \mid A)}{p(a_0 \mid A)} = v(a). \tag{8}$$

Similarly, for each $b \in A$, we have

$$\frac{p(b \mid A)}{p(a_0 \mid A)} = v(b). \tag{9}$$

Dividing (8) by (9), we conclude that for each set A containing a_0, we have

$$\frac{p(a \mid A)}{p(b \mid A)} = \frac{v(a)}{v(b)}. \tag{10}$$

Let us now consider a set B that contains a and b but that does not necessarily contain a_0. Then, by definition of a choice function, we have

$$p(a \mid B) = p(a \mid \{a, b\}) \cdot p(\{a, b\} \mid B) \tag{11}$$

and

$$p(b \mid B) = p(b \mid \{a, b\}) \cdot p(\{a, b\} \mid B). \tag{12}$$

Dividing (11) by (12), we conclude that

$$\frac{p(a \mid B)}{p(b \mid B)} = \frac{p(a \mid \{a, b\})}{p(b \mid \{a, b\})}. \tag{13}$$

The right-hand side of this equality does not depend on the set B. So the left-hand side, i.e., the ratio

$$\frac{p(a \mid B)}{p(b \mid B)} \tag{14}$$

also does not depend on the set B. In particular, for the sets B that contain a_0, this ratio – according to the formula (10) – is equal to $v(a)/v(b)$. Thus, the same equality (10) holds for all sets A – not necessarily containing the element a_0.

From the formula (10), we conclude that

$$\frac{p(a \mid A)}{v(a)} = \frac{p(b \mid A)}{v(b)}. \tag{15}$$

In other words, for all elements $a \in A$, the ratio

$$\frac{p(a \mid A)}{v(a)} \tag{16}$$

is the same. Let us denote this ratio by c_A; then, for each $a \in A$, we have:

$$p(a \mid A) = c_A \cdot v(a). \tag{17}$$

From $\sum\limits_{b \in A} p(b \mid A) = 1$, we can now conclude that: $c_A \cdot \sum\limits_{b \in A} v(b) = 1$, thus

$$c_A = \frac{1}{\sum\limits_{b \in A} v(b)}. \tag{18}$$

Substituting this expression (18) into the formula (17), we get the desired expression (5).

The proposition is proven.

Comment. This proof is similar to the proofs from [4,8].

Discussion. As we have mentioned earlier, a choice is rarely a stand-alone event. Usually, we make several choices – and often, at the same time. Let us consider a simple situation. Suppose that we need to make two independent choices:

- in the first choice, we must select one the alternatives $a_1, \ldots, a_n$, and
- in the second choice, we must select one of the alternatives $b_1, \ldots, b_m$.

We can view this as two separate selection processes. In this case, in the first process, we select each alternative a_i with probability $v(a_i) \Big/ \left(\sum\limits_{k=1}^{n} v(a_k) \right)$ and, in the second process, we select each alternative b_j with probability $v(b_j) \Big/ \left(\sum\limits_{\ell=1}^{m} v(b_\ell) \right)$. Since the two processes are independent, for each pair $\langle a_i, b_j \rangle$, the probability of selecting this pair is equal to the product of the corresponding probabilities:

$$\frac{v(a_i)}{\sum\limits_{k=1}^{n} v(a_k)} \cdot \frac{v(b_j)}{\sum\limits_{\ell=1}^{m} v(b_\ell)}. \tag{19}$$

Alternatively, we can view the whole two-stage selection as a single selection process, in which we select a pair $\langle a_i, b_j \rangle$ of alternatives out of all $n \cdot m$ possible pairs. In this case, the probability of selecting a pair is equal to

$$\frac{v(\langle a_i, b_j \rangle)}{\sum\limits_{k=1}^{n} \sum\limits_{\ell=1}^{m} v(\langle a_k, b_\ell \rangle)}. \tag{20}$$

The probability of selecting a pair should be the same in both cases, so the values (19) and (20) must be equal to each other. This equality limits possible functions $v(a)$.

Indeed, if all we know about each alternative a is the interval $\left[\underline{f}(a), \overline{f}(a) \right]$ of possible values of the equivalent monetary gain, then the value v should depend only on this information, i.e., we should have $v(a) = V\left(\underline{f}(a), \overline{f}(a) \right)$ for some function $V(x, y)$. Which functions $V(x, y)$ guarantee the above equality?

To answer this question, let us analyze how the gain corresponding to selecting a pair $\langle a_i, b_j \rangle$ is related to the gains corresponding to individual selections of a_i and b_j. Suppose that for the alternative a_i, our gain $f_i \overset{\text{def}}{=} f(a_i)$ can take any value from the interval $\left[\underline{f}_i, \overline{f}_i \right] \overset{\text{def}}{=} \left[\underline{f}(a_i), \overline{f}(a_i) \right]$, and for the alternative b_j, our gain $g_j \overset{\text{def}}{=} f(b_j)$ can take any value from the interval $\left[\underline{g}_j, \overline{g}_j \right] \overset{\text{def}}{=} \left[\underline{f}(b_j), \overline{f}(b_j) \right]$. These selections are assumed to be independent. This means that we can have all possible combinations of values $f_i \in \left[\underline{f}_i, \overline{f}_i \right]$ and $g_j \in \left[\underline{g}_j, \overline{g}_j \right]$.

The smallest possible value of the overall gain $f_i + g_j$ is when both gains are the smallest. In this case, the overall gain is $\underline{f}_i + \underline{g}_j$. The largest possible value of the overall gain $f_i + g_j$ is when both gains are the largest. In this case, the overall gain is $\overline{f}_i + \overline{g}_j$. Thus, the interval of possible values of the overall gain is

$$\left[\underline{f}(\langle a_i, b_j \rangle), \overline{f}(\langle a_i, b_j \rangle) \right] = \left[\underline{f}_i + \underline{g}_j, \overline{f}_i + \overline{g}_j \right]. \tag{21}$$

In these terms, the requirement that the expressions (19) and (20) coincide takes the following form:

Definition 2. *We say that a function $V : \mathbb{R} \times \mathbb{R} \to \mathbb{R}^+$ is consistent if for every two sequences of intervals $\left[\underline{f}_1, \overline{f}_1 \right], \ldots, \left[\underline{f}_n, \overline{f}_n \right]$, and $\left[\underline{g}_1, \overline{g}_1 \right], \ldots, \left[\underline{g}_m, \overline{g}_m \right]$, for every i and j, we have*

$$\frac{V\left(\underline{f}_i, \overline{f}_i \right)}{\sum\limits_{k=1}^{n} V\left(\underline{f}_k, \overline{f}_k \right)} \cdot \frac{V\left(\underline{g}_j, \overline{g}_j \right)}{\sum\limits_{\ell=1}^{m} V\left(\underline{g}_\ell, \overline{g}_\ell \right)} = \frac{V\left(\underline{f}_i + \underline{g}_j, \overline{f}_i + \overline{g}_j \right)}{\sum\limits_{k=1}^{n} \sum\limits_{\ell=1}^{m} V\left(\underline{f}_k + \underline{g}_\ell, \overline{f}_k + \overline{g}_\ell \right)}. \tag{22}$$

Monotonicity. Another reasonable requirement is that the larger the expected gain, the more probable that the corresponding alternative is selected.

The notion of "larger" is easy when gains are exact, but for intervals, we can provide the following definition.

Definition 3. *We say that an interval A is* smaller than or equal to *an interval B (and denote it by $A \leq B$) if the following two conditions hold:*

- *for every element $a \in A$, there is an element $b \in B$ for which $a \leq b$, and*
- *for every element $b \in B$, there is an element $a \in A$ for which $a \leq b$.*

Proposition 2. $[\underline{a}, \overline{a}] \leq [\underline{b}, \overline{b}] \Leftrightarrow (\underline{a} \leq \underline{b} \,\&\, \overline{a} \leq \overline{b})$.

Proof is straightforward.

Definition 4. *We say that a function $V : \mathbb{R} \times \mathbb{R} \to \mathbb{R}^+$ is* monotonic *if for every two intervals $[\underline{a}, \overline{a}]$ and $[\underline{b}, \overline{b}]$, if $[\underline{a}, \overline{a}] \leq [\underline{b}, \overline{b}]$ then $V(\underline{a}, \overline{a}) \leq V(\underline{b}, \overline{b})$.*

Proposition 3. *For each function $V : \mathbb{R} \times \mathbb{R} \to \mathbb{R}^+$, the following two conditions are equivalent to each other:*

- *the function V is consistent and monotonic;*
- *the function $V(\underline{f}, \overline{f})$ has the form*

$$V(\underline{f}, \overline{f}) = C \cdot \exp\left(k \cdot \left(\alpha_H \cdot \overline{f} + (1 - \alpha_H) \cdot \underline{f}\right)\right) \tag{23}$$

for some values $C > 0$, $k > 0$, and $\alpha_H \in [0, 1]$.

Conclusion. Thus, if we have n alternatives $a_1, \ldots, a_n$, and for each alternative a_i, we know the interval $\left[\underline{f}_i, \overline{f}_i\right]$ of possible values of the gain, we should select each alternative i with the probability

$$p_i = \frac{\exp\left(k \cdot \left(\alpha_H \cdot \overline{f}_i + (1 - \alpha_H) \cdot \underline{f}_i\right)\right)}{\sum\limits_{j=1}^{n} \exp\left(k \cdot \left(\alpha_H \cdot \overline{f}_j + (1 - \alpha_H) \cdot \underline{f}_j\right)\right)}. \tag{24}$$

So, *we have extended the softmax/McFadden's discrete choice formula to the case of interval uncertainty.*

Comment 1. Proposition 2 justifies the formula (24). It should be mentioned that the formula (24) coincides with what we would have obtained from the original McFadden's formula if, instead of the exact gain f_i, we substitute into this original formula, the expression $f_i = \alpha_H \cdot \overline{f}_i + (1 - \alpha_H) \cdot \underline{f}_i$ for some $\alpha_H \in [0, 1]$. This expression was first proposed by a future Nobelist Leo Hurwicz and is thus known as Hurwicz optimism-pessimism criterion [3, 5, 6, 9].

Comment 2. For the case when we know the exact values of the gain, i.e., when we have a degenerate interval $[f, f]$, we get a *new justification for the original McFadden's formula.*

Comment 3. Similar ideas can be used to extend softmax and McFadden's formula to other types of uncertainty. As one can see from the proof, by taking logarithm of V, we reduce the consistency condition to additivity, and additive functions are known; see, e.g., [6]. For example, for probabilities, the equivalent gain is the expected value – since, due to large numbers theorem, the sum of many independent copies of a random variable is practically a deterministic number. Similarly, a class of probability distributions is equivalent to the interval of mean values corresponding to different distributions, and specific formulas are known for the fuzzy case.

Proof of Proposition 3. It is easy to check that the function (24) is consistent and monotonic. So, to complete the proof, it is sufficient to prove that every consistent monotonic function has the desired form.

Indeed, let us assume that the function V is consistent and monotonic. Then, due to consistency, it satisfies the formula (22). Taking logarithm of both sides of the formula (22), we conclude that for the auxiliary function $u(\underline{a}, \overline{a}) \overset{\text{def}}{=} \ln(V(\underline{a}, \overline{a}))$, for every two intervals $[\underline{a}, \overline{a}]$ and $[\underline{b}, \overline{b}]$, we have

$$u(\underline{a}, \overline{a}) + u\left(\underline{b}, \overline{b}\right) = u\left(\underline{a} + \underline{b}, \overline{a} + \overline{b}\right) + c \tag{25}$$

for an appropriate constant c. Thus, for $U(\underline{a}, \overline{a}) \overset{\text{def}}{=} u(\underline{a}, \overline{a}) - c$, substituting $u(\underline{a}, \overline{a}) = U(\underline{a}, \overline{a}) + c$ into the formula (25), we conclude that

$$U(\underline{a}, \overline{a}) + U\left(\underline{b}, \overline{b}\right) = U\left(\underline{a} + \underline{b}, \overline{a} + \overline{b}\right), \tag{26}$$

i.e., that the function U is additive. Similarly to [6], we can use the general classification of additive locally bounded functions (and every monotonic function is locally bounded) from [1] to conclude that $U(\underline{a}, \overline{a}) = k_1 \cdot \overline{a} + k_2 \cdot \underline{a}$. Monotonicity with respect to changes in $\underline{a}$ and $\overline{a}$ imply that $k_1 \geq 0$ and $k_2 \geq 0$. Thus, for

$$V(\underline{a}, \overline{a}) = \exp(u(\underline{a}, \overline{a})) = \exp(U(\underline{a}, \overline{a}) + c) = \exp(c) \cdot \exp(U(\underline{a}, \overline{a})), \tag{27}$$

we get the desired formula, with $C = \exp(c)$, $k = k_1 + k_2$ and $\alpha_H = k_1/(k_1 + k_2)$.
The proposition is proven.

3   Conclusion

Currently, one of the most promising Artificial Intelligence techniques is deep learning. The successes of using deep learning are spectacular – from winning over human champions in Go (a very complex game that until recently resisted computer efforts) to efficient algorithms for self-driving cars. All these successes require a large amount of computations on high performance computers.

While deep learning has been very successful, there is a lot of room for improvement. For example, the existing deep learning algorithms implicitly assume that all the input data are exact, while in reality, data comes from measurements and measurement are never absolutely accurate. The simplest situation is when we know the upper bound Δ on the measurement error. In this case, based on the measurement result $\tilde{x}$, the only thing that we can conclude about the actual value x is that x is in the interval $[\tilde{x} - \Delta, \tilde{x} + \Delta]$.

In this paper, we have shown how computing softmax – one of the important steps in deep learning algorithms – can be naturally extended to the case of such interval uncertainty. The resulting formulas are almost as simple as the original ones, so their implementation will take about the same time on the same high performance computers.

References

1. Aczél, J., Dhombres, J.: Functional Equations in Several Variables. Camridge University Press, Cambridge (2008)
2. Goodfellow, I., Bengio, Y., Courville, A.: Deep Learning. MIT Press, Cambridge (2016)
3. Hurwicz, L.: Optimality Criteria for Decision Making Under Ignorance, Cowles Commission Discussion Paper, Statistics, no. 370 (1951)
4. Kosheleva, O., Kreinovich, V., Sriboonchitta, S.: Econometric models of probabilistic choice: beyond McFadden's formulas. In: Kreinovich, V., Sriboonchitta, S., Huynh, V.-N. (eds.) Robustness in Econometrics. SCI, vol. 692, pp. 79–87. Springer, Cham (2017). https://doi.org/10.1007/978-3-319-50742-2_5
5. Kreinovich, V.: Decision making under interval uncertainty (and beyond). In: Guo, P., Pedrycz, W. (eds.) Human-Centric Decision-Making Models for Social Sciences. SCI, vol. 502, pp. 163–193. Springer, Heidelberg (2014). https://doi.org/10.1007/978-3-642-39307-5_8
6. Kreinovich, V.: Decision making under interval (and more general) uncertainty: monetary vs. utility approaches. J. Comput. Technol. **22**(2), 37–49 (2017)
7. Kreinovich, V.: From traditional neural networks to deep learning: towards mathematical foundations of empirical successes. In: Shahbazova, S.N., et al. (eds.) Proceedings of the World Conference on Soft Computing, Baku, Azerbaijan, 29–31 May 2018 (2018)
8. Luce, D.: Inividual Choice Behavior: A Theoretical Analysis. Dover, New York (2005)
9. Luce, R.D., Raiffa, R.: Games and Decisions: Introduction and Critical Survey. Dover, New York (1989)
10. McFadden, D.: Conditional logit analysis of qualitative choice behavior. In: Zarembka, P. (ed.) Frontiers in Econometrics, pp. 105–142. Academic Press, New York (1974)
11. McFadden, D.: Economic choices. Am. Econ. Rev. **91**, 351–378 (2001)
12. Sutton, R.S., Barto, A.G.: Reinforcement Learning: An Introduction, 2nd edn. MIT Press, Cambridge (2018)
13. Train, K.: Discrete Choice Methods with Simulation. Cambridge University Press, Cambridge (2003)

Improvements of Monotonicity Approach to Solve Interval Parametric Linear Systems

Iwona Skalna[1]([⊠]) [iD], Marcin Pietroń[1], and Milan Hladík[2] [iD]

[1] AGH University of Science and Technology, Krakow, Poland
{skalna,pietron}@agh.edu.pl
[2] Charles University, Prague, Czech Republic
milan.hladik@matfyz.cz

Abstract. Recently, we have proposed several improvements of the standard monotonicity approach to solving parametric interval linear systems. The obtained results turned out to be very promising; i.e., we have achieved narrower bounds, while generally preserving the computational time. In this paper we propose another improvements, which aim to further decrease the widths of the interval bounds.

Keywords: Parametric linear systems · Monotonicity approach · Revised affine forms · Matrix equation

1 Introduction

Let us start by introducing some interval notation and preliminary theory of revised affine forms[1] (RAF). An interval is a closed and connected (in a topological sense) subset of the real number line, i.e.,

$$\boldsymbol{x} = \{x \in \mathbb{R} \mid \underline{x} \leqslant x \leqslant \overline{x}\}, \tag{1}$$

where $\underline{x}, \overline{x} \in \mathbb{R}$, $\underline{x} \leq \overline{x}$, are given. The midpoint of an interval $\boldsymbol{x}$ is denoted by $x^c := \frac{1}{2}(\underline{x} + \overline{x})$, and its radius by $x^\Delta := \frac{1}{2}(\overline{x} - \underline{x})$. The set of all closed and connected intervals is denoted by $\mathbb{IR}$. The set of $n \times m$ interval matrices is denoted by $\mathbb{IR}^{n \times m}$. Vectors are considered as single-column matrices, and a matrix can be represented as a set of column vectors $A = (A_{*1}, \ldots, A_{*m})$. The smallest (w.r.t. inclusion) interval vector containing $S \subseteq \mathbb{R}^n$ is called the interval hull of S and is denoted by $\square S$.

A revised affine form (cf. [9,15,16,20]) of length n is defined as a sum of a standard affine form (see, e.g., [1]) and a term that represents all errors introduced during a computation (including rounding errors), i.e.,

$$\hat{x} = x_0 + e^T x + x_r[-1, 1], \tag{2}$$

---

[1] All the methods used in the computation are based on revised affine forms.

© Springer Nature Switzerland AG 2020
R. Wyrzykowski et al. (Eds.): PPAM 2019, LNCS 12044, pp. 374–383, 2020.
https://doi.org/10.1007/978-3-030-43222-5_33

where $e = (\varepsilon_1, \ldots, \varepsilon_n)^T$ and $x = (x_1, \ldots, x_n)^T$. The *central value* x_0 and the *partial deviations* x_i are finite floating-point numbers, the *noise symbols* ε_i, $i = 1, \ldots, n$, are unknown, but assumed to vary within the interval $[-1, 1]$, and $x_r \geqslant 0$ is the radius of the *accumulative error* $x_r[-1, 1]$. Notice that the interval domain $[-1, 1]$ for ε_i is without loss of generality and can be obtained by a suitable scaling and shifting.

Remark. A revised affine form $\hat{x}$ is an interval-affine function of the noise symbols, so it can be written as $\boldsymbol{x}(e) = e^T x + \boldsymbol{x}$, where $\boldsymbol{x} = x_0 + x_r[-1, 1]$.

One of the basic problems of interval computation is the problem of solving *interval parametric linear systems* (see, e.g., [3, 15]), i.e., linear systems with coefficients being functions of parameters varying within prescribed intervals. Such systems are encountered, e.g., in structural mechanics, electrical engineering and various optimization problems. An interval parametric linear system is defined as the following family of real parametric linear systems

$$A(p)x = b(p), \; p \in \boldsymbol{p} \in \mathbb{IR}^K, \tag{3}$$

and is often denoted as

$$A(\boldsymbol{p})x = b(\boldsymbol{p}). \tag{4}$$

The coefficients of the system are, in the general case, some nonlinear real-valued functions $A_{ij}, b_i : \mathbb{R}^K \to \mathbb{R}$ of variables $p_1, \ldots, p_K$, usually described by closed-form expressions.

The (so-called united) solution set Σ of (4) is defined as the set of all solutions that can be obtained by solving systems from the family (3), i.e.,

$$\Sigma = \{x \in \mathbb{R}^n \mid \exists p \in \boldsymbol{p} : A(p)x = b(p)\}. \tag{5}$$

The solution set (5) is difficult to characterize (cf. [4, 8]) and handling it is computationally very hard. Many questions, such as nonemptiness, boundedness or approximation, are NP-hard to answer, even for very special subclasses of problems; see, e.g., [5]. In, particular, the problem of computing the hull $\square\Sigma$ is NP-hard. One of the possible approaches to approximate the hull in a polynomial time is the so-called *monotonicity approach* (MA).

2 Monotonicity Approach

In this section we briefly outline the main idea of the standard monotonicity approach (cf. [11, 13, 14, 17]) and we present some improvements of this approach.

If $A(p)$ is non-singular for every $p \in \boldsymbol{p}$, then the solution of the system $A(p)x = b(p)$ has the form of $x = A(p)^{-1}b(p)$. That is, each x_i, $i = 1, \ldots, n$, is a real valued function of p, i.e., $x_i = x_i(p)$. If $x_i(p)$ is monotonic on $\boldsymbol{p}$ with respect to all parameters, then the lowest and largest values of $x_i(p)$ on $\boldsymbol{p}$ (i.e., minimum and maximum of Σ in ith coordinate) are attained at the respective endpoints of $\boldsymbol{p}$. If $x_i(p)$ is monotonic with respect to some parameters only, then we can fix these parameters at the respective endpoints and then bound the range of $x_i(p)$ on a box of a lower dimension. Since the overestimation increases along with the number of parameters, we can expect that MA will produce narrower bounds. Suppose that

- $x_i(p)$ is nondecreasing on p in variables p_k, $k \in K_1$,
- $x_i(p)$ is nonincreasing on p in variables p_k, $k \in K_2$,
- $x_i(p)$ is non-monotonic on p in variables p_k, $k \in K_3$.

Define the restricted set of parameters p^1 and p^2 as follows

$$
p_k^1 = \begin{cases} \underline{p}_k & \text{if } k \in K_1, \\ \overline{p}_k & \text{if } k \in K_2, \\ \boldsymbol{p}_k & \text{if } k \in K_3, \end{cases} \qquad p_k^2 = \begin{cases} \overline{p}_k & \text{if } k \in K_1, \\ \underline{p}_k & \text{if } k \in K_2, \\ \boldsymbol{p}_k & \text{if } k \in K_3, \end{cases}
$$

$k = 1, \ldots, K$. Then

$$
\underline{(\Box \Sigma)}_i = \min\{x_i \mid x \in \Sigma\} = \min\{x_i \mid \exists p \in p^1 : A(p)x = b(p)\},
$$
$$
\overline{(\Box \Sigma)}_i = \max\{x_i \mid x \in \Sigma\} = \max\{x_i \mid \exists p \in p^2 : A(p)x = b(p)\}.
$$

In this way, the computation reduces to two problems of smaller dimension.

Now, what remains is to check whether $x_i(p)$ is monotonic on p in a parameter p_k. The standard way is to determine the bounds of partial derivative $\frac{\partial x_i(p)}{\partial p_k}$ on p. We can compute $\frac{\partial x_i(p)}{\partial p_k}$ for $i = 1, \ldots, n$ by solving the parametric system

$$
A(p)\frac{\partial x(p)}{\partial p_k} = \frac{\partial b(p)}{\partial p_k} - \frac{\partial A(p)}{\partial p_k}x(p), \quad p \in \boldsymbol{p}. \tag{6}
$$

The problem is that the vector $x(p)$ is not known *a priori*. Therefore, it is usually estimated by an outer enclosure. That is, let $\boldsymbol{x} \supseteq \Sigma$, and consider the interval parametric linear system

$$
A(p)\frac{\partial x(p)}{\partial p_k} = \frac{\partial b(p)}{\partial p_k} - \frac{\partial A(p)}{\partial p_k}x, \quad x \in \boldsymbol{x}, \ p \in \boldsymbol{p}, \tag{7}
$$

with $K + n$ interval parameters.

Let $\boldsymbol{d}$ be an enclosure of the solution set of this system. If $\underline{d}_i \geqslant 0$, then $x_i(p)$ is nondecreasing in p_k, and similarly if $\overline{d}_i \leqslant 0$, then $x_i(p)$ is nonincreasing in p_k.

By solving (7) for each $k = 1, \ldots, K$, we get interval vectors $\boldsymbol{d}^1, \ldots, \boldsymbol{d}^K$. Provided that $0 \notin \boldsymbol{d}_i^k$ for every $k = 1, \ldots, K$ and $i = 1, \ldots, n$, we can compute the exact range of the solution set Σ as follows. For every $k = 1, \ldots, K$ and $i = 1, \ldots, n$ define

$$
p_k^{1,i} = \begin{cases} \underline{p}_k, & \text{if } \underline{d}_i^k \geqslant 0, \\ \overline{p}_k, & \text{if } \overline{d}_i^k \leqslant 0, \end{cases} \qquad p_k^{2,i} = \begin{cases} \overline{p}_k, & \text{if } \underline{d}_i^k \geqslant 0, \\ \underline{p}_k, & \text{if } \overline{d}_i^k \leqslant 0. \end{cases}
$$

By solving a pair of real-valued linear systems

$$A(p^{1,i})x^1 = b(p^{1,i}),\tag{8a}$$

$$A(p^{2,i})x^2 = b(p^{2,i}),\tag{8b}$$

we obtain

$$\square\Sigma_i = [x^1_i, x^2_i].$$

Obviously, by solving n pairs of real-valued linear systems (8), we obtain the range of the solution set in all coordinates, that is, $\square\Sigma$. The number of equations to be solved can be decreased by removing redundant vectors from the list

$$\mathcal{L} = \{p^{1,1}, \ldots, p^{1,n}, p^{2,1}, \ldots, p^{2,n}\}.$$

If only some of the partial derivatives have constant sign on p, then, in the worst case, instead of $2n$ real-valued systems, we must solve $2n$ interval parametric linear systems with a smaller number of interval parameters.

The main deficiency of the approach described above follows from replacing x for $x(p)$ in (7). In [19], we have proposed the following two approaches to overcome this shortcoming.

Double System. For each $k = 1, \ldots, K$, we create the following $2n \times 2n$ interval parametric linear system

$$\begin{pmatrix} A(p) & 0 \\ \frac{\partial A(p)}{\partial p_k} & A(p) \end{pmatrix} \begin{pmatrix} x \\ \frac{\partial x}{\partial p_k} \end{pmatrix} = \begin{pmatrix} b(p) \\ \frac{\partial b(p)}{\partial p_k} \end{pmatrix}, \quad p \in \boldsymbol{p},\tag{9}$$

and solve it in order to obtain bounds for $\frac{\partial x}{\partial p_k}(p)$ over $\boldsymbol{p}$. This approach eliminates the problem of the "loss of information", but instead is more expensive than the standard approach.

p-Solution. For each $k = 1, \ldots, K$, we solve the following interval parametric linear system

$$A(p)\frac{\partial x}{\partial p_k} = \frac{\partial b(p)}{\partial p_k} - \frac{\partial A(p)}{\partial p_k}x(p), \quad p \in \boldsymbol{p},\tag{10}$$

where unknown $x(p)$ is replaced by the so-called p-solution form of an enclosure $x(p) = Lp + a$ (see, e.g., [6,15,18]). In our approach, the p-solution is represented by a vector of revised affine forms[2]. Since $x(p)$ partially preserves linear dependencies between $x(p)$ and p, the solutions of the systems (10) are usually narrower than the solution of the system (7), and hence we have better bounds for partial derivatives.

---

[2] We first substitute interval parameters by respective revised affine forms and then we performs computation of them.

Novel Approaches. Another variant of the MA that we consider in this paper is based on solving interval parametric linear system with multiple right-hand side (cf. [2,12]). The solution set of such system is defined as

$$\Sigma' = \{X \in \mathbb{R}^{n \times K} \mid (\forall k \in K)(\exists p \in \boldsymbol{p}) : A(p)X_{*k} = B(p)_{*k}\}. \tag{11}$$

Now, consider the following interval parametric matrix equation

$$A(p)\left(\frac{\partial x}{\partial p_1}, \dots, \frac{\partial x}{\partial p_K}\right) = B(p), \ p \in \boldsymbol{p}, \tag{12}$$

where $B(p)_{*k} = \frac{\partial b(p)}{\partial p_k} - \frac{\partial A(p)}{\partial p_k}X(p)$, $k = 1, \dots, K$. The solution set of the system (12) is defined as

$$\Sigma'' = \{X \in \mathbb{R}^{n \times K} \mid \exists p \in \boldsymbol{p} : A(p)X = B(p)\}. \tag{13}$$

The system (12) can be solved by an arbitrary, however accordingly modified, method for solving interval parametric linear systems. In our experiments we use modified Interval-affine Gauss-Seidel iteration[3].

As pointed in [2], it holds that $\Sigma'' \subseteq \Sigma'$, however, we have that $\square \Sigma'' = \square \Sigma'$.

Proposition 1. *It holds that*

$$\square \Sigma'' = \square \Sigma'.$$

Proof. If $A(p)$ is nonsingular for each $p \in \boldsymbol{p}$ then $\square \Sigma'$ and $\square \Sigma''$ are bounded and the respective bounds of the hull can be computed by solving $2nK$ optimization problems: $i = 1, \dots, n$, $k = 1, \dots, K$

$$\underline{\square \Sigma'_{ik}} = \min\{X(p)_{ik} \mid p \in \boldsymbol{p}\} = \underline{\square \Sigma''_{ik}},$$
$$\overline{\square \Sigma'_{ik}} = \max\{X(p)_{ik} \mid p \in \boldsymbol{p}\} = \overline{\square \Sigma''_{ik}},$$

where $X(p)_{ik}$ is the ith component of $X(p)_{*k} = A(p)^{-1}B(p)_{*k}$. $\square$

Due to the inclusion $\Sigma'' \subseteq \Sigma'$, focusing on enclosing the solution set Σ'' may provide tighter results.

Another possibility is to put together all the systems (9) in one and solve the following interval parametric linear system:

$$\begin{pmatrix} A(p) & 0 & \cdots & 0 \\ \frac{\partial A(p)}{\partial p_1} & A(p) & 0 & 0 \\ \vdots & & \ddots & \\ \frac{\partial A(p)}{\partial p_K} & 0 & 0 & A(p) \end{pmatrix} \begin{pmatrix} x(p) \\ \frac{\partial x}{\partial p_1} \\ \vdots \\ \frac{\partial x}{\partial p_K} \end{pmatrix} = \begin{pmatrix} b(p) \\ \frac{\partial b(p)}{\partial p_1} \\ \vdots \\ \frac{\partial b(p)}{\partial p_K} \end{pmatrix}, \quad p \in \boldsymbol{p}. \tag{14}$$

---

[3] For details on Interval-affine Gauss-Seidel iteration see [18].

3   Numerical Experiments

We use the following abbreviations for the discussed monotonicity based methods: MA1 for the standard MA, MA2 for the p-solution based MA, MA3 for the doubled system based MA, MA4 for the matrix equation based approach, and MA5 for the MA based on the formula (14).

Example 1. Consider the following interval parametric linear system with affine-linear dependencies

$$\begin{pmatrix} 1 + p_1 + p_2 & p_1 & p_2 \\ 0 & p_1 + p_2 & p_2 \\ 0.1 & 0 & 3p_1 + p_2 \end{pmatrix} x = \begin{pmatrix} p_1 + 5p_2 \\ 2 + p_1 + 3p_2 \\ 1 + 2p_1 + p_2, \end{pmatrix},$$

where $p_1 \in [0.4 - 2\delta, 0.5 + 2\delta], p_2 \in [0.2 - \delta, 0.3 + \delta]$. We solve the system with $\delta = 0.05$ and $\delta = 0.1$. The overestimation of the monotonicity based methods over the hull is presented in Table 1. The overestimation is computed using the following formula:

$$O_\omega(a, b) = \left(1 - \frac{a^\Delta}{b^\Delta}\right) \cdot 100\%, \tag{15}$$

where $a, b \in \mathbb{IR}$, $a \subseteq b$, and a^Δ (b^Δ) is the radius of a (b).

Table 1. Results for Example 1: overestimation of monotonicity based bounds over interval hull

	MA1	MA2	MA3	MA4	MA5
$\delta = 0.05$					
x_1	3%	3%	3%	3%	3%
x_2	15%	1%	1%	0%	0%
x_3	14%	0%	0%	0%	0%
Time[s]	0.005	0.006	0.009	0.004	0.008
$\delta = 0.1$					
x_1	36%	36%	4%	4%	4%
x_2	22%	22%	22%	22%	22%
x_3	19%	19%	19%	19%	19%
Time[s]	0.006	0.006	0.11	0.005	0.008

As can be seen from Table 1, for $\delta = 0.05$, the MA4 and MA5 methods produced the best results; whereas for $\delta = 0.1$ the best results were produced by the MA3, MA4 and MA5 methods. However, MA4 turned out to be less time consuming than MA3 and MA5.

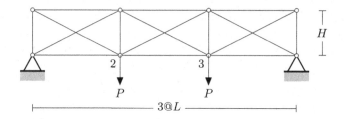

Fig. 1. 3-bay 1-floor truss structure

Example 2. Consider the 3-bay 1-floor (statically indeterminate) planar truss structure depicted in Fig. 1.

The nominal values of the truss parameters are as follows: Young modulus $E = 200$ [GPa], cross-sectional area $A = 0.01$ [m^2]. The length of the horizontal bars $L = 10$ [m], and the length of the vertical bars $H = 5$ [m]. The truss is subjected to external downward load $P = 10$ [kN] at node 2 and 3. The loads and the stiffness of each bar are assumed to be uncertain. The uncertainty considered is 10% corresponding to a variation of ±5% about the nominal values. In order to find the displacements of the truss nodes, we must solve the following interval parametric linear system

$$K(p)d = F(p), \ p \in \boldsymbol{p}, \tag{16}$$

where $K(p)$ is the stiffness matrix, $F(p)$ is the external loads vector and $\boldsymbol{p}$ is a vector of interval parameters. We compare the results of the considered

Table 2. Results for Example 2: overestimation of monotonicity based bounds by NeP bounds

Displ.	MA1	MA2	MA3	MA4	MA5
d_2^x	−6%	3%	11%	3%	3%
d_2^y	1%	4%	6%	5%	6%
d_3^x	−6%	3%	11%	3%	2%
d_3^y	1%	4%	6%	4%	6%
d_5^x	3%	4%	9%	4%	−13%
d_5^y	4%	6%	10%	6%	11%
d_6^x	7%	7%	9%	7%	−6%
d_6^y	4%	5%	6%	5%	6%
d_7^x	7%	7%	9%	7%	8%
d_7^y	4%	5%	6%	5%	6%
d_8^x	3%	4%	9%	4%	8%
d_8^y	4%	6%	10%	6%	7%
Time[s]	0.15	0.14	0.32	0.11	0.48

methods with the result of the iterative method (for the comparison purposes we will refer to this method as the NeP method) proposed in [10], which is considered as one of the best methods for solving interval parametric linear systems having the specific form (see [10]) (the truss system (16) can be transformed to this form, therefore the NeP is applicable). The comparison of the results of the MAs and NeP is presented in Table 2.

As we can see from the table, the MA2, MA3 and MA4 improved the NeP bounds for all the displacements. The best results were produced by the MA3 method, and the MA4 method was slightly better than the MA2 method. The results of the MA5 method are not straightforward. The method produced the best bounds for the vertical displacement of the node 5, for some displacements it produced better bounds than MA1, MA2 and MA4 methods, however, it produced the worst bounds for the horizontal displacements of the nodes 5 and 6. Similarly as in Example 1, the MA4 was computationally the most efficient.

4 Conclusions

We proposed some improvements of the monotonicity approach. In general, the monotonicity approach based on solving interval parametric matrix equation is a very promising approach, and our improvements can in certain situations further decrease the overestimation caused by the solving method. Nevertheless, the monotonicity approach is an expensive technique which is the main obstacle in broader exploitation of this method in solving practical problems.

Therefore, in the future we will try to decrease the computational time of the monotonicity approach by using parallelization[4]. Indeed, the monotonicity approach for solving interval parametric systems is suitable for parallelization in general. Particularly the systems (9) and (10) can be solved independently for each $k = 1, \ldots, K$. For systems (12) and (14) there is no direct way for parallelization, and its utilization heavily depends on the particular method chosen to solve the systems.

Acknowledgments. M. Hladík was supported by the Czech Science Foundation Grant P403-18-04735S.

References

1. Comba, J., Stolfi, J.: Affine arithmetic and its applications to computer graphics. In: Proceedings of SIBGRAPI 1993 VI Simpósio Brasileiro de Computação Gráfica e Processamento de Imagens (Recife, BR), pp. 9–18 (1993)
2. Dehghani-Madiseh, M., Dehghan, M.: Parametric AE-solution sets to the parametric linear systems with multiple right-hand sides and parametric matrix equation $A(p)X = B(p)$. Numer. Algorithms **73**(1), 245–279 (2016). https://doi.org/10.1007/s11075-015-0094-3

---

[4] Some attempts to use parallelization for increasing the efficiency of a few methods for solving interval parametric linear systems are described in [7].

3. Hladík, M.: Enclosures for the solution set of parametric interval linear systems. Int. J. Appl. Math. Comput. Sci. **22**(3), 561–574 (2012)
4. Hladík, M.: Description of symmetric and skew-symmetric solution set. SIAM J. Matrix Anal. Appl. **30**(2), 509–521 (2008)
5. Horáček, J., Hladík, M., Černý, M.: Interval linear algebra and computational complexity. In: Bebiano, N. (ed.) MAT-TRIAD 2015. SPMS, vol. 192, pp. 37–66. Springer, Cham (2017). https://doi.org/10.1007/978-3-319-49984-0_3
6. Kolev, L.: Parameterized solution of linear interval parametric systems. Appl. Math. Comput. **246**, 229–246 (2014)
7. Král, O., Hladík, M.: Parallel computing of linear systems with linearly dependent intervals in MATLAB. In: Wyrzykowski, R., Dongarra, J., Deelman, E., Karczewski, K. (eds.) PPAM 2017. LNCS, vol. 10778, pp. 391–401. Springer, Cham (2018). https://doi.org/10.1007/978-3-319-78054-2_37
8. Mayer, G.: An Oettli-Prager-like theorem for the symmetric solution set and for related solution sets. SIAM J. Matrix Anal. Appl. **33**(3), 979–999 (2012)
9. Messine, F.: New affine forms in interval branch and bound algorithms. Technical report, R2I 99–02, Université de Pau et des Pays de l'Adour (UPPA), France (1999)
10. Neumaier, A., Pownuk, A.: Linear systems with large uncertainties, with applications to truss structures. Reliab. Comput. **13**, 149–172 (2007). https://doi.org/10.1007/s11155-006-9026-1
11. Popova, E.: Computer-assisted proofs in solving linear parametric problems. In: 12th GAMM/IMACS International Symposium on Scientific Computing, Computer Arithmetic and Validated Numerics, SCAN 2006, Duisburg, Germany, p. 35 (2006)
12. Popova, E.D.: Enclosing the solution set of parametric interval matrix equation $A(p)X = B(p)$. Numer. Algorithms **78**(2), 423–447 (2018). https://doi.org/10.1007/s11075-017-0382-1
13. Rohn, J.: A method for handling dependent data in interval linear systems. Technical report, 911, Institute of Computer Science, Academy of Sciences of the Czech Republic, Prague (2004). https://asepactivenode.lib.cas.cz/arl-cav/en/contapp/?repo=crepo1&key=20925094170
14. Skalna, I.: On checking the monotonicity of parametric interval solution of linear structural systems. In: Wyrzykowski, R., Dongarra, J., Karczewski, K., Wasniewski, J. (eds.) PPAM 2007. LNCS, vol. 4967, pp. 1400–1409. Springer, Heidelberg (2008). https://doi.org/10.1007/978-3-540-68111-3_148
15. Skalna, I.: Parametric Interval Algebraic Systems. Studies in Computational Intelligence. Springer, Cham (2018). https://doi.org/10.1007/978-3-319-75187-0
16. Skalna, I., Hladík, M.: A new algorithm for Chebyshev minimum-error multiplication of reduced affine forms. Numer. Algorithms **76**(4), 1131–1152 (2017). https://doi.org/10.1007/s11075-017-0300-6
17. Skalna, I., Duda, J.: A study on vectorisation and paralellisation of the monotonicity approach. In: Wyrzykowski, R., Deelman, E., Dongarra, J., Karczewski, K., Kitowski, J., Wiatr, K. (eds.) PPAM 2015. LNCS, vol. 9574, pp. 455–463. Springer, Cham (2016). https://doi.org/10.1007/978-3-319-32152-3_42
18. Skalna, I., Hladík, M.: A new method for computing a p-solution to parametric interval linear systems with affine-linear and nonlinear dependencies. BIT Numer. Math. **57**(4), 1109–1136 (2017). https://doi.org/10.1007/s10543-017-0679-4

19. Skalna, I., Hladík, M.: Enhancing monotonicity checking in parametric interval linear systems. In: Martel, M., Damouche, N., Sandretto, J.A.D. (eds.) Trusted Numerical Computations, TNC 2018. Kalpa Publications in Computing, vol. 8, pp. 70–83. EasyChair (2018)
20. Vu, X.H., Sam-Haroud, D., Faltings, B.: A generic scheme for combining multiple inclusion representations in numerical constraint propagation. Technical report no. IC/2004/39, Swiss Federal Institute of Technology in Lausanne (EPFL), Lausanne, Switzerland, April 2004. http://liawww.epfl.ch/Publications/Archive/vuxuanha2004a.pdf

The First Approach to the Interval Generalized Finite Differences

Malgorzata A. Jankowska[1(✉)] and Andrzej Marciniak[2,3]

[1] Institute of Applied Mechanics, Poznan University of Technology,
Jana Pawla II 24, 60-965 Poznan, Poland
malgorzata.jankowska@put.poznan.pl
[2] Institute of Computing Science, Poznan University of Technology,
Piotrowo 2, 60-965 Poznan, Poland
andrzej.marciniak@put.poznan.pl
[3] Department of Computer Science, State University of Applied Sciences in Kalisz,
Poznanska 201-205, 62-800 Kalisz, Poland

Abstract. The paper concerns the first approach to interval generalized finite differences. The conventional generalized finite differences are of special interest due to the fact that they can be applied to irregular grids (clouds) of points. Based on these finite differences we can compute approximate values of some derivatives (ordinary or partial). Furthermore, one can formulate the generalized finite difference method for solving some boundary value problems with a complicated boundary of a domain. The aim of this paper is to propose the interval counterparts of generalized finite differences. Under the appropriate assumptions the exact values of the derivatives are included in the interval values obtained.

Keywords: Conventional and interval generalized finite differences · Interval arithmetic · Interval enclosure of derivatives

1 Introduction

The development of interval methods for solving initial-boundary value problems for ordinary and partial differential equations can be made on the basis of finite difference methods. Such methods utilize the so called finite differences that are used to approximate a value of some derivative at a given point. As a result the finite difference method provides an approximate solution of the initial-boundary value problem in the points of a region.

In the area of conventional finite differences we can indicate two main classes that differ, in particular, in the way the points of a grid are located in the domain of interest. In the case of well-known classical finite differences (FD) we generate a regular grid of points such that the distances between two neighbouring points in a given direction (e.g., horizontal and/or vertical) are equal. Such an approach is very useful and efficient when we formulate some finite difference method

ⓒ Springer Nature Switzerland AG 2020
R. Wyrzykowski et al. (Eds.): PPAM 2019, LNCS 12044, pp. 384–394, 2020.
https://doi.org/10.1007/978-3-030-43222-5_34

(FDM) in the area of a rectangular/square region. Otherwise, when an irregular boundary occurs, these finite differences can be also applied, but we have to modify their main formula for the points located near the boundary to take into account different values of mesh increments in some direction. In such a case we have to apply different formulas of finite differences depending on the location of points in the domain. To another class of finite differences we include the so called generalized finite differences (GFD) [1,6,9]. In the case of this approach we generate a grid (cloud) of points with an arbitrary (irregular) arrangement in the region, although the regular distribution can be also applied. Similarly as in the classical case the technique is based on the local Taylor series expansion. The concept of the generation of finite difference formulas at irregular grid of points leads to a complete set of derivatives up to the order n. This feature differs the FDs from the GFDs, as in the first approach only an approximation of one particular derivative at a given point was obtained at one time.

A number of interval finite difference methods based on their conventional counterparts was recently formulated in the literature [3,5,7] and applied to some initial-boundary value problems defined in regions of regular shapes. To the best knowledge of the authors, neither the interval generalized finite differences (IGFD) nor the appropriate interval generalized finite difference methods (IGFDM) have been proposed yet. Hence, as a first step towards this task, we focus on the interval counterparts of the GFDs. The construction of some interval generalized finite difference methods is not considered in the paper. We plan to take it into account in our future research.

The paper is organized in the following way. After a short introduction to the conventional GFDs presented in Sect. 2, we formulate the concept of their interval counterparts in Sect. 3. The interval approach has many significant advantages. First of all, the remainder term of the Taylor series is included in the final solution obtained. Hence, the use of interval methods in the floating-point interval arithmetic allows to perform computations such that the interval solution contains an exact solution of the problem (see Sect. 4). If the endpoints of the error term intervals are approximated, then the guaranteed nature of results is lost. However, the numerical experiments show that values of derivatives are still contained in the intervals obtained. We compute interval values of derivatives up to the second order for four example functions and regular/irregular grids of points. A short result discussion is given Sect. 5.

2   Conventional Generalized Finite Differences

Consider the following derivatives of a function $u = u(x, y)$

$$\frac{\partial u}{\partial x}(p_0), \ \frac{\partial u}{\partial y}(p_0), \ \frac{\partial^2 u}{\partial x^2}(p_0), \ \frac{\partial^2 u}{\partial y^2}(p_0), \ \frac{\partial^2 u}{\partial x \partial y}(p_0). \tag{1}$$

The approximate values of these derivatives at some point p_0 can be computed with generalized finite differences described in detail in, e.g., [1,6,9,10].

We assume that the function u has continuous derivatives up to the third order with respect to x and y in a region $\Omega \subset \mathbb{R}^2$.

First we generate a grid (cloud) of points such that the point $p_0 = (x_0, y_0)$ is the central node and the points $p_i = (x_i, y_i)$, $i = 1, 2, \ldots n$ (located in the surroundings of p_0) are the ith nodes of the star obtained. We also have $h_i = x_i - x_0$, $k_i = y_i - y_0$. We expand the function u in the Taylor series about the point p_0 and evaluate it at the points p_i, $i = 1, 2, \ldots n$. For each point p_i we have

$$
\begin{aligned}
u(p_i) = {} & u(p_0) + h_i \frac{\partial u}{\partial x}(p_0) + k_i \frac{\partial u}{\partial y}(p_0) \\
& + \frac{1}{2!}\left(h_i^2 \frac{\partial^2 u}{\partial x^2}(p_0) + k_i^2 \frac{\partial^2 u}{\partial y^2}(p_0) + 2h_i k_i \frac{\partial^2 u}{\partial x \partial y}(p_0) \right) \\
& + \frac{1}{3!}\left(h_i^3 \frac{\partial^3 u}{\partial x^3}(q_i) + k_i^3 \frac{\partial^3 u}{\partial y^3}(q_i) + 3h_i^2 k_i \frac{\partial^3 u}{\partial x^2 \partial y}(q_i) + 3h_i k_i^2 \frac{\partial^3 u}{\partial x \partial y^2}(q_i) \right).
\end{aligned}
\tag{2}
$$

Note that $q_i = (\xi_i, \eta_i)$ is an intermediate point of the remainder term such that $\xi_i \in \left(\xi_i^{\min}, \xi_i^{\max}\right)$, $\eta_i \in \left(\eta_i^{\min}, \eta_i^{\max}\right)$. Furthermore, we have $\xi_i^{\min} = \min\{x_i, x_0\}$, $\xi_i^{\max} = \max\{x_i, x_0\}$ and $\eta_i^{\min} = \min\{y_i, y_0\}$, $\eta_i^{\max} = \max\{y_i, y_0\}$. If we add the expressions (2), we obtain

$$
\begin{aligned}
\sum_{i=1}^{N}(u(p_i) - u(p_0)) = {} & \sum_{i=1}^{N} h_i \frac{\partial u}{\partial x}(p_0) + \sum_{i=1}^{N} k_i \frac{\partial u}{\partial y}(p_0) + \frac{1}{2}\left(\sum_{i=1}^{N} h_i^2 \frac{\partial^2 u}{\partial x^2}(p_0) \right. \\
& \left. + \sum_{i=1}^{N} k_i^2 \frac{\partial^2 u}{\partial y^2}(p_0) + 2 \sum_{i=1}^{N} h_i k_i \frac{\partial^2 u}{\partial x \partial y}(p_0) \right) + \frac{1}{6} \sum_{i=1}^{N} r(q_i),
\end{aligned}
\tag{3}
$$

where

$$
r(q_i) = h_i^3 \frac{\partial^3 u}{\partial x^3}(q_i) + k_i^3 \frac{\partial^3 u}{\partial y^3}(q_i) + 3h_i^2 k_i \frac{\partial^3 u}{\partial x^2 \partial y}(q_i) + 3h_i k_i^2 \frac{\partial^3 u}{\partial x \partial y^2}(q_i).
\tag{4}
$$

We further define the function $\mathcal{F}$ as follows

$$
\begin{aligned}
\mathcal{F}(u) = \sum_{i=1}^{N} \Bigg\{ & \bigg[u(p_0) - u(p_i) + h_i \frac{\partial u}{\partial x}(p_0) + k_i \frac{\partial u}{\partial y}(p_0) + \frac{1}{2} h_i^2 \frac{\partial^2 u}{\partial x^2}(p_0) \\
& + \frac{1}{2} k_i^2 \frac{\partial^2 u}{\partial y^2}(p_0) + h_i k_i \frac{\partial^2 u}{\partial x \partial y}(p_0) + \frac{1}{6} r(q_i) \bigg] w(h_i, k_i) \Bigg\}^2,
\end{aligned}
\tag{5}
$$

where $w = w(h_i, k_i)$ are the weight functions simply denoted by w_i. We minimize $\mathcal{F}$ with respect to the values of the derivatives at the point p_0 (1). We have

$$
\frac{\partial \mathcal{F}(u)}{\partial \mathcal{A}} = \frac{\partial \mathcal{F}(u)}{\partial \mathcal{B}} = \frac{\partial \mathcal{F}(u)}{\partial \mathcal{C}} = \frac{\partial \mathcal{F}(u)}{\partial \mathcal{D}} = \frac{\partial \mathcal{F}(u)}{\partial \mathcal{E}} = 0,
\tag{6}
$$

where

$$
\mathcal{A} = \frac{\partial u}{\partial x}(p_0), \quad \mathcal{B} = \frac{\partial u}{\partial y}(p_0), \quad \mathcal{C} = \frac{\partial^2 u}{\partial x^2}(p_0), \quad \mathcal{D} = \frac{\partial^2 u}{\partial y^2}(p_0), \quad \mathcal{E} = \frac{\partial^2 u}{\partial x \partial y}(p_0).
\tag{7}
$$

Finally, we obtain a linear system of equations of the form

$$\widehat{A}\widehat{D} = \widehat{B} + \widehat{E}, \tag{8}$$

where

$$\widehat{A} = \begin{bmatrix} \sum\limits_{i=1}^{N} h_i^2 w_i^2 & \sum\limits_{i=1}^{N} h_i k_i w_i^2 & \sum\limits_{i=1}^{N} \frac{1}{2} h_i^3 w_i^2 & \sum\limits_{i=1}^{N} \frac{1}{2} h_i k_i^2 w_i^2 & \sum\limits_{i=1}^{N} h_i^2 k_i w_i^2 \\[2mm] \sum\limits_{i=1}^{N} h_i k_i w_i^2 & \sum\limits_{i=1}^{N} k_i^2 w_i^2 & \sum\limits_{i=1}^{N} \frac{1}{2} h_i^2 k_i w_i^2 & \sum\limits_{i=1}^{N} \frac{1}{2} k_i^3 w_i^2 & \sum\limits_{i=1}^{N} h_i k_i^2 w_i^2 \\[2mm] \sum\limits_{i=1}^{N} \frac{1}{2} h_i^3 w_i^2 & \sum\limits_{i=1}^{N} \frac{1}{2} h_i^2 k_i w_i^2 & \sum\limits_{i=1}^{N} \frac{1}{4} h_i^4 w_i^2 & \sum\limits_{i=1}^{N} \frac{1}{4} h_i^2 k_i^2 w_i^2 & \sum\limits_{i=1}^{N} \frac{1}{2} h_i^3 k_i w_i^2 \\[2mm] \sum\limits_{i=1}^{N} \frac{1}{2} h_i k_i^2 w_i^2 & \sum\limits_{i=1}^{N} \frac{1}{2} k_i^3 w_i^2 & \sum\limits_{i=1}^{N} \frac{1}{4} h_i^2 k_i^2 w_i^2 & \sum\limits_{i=1}^{N} \frac{1}{4} k_i^4 w_i^2 & \sum\limits_{i=1}^{N} \frac{1}{2} h_i k_i^3 w_i^2 \\[2mm] \sum\limits_{i=1}^{N} h_i^2 k_i w_i^2 & \sum\limits_{i=1}^{N} h_i k_i^2 w_i^2 & \sum\limits_{i=1}^{N} \frac{1}{2} h_i^3 k_i w_i^2 & \sum\limits_{i=1}^{N} \frac{1}{2} h_i k_i^3 w_i^2 & \sum\limits_{i=1}^{N} h_i^2 k_i^2 w_i^2 \end{bmatrix}, \tag{9}$$

$$\widehat{D} = \left[\frac{\partial u}{\partial x}(p_0), \frac{\partial u}{\partial y}(p_0), \frac{\partial^2 u}{\partial x^2}(p_0), \frac{\partial^2 u}{\partial y^2}(p_0), \frac{\partial^2 u}{\partial x \partial y}(p_0) \right]^{\mathrm{T}}, \tag{10}$$

$$\widehat{B} = \begin{bmatrix} \sum\limits_{i=1}^{N} (-u(p_0) + u(p_i)) h_i w_i^2 \\[2mm] \sum\limits_{i=1}^{N} (-u(p_0) + u(p_i)) k_i w_i^2 \\[2mm] \sum\limits_{i=1}^{N} (-u(p_0) + u(p_i)) \frac{1}{2} h_i^2 w_i^2 \\[2mm] \sum\limits_{i=1}^{N} (-u(p_0) + u(p_i)) \frac{1}{2} k_i^2 w_i^2 \\[2mm] \sum\limits_{i=1}^{N} (-u(p_0) + u(p_i)) k_i h_i w_i^2 \end{bmatrix}, \quad \widehat{E} = \begin{bmatrix} -\sum\limits_{i=1}^{N} r(q_i) h_i w_i^2 \\[2mm] -\sum\limits_{i=1}^{N} r(q_i) k_i w_i^2 \\[2mm] -\sum\limits_{i=1}^{N} r(q_i) \frac{1}{2} h_i^2 w_i^2 \\[2mm] -\sum\limits_{i=1}^{N} r(q_i) \frac{1}{2} k_i^2 w_i^2 \\[2mm] -\sum\limits_{i=1}^{N} r(q_i) k_i h_i w_i^2 \end{bmatrix}. \tag{11}$$

Note that as the weight functions w_i we choose (see, e.g., [1,10]) $w(h_i, k_i) = 1/d_i^3$, where $d_i = ((x_0 - x_i)^2 + (y_0 - y_i)^2)^{1/2} = (h_i^2 + k_i^2)^{1/2}$.

Remark 1. Let $u_0, u_i, i = 1, 2, \ldots n$ approximate the exact values $u(p_0), u(p_i)$ of the function u at the central and surrounding nodes. If we also ignore the remaining terms of the Taylor series expansion given in the components of a vector $\widehat{E}$, we obtain the linear system of equations whose solution provides approximate values of a complete set of the first and second order derivatives of u at the central node p_0. Such an approach utilizes the conventional generalized finite differences considered. The matrix $\widehat{A}$ of the linear system of equations (8) is symmetrical. As proposed in, e.g., [1,10], this system of equations can be efficiently solved with the Cholesky method.

3   Interval Generalized Finite Differences

Let us denote by X_i, Y_i, $i = 0, 1, \ldots, n$, the intervals such that $x_i \in X_i$, $y_i \in Y_i$ and by $U = U(X, Y)$ the interval extension of $u = u(x, y)$. Hence, we have $H_i = X_i - X_0$, $K_i = Y_i - Y_0$ and $W(H_i, K_i) = 1/D_i^3$, where $D_i = (H_i^2 + K_i^2)^{1/2}$.

Finally, we make the following assumptions about values in the midpoints of the derivatives included in the remainder term of the Taylor series. We denote by $D^{(3,1)} = D^{(3,1)}(X, Y)$, $D^{(3,2)} = D^{(3,2)}(X, Y)$, $D^{(3,3)} = D^{(3,3)}(X, Y)$, $D^{(3,4)} = D^{(3,4)}(X, Y)$ the interval extensions of the appropriate derivatives of u, i.e., $d^{(3,1)} = \partial^3 u / \partial x^3(x, y)$, $d^{(3,2)} = \partial^3 u / \partial y^3(x, y)$, $d^{(3,3)} = \partial^3 u / \partial x^2 \partial y(x, y)$, $d^{(3,4)} = \partial^3 u / \partial x \partial y^2(x, y)$, respectively. Then, for the midpoints ξ_i, η_i, we assume that there exist the intervals such that $\xi_i \in \Xi_i$, $\eta_i \in H_i$, $i = 1, 2, \ldots, n$. Based on that we can define the interval extension $R = R(X, Y)$ of the error term function $r = r(x, y)$ (4) and compute its value at the point (Ξ_i, H_i). We have

$$R_i = H_i^3 D_i^{(3,1)} + K_i^3 D_i^{(3,2)} + 3H_i^2 K_i D_i^{(3,3)} + 3H_i K_i^2 D_i^{(3,4)}, \qquad (12)$$

where $R_i = R(\Xi_i, H_i)$, $D_i^{(3,j)} = D^{(3,j)}(\Xi_i, H_i)$, $i = 1, 2, \ldots, n$, $j = 1, 2, 3, 4$.

Based on the above notations and assumptions if we replace all real values used in (8)–(11) by the appropriate intervals and all functions by their interval extensions, then we obtain an interval linear system of equations of the form

$$AD = B + E, \qquad (13)$$

where

$$A = \begin{bmatrix}
\sum\limits_{i=1}^{N} H_i^2 W_i^2 & \sum\limits_{i=1}^{N} H_i K_i W_i^2 & \sum\limits_{i=1}^{N} \frac{1}{2} H_i^3 W_i^2 & \sum\limits_{i=1}^{N} \frac{1}{2} H_i K_i^2 W_i^2 & \sum\limits_{i=1}^{N} H_i^2 K_i W_i^2 \\
\sum\limits_{i=1}^{N} H_i K_i W_i^2 & \sum\limits_{i=1}^{N} K_i^2 W_i^2 & \sum\limits_{i=1}^{N} \frac{1}{2} H_i^2 K_i W_i^2 & \sum\limits_{i=1}^{N} \frac{1}{2} K_i^3 W_i^2 & \sum\limits_{i=1}^{N} H_i K_i^2 W_i^2 \\
\sum\limits_{i=1}^{N} \frac{1}{2} H_i^3 W_i^2 & \sum\limits_{i=1}^{N} \frac{1}{2} H_i^2 K_i w_i^2 & \sum\limits_{i=1}^{N} \frac{1}{4} H_i^4 W_i^2 & \sum\limits_{i=1}^{N} \frac{1}{4} H_i^2 K_i^2 W_i^2 & \sum\limits_{i=1}^{N} \frac{1}{2} H_i^3 K_i W_i^2 \\
\sum\limits_{i=1}^{N} \frac{1}{2} H_i K_i^2 W_i^2 & \sum\limits_{i=1}^{N} \frac{1}{2} K_i^3 W_i^2 & \sum\limits_{i=1}^{N} \frac{1}{4} H_i^2 K_i^2 W_i^2 & \sum\limits_{i=1}^{N} \frac{1}{4} K_i^4 W_i^2 & \sum\limits_{i=1}^{N} \frac{1}{2} H_i K_i^3 W_i^2 \\
\sum\limits_{i=1}^{N} H_i^2 K_i W_i^2 & \sum\limits_{i=1}^{N} H_i K_i^2 W_i^2 & \sum\limits_{i=1}^{N} \frac{1}{2} H_i^3 K_i W_i^2 & \sum\limits_{i=1}^{N} \frac{1}{2} H_i K_i^3 W_i^2 & \sum\limits_{i=1}^{N} H_i^2 K_i^2 W_i^2
\end{bmatrix}, \qquad (14)$$

$$D = \left[D_0^{(X)}, \ D_0^{(Y)}, \ D_0^{(XX)}, \ D_0^{(YY)}, \ D_0^{(XY)} \right]^T, \qquad (15)$$

$$B = \begin{bmatrix}
\sum\limits_{i=1}^{N} (-U_0 + U_i) H_i W_i^2 \\
\sum\limits_{i=1}^{N} (-U_0 + U_i) K_i W_i^2 \\
\sum\limits_{i=1}^{N} (-U_0 + U_i) \frac{1}{2} H_i^2 W_i^2 \\
\sum\limits_{i=1}^{N} (-U_0 + U_i) \frac{1}{2} K_i^2 W_i^2 \\
\sum\limits_{i=1}^{N} (-U_0 + U_i) K_i H_i W_i^2
\end{bmatrix}, \quad
E = \begin{bmatrix}
-\sum\limits_{i=1}^{N} R_i H_i W_i^2 \\
-\sum\limits_{i=1}^{N} R_i K_i W_i^2 \\
-\sum\limits_{i=1}^{N} R_i \frac{1}{2} H_i^2 W_i^2 \\
-\sum\limits_{i=1}^{N} R_i \frac{1}{2} K_i^2 W_i^2 \\
-\sum\limits_{i=1}^{N} R_i K_i H_i W_i^2
\end{bmatrix}. \qquad (16)$$

Note that we can solve the interval linear system of equations (13) with the interval Cholesky method (similarly as in the case of the linear system of equations (8) when the conventional Cholesky method is proposed). Such a choice has an important consequence that can be easily noticed when we follow the assumptions and the theorem provided by Moore, Kearfott and Cloud in [8].

Consider a finite system of linear algebraic equations of the form $Ax = b$, where A is an n-by-n matrix, b is an n-dimensional vector and the coefficients of A and b are real or interval values. The existence of the solution to $Ax = b$ is provided by the following theorem (see [8]).

Theorem 1 (Moore et al. [8]). *If we can carry out all the steps of a direct method for solving $Ax = b$ in the interval arithmetic (if no attempted division by an interval containing zero occurs, nor any overflow or underflow), then the system has a unique solution for every real matrix in A and every real matrix in b, and the solution is contained in the resulting interval vector X.*

Remark 2. In the interval approach proposed, all real value coefficients of the matrix $\widehat{A}$ and the vectors $\widehat{B}$, $\widehat{E}$ of the linear system of equations (8) are included in the interval value coefficients of the matrix A and the vectors B, E of the interval linear system of equations (13). Hence, based on Theorem 1 we can conclude the following. If we solve the interval linear system of equations (13) with the interval Cholesky method (i.e., the interval counterpart of the direct Cholesky method), then the exact values of the derivatives given in $\widehat{D}$ (10) at the node p_0 are included in the interval values of the vector D (15) and we have

$$\frac{\partial u}{\partial x}(p_0) \in D_0^{(X)}, \quad \frac{\partial u}{\partial y}(p_0) \in D_0^{(Y)},$$

$$\frac{\partial^2 u}{\partial x^2}(p_0) \in D_0^{(XX)}, \quad \frac{\partial^2 u}{\partial y^2}(p_0) \in D_0^{(YY)}, \quad \frac{\partial^2 u}{\partial x \partial y}(p_0) \in D_0^{(XY)}. \tag{17}$$

4   Numerical Examples

We consider the following functions

$$u_1(x,y) = \exp(xy), \qquad u_2(x,y) = \left(x^2 + y^2 + 0.5\right)^2 \exp(xy),$$
$$u_3(x,y) = \cos(x)\cos(y), \quad u_4(x,y) = u_a(x,y) + u_b(x,y), \tag{18}$$

where the functions u_a, u_b are defined in [2] and given in the form

$$u_a(x,y) = \frac{3}{4}\exp\left(-\frac{(9x-2)^2}{4} - \frac{(9y-2)^2}{4}\right) + \frac{3}{4}\exp\left(-\frac{(9x+1)^2}{49} - \frac{(9y+1)^2}{10}\right),$$

$$u_b(x,y) = \frac{1}{2}\exp\left(-\frac{(9x-7)^2}{4} - \frac{(9y-3)^2}{4}\right) - \frac{1}{5}\exp\left(-(9x-4)^2 - (9y-7)^2\right).$$

We use the approach proposed in the above section to compute the interval enclosures of the second order derivatives (1) of the functions u_1, u_2, u_3, u_4 at the point $p_0(0.5, 0.5)$. We assume that we know a set of points $p_i = (x_i, y_i)$ and our task is to compute a value of the derivative of some function u at a given point $p_0 = (x_0, y_0)$. We take into account the regular and irregular grids of 9, 17 and 25 points (see Fig. 1). To examine the influence of the distances between the points in the cloud, we define the distances ρ_x, ρ_y that are further used to determine the position of the points p_0, p_i, $i = 1, 2, \ldots, n$.

We performed two kinds of numerical tests. In Example 1 we assume that the analytical formula of u is known. In such a case we can compute the analytical formula of all the derivatives of u given in (4) and then their interval extensions required in (12). Under this assumption the exact values of all derivatives at the point p_0 are included in the interval enclosures obtained. Note that in general, the analytical formula of u is unknown. Hence, in Example 2, we propose some kind of method that shows how we can approximate values of the endpoints of the error term intervals in the formula (12). Note that we performed computations with the C++ libraries for the floating-point conversions and the floating-point interval arithmetic dedicated for the Intel C++ compiler [4].

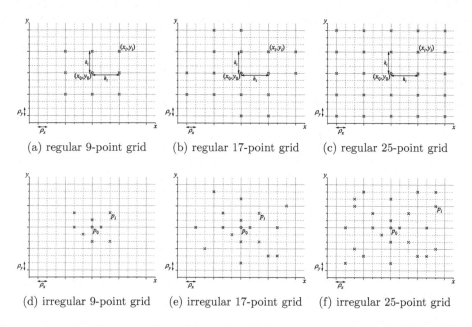

(a) regular 9-point grid   (b) regular 17-point grid   (c) regular 25-point grid

(d) irregular 9-point grid   (e) irregular 17-point grid   (f) irregular 25-point grid

Fig. 1. Examples of regular and irregular grids of n-nodes.

Example 1. Consider the derivatives (1) of the functions u_1, u_3, u_4 at the point $p_0(0.5, 0.5)$. We computed the exact values of these derivatives and then their interval enclosures for all regular/irregular clouds of 9, 17 and 25 points, and a sequence of the grid parameters $\rho_x = \rho_y$ equal to $1E-12, 5E-12, \ldots, 5E-2$. We found the results such that the widths of the interval enclosures were the

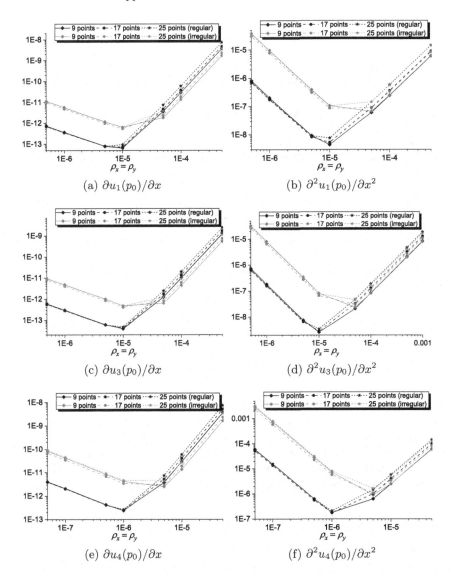

Fig. 2. Widths of interval enclosures of the derivatives $\partial u(p_0)/\partial x$ and $\partial^2 u(p_0)/\partial x^2$ of the functions u_1, u_3, u_4 for different values of the grid parameter $\rho_x = \rho_y$.

smallest in the case of regular and irregular grids. These widths occurred to be of the same order. The comparison in the case of the first and second order derivatives with respect to the x coordinate is shown in Fig. 2.

Example 2. Let us compute the derivatives (1) of the function u_2 at the point $p_0(0.5, 0.5)$ using a method of approximation of the error term intervals. In the formula (12), we assume that we know the interval enclosures of the third order

Table 1. Exact values of the derivatives, their interval enclosures obtained with the approximation of the endpoints of the error term intervals and the widths of intervals

Deriv.	Interval enclosure of the derivative	Width
$\partial u/\partial x$	[3.21006354171910379E+0000, 3.21006354171960945E+0000] exact $\approx$ 3.21006354171935371E+0000	5.0565E−13
$\partial u/\partial y$	[3.21006354171910369E+0000, 3.21006354171960958E+0000] exact $\approx$ 3.21006354171935371E+0000	5.0587E−13
$\partial^2 u/\partial x^2$	[1.05932096499612316E+0001, 1.05932097153302133E+0001] exact $\approx$ 1.05932096876738672E+0001	6.5368E−08
$\partial^2 u/\partial y^2$	[1.05932096499385176E+0001, 1.05932097153178944E+0001] exact $\approx$ 1.05932096876738672E+0001	6.5379E−08
$\partial^2 u/\partial x\partial y$	[6.74113340305148239E+0000, 6.74113345648631275E+0000] exact $\approx$ 6.74113343761064279E+0000	5.3434E−08

derivatives such that for a given point q_i the following relations hold

$$\frac{\partial^3 u}{\partial x^3}(q_i) \in D_i^{(3,1)} = \left[\underline{D}_i^{(3,1)}, \overline{D}_i^{(3,1)}\right], \quad \frac{\partial^3 u}{\partial y^3}(q_i) \in D_i^{(3,2)} = \left[\underline{D}_i^{(3,2)}, \overline{D}_i^{(3,2)}\right],$$

$$\frac{\partial^3 u}{\partial x^2 \partial y}(q_i) \in D_i^{(3,3)} = \left[\underline{D}_i^{(3,3)}, \overline{D}_i^{(3,3)}\right], \quad \frac{\partial^3 u}{\partial x \partial y^2}(q_i) \in D_i^{(3,4)} = \left[\underline{D}_i^{(3,4)}, \overline{D}_i^{(3,4)}\right].$$

If the analytical formulas of the third order derivatives are not known, we have to approximate the endpoints of the error term intervals. One approach assumes that we compute the derivatives up to the third order using the conventional generalized finite differences of higher order (see, e.g., [10]) and then we use the results obtained to approximate the endpoints considered in the way similar to the one proposed in, e.g., [5]. For $k = 1, 2, 3, 4$, we choose

$$\underline{D}_i^{(3,k)} \approx \min\left\{D_i^{(3,k)*}, D_0^{(3,k)*}\right\}, \quad \underline{D}_i^{(3,k)} \approx \max\left\{D_i^{(3,k)*}, D_0^{(3,k)*}\right\}, \quad (19)$$

where, for $s = i$ and $s = 0$, we take

$$D_s^{(3,1)*} = \frac{\partial^3 u(p_s)}{\partial x^3}, \quad D_s^{(3,2)*} = \frac{\partial^3 u(p_s)}{\partial y^3}, \quad D_s^{(3,3)*} = \frac{\partial^3 u(p_s)}{\partial x^2 \partial y}, \quad D_s^{(3,4)*} = \frac{\partial^3 u(p_s)}{\partial x \partial y^2}.$$

We chose the regular 25-grid of points with $\rho_x = \rho_y = 5\text{E}{-6}$. We computed the second order derivatives of u_2 at the point $p_0(0.5, 0.5)$ with the approximated values of the endpoints of the error term intervals. As we can see in Table 1, the exact values of the derivatives are included in the corresponding interval results. The similar situation was observed also in the case of other examples considered. Nevertheless, in this approach, we cannot formally guarantee the inclusion of the truncation error in the resultant interval solutions.

5   Results Discussion and Final Conclusions

The results obtained with the IGFDs lead to the following general conclusions.

- The interval solution includes the exact value of the derivative in the case of all functions and each numerical experiment. For each example function u and each number of grid points 9, 17, 25, we can find the grid parameter $\rho_x = \rho_y$ such that the widths of interval solutions are the smallest. Their further decrease does not improve the results or even makes them worse.
- The smallest widths of interval solutions are usually obtained with the 9-point grid of points. The larger number of the grid points improves the results when we take $\rho_x = \rho_y$ much smaller than the optimal one. Hence, it seems that there is no reason to perform computations with 17 or even 25 points.
- The widths of interval solutions are smaller in the case of the regular arrangement of points than the irregular one (in the case of each example function such a difference is equal to about one order of accuracy). Nevertheless, the regular distribution is rarely possible near irregular and complicated boundary. In such a case the interval GFDs are very useful.

The results seem to be a promising starting point towards development of interval generalized finite difference methods for solving the boundary value problems.

Acknowledgments. The paper was supported by the Poznan University of Technology (Poland) through Grants No. 02/21/DSPB/3544, 09/91/DSPB/1649.

References

1. Benito, J.J., Urena, F., Gavete, L.: Solving parabolic and hyperbolic equations by the generalized finite diference method. J. Comput. Appl. Math. **209**(2), 208–233 (2007)
2. Chen, W., Fu, Z.J., Chen, C.S.: Recent Advances in Radial Basis Function Collocation Methods. Springer, Heidelberg (2014). https://doi.org/10.1007/978-3-642-39572-7
3. Hoffmann, T., Marciniak, A., Szyszka, B.: Interval versions of central difference method for solving the Poisson equation in proper and directed interval arithmetic. Found. Comput. Decis. Sci. **38**(3), 193–206 (2013)
4. Jankowska, M.A.: Remarks on algorithms implemented in some C++ libraries for floating-point conversions and interval arithmetic. In: Wyrzykowski, R., Dongarra, J., Karczewski, K., Wasniewski, J. (eds.) PPAM 2009. LNCS, vol. 6068, pp. 436–445. Springer, Heidelberg (2010). https://doi.org/10.1007/978-3-642-14403-5_46
5. Jankowska, M.A., Sypniewska-Kaminska, G.: Interval finite difference method for solving the one-dimensional heat conduction problem with heat sources. In: Manninen, P., Öster, P. (eds.) PARA 2012. LNCS, vol. 7782, pp. 473–488. Springer, Heidelberg (2013). https://doi.org/10.1007/978-3-642-36803-5_36
6. Jensen, P.: Finite difference techniques for variable grids. Comput. Struct. **2**(1–2), 17–29 (1972)
7. Mochnacki, B., Piasecka-Belkhayat, A.: Numerical modeling of skin tissue heating using the interval finite difference method. MCB Mol. Cell. Biomech. **10**(3), 233–244 (2013)

8. Moore, R.E., Kearfott, R.B., Cloud, M.J.: Introduction to Interval Analysis. SIAM, Philadelphia (2009)

9. Orkisz, J.: Meshless finite difference method I - basic approach. Meshless finite difference method II - adaptative approach. In: Idelsohn, S.R., Oñate, E.N., Dvorkin, E. (eds.) Computational Mechanics. New Trends and Applications. CIMNE (1998)

10. Urena, F., Salete, E., Benito, J.J., Gavete, L.: Solving third- and fourth-order partial differential equations using GFDM: application to solve problems of plates. Int. J. Comput. Math. **89**(3), 366–376 (2012)

An Interval Calculus Based Approach to Determining the Area of Integration of the Entropy Density Function

Bartłomiej Jacek Kubica[(⊠)] and Arkadiusz Janusz Orłowski

Institute of Information Technology, Warsaw University of Life Sciences – SGGW,
ul. Nowoursynowska 159, 02-776 Warsaw, Poland
{bartlomiej_kubica,arkadiusz_orlowski}@sggw.pl

Abstract. This paper considers the problem of numerical computation
of a definite integral of the entropy density function over a wide, potentially unbounded, domain. Despite there are efficient approaches to compute the quadrature over a finite (hyper)rectangle, it may be challenging
to bound the whole domain, out of which the function is negligible. An
approach based on the interval analysis is proposed in this paper. Preliminary numerical results are also discussed.

Keywords: Interval computations · Entropy · Integration · Numerical
quadrature · Heuristics

1   Introduction

Several integrals, found in various branches of science and engineering, cannot
be computed exactly, but only approximated using some numerical procedures.
Numerical integration is a well-known problem and several approaches exist to
solving it.

In many applications, the area of integration is unbounded. To perform any
kind of numerical algorithm, we need to provide some bounds or other kind
of enclosure for the support of the function. In particular, this is the case for
applications, where entropy density functions have to be integrated.

The paper is organized as follows. Problems of integrating the entropy function are briefly presented in Sect. 2. In Sect. 3, the numerical procedure for this
purpose is discussed. In Sect. 4, basics of the interval calculus are introduced.
Also operations on unbounded intervals are presented there. In Sect. 5, an interval branch-and-bound type algorithm to enclose the approximate support of the
integrated function is proposed. Its implementation is considered in Sect. 6. Many
issues, including the use of algorithmic differentiation, are given there. Section 7
presents and discusses the preliminary numerical results and the conclusions are
given in Sect. 8.

© Springer Nature Switzerland AG 2020
R. Wyrzykowski et al. (Eds.): PPAM 2019, LNCS 12044, pp. 395–406, 2020.
https://doi.org/10.1007/978-3-030-43222-5_35

2 Integrating the Entropy Density Function

Majority of interesting and nontrivial problems in physics cannot be solved analytically. It is necessary to use some kind of numerical procedures. Even computing definite integrals can be a challenging problem, especially in multidimensional
cases with nontrivial boundary conditions. In this paper we focus our attention
on computation of the values of phase-space entropy functions

$$S = - \iint\limits_{A} \mathrm{d}\mu(x_1, x_2) f(x_1, x_2) \ln f(x_1, x_2) \,, \tag{1}$$

where $f(x_1, x_2)$ is a probability distribution defined over a phase-space of the
physical system under consideration, $\mu(x_1, x_2)$ is a proper integration measure,
and A – integration area.

As an example, let us consider so-called Husimi function or Glauber representation of a quantum state of the one-dimensional harmonic oscillator. In
contrast to its more famous cousin, the Wigner function, the Husimi function is
always non-negative and can be used as an argument of the logarithm. As both
entropy and harmonic oscillator are ubiquitous concepts in so many different
physical contexts (cf., e.g., [8,15–17]), we find this example particularly enlight-
ening and valuable. Let us consider the Husimi function of the nth excited state
of harmonic oscillator, which in proper units reads

$$f_n(x_1, x_2) = \frac{1}{n!}(x_1^2 + x_2^2)^n \exp(-x_1^2 - x_2^2) \,. \tag{2}$$

In the following we concentrate on three different quantum states, presenting
both $f_n(x_1, x_2)$ and $g_n(x_1, x_2) = -f_n(x_1, x_2) \ln f_n(x_1, x_2)$ for the first, second,
and third excited state. We take $\mathrm{d}\mu(x_1, x_2) = \frac{1}{\pi}\mathrm{d}x_1\mathrm{d}x_2$ for convenience.

Graphs of these functions are presented in Fig. 1 and their corresponding
entropy densities – in Fig. 2.

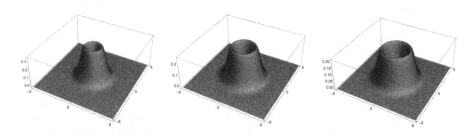

Fig. 1. Graphs of functions (2), for $n = 1, 2, 3$.

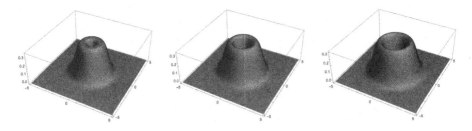

Fig. 2. Graphs of entropy densities of functions from Fig. 1.

3 Numerical Integration

Various approaches to numerically approximating the definite integral exist, and
their various efficient and well-tuned implementations are commonly used. In
particular, in the Python library numpy for numerical computations, we have
functions quad(), dblquad(), tplquad() and nquad() for numerical integration
over domains of various dimensionalities [1].

These functions are very efficient. They have not been implemented in
Python, but they are Python wrappers for functions written in C or Fortran,
basing on highly-tuned numerical libraries. Nevertheless, they require explicitly
giving bounds of integration.

It may seem a sane approach to simply provide wide approximate bounds.
Yet this can cause two kinds of problems:

- we can miss a non-negligible region of the function's support,
- we can make our procedure perform superfluous computations on relatively
  wide regions, where the function is negligible.

Yet, to determine where the function has non-negligible values we need either to
compute its values in an outstanding number of points or determine ranges of
its values on some regions. The latter can be achieved using interval analysis.

4 The Interval Calculus

The idea of interval analysis is to perform computations using intervals, instead
of specific numbers. It can be found in several textbooks; e.g., [4, 7, 9, 13, 14, 19].

We define the (closed) interval $[\underline{x}, \overline{x}]$ as a set $\{x \in \mathbb{R} \mid \underline{x} \le x \le \overline{x}\}$. It
is assumed that all considered interval are closed (why such an assumption is
made has been discussed, i.a., in Sect. 2.8 of [11]).

Following [10], we use boldface lowercase letters to denote interval variables,
e.g., **x**, **y**, **z**, and $\mathbb{IR}$ denotes the set of all real intervals.

4.1 Basics

We design arithmetic (and other) operations on intervals so that the following
condition was fulfilled:

$$\odot \in \{+, -, \cdot, /\}, \ a \in \mathbf{a}, \ b \in \mathbf{b} \ \text{ implies } \ a \odot b \in \mathbf{a} \odot \mathbf{b} . \tag{3}$$

The actual formulae for arithmetic operations are as follows:

$$[\underline{a}, \overline{a}] + [\underline{b}, \overline{b}] = [\underline{a} + \underline{b}, \overline{a} + \overline{b}] ,$$
$$[\underline{a}, \overline{a}] - [\underline{b}, \overline{b}] = [\underline{a} - \overline{b}, \overline{a} - \underline{b}] ,$$
$$[\underline{a}, \overline{a}] \cdot [\underline{b}, \overline{b}] = [\min(\underline{a}\underline{b}, \underline{a}\overline{b}, \overline{a}\underline{b}, \overline{a}\overline{b}), \max(\underline{a}\underline{b}, \underline{a}\overline{b}, \overline{a}\underline{b}, \overline{a}\overline{b})] ,$$
$$[\underline{a}, \overline{a}] / [\underline{b}, \overline{b}] = [\underline{a}, \overline{a}] \cdot [1 / \overline{b}, 1 / \underline{b}] , \qquad 0 \notin [\underline{b}, \overline{b}].$$

(4)

Other operations can be defined in a similar manner. For instance, we have a formula for the power of an interval:

$$[\underline{a}, \overline{a}]^n = \begin{cases} [\underline{a}^n, \overline{a}^n] & \text{for odd } n \\ [\min\{\underline{a}^n, \overline{a}^n\}, \max\{\underline{a}^n, \overline{a}^n\}] & \text{for even } n \text{ and } 0 \notin [\underline{a}, \overline{a}] \\ [0, \max\{\underline{a}^n, \overline{a}^n\}] & \text{for even } n \text{ and } 0 \in [\underline{a}, \overline{a}] \end{cases} .$$

(5)

and for other transcendental functions. For instance:

$$\exp\left([\underline{a}, \overline{a}]\right) = [\exp(\underline{a}), \exp(\overline{a})],$$
$$\log\left([\underline{a}, \overline{a}]\right) = [\log(\underline{a}), \log(\overline{a})], \text{ for } \underline{a} > 0 .$$

Links between real and interval functions are set by the notion of an *inclusion function*: see; e.g., [7]; also called an *interval extension*; e.g., [9].

Definition 1. *A function* $f: \mathbb{IR} \to \mathbb{IR}$ *is an* inclusion function *of* $f: \mathbb{R} \to \mathbb{R}$, *if for each interval* **x** *within the domain of* f *the following condition is satisfied:*

$$\{f(x) \mid x \in \mathbf{x}\} \subseteq f(\mathbf{x}) .$$

The definition is analogous for functions $f: \mathbb{R}^n \to \mathbb{R}^m$.

For each interval, we can define its midpoint:

$$\operatorname{mid} \mathbf{x} = \frac{\underline{x} + \overline{x}}{2} .$$

(6)

This operation is of particular importance, as it allows us to subdivide intervals into subintervals. This is a crucial operation for all branch-and-bound type methods [11].

4.2   Unbounded Intervals

Formulae (4), although well-known, are less universal than they would seem. If the endpoints are represented using the IEEE 754 Standard for floating-point numbers [5], these endpoints can have infinite values. So what would be the result of the following multiplication: $[0, 2] \cdot [2, +\infty]$? According to Formulae (4), we obtain a NaN (Not a Number) for the right endpoint, but we can simply bound the set:

$$\{z = x \cdot y \mid x \in [0, 2], \ y \in [2, +\infty]\} ;$$

its bounds are obviously: $[0, +\infty]$. Various interval libraries and packages implement such operations in different manners; some unification has been provided by the IEEE Standard 1788-2015 [6].

Subdivision of an unbounded interval is another issue. According to (6), for unbounded intervals we would obtain:

$$\text{mid}\,[a, +\infty] = +\infty \ ,$$
$$\text{mid}\,[-\infty, a] = -\infty \ ,$$
$$\text{mid}\,[-\infty, +\infty] = \text{NaN},$$

which would not be very useful.

In [18], Ratschek and Voller present another approach. A global parameter λ is chosen, such that solutions are supposed to lie in the range $[-\lambda, +\lambda]$. For intervals contained in this range, the midpoint and width are defined in a traditional manner.

For intervals exceeding this range, we have:

$$\text{mid}\,[a, +\infty] = \begin{cases} \lambda \text{ for } a < \lambda, \\ 2 \cdot a \text{ for } a \geq \lambda, \end{cases} \cdot \tag{7}$$

Obviously, $\text{mid}\,[-\infty, a] = -\text{mid}\,[-a, +\infty]$.

Please note that a in formula (7) can be both finite or infinite. Also, it is worth noting that, according to the above equation:

$$\text{mid}\,[-\infty, +\infty] = +\lambda \ ,$$

so the interval $[-\lambda, +\lambda]$ will not get subdivided too early; we shall start with "amputating" the regions from its outside.

Ratschek and Voller have proposed alternative formulae for the diameter of an unbounded interval, as well. These formulae seem more controversial and we shall not use them. Fortunately, we do not need to distinguish, which of the unbounded intervals is "wider" – as long, as we can subdivide any of them, and formula (7) is sufficient for that.

5   Proposed Algorithms

We need to find the approximate support of the non-negative function g (the entropy of f) over its domain: $\mathbb{R}^2$, in our current implementation or $\mathbb{R}^n$, in general. This is equivalent to seeking solutions of the inequality:

$$g(x) \geq \delta, \quad x \in \mathbb{R}^n \ . \tag{8}$$

According to the general approach presented in [11], and making use of the operations defined in Sect. 4.2, solution to (8) can be done by simple procedure, presented in Algorithm 1.

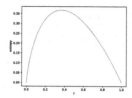

Fig. 3. Entropy of function f, for $f \in [0,1]$

Yet solving (8) directly, would not be the optimal approach. Please note that we know the structure of g: $g(x) = -f(x) \cdot \ln f(x)$. So, we can reduce solving the inequality on g to solving an equivalent inequality (or inequalities system) on f.

Please not that we know the relation between f and g – it is the function $g = -f \cdot \ln f$ and it is shown on Fig. 3.

Consequently, $g(x) \geq \delta$ is equivalent to:

$$f(x) \geq f_l \text{ and } f(x) \leq f_h , \tag{9}$$

for f_l and f_h being the two solutions to the equation $-f \cdot \ln f = \delta$.

These values can easily be computed by any interval or non-interval procedure. Even as simple linear procedure, checking subsequent values of $0 + k \cdot \varepsilon$ and $1 - k \cdot \varepsilon$ for $k = 0, 1, 2, \ldots$ is efficient enough.

Actually, in the considered application, we could neglect the distinction between verified and possible solutions, as all of them are going to be treated in the same manner: fed as regions of the domain to the integration algorithm.

Also, in contrast to other typical applications of the branch-and-bound type methods, ε in Algorithm 1 does not have to be small. The algorithms is supposed to produce the set of boxes that will enclose the support of some function, but the overestimation of this region is not dangerous.

A parameter that should have a smaller value is δ. Actually, using any $\delta > 0$ value thwarts us from computing the verified enclosure of the integral, as some parts of the support region will not be taken into account. Possibly, for some problems, it could be repaired by adding some additional margin $[-\rho, +\rho]$ to the computed integral. In our current implementation, we simply accept that the computed integral will not be verified; actually, non-interval algorithms are used for its computation.

An even more serious issue is to verify the conditions in lines 9 and 11.

5.1   The Initial Box

What initial box $\mathbf{x}^{(0)}$ should we consider? The optimal approach would be to use the whole real plane: $\mathbf{x}^{(0)} = \mathbb{R}^n$.

While techniques from Subsect. 4.2 make it technically possible, such an approach might be cumbersome for several forms of the function f.

Let us consider a simple example. For the function $f_1(x) = \exp(-x)$, we can easily get finite bounds, on any unbounded interval $\mathbf{x} = [\underline{x}, +\infty]$. But for a

---

Algorithm 1. Interval branch-and-bound type algorithm for Problem (8)

Require: $\mathbf{x}^{(0)}, \mathbf{f}, \delta, \varepsilon$

1: {$\mathbf{x}^{(0)}$ is the initial box (cf. Subsection 5.1)}
2: {$\mathbf{f}(\cdot)$ is the interval extension of the function $f : \mathbb{R}^n \to \mathbb{R}$}
3: {L_{ver} – verified solution boxes, L_{pos} – possible solution boxes}
4: $L_{ver} = L_{pos} = \emptyset$
5: find f_l and f_h, as the two solutions of $-f \cdot \ln f = \delta$ for $f \in [0, 1]$
6: $\mathbf{x} = \mathbf{x}^{(0)}$
7: **loop**
8:     compute $[\underline{y}, \overline{y}] = \mathbf{f}(\mathbf{x})$
9:     **if** ($\overline{y} < f_l$ or $\underline{y} > f_h$) **then**
10:         discard $\mathbf{x}$
11:     **else if** ($\underline{y} \geq f_l$ and $\overline{y} \leq f_h$) **then**
12:         push (L_{ver}, $\mathbf{x}$)
13:     **else if** (wid $\mathbf{x} < \varepsilon$) **then**
14:         push (L_{pos}, $\mathbf{x}$) {The box $\mathbf{x}$ is too small for bisection}
15:     **if** ($\mathbf{x}$ was discarded **or** $\mathbf{x}$ was stored) **then**
16:         **if** ($L == \emptyset$) **then**
17:             **return** L_{ver}, L_{pos} {All boxes have been considered}
18:         $\mathbf{x} = \text{pop}(L)$
19:     **else**
20:         bisect ($\mathbf{x}$), obtaining $\mathbf{x}^{(1)}$ and $\mathbf{x}^{(2)}$
21:         $\mathbf{x} = \mathbf{x}^{(1)}$
22:         push (L, $\mathbf{x}^{(2)}$)

---

pretty similar function $f_2(x) = x \cdot \exp(-x)$, on a similar interval, the interval evaluation will result in:

$$\mathbf{f}_2\left([\underline{x}, +\infty]\right) = [\underline{x}, +\infty] \cdot \exp\left([-\infty, -\underline{x}]\right) = [\underline{x}, +\infty] \cdot [0, \exp(-\underline{x})] = [0, +\infty] \,,$$

regardless of the value of $\underline{x}$!

Another version of f_2:

$$f_{2a}(x) = \exp\left(\ln(x) - x\right) \,,$$

is of little help, as it results in:

$$\mathbf{f}_{2a}\left([\underline{x}, +\infty]\right) = \exp\left(\ln\left([\underline{x}, +\infty]\right) - [\underline{x}, +\infty]\right)$$

$$= \exp\left([\ln \underline{x}, +\infty] - [\underline{x}, +\infty]\right) = \exp[-\infty, +\infty] = [0, +\infty] \,,$$

Obtaining a more useful form is not easy. Using derivatives will be of some help, as we shall see in next subsection.

Also, please note that for any of the forms, bounding the function is of no problem for a bounded interval, even for a wide one.

5.2   How to Verify that $f(x) \geq f_l$ and $f(x) \leq f_h$

The very objective of the interval calculus is to bound various quantities. Consequently, it is a proper tool to bound $f(x)$ and allow checking conditions of type (9). For bounded intervals, it works perfectly. But what about unbounded domains?

For some functions, their interval extensions allow checking (9) even over unbounded domains. Assume, we have $x = (x_1, x_2)$ and $f(x) = \exp(-x_1^2 - x_2^2)$. Now, for large enough arguments, the condition that $f(x) \geq f_l$ can easily be falsified.

But checking it will be much more tedious for a more sophisticated function. A function described by an infinite series, that has to be truncated at some point, seems particularly challenging.

What can we do in such a case?

We know that f is non-negative. So, we can make use of its derivatives to make sure it is non-increasing over (or non-decreasing under) some threshold value.

Derivatives can be computed using algorithmic differentiation, as we shall see below.

Hence, we can formulate three conditions to discard a box $\mathbf{x}$:

1. It is verified that $\mathsf{f}(\mathbf{x}) \subseteq]-\infty, +f_l[$.
2. It is verified that $\mathsf{f}(\underline{x}) \subseteq]-\infty, +f_l[$ and $\nabla \mathsf{f}(\mathbf{x}) \in]-\infty, 0]$ (i.e., the function is decreasing).
3. It is verified that $\mathsf{f}(\overline{x}) \subseteq]-\infty, +f_l[$ and $\nabla \mathsf{f}(\mathbf{x}) \in]0, +\infty]$ (the function is increasing).

Unfortunately, none of the above conditions turned out to be easy to verify on unbounded intervals – at least for functions we have considered in this paper. Consequently, in our current implementation we do not use derivatives, but natural interval extensions over (wide but) finite intervals only.

We are going to test other conditions in the future. In particular, conditions basing on the 2nd (or even higher) derivative can be formulated. Possibly, they will be considered in our further research.

6   Implementation

The program computing numerical integrals is currently implemented in Python 3. The quadrature uses function `dblquad()` from module `numpy.integrate`.

The implementation has been parallelized using processes. Today Python translators do not allow threads to run in parallel, because of the presence of the so-called Global Interpreter Lock (GIL), that allows executing only one Python instruction at a time. So, several processes are applied instead, to perform parallel computations. Specifically, the class `ProcessPoolExecutor` from module `concurrent.futures` is used. This class represents a scheduler that allows assigning tasks to a process pool.

Parallelization is straightforward as we already have the domain subdivided into several boxes. Simply the tasks of computing `dblquad()` on many of them are submitted to the `ProcessPoolExecutor` scheduler, it assigns them to various processes and the main program waits for the completion and adds up the results.

As for bounding the support of f, it has been implemented in C++ and not Python, as some libraries have to be used for interval computations (C-XSC [2]) and algorithmic differentiation (ADHC [3]). Communication between the two components is performed using a file; an unnamed pipe or another IPC mechanism would be equally appropriate.

Algorithmic differentiation is based on the following observation: each function evaluated by a computer is described by a computer program, that consists of several elementary operations (arithmetic operations, transcendental functions, etc.). And it is known, how to compute derivatives of such elementary operations. So, we can enhance this program, to compute derivative(s) together with the original function.

This can be done in a few manners, but the ADHC library, described below, overloads the basic arithmetic operations to compute the derivative(s) together with basic values.

For instance, for basic arithmetic operations, we have:

$$\langle u, u' \rangle + \langle v, v' \rangle = \langle u + v, \ u' + v' \rangle \ ,$$
$$\langle u, u' \rangle - \langle v, v' \rangle = \langle u - v, \ u' - v' \rangle \ ,$$
$$\langle u, u' \rangle \cdot \langle v, v' \rangle = \langle u \cdot v, \ u \cdot v' + u' \cdot v \rangle \ ,$$
$$\langle u, u' \rangle / \langle v, v' \rangle = \langle u/v, \ (u' \cdot v - u \cdot v')/v^2 \rangle \ .$$

Transcendental functions, can be extended in an analogous manner.

The library, called above ADHC (Algorithmic Differentiation and Hull Consistency enforcing), has been developed by the first author [3]. It has been described, i.a., in [12] and in Chapter 3 of [11].

Version 1.0 has been used in our experiments. This version has all necessary operations, including the exp function and division of terms (that was not implemented in earlier versions of the package).

7   Numerical Results

In our current implementation, the initial box $\mathbf{x}^{(0)} = [-2000, 2000] \times [-2000, 2000]$ has been used.

All solutions get found so quickly that parallelization of the algorithms has not been considered, yet. In Fig. 4 we can see results for various values of δ and relatively large $\varepsilon = 1$, for function $f(x_1, x_2) = \exp(-x_1^2 - x_2^2)$.

The numbers of verified and possible solution boxes are: 16 and 32, 20 and 32, and 28 and 40, respectively.

In Fig. 5, an analogous result is presented for $\varepsilon = 0.1$. For these parameters, the numbers of boxes are: 216 and 288.

But a large ε is sufficient in most applications.

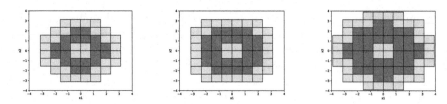

Fig. 4. Approximated support of entropy density of function $f(x) = \exp(-x_1^2 - x_2^2)$, for $\varepsilon = 1$ and $\delta = 10^{-2}, 10^{-3}, 10^{-4}$

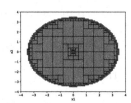

Fig. 5. Approximated support of entropy density of function $f(x) = \exp(-x_1^2 - x_2^2)$, for $\varepsilon = 0.1$ and $\delta = 10^{-4}$

For the family (2) of functions, we present results for $n = 1, 2, 3$ in Fig. 6. Parameters $\varepsilon = 1.0$ and $\delta = 10^{-4}$ have been used there.

The numbers of verified and possible solution boxes have been: 20 and 52, 16 and 92, and 32 and 92, respectively.

Computation times have been negligible, in all above cases.

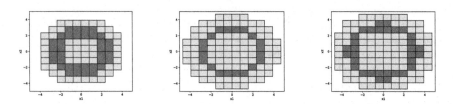

Fig. 6. Approximated support of entropy densities of functions (2), for $\varepsilon = 1, \delta = 10^{-4}$, and $n = 1, 2, 3$

8 Summary

We have provided a simple interval algorithm to compute a relatively narrow enclosure for the support of a function. The algorithm is used for numerical integration of the entropy density function, in some physical applications.

The approach described in this paper, while pretty simple, seems interesting, at least because of two reasons. Firstly, it has a useful practical application in quantum physics. Secondly, it shows a non-obvious manner of using the interval calculus, which has been applied not for validated computations, resulting in a verified enclosure of some solution, but to provide a heuristic for a non-interval procedure.

Further development of the obtained program is planned. In particular, some symbolic transformations will hopefully allow us to use unbounded domains for non trivial functions. Also, we are going to test it on more serious test problems.

An interesting side observation is that operations on unbounded intervals turn out to be less useful than they may seem. Using wide but bounded intervals is much more likely to provide meaningful results.

References

1. Documentation of scipy.integrate Python module (2019). https://docs.scipy.org/doc/scipy/reference/integrate.html
2. C++ eXtended Scientific Computing library (2017). http://www.xsc.de
3. ADHC, C++ library (2018). https://www.researchgate.net/publication/3166104 15_ADHC_Algorithmic_Differentiation_and_Hull_Consistency_Alfa-05
4. Hansen, E., Walster, W.: Global Optimization Using Interval Analysis. Marcel Dekker, New York (2004)
5. IEEE. 754–2008 - IEEE standard for floating-point arithmetic (2008). http://ieeexplore.ieee.org/document/4610935/
6. IEEE. 1788–2015 - IEEE standard for interval arithmetic (2015). http://standards.ieee.org/findstds/standard/1788-2015.html
7. Jaulin, L., Kieffer, M., Didrit, O., Walter, E.: Applied Interval Analysis. Springer, London (2001). https://doi.org/10.1007/978-1-4471-0249-6_2
8. Jex, I., Orłowski, A.: Wehrl's entropy dynamics in a Kerr-like medium. J. Mod. Opt. **41**(12), 2301–2306 (1994)
9. Kearfott, R.B.: Rigorous Global Search: Continuous Problems. Kluwer, Dordrecht (1996)
10. Kearfott, R.B., Nakao, M.T., Neumaier, A., Rump, S.M., Shary, S.P., van Hentenryck, P.: Standardized notation in interval analysis. Vychislennyie Tiehnologii (Comput. Technol.) **15**(1), 7–13 (2010)
11. Kubica, B.J.: Interval Methods for Solving Nonlinear Constraint Satisfaction, Optimization and Similar Problems: From Inequalities Systems to Game Solutions. Studies in Computational Intelligence, vol. 805. Springer, Heidelberg (2019). https://doi.org/10.1007/978-3-030-13795-3
12. Kubica, B.J., Kurek, J.: Interval arithmetic, hull-consistency enforcing and algorithmic differentiation using a template-based package. In: CPEE 2018 Proceedings (2018)
13. Kulisch, U.: Computer Arithmetic and Validity - Theory Implementation and Applications. De Gruyter, Berlin (2008)
14. Moore, R.E., Kearfott, R.B., Cloud, M.J.: Introduction to Interval Analysis. SIAM, Philadelphia (2009)
15. Orłowski, A.: Classical entropy of quantum states of light. Phys. Rev. A **48**, 727–731 (1993)
16. Orłowski, A.: Information entropy and squeezing of quantum fluctuations. Phys. Rev. A **56**(4), 2545 (1997)
17. Orłowski, A.: Wehrl's entropy and classification of states. Rep. Math. Phys. **43**(1–2), 283–289 (1999)

18. Ratschek, H., Voller, R.L.: Global optimization over unbounded domains. SIAM J. Control Optim. **28**(3), 528–539 (1990)
19. Shary, S.P.: Finite-dimensional Interval Analysis. Institute of Computational Technologies, SB RAS, Novosibirsk (2013)

An Interval Difference Method of Second Order for Solving an Elliptical BVP

Andrzej Marciniak[1,2]([⊠]), Malgorzata A. Jankowska[3], and Tomasz Hoffmann[4]

[1] Institute of Computing Science, Poznan University of Technology,
Piotrowo 2, 60-965 Poznan, Poland
andrzej.marciniak@put.poznan.pl
[2] Department of Computer Science, State University of Applied Sciences in Kalisz,
Poznanska 201-205, 62-800 Kalisz, Poland
[3] Institute of Applied Mechanics, Poznan University of Technology,
Jana Pawla II 24, 60-965 Poznan, Poland
malgorzata.jankowska@put.poznan.pl
[4] Poznan Supercomputing and Networking Center,
Jana Pawla II 10, 61-139 Poznan, Poland
tomhof@man.poznan.pl

Abstract. In the article we present an interval difference scheme for solving a general elliptic boundary value problem with Dirichlet' boundary conditions. The obtained interval enclosure of the solution contains all possible numerical errors. A numerical example we present confirms that the exact solution belongs to the resulting interval enclosure.

Keywords: Interval difference methods · Elliptic boundary value problem · Floating-point interval arithmetic

1 Introduction

It is well-known that floating-point arithmetic causes rounding errors, both for the representation of real numbers and for the result of operations. Applying approximate methods to solve problems on a computer we introduce also the error of methods (usually called the truncation errors). Using interval methods realized in interval floating-point arithmetic we can obtain interval enclosures of solutions which are guaranteed to contain the actual solution.

The first monograph on interval arithmetic has been written by Moore in 1966 [13]. Other researchers has been extended this arithmetic in the following years (see, e.g., [1–3,14,16]). In the so called proper interval arithmetic we have four rules for four basic operations, and the realization of them on computers is based on a simple rule, where left and right end-points of intervals are calculated by using downward and upward roundings (see, e.g., [2])[1].

---

[1] There is also known directed interval arithmetic in which the left-ends of intervals may be greater than the right-end of ones. But it is not the case of our paper – we use only proper intervals.

© Springer Nature Switzerland AG 2020
R. Wyrzykowski et al. (Eds.): PPAM 2019, LNCS 12044, pp. 407–417, 2020.
https://doi.org/10.1007/978-3-030-43222-5_36

In this paper we consider more general elliptic equations with Dirichlet's boundary conditions than in our previous papers [4–8,12]. The generalization consists in taking into account some continuous functions about the second order partial derivatives and in adding a term $c(x, y) \cdot u(x, y)$ into equation, where $c(x, y)$ denotes also such a function.

The paper is divided into four sections. In Sect. 2 we recall shortly the well-known elliptic equation with Dirichlet's boundary conditions which is of our interest. In Sect. 3 (the main section of this paper) we present an interval difference scheme of second order for solving a special problem of this kind. Finally, in Sect. 4, a numerical example is presented. This example is only one of the numerous examples we have solved; in every case the exact solution belongs to the interval enclosure obtained by the method. Since, in our opinion, it is rather impossible to obtain a theoretical proof of this fact, the paper can be treated as an experimental one.

2   Boundary Value Problem for Elliptic Equations

The well-known general form of elliptic partial differential equation is as follows:

$$
\begin{aligned}
a\left(x, y\right) \frac{\partial^{2} u}{\partial x^{2}} + 2g\left(x, y\right) \frac{\partial^{2} u}{\partial x \partial y} + b\left(x, y\right) \frac{\partial^{2} u}{\partial y^{2}} \\
+ 2d\left(x, y\right) \frac{\partial u}{\partial x} + 2e\left(x, y\right) \frac{\partial u}{\partial y} + c\left(x, y\right) u = f\left(x, y\right),
\end{aligned}
\tag{1}
$$

where $u = u(x, y)$, $0 \leq x \leq \alpha$, $0 \leq y \leq \beta$. The functions $a = a(x, y)$, $b = b(x, y)$, $c = c(x, y)$, $d = d(x, y)$, $e = e(x, y)$, $f = f(x, y)$ and $g = g(x, y)$ are arbitrary continuous functions determined in the rectangle $\Omega = (x, y) : 0 \leq x \leq \alpha, 0 \leq y \leq \beta$ fulfilling in the interior of Ω the condition

$$
a\left(x, y\right) b\left(x, y\right) - g^{2}\left(x, y\right) > 0.
$$

For (1) we can consider the Dirichlet boundary conditions of the form

$$
u|_{\Gamma} = \varphi(x, y) = \begin{cases} \varphi_1\left(y\right), & \text{for } x = 0, \\ \varphi_2\left(x\right), & \text{for } y = 0, \\ \varphi_3\left(y\right), & \text{for } x = \alpha, \\ \varphi_4\left(x\right), & \text{for } y = \beta, \end{cases}
\tag{2}
$$

where

$$
\varphi_1\left(0\right) = \varphi_2\left(0\right), \quad \varphi_2\left(\alpha\right) = \varphi_3\left(0\right), \quad \varphi_3\left(\beta\right) = \varphi_4\left(\alpha\right), \quad \varphi_4\left(0\right) = \varphi_1\left(\beta\right).
$$

and $\Gamma = \{(x, y) : x = 0, \alpha \text{ and } 0 \leq y \leq \beta \text{ or } 0 \leq x \leq \alpha \text{ and } y = 0, \beta\}$.

If in (1) we take $a(x, y) = b(x, y) = 1$, $c(x, y) = d(x, y) = e(x, y) = g(x, y) = 0$, then we have the following well-known Poisson equation:

$$
\frac{\partial^{2} u}{\partial x^{2}} + \frac{\partial^{2} u}{\partial y^{2}} = f\left(x, y\right).
$$

Interval difference methods for solving this equation with boundary conditions (2) we have presented in [4–8, 12].

The equation

$$\frac{\partial^2 u}{\partial x^2} + \frac{\partial^2 u}{\partial y^2} + c(x,y)u = f(x,y)$$

is another special kind of elliptic equation of the form (1). In [11] we have constructed interval difference scheme for solving this equation with conditions (2) and compared with Nakao's method [15] based on Galerkin's approximation. As we have shown in [11], our method gives better interval enclosures of the exact solution than the method proposed by Nakao. Note that the Nakao method is not applicable to the Poisson equation (see [11] for details).

In this paper we consider the elliptic differential equation of the form

$$a(x,y)\frac{\partial^2 u}{\partial x^2} + b(x,y)\frac{\partial^2 u}{\partial y^2} + c(x,y)u = f(x,y), \tag{3}$$

in which

$$a(x,y)b(x,y) > 0$$

in the interior of rectangle Ω.

3   An Interval Difference Scheme

Partitioning the interval $[0, \alpha]$ into n equal parts of width h and interval $[0, \beta]$ into m equal parts of width k provides a mean of placing a grid on the rectangle $[0, \alpha] \times [0, \beta]$ with mesh points (x_i, y_j), where $h = \alpha/n$, $k = \beta/m$. Assuming that the fourth order partial derivatives of u exist and using Taylor series in the variable x about x_i and in the variable y about y_j, we can express the Eq. (3) at the points (x_i, y_j) as

$$a_{ij}\left[\delta_x^2 u_{ij} - \frac{h^2}{12}\frac{\partial^4 u}{\partial x^4}(\xi_i, y_j)\right] + b_{ij}\left[\delta_y^2 u_{ij} - \frac{k^2}{12}\frac{\partial^4 u}{\partial y^4}(x_i, \eta_j)\right] + c_{ij}u_{ij} = f_{ij}, \tag{4}$$

where

$$\delta_x^2 u_{ij} = \frac{u_{i+1,j} - 2u_{ij} + u_{i-1,j}}{h^2}, \quad \delta_y^2 u_{ij} = \frac{u_{i,j+1} - 2u_{ij} + u_{i,j-1}}{k^2},$$

$i = 1, 2, \ldots, n-1$; $j = 1, 2, \ldots, m-1$, $v_{ij} = v(x_i, y_j)$ for $v \in \{u, a, b, c, f\}$, and where $\xi_i \in (x_{i-1}, x_{i+1})$, $\eta_j \in (y_{j-1}, y_{j+1})$ are intermediate points, and the boundary conditions (2) as

$$\begin{aligned}
u(0, y_j) &= \varphi_1(y_j), && \text{for } j = 0, 1, \ldots, m, \\
u(x_i, 0) &= \varphi_2(x_i), && \text{for } i = 1, 2, \ldots, n-1, \\
u(\alpha, y_j) &= \varphi_3(y_j), && \text{for } j = 0, 1, \ldots, m, \\
u(x_i, \beta) &= \varphi_4(x_i), && \text{for } i = 1, 2, \ldots, n-1.
\end{aligned} \tag{5}$$

Differentiating (3) with respect to x and y, we have

$$a\frac{\partial^3 u}{\partial x^3} = \frac{\partial f}{\partial x} - \frac{\partial a}{\partial x}\frac{\partial^2 u}{\partial x^2} - \frac{\partial b}{\partial x}\frac{\partial^2 u}{\partial y^2} - b\frac{\partial^3 u}{\partial x \partial y^2} - \frac{\partial c}{\partial x}u - c\frac{\partial u}{\partial x},$$
$$b\frac{\partial^3 u}{\partial y^3} = \frac{\partial f}{\partial y} - \frac{\partial a}{\partial y}\frac{\partial^2 u}{\partial x^2} - a\frac{\partial^3 u}{\partial x^2 \partial y} - \frac{\partial b}{\partial y}\frac{\partial^2 u}{\partial x^2} - \frac{\partial c}{\partial y}u - c\frac{\partial u}{\partial y},$$
(6)

and differentiating again with respect to x and y, we get

$$a\frac{\partial^4 u}{\partial x^4} = \frac{\partial^2 f}{\partial x^2} - \frac{\partial^2 a}{\partial x^2}\frac{\partial^2 u}{\partial x^2} - 2\frac{\partial a}{\partial x}\frac{\partial^3 u}{\partial x^3} - \frac{\partial^2 b}{\partial x^2}\frac{\partial^2 u}{\partial y^2} - 2\frac{\partial b}{\partial x}\frac{\partial^3 u}{\partial x \partial y^2} - b\frac{\partial^4 u}{\partial x^2 \partial y^2}$$
$$- \frac{\partial^2 c}{\partial x^2}u - 2\frac{\partial c}{\partial x}\frac{\partial u}{\partial x} - c\frac{\partial^2 u}{\partial x^2},$$
$$b\frac{\partial^4 u}{\partial y^4} = \frac{\partial^2 f}{\partial y^2} - \frac{\partial^2 a}{\partial y^2}\frac{\partial^2 u}{\partial x^2} - 2\frac{\partial a}{\partial y}\frac{\partial^3 u}{\partial x^2 \partial y} - a\frac{\partial^4 u}{\partial x^2 \partial y^2} - \frac{\partial^2 b}{\partial y^2}\frac{\partial^2 u}{\partial y^2} - 2\frac{\partial b}{\partial y}\frac{\partial^3 u}{\partial y^3}$$
$$- \frac{\partial^2 c}{\partial y^2}u - 2\frac{\partial c}{\partial y}\frac{\partial u}{\partial y} - c\frac{\partial^2 u}{\partial y^2},$$
(7)

Taking into account in (7) the relations (6), we obtain

$$a\frac{\partial^4 u}{\partial x^4} = \frac{\partial^2 f}{\partial x^2} - \frac{2}{a}\frac{\partial a}{\partial x}\frac{\partial f}{\partial x}$$
$$- \left[\frac{\partial^2 a}{\partial x^2} - \frac{2}{a}\left(\frac{\partial a}{\partial x}\right)^2 + c\right]\frac{\partial^2 u}{\partial x^2} - \left(\frac{\partial^2 b}{\partial x^2} - \frac{2}{a}\frac{\partial a}{\partial x}\frac{\partial b}{\partial x}\right)\frac{\partial^2 u}{\partial y^2}$$
$$- 2\left(\frac{\partial b}{\partial x} - \frac{b}{a}\frac{\partial a}{\partial x}\right)\frac{\partial^3 u}{\partial x \partial y^2} - b\frac{\partial^4 u}{\partial x^2 \partial y^2}$$
$$- \left(\frac{\partial^2 c}{\partial x^2} - \frac{2}{a}\frac{\partial a}{\partial x}\frac{\partial c}{\partial x}\right)u - 2\left(\frac{\partial c}{\partial x} - \frac{c}{a}\frac{\partial a}{\partial x}\right)\frac{\partial u}{\partial x}$$
(8)

and

$$b\frac{\partial^4 u}{\partial y^4} = \frac{\partial^2 f}{\partial y^2} - \frac{2}{b}\frac{\partial a}{\partial y}\frac{\partial f}{\partial y}$$
$$- \left[\frac{\partial^2 b}{\partial y^2} - \frac{2}{b}\left(\frac{\partial b}{\partial y}\right)^2 + c\right]\frac{\partial^2 u}{\partial y^2} - \left(\frac{\partial^2 a}{\partial y^2} - \frac{2}{b}\frac{\partial a}{\partial y}\frac{\partial b}{\partial y}\right)\frac{\partial^2 u}{\partial x^2}$$
$$- 2\left(\frac{\partial a}{\partial y} - \frac{a}{b}\frac{\partial b}{\partial y}\right)\frac{\partial^3 u}{\partial x^2 \partial y} - a\frac{\partial^4 u}{\partial x^2 \partial y^2}$$
$$- \left(\frac{\partial^2 c}{\partial y^2} - \frac{2}{b}\frac{\partial b}{\partial y}\frac{\partial c}{\partial y}\right)u - 2\left(\frac{\partial c}{\partial y} - \frac{c}{b}\frac{\partial b}{\partial y}\right)\frac{\partial u}{\partial y}$$
(9)

The Eq. (8) should be considered at (ξ_i, y_j) and the Eq. (9) – at (x_i, η_j).

It is obvious that

$$b\left(\xi_i, y_j\right) = b_{ij} + O\left(h\right), \quad c\left(\xi_i, y_j\right) = c_{ij} + O\left(h\right), \quad \frac{1}{a\left(\xi_i, y_j\right)} = \frac{1}{a_{ij}} + O\left(h\right),$$

$$\frac{\partial^p v}{\partial x^p}\left(\xi_i, y_j\right) = \frac{\partial^p v}{\partial x^p}\left(x_i, y_j\right) + O\left(h\right) = \frac{\partial^p v_{ij}}{\partial x^p} + O\left(h\right),$$

$$a\left(x_i, \eta_j\right) = a_{ij} + O\left(k\right), \quad c\left(x_i, \eta_j\right) = c_{ij} + O\left(k\right), \quad \frac{1}{b\left(x_i, \eta_j\right)} = \frac{1}{b_{ij}} + O\left(k\right),$$

$$\frac{\partial^p v}{\partial y^p}\left(x_i, \eta_j\right) = \frac{\partial^p v}{\partial y^p}\left(x_i, y_j\right) + O\left(k\right) = \frac{\partial^p v_{ij}}{\partial y^p} + O\left(k\right) \tag{10}$$

for $p = 1, 2$ and $v = a, b, c$. Moreover, we have

$$\frac{\partial u}{\partial x}\left(\xi_i, y_j\right) = \frac{\partial u}{\partial x}\left(x_i, y_j\right) + O\left(h\right) = \delta_x u_{ij} + O\left(h\right),$$

$$\frac{\partial^2 u}{\partial x^2}\left(\xi_i, y_j\right) = \frac{\partial^2 u}{\partial x^2}\left(x_i, y_j\right) + O\left(h\right) = \delta_x^2 u_{ij} + O\left(h\right),$$

$$\frac{\partial^2 u}{\partial y^2}\left(\xi_i, y_j\right) = \frac{\partial^2 u}{\partial y^2}\left(x_i, y_j\right) + O\left(h\right) = \delta_y^2 u_{ij} + O\left(k^2\right) + O\left(h\right),$$

$$\frac{\partial u}{\partial y}\left(x_i, \eta_j\right) = \frac{\partial u}{\partial y}\left(x_i, y_j\right) + O\left(k\right) = \delta_y u_{ij} + O\left(k\right), \tag{11}$$

$$\frac{\partial^2 u}{\partial y^2}\left(x_i, \eta_j\right) = \frac{\partial^2 u}{\partial y^2}\left(x_i, y_j\right) + O\left(k\right) = \delta_y^2 u_{ij} + O\left(k\right),$$

$$\frac{\partial^2 u}{\partial x^2}\left(x_i, \eta_j\right) = \frac{\partial^2 u}{\partial x^2}\left(x_i, y_j\right) + O\left(k\right) = \delta_x^2 u_{ij} + O\left(h^2\right) + O\left(k\right),$$

where

$$\delta_x u_{ij} = \frac{u_{i+1,j} - u_{i-1,j}}{2h}, \quad \delta_y u_{ij} = \frac{u_{i,j+1} - u_{i,j-1}}{2k}.$$

Substituting (10) and (11) into (8) and (9), and then substituting the resulting formulas into (4), after some transformations we finally obtain

$$\left(\frac{w_{1ij}}{h^2} - \frac{w_{3ij}}{2h}\right) u_{i-1,j} + \left(\frac{w_{2ij}}{k^2} - \frac{w_{4ij}}{2k}\right) u_{i,j-1}$$

$$- \left(2\frac{w_{1ij}}{h^2} + 2\frac{w_{2ij}}{k^2} - w_{5ij} - w_{6ij} - c_{ij}\right) u_{ij}$$

$$+ \left(\frac{w_{2ij}}{k^2} + \frac{w_{4ij}}{2k}\right) u_{i,j+1} + \left(\frac{w_{1ij}}{h^2} + \frac{w_{3ij}}{2h}\right) u_{i+1,j} \tag{12}$$

$$= f_{ij} + w_{7ij} + O\left(h^3\right) + O\left(k^3\right) + O\left(h^2 k^2\right),$$

where

$$w_{1ij} = a_{ij} + \frac{h^2}{12}\left[\frac{\partial^2 a_{ij}}{\partial x^2} - \frac{2}{a_{ij}}\left(\frac{\partial a_{ij}}{\partial x}\right)^2 + c_{ij}\right] + \frac{k^2}{12}\left(\frac{\partial^2 a_{ij}}{\partial y^2} - \frac{2}{b_{ij}}\frac{\partial a_{ij}}{\partial y}\frac{\partial b_{ij}}{\partial y}\right),$$

$$w_{2ij} = b_{ij} + \frac{h^2}{12}\left(\frac{\partial^2 b_{ij}}{\partial x^2} - \frac{2}{a_{ij}}\frac{\partial a_{ij}}{\partial x}\frac{\partial b_{ij}}{\partial x}\right) + \frac{k^2}{12}\left[\frac{\partial^2 b_{ij}}{\partial y^2} - \frac{2}{b_{ij}}\left(\frac{\partial b_{ij}}{\partial y}\right)^2 + c_{ij}\right],$$

$$w_{3ij} = \frac{h^2}{6}\left(\frac{\partial c_{ij}}{\partial x} - \frac{c_{ij}}{a_{ij}}\frac{\partial a_{ij}}{\partial x}\right), \qquad w_{4ij} = \frac{k^2}{6}\left(\frac{\partial c_{ij}}{\partial y} - \frac{c_{ij}}{b_{ij}}\frac{\partial b_{ij}}{\partial y}\right),$$

$$w_{5ij} = \frac{h^2}{12}\left(\frac{\partial^2 c_{ij}}{\partial x^2} - \frac{2}{a_{ij}}\frac{\partial a_{ij}}{\partial x}\frac{\partial c_{ij}}{\partial x}\right), \qquad w_{6ij} = \frac{k^2}{12}\left(\frac{\partial^2 c_{ij}}{\partial y^2} - \frac{2}{b_{ij}}\frac{\partial b_{ij}}{\partial y}\frac{\partial c_{ij}}{\partial y}\right),$$

$$\begin{aligned}
w_{7ij} = &\ \frac{h^2}{12}\left[\frac{\partial^2 f}{\partial x^2}(\xi_i, y_j) - \frac{2}{a_{ij}}\frac{\partial a_{ij}}{\partial x}\frac{\partial f}{\partial x}(\xi_i, y_j)\right. \\
&\ \left. -2\left(\frac{\partial b_{ij}}{\partial x} - \frac{b_{ij}}{a_{ij}}\frac{\partial a_{ij}}{\partial x}\right)\frac{\partial^3 u}{\partial x\partial y^2}(\xi_i, y_j) - b_{ij}\frac{\partial^4 u}{\partial x^2\partial y^2}(\xi_i, y_j)\right] \\
&\ +\frac{k^2}{12}\left[\frac{\partial^2 f}{\partial x^2}(x_i, \eta_j) - \frac{2}{b_{ij}}\frac{\partial b_{ij}}{\partial y}\frac{\partial f}{\partial y}(x_i, \eta_j)\right. \\
&\ \left. -2\left(\frac{\partial a_{ij}}{\partial y} - \frac{a_{ij}}{b_{ij}}\frac{\partial b_{ij}}{\partial y}\right)\frac{\partial^3 u}{\partial x^2\partial y}(x_i, \eta_j) - a_{ij}\frac{\partial^4 u}{\partial x^2\partial y^2}(x_i, \eta_j)\right].
\end{aligned}$$

From (12) we can obtain an interval method. Let us assume that

$$\left|\frac{\partial^4 u}{\partial x^2\partial y^2}(x, y)\right| \le M, \quad \left|\frac{\partial^3 u}{\partial x^2\partial y}(x, y)\right| \le P, \quad \left|\frac{\partial^3 u}{\partial x\partial y^2}(x, y)\right| \le Q$$

for all (x, y) in Ω, and let $\Psi_1(X, Y)$, $\Psi_2(X, Y)$, $\Xi_1(X, Y)$, $\Xi_2(X, Y)$ denote interval extensions of $\partial f/\partial x(x, y)$, $\partial^2 f/\partial x^2(x, y)$, $\partial f/\partial y(x, y)$, $\partial^2 f/\partial y^2(x, y)$, respectively. Then,

$$\frac{\partial^4 u}{\partial x^2\partial y^2}(x, y) \in [-M, M], \quad \frac{\partial^3 u}{\partial x^2\partial y}(x, y) \in [-P, P], \quad \frac{\partial^3 u}{\partial x\partial y^2}(x, y) \in [-Q, Q]$$

for each (x, y), and

$$\frac{\partial f}{\partial x}(\xi_i, y_j) \in \Psi_1(X_i + [-h, h], Y_j), \quad \frac{\partial^2 f}{\partial x^2}(\xi_i, y_j) \in \Psi_2(X_i + [-h, h], Y_j),$$

$$\frac{\partial f}{\partial y}(x_i, \eta_j) \in \Xi_1(X_i, Y_j + [-k, k]), \quad \frac{\partial^2 f}{\partial y^2}(x_i, \eta_j) \in \Xi_2(X_i, Y_j + [-k, k]),$$

since $\xi_i \in (x_i - h, x_i + h)$ and $\eta_j \in (y_j - k, y_j + k)$. Thus, we have $w_{7ij} \in W_{7ij}$, where

$$
W_{7ij} = \frac{h^2}{12}\left\{ \Psi_2\left(X_i + [-h,h], Y_j\right) - \frac{2}{A_{ij}} D_x A_{ij} \Psi_1\left(X_i + [-h,h], Y_j\right)\right.
$$
$$
\left. -2\left(D_x B_{ij} - \frac{B_{ij}}{A_{ij}} D_x A_{ij}\right)[-Q,Q] - B_{ij}[-M,M]\right\}
$$
$$
+ \frac{k^2}{12}\left\{ \Xi_2\left(X_i, Y_j + [-k,k]\right) - \frac{2}{B_{ij}} D_y B_{ij} \Xi_1\left(X_i, Y_j + [-k,k]\right)\right.
$$
$$
\left. -2\left(D_y A_{ij} - \frac{A_{ij}}{B_{ij}} D_y B_{ij}\right)[-P,P] - A_{ij}[-M,M]\right\}, \tag{13}
$$

where V_{ij} and $D_z V_{ij}$ for $V \in \{A,B\}$ and $z \in \{x,y\}$ denote interval extensions of v_{ij} and $\partial v_{ij}/\partial z$ for $v \in \{a,b\}$, respectively. If we denote interval extensions of f_{ij}, c_{ij} and w_{pij} by F_{ij}, C_{ij} and W_{pij}, respectively ($p = 1,2,\ldots,6$), then from the above considerations and (12) it follows an interval method of the form

$$
\left(\frac{W_{1ij}}{h^2} - \frac{W_{3ij}}{2h}\right)U_{i-1,j} + \left(\frac{W_{2ij}}{k^2} - \frac{W_{4ij}}{2k}\right)U_{i,j-1}
$$
$$
-\left(2\frac{W_{1ij}}{h^2} + 2\frac{W_{2ij}}{k^2} - W_{5ij} - W_{6ij} - C_{ij}\right)U_{ij} \tag{14}
$$
$$
+\left(\frac{W_{2ij}}{k^2} + \frac{W_{4ij}}{2k}\right)U_{i,j+1} + \left(\frac{W_{1ij}}{h^2} + \frac{W_{3ij}}{2h}\right)U_{i+1,j}
$$
$$
= F_{ij} + W_{7ij} + [-\delta,\delta], \quad i = 1,2,\ldots,n-1, \quad j = 1,2,\ldots,m-1,
$$

where the interval $[-\delta,\delta]$, called the δ–extension, represents $O\left(h^3\right) + O\left(k^3\right) + O\left(h^2 k^2\right)$, and where

$$
U_{0j} = \Phi_1\left(Y_j\right), \quad U_{i0} = \Phi_2\left(X_i\right), \quad U_{nj} = \Phi_3\left(Y_j\right), \quad U_{im} = \Phi_4\left(X_i\right),
$$
$$
j = 0,1,\ldots,m, \quad i = 1,2,\ldots,n-1. \tag{15}
$$

Here, $\Phi_1\left(Y\right)$, $\Phi_2\left(X\right)$, $\Phi_3\left(Y\right)$ and $\Phi_4\left(X\right)$ denote interval extensions of $\varphi_1\left(y\right)$, $\varphi_2\left(x\right)$, $\varphi_3\left(y\right)$ and $\varphi_4\left(x\right)$, respectively. The system of linear interval Eq. (14) with (15), with unknowns U_{ij} can be solved in conventional (proper) floating-point interval arithmetic, because all intervals are proper.

It should be added a remark concerning the constants M, P and Q occurring in (13). If nothing can be concluded about M, P and Q from physical or technical properties or characteristics of the problem considered, we proposed to find these constants taking into account that

$$\frac{\partial^4 u}{\partial x^2 \partial y^2}(x_i, y_j) = \lim_{h\to\infty} \lim_{k\to\infty} \left(\frac{u_{i-1,j-1} + u_{i-1,j+1} + u_{i+1,j-1} + u_{i+1,j+1}}{h^2 k^2} \right.$$
$$\left. + \frac{4u_{ij} - 2\left(u_{i-1,j} + u_{i,j-1} + u_{i,j+1} + u_{i+1,j}\right)}{h^2 k^2} \right),$$

$$\frac{\partial^3 u}{\partial x^2 \partial y}(x_i, y_j) = \lim_{h\to\infty} \lim_{k\to\infty} \left(\frac{u_{i-1,j+1} - u_{i-1,j-1} - 2\left(u_{i,j+1} - u_{i,j-1}\right)}{2h^2 k} \right.$$
$$\left. + \frac{u_{i+1,j+1} - u_{i+1,j-1}}{2h^2 k} \right),$$

$$\frac{\partial^3 u}{\partial x \partial y^2}(x_i, y_j) = \lim_{h\to\infty} \lim_{k\to\infty} \left(\frac{u_{i+1,j-1} - u_{i-1,j-1} - 2\left(u_{i+1,j} - u_{i-1,j}\right)}{2h k^2} \right.$$
$$\left. + \frac{u_{i+1,j+1} - u_{i-1,j+1}}{2h k^2} \right).$$

We can calculate

$$M_{nm} = \frac{1}{h^2 k^2} \max_{\substack{i=1,2,\ldots,n-1 \\ j=1,2,\ldots,m-1}} |u_{i-1,j-1} + u_{i-1,j+1} + u_{i+1,j-1} + u_{i+1,j+1}$$
$$+ 4u_{ij} - 2\left(u_{i-1,j} + u_{i,j-1} + u_{i,j+1} + u_{i+1,j}\right)|,$$

$$P_{nm} = \frac{1}{2h^2 k} \max_{\substack{i=1,2,\ldots,n-1 \\ j=1,2,\ldots,m-1}} |u_{i-1,j+1} - u_{i-1,j-1} - 2\left(u_{i,j+1} - u_{i,j-1}\right)$$
$$+ u_{i+1,j+1} - u_{i+1,j-1}|,$$

$$Q_{nm} = \frac{1}{2h k^2} \max_{\substack{i=1,2,\ldots,n-1 \\ j=1,2,\ldots,m-1}} |u_{i+1,j-1} - u_{i-1,j-1} - 2\left(u_{i+1,j} - u_{i-1,j}\right)$$
$$+ u_{i+1,j+1} - u_{i-1,j+1}|,$$

where u_{ij} are obtained by a conventional method for a variety of n and m. Then we can plot M_{nm}, P_{nm} and Q_{nm} against different n and m. The constants M, P and Q can be easy determined from the obtained graphs, since

$$\lim_{\substack{h\to\infty \\ k\to\infty}} M_{nm} \leq M, \quad \lim_{\substack{h\to\infty \\ k\to\infty}} P_{nm} \leq P, \quad \lim_{\substack{h\to\infty \\ k\to\infty}} Q_{nm} \leq Q.$$

4   A Numerical Example

In the example presented we have used our own implementation of floating-point interval arithmetic written in Delphi Pascal. This implementation has been written as a unit called *IntervalArithmetic32and64*, which current version may be found in [9]. The program written in Delphi Pascal for the example considered one can find in [10]. We have run this program on Lenovo® Z51 computer with Intel® Core i7 2.4 GHz processor.

Let $\Omega = [0,1] \times [0,1]$ and consider the following problem:

$$x^2 \sin \pi y \frac{\partial^2 u}{\partial x^2} + y^2 \sin \pi x \frac{\partial^2 u}{\partial y^2} - xyu$$
$$= xy \exp(xy) \left[2xy \sin \frac{\pi(x+y)}{2} \cos \frac{\pi(x+y)}{2} - 1 \right],$$
$$\varphi_1(y) = 1, \quad \varphi_2(x) = 1, \quad \varphi_3(y) = \exp(y), \quad \varphi_4(x) = \exp(x).$$

This problem has the exact solution $u(x,y) = \exp(xy)$. Since the exact solution is known, we can calculate the constants M, P and Q and take $M = 19.03$, $P = Q = 8.16$. These constants can be also estimates from the graphs presented in Fig. 1 (the method still succeeds for less accurate bounds). For $h = k = 0.01$, i.e., $n = m = 100$, and $\delta = 10^{-6}$, using an interval version of LU decomposition, after 6 min we have obtained by our program [10] the results presented in Table 1. Note that using the interval version of full Gauss elimination (with pivoting) we need about 200 days (!) of CPU time to obtain such results.

One can observe that for each (x_i, y_j) the exact solution is within the interval enclosures obtained. It should be added that CPU time grows significantly for greater values of n and m.

The restricted size of this paper does not allow to present other numerical experiments carried out by us. But all these experiments confirm the fact that exact solutions are within interval enclosures obtained by the method (14)–(15).

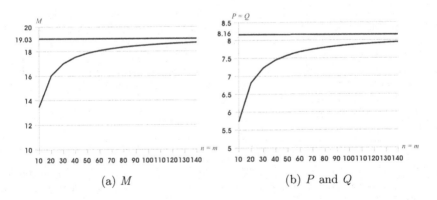

(a) M (b) P and Q

Fig. 1. Estimations of the constants M, P and Q.

Table 1. Enclosures of the exact solution obtained by the method (14)–(15)

(i,j)	U_{ij}	Width
$(0.1, 0.5)$	[1.0509815441511615E+0000, 1.0515636639134466E+0000] exact ≈1.0512710963760240E+0000	≈5.82 · 10⁻⁴
$(0.3, 0.5)$	[1.1616066450304161E+0000, 1.1620625933013910E+0000] exact ≈1.1618342427282831E+0000	≈4.56 · 10⁻⁴
$(0.5, 0.1)$	[1.0509815441511614E+0000, 1.0515636639134467E+0000] exact ≈1.0512710963760240E+0000	≈5.82 · 10⁻⁴
$(0.5, 0.3)$	[1.1616066450304160E+0000, 1.1620625933013911E+0000] exact ≈1.1618342427282831E+0000	≈4.56 · 10⁻⁴
$(0.5, 0.5)$	[1.2838256641451216E+0000, 1.2842221958365290E+0000] exact ≈1.2840254166877415E+0000	≈3.97 · 10⁻⁴
$(0.5, 0.7)$	[1.4189191705563123E+0000, 1.4192105027251595E+0000] exact ≈1.4190675485932573E+0000	≈2.91 · 10⁻⁴
$(0.5, 0.9)$	[1.5682528588111690E+0000, 1.5683682637095705E+0000] exact ≈1.5683121854901688E+0000	≈1.15 · 10⁻⁴
$(0.7, 0.5)$	[1.4189191705563122E+0000, 1.4192105027251595E+0000] exact ≈1.4190675485932573E+0000	≈2.91 · 10⁻⁴
$(0.9, 0.5)$	[1.5682528588111689E+0000, 1.5683682637095705E+0000] exact ≈1.5683121854901688E+0000	≈1.15 · 10⁻⁴

Acknowledgments. The paper was supported by the Poznan University of Technology (Poland) through the Grants No. 09/91/DSPB/1649 and 02/21/ SBAD/3558.

References

1. Alefeld, G., Herzberger, J.: Introduction to Interval Computations. Academic Press, New York (1983)
2. Hammer, R., Hocks, M., Kulisch, U., Ratz, D.: Numerical Toolbox for Verified Computing I. Basic Numerical Problems, Theory, Algorithms, and Pascal-XSC Programs. Springer, Berlin (1993). https://doi.org/10.1007/978-3-642-78423-1
3. Hansen, E.R.: Topics in Interval Analysis. Oxford University Press, London (1969)
4. Hoffmann, T., Marciniak, A.: Finding optimal numerical solutions in interval versions of central-difference method for solving the poisson equation. In: Latuszyńska, M., Nermend, K. (eds.) Data Analysis - Selected Problems, pp. 79–88. Scientific Papers of the Polish Information Processing Society Scientific Council, Szczecin-Warsaw (2013). Chapter 5
5. Hoffmann, T., Marciniak, A.: Solving the Poisson equation by an interval method of the second order. Comput. Methods Sci. Technol. **19**(1), 13–21 (2013)
6. Hoffmann, T., Marciniak, A.: Solving the generalized poisson equation in proper and directed interval arithmetic. Comput. Methods Sci. Technol. **22**(4), 225–232 (2016)
7. Hoffmann, T., Marciniak, A., Szyszka, B.: Interval versions of central difference method for solving the Poisson equation in proper and directed interval arithmetic. Found. Comput. Decis. Sci. **38**(3), 193–206 (2013)
8. Marciniak, A.: An interval difference method for solving the Poisson equation - the first approach. Pro Dialog **24**, 49–61 (2008)
9. Marciniak, A.: Interval Arithmetic Unit (2016). http://www.cs.put.poznan.pl/amarciniak/IAUnits/IntervalArithmetic32and64.pas

10. Marciniak, A.: Delphi Pascal Programs for Elliptic Boundary Value Problem (2019). http://www.cs.put.poznan.pl/amarciniak/IDM-EllipticEqn-Example
11. Marciniak, A.: Nakao's method and an interval difference scheme of second order for solving the elliptic BVS. Comput. Methods Sci. Technol. **25**(2), 81–97 (2019)
12. Marciniak, A., Hoffmann, T.: Interval difference methods for solving the Poisson equation. In: Pinelas, S., Caraballo, T., Kloeden, P., Graef, J.R. (eds.) ICDDEA 2017. SPMS, vol. 230, pp. 259–270. Springer, Cham (2018). https://doi.org/10.1007/978-3-319-75647-9_21
13. Moore, R.E.: Interval Analysis. Prentice-Hall, Englewood Cliffs (1966)
14. Moore, R.E.: Methods and Applications of Interval Analysis. SIAM, Philadelphia (1979)
15. Nakao, M.T.: A numerical approach to the proof of existence of solutions for elliptic problems. Japan J. Appl. Math. **5**, 313–332 (1988)
16. Shokin, Y.I.: Interval Analysis. Nauka, Novosibirsk (1981)

A Parallel Method of Verifying Solutions for Systems of Two Nonlinear Equations

Bartłomiej Jacek Kubica$^{(\boxtimes)}$ and Jarosław Kurek

Department of Applied Informatics, Warsaw University of Life Sciences,
ul. Nowoursynowska 159, 02-776 Warsaw, Poland
bartlomiej_kubica@sggw.pl

Abstract. The paper describes a new algorithm for verifying solutions of nonlinear systems of equations. Interval methods provide us a few tools for such verification. Some of them are based on topological theorems. Also our new test is based on checking the extendability of the function from a subspace of the boundary of the box to its interior. For a system of two equations, we can provide an efficient implementation. Generalization to a higher number of equations is also theoretically possible, yet cumbersome. Some numerical results are presented.

Keywords: Interval computations · Nonlinear systems · Verification · Algebraic topology · Extendability · Multithreading

1 Introduction

Let us consider the problem of solving the nonlinear system of equations $f(x) = 0$, i.e., finding zeros of the function:

$$f\colon X \to \mathbb{R}^m, \text{ where } X \subseteq \mathbb{R}^n \text{ and } n \geq m. \tag{1}$$

Such problems are hard, in general. We can apply Monte Carlo-type methods to seek solutions (e.g., various kinds of genetic algorithms, evolutionary algorithms, evolutionary strategies, differential evolution, etc.). All of them can find the solution (or solutions) with some probability, but are not bound to be successful.

Interval methods (see, e.g., [8,10,19]) are an alternative approach. Unlike randomized methods, they are robust, guaranteed to enclose *all* solutions, even if they are computationally intensive and memory demanding. Their important advantage is allowing not only to locate solutions of well-determined and under-determined systems, but also to *verify* them, i.e., prove that in a given box there is a solution point (resp. a segment of the solution manifold).

2 Generic Algorithm

There are several interval solvers; let us focus on HIBA_USNE [2], developed by the first author. The name HIBA_USNE stands for Heuristical Interval Branch-and-prune Algorithm for Underdetermined and well-determined Systems of Nonlinear Equations and it has been described in a series of papers (including [12–16] and [17]; cf. Chapter 5 of [18] and the references therein).

R. Wyrzykowski et al. (Eds.): PPAM 2019, LNCS 12044, pp. 418–430, 2020.
https://doi.org/10.1007/978-3-030-43222-5_37

Let us present it (the standard interval notation, described in [11], will be used). The solver is based on the branch-and-prune (B&P) schema that can be expressed by pseudocode presented in Algorithm 1.

Algorithm 1. Interval branch-and-prune algorithm

Require: $L, \mathsf{f}, \varepsilon$
1: $\{L$ – the list of initial boxes, often containing a single box $\mathbf{x}^{(0)}\}$
2: $\{L_{ver}$ – verified solution boxes, L_{pos} – possible solution boxes$\}$
3: $L_{ver} = L_{pos} = \emptyset$
4: $\mathbf{x} = \text{pop}\,(L)$
5: **loop**
6:     process the box $\mathbf{x}$, using the rejection/reduction tests
7:     **if** ($\mathbf{x}$ does not contain solutions) **then**
8:         discard $\mathbf{x}$
9:     **else if** ($\mathbf{x}$ is verified to contain a segment of the solution manifold) **then**
10:        push $(L_{ver}, \mathbf{x})$
11:     **else if** (the tests resulted in two subboxes of $\mathbf{x}$: $\mathbf{x}^{(1)}$ and $\mathbf{x}^{(2)}$) **then**
12:        $\mathbf{x} = \mathbf{x}^{(1)}$
13:        push $(L, \mathbf{x}^{(2)})$
14:        **cycle loop**
15:     **else if** ($\text{wid}\,\mathbf{x} < \varepsilon$) **then**
16:        push $(L_{pos}, \mathbf{x})$ {The box $\mathbf{x}$ is too small for bisection}
17:     **if** ($\mathbf{x}$ was discarded **or** $\mathbf{x}$ was stored) **then**
18:        **if** ($L == \emptyset$) **then**
19:           **return** L_{ver}, L_{pos} {All boxes have been considered}
20:        $\mathbf{x} = \text{pop}\,(L)$
21:     **else**
22:        bisect $(\mathbf{x})$, obtaining $\mathbf{x}^{(1)}$ and $\mathbf{x}^{(2)}$
23:        $\mathbf{x} = \mathbf{x}^{(1)}$
24:        push $(L, \mathbf{x}^{(2)})$

---

3   Verification Tools

Interval methods allow Algorithm 1 not only to enclose the solutions, but also to verify them, under proper conditions.

Where are these verification tools used in the solver? They belong to the "rejection/reduction tests" in line 6. Some of such tests can only narrow a box (possibly discarding it completely), but other ones can verify it to contain a solution (or solutions). The interval Newton operator is the most famous of such verification tools, but other ones – like the quadratic approximation test [14] – can be used as well. Other tests are applied in line 16, before putting the box to the list of possible (yet non-verified) solutions. So, line 16 takes the form:

if (**x** gets verified by one of the advanced tests) **then**
    push $(L_{adv\_ver}, \mathbf{x})$
else
    push $(L_{pos}, \mathbf{x})$

We use a third list, different from L_{ver}, to store boxes verified by these new tests (other than the Newton and quadratic tests). Now, let us describe the verification tests.

The most celebrated tool allowing such verification is the aforementioned interval Newton operator; we omit its description as it has already been done in several textbooks (including [10] and [18]).

3.1 Miranda Test

This test, presented, e.g., in [10] is one of the simplest. Consider a continuous function $f: X \to \mathbb{R}$. Assume, we have found two points $a, b \in X$ such that $f(a) > 0$ and $f(b) < 0$. It is well known (from Bolzano's intermediate value theorem), that any curve connecting a and b and lying inside X, on which f is continuous, will contain a point x such that $f(x) = 0$.

Miranda's theorem generalizes this result to a function $f: \mathbb{R}^n \to \mathbb{R}^n$.

Another similar test is also presented in [7].

3.2 Using Quadratic Approximation

This test has been proposed by the author in [14]. As Newton test is based on the linear approximation of f, a quadratic approximation can also be used. Yet, unlike in the Newton case, we do not approximate the whole $f = (f_1, \ldots, f_m)$, but only one of its components f_i. We obtain a representation in the form of a quadratic function; the resulting quadratic equation is used as a constraint. Experiments show that this test can dramatically improve the performance on some problems, but is rather inefficient on other ones. An adequate heuristic for choosing when to use the test is proposed in [14]. Many other details are described there, as well.

3.3 Borsuk Test

This tool, proposed in [6] is based on one of the theorems of Karol Borsuk. The theorem states (slightly simplifying) that the function $f(\cdot)$ must have a zero on the box $\mathbf{x}$, if:

$$f(\text{mid}\,\mathbf{x} + r) \neq \lambda \cdot f(\text{mid}\,\mathbf{x} - r), \text{ for all } \lambda > 0 \text{ and } r \text{ such that } \text{mid}\,\mathbf{x} + r \in \partial\mathbf{x}. \tag{2}$$

For each pair of faces $\mathbf{x}_{i+}$, $\mathbf{x}_{i-}$ of $\mathbf{x}$, we have to compute the intersection of interval expressions:

$$]0, +\infty[\cap \left(\cap_{j=1}^{m} \frac{f_j(\mathbf{x}_{i+})}{f_j(\mathbf{x}_{i-})} \right). \tag{3}$$

If the intersection is empty for at least m pairs of faces, then there is no λ for which the inequality (2) becomes an equality; hence, according to the theorem, f has a zero in $\mathbf{x}$.

In its original formulation [6], the test is used for well-determined problems ($n = m$). Hence, the intersection (3) must be nonempty for all $i = 1, \ldots, n$. In the underdetermined case, it suffices that the intersection is nonempty for m arbitrary values of i. We get back to this topic later in the paper.

3.4 Computing Topological Degree

Topological degree is one of the most general tools to verify existence of zeros of equations.

Let us consider the quantity $deg(f, y, B)$, where $f \colon \mathbb{R}^n \to \mathbb{R}^n$ is a function, y a value from its co-domain (in all our considerations we shall take $y = 0$) and B is a closed connected n-dimensional subset of the f's domain.

Due to the lack of space, we shall not present definition of the degree; it is relatively complicated, but a few algorithms have been developed to compute the topological degree: [4,5,9]. An important property of the degree is its composability: if we have two sets B_1, B_2 such that $B_1 \cap B_2 = \emptyset$ or even $\operatorname{int} B_1 \cap \operatorname{int} B_2 = \emptyset$, then:

$$deg(f, y, B_1 \cup B_2) = deg(f, y, B_1) + deg(f, y, B_2). \tag{4}$$

The algorithm of Franek and Ratschan [5] is based on this feature. It subdivides the set under consideration to obtain sets for which the degrees can be computed more easily.

Why would we want to compute this degree, anyway? Because of its most important property:

$$\text{If } deg(f, y, B) \neq 0, \text{ then } \exists x_0 \in B \ f(x_0) = 0. \tag{5}$$

So, by computing the deg for a specific box, we can obtain the information about the possibility of a solution lying in it.

4   How to Deal with Underdetermined Problems?

All tests considered so far were aimed at verifying zeros of well-determined systems, i.e., such that $f \colon \mathbb{R}^n \to \mathbb{R}^n$. Such systems have discrete solution sets, consisting of isolated points (unless they are ill-posed, e.g., there is a dependency between some equations).

What about underdetermined systems? If $f \colon \mathbb{R}^n \to \mathbb{R}^m$, where $m < n$, then the solution set is typically a manifold, consisting of uncountably many points. Enclosing such a manifold with a collection of boxes is a natural operation and hence interval methods are well-suited to solving such problems.

Hence, we can enclose the solution set, but what about its verification?

In [12], the author described how the interval Newton operator can be used to verify an underdetermined system of equations. Succinctly, if we have m equations in $n > m$ variables, we need to choose a square submatrix of the Jacobi matrix. Treating the chosen m variables normally and the remaining $(n - m)$ ones as parameters, we can perform an normal Newton iteration to narrow, discard or verify the existence of solution *for all* values of the $(n - m)$ "parameters".

A similar procedure can be performed for other above tests, but preliminary experiments of the author have not proved it very useful.

In any case, due to the nature of interval calculus, such an approach can verify boxes where the solution exists *for all* values of the other $(n - m)$ variables. For instance, when we verify a single equation in two variables, such methods can verify a solution segment on the left in Fig. 1, but not the one on the right. For the Borsuk test, as we already have indicated, we can compute (3) for all n variables and check if the condition is fulfilled for at least m of them. This seems a much better approach than choosing m variables in advance.

What about other approaches?

Fig. 1. Left: a solution segment relatively easy to verify using interval tests, right: a harder one

4.1   Obstruction Theory Test

In a series of papers (see, e.g., [4,5] and the references therein), Franek et al. propose a fascinating family of methods targeted specifically at underdetermined systems.

Assume, we have $f \colon \mathbb{R}^n \to \mathbb{R}^m$, where $m < n$ and we want to verify that a box $\mathbf{x}$ contains a segment of the solution manifold. Assume the boundary region of $\mathbf{x}$ does not contain zeros: $\forall x \in \partial \mathbf{x} \; f(x) \neq 0$.

The question is, whether f can be extended from $\partial \mathbf{x}$ to the whole $\mathbf{x}$ without containing a zero.

Let us formulate it differently. As the boundary region of $\mathbf{x}$ does not contain zeros, we can consider the image of f on the boundary as a space homeomorphic to a subset of the $(m-1)$-dimensional sphere S^{m-1}. The boundary $\partial \mathbf{x}$ of $\mathbf{x} \subset \mathbb{R}^n$ itself, is obviously homeomorphic to S^{n-1}.

So, the problem boils down to checking the *extendability* of some function $f \colon S^{n-1} \to S^{m-1}$ from S^{n-1} to the whole disk D^n. Abusing the notation, we do not distinguish between the original f and $f \colon S^{n-1} \to S^{m-1}$. This should not lead to any confusion.

To be succinct, the methods of Franek et al. try to approximate the boundary $\partial \mathbf{x}$ as a cell complex or a simplicial set and they construct a so-called Postnikov complex, built of Eilenberg-MacLane spaces [4]. Basing on this representation, we can check possible extendability of a function for subsequent skeletons of the complex.

This test seems a pretty general tool, suitable for underdetermined problems as well as well-determined ones (in the latter case it is equivalent to using the topological degree). Unfortunately, it is not only based on complicated mathematical notions, but also it seems extremely cumbersome to implement and usually requiring high computational effort. Eilenberg-MacLane spaces have often infinite dimensionality and thus they can only be represented implicitly. And they are only a building block of the algorithm!

No full implementation of this approach is known, at the moment.

Ironically, though the algorithm is so hard (or impossible) to implement, its parallelization should be natural; operations on various simplices (or cells) of the complex can be performed concurrently.

4.2   A New Idea

Instead of proving non-extendability of a function from the whole boundary $\partial \mathbf{x}$ of a box, we can apply another approach. Let us find a subspace $C \subset \partial \mathbf{x}$ of dimension $(m - 1)$ such that f cannot be extended from C to D^n.

What does it simplify? Let us consider the restriction of f to C. Now, we have the domain C and the co-domain homeomorphic to S^{m-1} of f, having the same dimension. So, the problem boils down to checking if the degree of some mapping is nonzero.

For instance if $C \cong S^{m-1}$, we just need to check if $f\colon S^{m-1} \to S^{m-1}$ is homotopic to the identity mapping.

 Obviously, checking this condition is still non-trivial, but we have a few important special cases.

The Case of a Single Equation. Actually, the case of a single equation $f(x) = 0$ is trivial. $S^0 = \partial D^1$ is a discrete set of two disjoint points $\{a, b\}$. If $f(a)$ and $f(b)$ have different signs, we cannot extend f to $[a, b]$ without having some $x_0 \in [a, b]$ such that $f(x_0) = 0$. This is a direct consequence of the intermediate value theorem.

So, if we find on $\partial \mathbf{x}$ two points x_a and x_b such that $f(x_a) \cdot f(x_b) < 0$, any connected line containing these points contains a solution; hence, the box $\mathbf{x}$ is guaranteed to contain it.

The Case of Two Equations. Let us consider the case of $m = 2$ and $n \geq 2$. Now, $\partial \mathbf{x} \cong S^{n-1}$ and range$(f, \partial \mathbf{x}) \cong S^1$.

We need to find C, a closed non-self-intersecting path in $\partial \mathbf{x}$, such that when moving around this path, we revolve around the center making a different number of circles in one than in the opposite direction. Again, this is equivalent to saying that the degree of the mapping (which can also be called the *winding number*, in this case) is different from zero.

How to check this condition? Actually, we can make use of the toolset pretty similar to the one of Franek et al. [5], but we arrange these tools in a quite different manner.

Firstly, let us assume that we have a *sign covering*, i.e., a subdivision of $\partial \mathbf{x}$ such that for each box $\mathbf{t}$ of this subdivision, we have a vector s of two (m in general) elements such that:

- $s_i = 1$, if $\forall t \in \mathbf{t}\ f_i(t) > 0$,
- $s_i = -1$, if $\forall t \in \mathbf{t}\ f_i(t) < 0$,
- $s_i = 0$, otherwise (i.e., when we cannot determine the unique sign of f_i in ti).

Such a sign covering can be obtained by a simple branch-and-bound type procedure. To seek paths, it will be useful to have the information about adjacency of various subboxes of $\partial \mathbf{x}$. Thus, we construct a graph representing the covering. This can be done by Algorithm 2.

Algorithm 2. Create box graph

Require: $X = X[1, \ldots, N]$ – vector of boxes $\mathbf{t}$
1: **for** $(i = 1$ **to** $N)$ **do**
2:   **for** $(j = i + 1$ **to** $N)$ **do**
3:     **if** $(X[i] \cap X[j] \neq \emptyset)$ **then**
4:       add to the graph an edge between nodes $X[i]$ and $X[j]$

Assume we have found in the function's domain, a cubical complex C that is homotopically equivalent to a one-dimensional circle S^1 (other words: a closed path in the boundary). Assume the image of C is homotopically equivalent to $S^1 \subset \mathbb{R}^2$, also; other words: it does not contain $(0,0)$ and it winds around this point (its winding number aka topological degree is different from zero). We can find such C, by seeking cycles in a graph derived from the cubical complex representing the subdivision of $\partial \mathbf{x}$ and checking the sign vector of adjacent boxes. Please note that segments of C can belong to arbitrary faces of the considered box, e.g., Fig. 2.

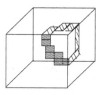

Fig. 2. Representation of C

Traversing the graph can be done in the DFS (depth-first-search), BFS (breadth-first-search) or any other manner.

In our implementation, the DFS approach has been adopted. In each step, we seek a neighboring subbox with the sign covering of the next phase of the cycle or of the same phase. If we have returned to the initial box, changing the phase at least once, it means we have done the whole cycle. Subsequent phases are presented on Fig. 3. Sign coverings in brackets are optional.

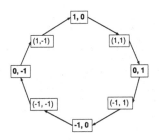

Fig. 3. Subsequent values of sign vectors for elements of C

The procedure for finding the cycle from a given node, is recursive and it can be expressed as Algorithm 3. This procedure is called by another one, named find_cycle_with_nonzero_degree(nd, graph). It marks all nodes in the graph as unread and than calls find_cycle_with_nonzero_degree_recur(nd, nd, graph). This procedure can be called directly for the graph generated by Algorithm 2. Yet, there is a preprocessing procedure that can dramatically improve its performance: adjacent boxes with identical sign coverings can be merged. After this operation, nodes of the considered graph are no longer boxes; they are unions of boxes, potentially quite irregular. But the important information is whether the areas with various sign vectors are direct neighbors and not their precise shape.

The overall 2D-cycle test can be described by Algorithm 4.

Several parts of the above procedures have been implemented in a multi-threaded manner. Details will be given in Sect. 5.

The Case of More than Two Equations. Algorithm from the previous paragraph can (probably) be generalized to $m > 2$, but such extension would be difficult and it might not be efficient. If $C \cong S^k$ for $k > 1$, we cannot move around it as around the path (because there is more than one dimension). Also, we would have to compute Betti numbers, to check if C does not have "holes". It could probably be possible to implement, but the total complexity and difficulty of the algorithm seems high.

Algorithm 3. Function find_cycle_with_nonzero_degree_recur

Require: nd, first, graph
1: mark nd
2: **for all** (neighbor **of** nd) **do**
3:    **if**   not   is_next_element_of_cycle(neighbor,   nd)   **and**   **not** is_same_element_of_cycle(neighbor, nd) **and** neighbor != first **then**
4:       {We shall not consider the path going through the "neighbor" node.}
5:       **cycle loop**
6:    **if** (neighbor == first) **then**
7:       **return true**
8:    **if** (neighbor is marked) **then**
9:       {The "neighbor" already in the path.}
10:      **cycle loop**
11:   result = find_cycle_with_nonzero_degree_recur(neighbor, first, graph)
12:   **if** (result == **true**) **then**
13:      **return true**
14:   **else**
15:      {No cycle going through the "neighbor" has been found. We need to consider another path.}
16:      unmark neighbor
17: {No cycle going through the "nd" has been found.}
18: **return false**

Algorithm 4. The 2D-cycle test

Require: x, f, ε
1: perform a branch-and-bound algorithm on ∂x to obtain boxes t with the sign covering not containing zeros; store these boxes in a vector X
2: create the graph for the set of nodes X, using Algorithm 2
3: merge neighboring nodes with identical s vectors
4: res = **false**
5: **for all** (node **in** graph) **do**
6:    **if** (find_cycle_with_nonzero_degree(node, graph)) **then**
7:       res = **true**
8: **return** res

---

5   Parallelization

Both implemented tests have been parallelized. Obviously, the HIBA_USNE solver is parallel itself, i.e., the B&P process is multithreaded, but parallelizing tools processing a single box, becomes more and more important with the development of multi- and many-core architectures (cf., e.g., [16]).

For the Borsuk test, the parallelization is obvious: formulae (3) for various variables i are computed concurrently, then the results are combined using the tbb::reduce procedure [3], performing the *reduction* operation – in this case, intersection of boxes.

Parallelizing the procedure for seeking 2D-cycle on the boundary of a box, we have more options: we can parallelize the construction process of the sign covering, we can seek cycles starting from several nodes, in parallel, and we can parallelize the cycle-seeking algorithm itself.

In our current implementation, we do not parallelize the cycle-seeking procedure itself: serial DFS method is used. Yet we seek them from several nodes, in parallel. Also, the sign covering is created by a parallelized branch-and-bound type algorithm [18].

Multithreaded Implementation of Algorithm 2 Checking if two boxes are adjacent (and if there should be an edge between them in the graph) is performed in parallel. The loop in line 1 has been parallelized (TBB allows us to use the `tbb::parallel_for` concept [3]).

There has to be some synchronization so that various threads do not maintain the same nodes at the same time. This is realized by adding a dedicated lock (`tbb::spin_mutex`) to each of the nodes. The lock associated with $X[i]$ is acquired as the first one and then the lock of $X[j]$. There is no deadlock, as $j > i$ (please cf. the construction of the loop in line 2 of the considered algorithm) and the locks are always acquired in the same order.

Multithreaded Implementation of Algorithm 4. The loop in line 5 is split between threads. It is worth noting, how marking of threads is realized: many threads perform the DFS search in parallel and each of them must have its own markings of already considered nodes. Indeed, the marking of the node is implemented as thread-specific; we use the `tbb::enumerable_thread_specific<bool>` class [3].

6   Numerical Experiments

Numerical experiments have been performed on a machine with two Intel Xeon E5-2695 v2 processors (2.4 GHz). Each of them has 12 cores and on each core two hyper-threads (HT) can run. So, $2 \times 12 \times 2 = 48$ HT can be executed in parallel. The machine runs under control of a 64-bit GNU/Linux operating system, with the kernel 3.10.0-123.e17.x86_64 and glibc 2.17. They have non-uniform turbo frequencies from range 2.9–3.2 GHz.

As there have been other users performing their computations also, we limited ourselves to using 24 threads only.

The Intel C++ compiler ICC 15.0.2 has been used.

The solver has been written in C++, using the C++11 standard. The C-XSC library (version 2.5.4) [1] was used for interval computations. The parallelization was done with the packaged version of TBB 4.3 [3].

The author's HIBA_USNE solver has been used in version Beta2.5. The additional procedures for the Borsuk test and the 2D-cycle-detecting procedure have been incorporated to the body of the solver, in additional source files.

We have performed experiments with three versions of the algorithm: "No additional test", "Borsuk test", and "2D-cycle test". Hopefully, these names are self-explanatory.

Two underdetermined test problems have been considered. Both have already been considered, i.a., in [15] (see the references therein for their sources' description). The first of the underdetermined ones is a set of two equations – a quadratic

Table 1. Computational results

Quantity	No additional test	Borsuk test	2D-cycle test
Academic problem			
Time (sec.)	4	12	175
Possible boxes	915,934	915,934	32,996
Newton verif. boxes	721	721	721
Adv. verif. boxes	—	0	882,938
Hippopede problem			
Time (sec.)	1	1	3
Possible boxes	170,411	165,084	78
Newton verif. boxes	20,494	20,494	20,494
Adv. verif. boxes	—	5,327	170,333

one and a linear one – in five variables. It is called the Academic problem (accuracy $\varepsilon = 0.05$ has been used). The second one is called the Hippopede problem – two equations in three variables (accuracy $\varepsilon = 10^{-7}$) (Table 1).

6.1 Analysis of the Results

The Borsuk test did not manage to verify any boxes (out of 915934) for the Academic problem, while verifying 5327 boxes (out of 170411) for the Hippopede problem. The 2D-cycle test turned out to be much more successful, but at a very high computational cost. It verified 882938 boxes for the Academic problem and 170333 for the Hippopede one. Unfortunately, performing this procedure turned out to be very intensive: the overall computational time increases from three to 175 seconds for the Academic problem.

The obtained results suggest that the Borsuk test is not very useful, at least for underdetermiend problems. And both tests are very time-consuming, so they should not be considered for execution, unless verifying as many of the possible solutions as we can, is really meaningful.

7   Conclusions

We have investigated a new test for checking if an underdetermined system of two nonlinear equations has a solution in a given box. The test has been incorporated to the HIBA_USNE interval solver. It occurred that the test is able to verify dramatically more boxes than alternative tests based on the Borsuk's theorem or the interval Newton operator. Unfortunately, the computational cost is also significantly higher for the developed test. It can be very useful, when we actually need to verify the solutions, but would not be worthwhile, if it suffices to enclose all *possible* solutions, without their *verification*.

Some effort on parallelizing graph algorithms is an interesting side effect of the research.

References

1. C++ eXtended Scientific Computing library (2017). http://www.xsc.de
2. HIBA_USNE, C++ library (2017). https://www.researchgate.net/publication/ 316687827_HIBA_USNE_Heuristical_Interval_Branch-and-prune_Algorithm_for_ Underdetermined_and_well-determined_Systems_of_Nonlinear_Equations_-_Beta_ 25
3. Intel TBB (2017). http://www.threadingbuildingblocks.org
4. Franek, P., Krčál, M.: Robust satisfiability of systems of equations. In: Proceedings of the Twenty-Fifth Annual ACM-SIAM Symposium on Discrete Algorithms, pp. 193–203. SIAM (2014)
5. Franek, P., Ratschan, S.: Effective topological degree computation based on interval arithmetic. Math. Comput. **84**(293), 1265–1290 (2015)
6. Frommer, A., Lang, B.: Existence tests for solutions of nonlinear equations using Borsuk's theorem. SIAM J. Numer. Anal. **43**(3), 1348–1361 (2005)
7. Goldsztejn, A.: Comparison of the Hansen-Sengupta and the Frommer-Lang-Schnurr existence tests. Computing **79**(1), 53–60 (2007)
8. Jaulin, L., Kieffer, M., Didrit, O., Walter, E.: Applied Interval Analysis. Springer, London (2001). https://doi.org/10.1007/978-1-4471-0249-6_2
9. Kearfott, R.B.: An efficient degree-computation method for a generalized method of bisection. Numerische Mathematik **32**(2), 109–127 (1979)
10. Kearfott, R.B.: Rigorous Global Search: Continuous Problems. Kluwer, Dordrecht (1996)
11. Kearfott, R.B., Nakao, M.T., Neumaier, A., Rump, S.M., Shary, S.P., van Hentenryck, P.: Standardized notation in interval analysis. Vychislennyie Tiehnologii (Comput. Technol.) **15**(1), 7–13 (2010)
12. Kubica, B.J.: Interval methods for solving underdetermined nonlinear equations systems. In: SCAN 2008 Proceedings Reliable Computing, vol. 15, pp. 207–217 (2011)
13. Kubica, B.J.: Tuning the multithreaded interval method for solving underdetermined systems of nonlinear equations. In: Wyrzykowski, R., Dongarra, J., Karczewski, K., Waśniewski, J. (eds.) PPAM 2011. LNCS, vol. 7204, pp. 467–476. Springer, Heidelberg (2012). https://doi.org/10.1007/978-3-642-31500-8_48
14. Kubica, B.J.: Using quadratic approximations in an interval method for solving underdetermined and well-determined nonlinear systems. In: Wyrzykowski, R., Dongarra, J., Karczewski, K., Waśniewski, J. (eds.) PPAM 2013. LNCS, vol. 8385, pp. 623–633. Springer, Heidelberg (2014). https://doi.org/10.1007/978-3-642-55195-6_59
15. Kubica, B.J.: Presentation of a highly tuned multithreaded interval solver for underdetermined and well-determined nonlinear systems. Numer. Algorithms **70**(4), 929–963 (2015)
16. Kubica, B.J.: Parallelization of a bound-consistency enforcing procedure and its application in solving nonlinear systems. J. Parallel Distrib. Comput. **107**, 57–66 (2017)
17. Kubica, B.J.: Role of hull-consistency in the HIBA_USNE multithreaded solver for nonlinear systems. In: Wyrzykowski, R., Dongarra, J., Deelman, E., Karczewski, K. (eds.) PPAM 2017. LNCS, vol. 10778, pp. 381–390. Springer, Cham (2018). https://doi.org/10.1007/978-3-319-78054-2_36

18. Kubica, B.J.: Interval Methods for Solving Nonlinear Constraint Satisfaction, Optimization and Similar Problems: From Inequalities Systems to Game Solutions. Studies in Computational Intelligence, vol. 805. Springer, Heidelberg (2019). https://doi.org/10.1007/978-3-030-13795-3
19. Shary, S.P.: Finite-dimensional Interval Analysis. Institute of Computational Technologies, SB RAS, Novosibirsk (2013)

Workshop on Complex Collective Systems

Experiments with Heterogenous Automata-Based Multi-agent Systems

Franciszek Seredyński[1], Jakub Gąsior[1(✉)], Rolf Hoffmann[2],
and Dominique Désérable[3]

[1] Department of Mathematics and Natural Sciences,
Cardinal Stefan Wyszyński University, Warsaw, Poland
`f.seredynski@uksw.edu.pl`, `j.gasior@uksw.edu.pl`
[2] Technische Universität Darmstadt, Darmstadt, Germany
`hoffmann@informatik.tu-darmstadt.de`
[3] Institut National des Sciences Appliquées, Rennes, France
`domidese@gmail.com`

Abstract. We present a theoretical framework and an experimental tool to study behavior of heterogeneous multi-agent systems composed of the two classes of automata-based agents: Cellular Automata (CA) and Learning Automata (LA). Our general aim is to use this framework to solve global optimization problems in a distributed way using the collective behavior of agents. The common feature of CA and LA systems is the ability to show a collective behavior which, however, is understood differently. It is natural for LA-based agents that are able to learn and adapt, but for CA-based agents, extra features have to be used like the second–order CA. We create a theoretical framework of the system based on a spatial Prisoner's Dilemma (PD) game in which both classes of players may participate. We introduce to the game some mechanisms like local profit sharing, mutation, and competition which stimulate the evolutionary process of developing collective behavior among players. We present some results of an experimental study showing the emergence of collective behavior in such systems.

Keywords: Collective behavior · Learning Automata · Multi-agent systems · Spatial Prisoner's Dilemma game · Second order cellular automata

1 Introduction

Fast development of sensor technology and massive appearance of the Internet–of–Things (IoT) devices generating big data volumes which need to be processed in real-time applications have resulted in setting new distributed computing paradigms. The concept of Fog Computing as an extension of Cloud Computing (CC) appeared a few years ago and edge computing becomes today a natural extension of Fog Computing [1]. Both these new computing paradigms assume

© Springer Nature Switzerland AG 2020
R. Wyrzykowski et al. (Eds.): PPAM 2019, LNCS 12044, pp. 433–444, 2020.
https://doi.org/10.1007/978-3-030-43222-5_38

an increasing degree of computational intelligence of a huge number of small heterogeneous IoT devices and their ability to perform collectively control, analytic and machine–learning tasks.

Very often solving these tasks can be reduced to an optimization problem. Solving optimization problems in a centralized way with a request of full information about the system resources and users' demands is intractable for realistic problems. One can rely rather on a distributed problem solving by several independent entities that can use only some local information but may face a conflict of local goals. Therefore, we propose a large scale multi-agent system approach, where agents are capable to solve optimization problems by local interaction to achieve, on one side, some compromise between their local goals but at the same time to show a certain ability of global collective behavior. Because of the simplicity and non-trivial computational possibilities of automata models they are good candidates to be applied in the context of these problems.

The purpose of this work is to develop a system to study phenomena of global collective behavior of heterogeneous systems composed of both Cellular Automata (CA) [2] and Learning Automata (LA) [3,4] based agents, where they act in an environment described in terms of non–cooperative game theory [5] with the use of Prisoner's Dilemma (PD) type game. These two classes of automata have different origins, features, and applications. In contrast to LA, classical CA does not have adaptability features and the notion of collective behavior is differently understood.

In recent years a new term –the *second–order* CA–has appeared [6] to express an attempt to design CA with adaptability features. We will be using this class of CA and it gives us an opportunity to interpret the notion of the global collective behavior in a unified way measured by the number of cooperating agent–players participating in a game or by the value of the average total payoff of all agents.

The structure of the paper is the following. In the next section works related to the subject of our study are discussed. Section 3 contains a description of a theoretical framework of the proposed heterogeneous multi-agent system including a Spatial Prisoner's Dilemma (SPD) game, CA and LA–based agents, and rules of interaction between them. Section 4 contains some results of the experimental study of the model from the point of view of the ability of collective behavior. Finally, the last section concludes the paper.

2   Related Work

The literature related to multi-agent collective behavior recognizes [7–9] three classes of systems with specific characteristic of behavior: (a) spatially-organizing behaviors, where agents interact a little with an environment but they coordinate themselves to achieve a desired spatial formation, (b) collective exploration, where agents interact a little between themselves but interact mainly with an environment to achieve some goal, and (c) cooperative decision making, where agents both coordinate their actions and interact with an environment to accomplish some complex tasks.

While CA belong to the first class according to this classification, LA belong to the second class. Achieving by CA a desired spatial formation that serves as a solution of a given problem is obtained by a collective behavior of CA cells which are governed by predefined rules assigned to each cell.

Classical CA are not adaptive systems. Rules used to obtain a desired spatial formation are either proposed by humans or found by optimization techniques like e.g., Genetic Algorithms (GA). The issue of self-optimization of CA has been recognized recently in the literature [6,10] and CA with such features are called the "second–order" CA.

In contrast to CA, LA are adaptive systems that interact with an environment and demonstrate an ability of a global collective behavior by finding actions enabling to achieve a global goal. The distinctive feature of this paper is that we combine these two approaches in such a way that we use a SPD game defined on 2D discrete space as a multi-agent framework, where we set both CA and LA types of agents, and we intend to study conditions of emerging a global collective behavior of agents, measured by the total number of cooperating players, i.e. an ability to maximize the average total payoff of all agents of the system like it is expected in LA–based systems.

PD game [5] is one of the most accepted game-theoretical models, where both *cooperation* (C) and *defection* (D) of rational players can be observed and it was a subject of study both for CA and LA models. Tucker formalized the game as the 2–person game and in the 1980's Axelrod organized the first tournament [11] to recognize competitive strategies in this game. The winner was a strategy Tit-For-Tat (TFT) which assumes cooperation of a player on the first move and subsequent repeating actions of the opponent player used in the previous move. Next, Axelrod proposed [12] to apply GA to discover strategies enabling cooperation in the 2-person PD game. Genetic Algorithms (GAs) were able to discover the TFT strategy and several interesting strategies specific for humans.

Discovering strategies of cooperation in N–person PD games ($N > 2$) is a more complex problem. Therefore, Yao and Darwen proposed in [13] another approach where GAs are still applied but the payoff function was simplified. The main idea was that a payoff of a given player depends on a number of cooperating players among the remaining $N - 1$ participants. Under these assumptions, GAs were able to find strategies enabling global cooperation for up to 10 players. For more players, such strategies were not discovered by GA. One of the main reasons for that is the form of the payoff function which assumes participation of a player with a "crowd" – a large number of anonymous players.

A concept of spatial games with a neighbor relation between players helps to solve the crowd problem. Among the first concepts related to SPD game was the game on the ring considered by Tsetlin [4] in the context of LA games, where a payoff of a given player depends on its action and actions of two immediate neighbors. Later such a game in the context of homogeneous LA and GA–based models was studied in [14].

A number of SPD games on 2D grids have been studied recently. Nowak and May proposed [15–17] an original SPD game on a 2D grid with only two types of players – players who always cooperate (*all–C*) and players who always defect (*all–D*). Players occupy cells of 2D space and each of them plays the PD game with all neighbors, and depending on the total score it is replaced by the best performing player in the neighborhood. The study was oriented on the first class of the behavior and it has shown that both types of players persist indefinitely in chaotically changing local structures.

Ishibuchi and Namikawa [18] studied a variant of SPD game with two neighborhoods: one for playing locally defined PD game with randomly selected neighbors, and the second one for matching strategies of the game by a locally defined GA. Howley and O'Riordan [19] considered a *N*–person PD game with *all–to–all* interactions between players and the existence of a tagging mechanism in subgroups of players. Katsumata and Ishida [6] extended the model of SPD game proposed in [15] by considering 2D space as the 2D CA and introducing an additional strategy called *k–D*, which tolerates at most *k* defections in a local neighborhood in the case of cooperation.

The study [6] was oriented on the first class of the system behavior and it has shown a possibility of the emergence of specific spatial structures called membranes created by players using *k–D* strategies, which separate cooperating and defecting players. The interesting issue in these works was an attempt to extend classical CA to a new class called the second–order CA, where rules assigned to CA cells could be changed during the evolution of cells in time. It opens the possibility to study the issue of collective system behavior from this novel point of view and our recent study [20] confirms it.

3   An Environment of Heterogenous Multi-agent System

3.1   Iterated Spatial Prisoner's Dilemma Game

We consider a 2D spatial array of size $n \times m$. We assume that a cell (i, j) will be alive with a probability $p_{i,j}^{alive}$. Each alive cell will be considered as an agent–player participating in the SPD game [6,15]. We assume that a neighborhood of a given player is defined in some way. Players from this neighborhood will be considered as his opponents in the game. At a given discrete moment, each cell can be in one of two states: C or D. The state of a given cell will be considered as an action C (cooperate) or D (defect) of its player against an opponent player from his neighborhood. Payoff function of the game is given in Table 1.

Table 1. Payoff function of a row player participating in SPD game.

Player's action	Opponent's action	
	Cooperate (C)	Defect (D)
Cooperate (C)	$R = 1$	$S = 0$
Defect (D)	$T = b$	$P = a$

Each player playing a game with an opponent in a single round (iteration) receives a payoff equal to R, T, S or P, where $T > R > P > S$. We will assume that $R = 1$, $S = 0$, $T = b$ and $P = a$, and the values of a and b can vary depending on the purpose of an experiment.

If a player takes the action C and the opponent also takes the action C then the player receives payoff $R = 1$. If a player takes the action D and the opponent player still keeps the action C, the defecting player receives payoff $T = b$. If a player takes the action C while the opponent takes the action D, the cooperating player receives payoff $S = 0$. When both players use the action D then both of them receive payoff $P = a$.

It is worth to notice that choosing by all players the action D corresponds to the Nash equilibrium point [5] and it is considered as a solution of the one–shot game. Indeed, if all players select the action D each of them receives a payoff equal to a and there is no reason for any of them to change the action to C while the others keep their actions unchanged, what would result in decreasing his payoff to value 0.

The average total payoff of all players in the Nash equilibrium point is also equal to a. Looking from the point of view of global collective behavior of players this average total payoff of all players is low. We would rather expect choosing by all players the action C which provides the value of the average total payoff of all players equal to 1. For this instance of the game, it is the maximal value of a possible average total payoff of all players and we are interested in studying conditions when such a behavior of players in iterated games is possible.

3.2   CA–Based Players

We will be using two classes of players in the game: CA–based and LA–based players. CA are spatially and temporally discrete computational systems (see, e.g. [2]) originally proposed by Ulam and von Neumann and today they are a powerful tool used in computer science, mathematics and natural science to solve problems and model different phenomena.

When a given alive cell (i, j) is considered as a CA–based player it will be assumed that it is a part of the 2D array and at a given discrete moment of time t, each cell is either in state D or in state C. The value of the state is used by CA–based player as an action with an opponent player. For each cell, a local 2D neighborhood of a radius r is defined. Because we employ a 2D finite space a cyclic boundary condition is applied. We will assume that the Moore neighborhood is used, with 8 immediate neighbors. A neighbor of a given CA or LA–based player will be either another CA or LA–based player and each of them is considered as his opponent in the game.

In discrete moments, CA–based players will select new actions according to local rules (called also strategies or transition functions) assigned to them, which will change the states of the corresponding cells. We will be using some number of rules among which one of them will be randomly assigned to each CA cell, so we deal with a non–uniform CA.

We will consider two types of CA–based players. To cells of the first type, one of the following rules: *all-C*, *all-D*, and *k-D* will be assigned. The *k-D* strategy is a generalized TFT strategy, and when $k = 0$ it is exactly the TFT strategy. The second type of CA–based player uses probabilistic CA. To cells of this type, the following rule will be assigned: *cooperate* with probability p_{coop} or *defect* with probability $1 - p_{coop}$, where p_{coop} is some predefined value.

One can see that CA–based player does not interact with an external environment represented by the payoff function, it does not have a memory and it is not supported by a learning algorithm. The first type of the CA–based player acts reactively on the basis of local information about the number of cooperating players when it uses *k-D* strategy or it does not pay any attention to the behavior of the neighbors when it uses strategies *all-C* or *all-D*. The second type of CA–based player uses probabilistic CA.

3.3   LA–Based Players

We will use a deterministic ϵ–LA [14,21] as LA–based players; ϵ–LA has $d = 2$ actions and acts in a deterministic environment $c = (c_1, c_2, ..., c_{2*d})$, where c_k stands for a reward defined by the payoff function from Table 1 obtained for its action and action of his opponent (CA or LA–based player) from the Moore neighborhood. It also has a memory of length h and a reinforcement learning algorithm which selects a new action. In our case, C and D are actions of an automaton and they are associated with states of the array cells occupied by LA–based players.

Whenever ϵ–LA generates an action, and its opponent from a neighborhood selects an action, the local environment (payoff function) sends it a payoff in a deterministic way. The objective of a reinforcement learning algorithm represented by ϵ–LA is to maximize its payoff in an environment where it operates.

The automaton remembers its last h actions and corresponding payoffs from the last h moments. As the next action ϵ–LA chooses its best action from the last h games (rounds) with the probability $1 - \epsilon$ ($0 < \epsilon \leq 1$), and with the probability ϵ/d any of its d actions.

3.4   Sharing, Mutation and Competition Mechanisms in the Game

To study a possibility of emergence of a global collective behavior of players in the sense of the second class of the collective behavior classification [7,9] we will introduce some local mechanisms of interaction between players, which can be potentially spread or dismissed during the evolution.

The first mechanism is the possibility of sharing locally profits obtained by players. Some kind of hard local sharing was successfully used [14] in the context of LA games. Here we will be using a soft version of sharing, where a player decides to use it or not. It is assumed that each player has a tag indicating whether he wishes (*on*) or not (*off*) to share his payoff with players from the neighborhood who also wish to share. The sharing works in such a way that if two players both wish to share, each of them receives half of the payoff from the sum. Before starting the iterated game each player turns *on* its tag with a predefined probability $p_{sharing}$.

The second mechanism which will be used is a mutation of a number of system parameters. With some predefined value of probability, a CA–based agent of the first type can change the currently assigned strategy (rule) to one of the two other strategies. Similarly, a CA–based agent of the second type can increase/decrease its probability of cooperation. Also, parameters h and ϵ of LA–based agents can be a subject of mutation.

The third mechanism which can be used is a competition which is a generalization of the idea proposed in [15]. Each player associated with a given cell plays in a single round a game with each of his neighbors and this way collects some total score. If the competition mechanism is turned *on*, after a q number of rounds (iterations) each agent compares its total payoff with the total payoffs of its neighbors. If a more successful player exists in the neighborhood of a given player, this player is replaced by the most successful one.

In the case when both players are CA–based players, a rule of a given player is replaced by a rule of the most successful players and also the value of the sharing tag is copied. If both players are LA–based players, then replacing happens only if the best player differs in at least one value of such parameters like h, ϵ or sharing tag. If one player is a CA–based player and the other one is a LA–based player, then a player of the most successful class replaces a given player.

4   Experimental Study

4.1   Graphical Interface

Monitoring the changes of some system parameters during an iterated game is possible with the use of visualization tools. Figure 1 presents snapshots in a 2D window of size 5×5. One can observe which cell is occupied at a given timestep by CA or LA–based player (a), what is a state of a cell (b), which strategies are assigned to players (c) or what are the specific k–D strategies of players (d).

4.2   Parameter Settings

Presented experimental tool has a wide variety of possibilities to study the phenomenon of collective behavior under different values parameters. Below we present some experimental results illustrating potentials of the system. We used the following system parameters: a 2D array of size 50×50 was used, with an initial state C or D (player action) set with probability 0.5. Initially, the rule k–D was assigned (if applied) to CA cells with probability 0.7, and the remaining three rules (*all–C*, *all–D*, probabilistic CA) with probability 0.1. When k–D was applied, k was considered randomly selected from the range $\{0,7\}$. When the competition mechanism was turned *on*, updating the array cells (by a winner in a local neighborhood) was conducted after each iteration. Parameters of the payoff function were set to $a = 0.3$ and $b = 1.4$.

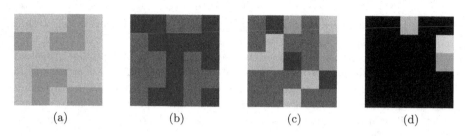

(a)   (b)   (c)   (d)

Fig. 1. Visualization (5 × 5 window): (a) CA players (in orange) and LA players (in green), (b) players' states: C (in red), D (in blue), (c) agents' strategies: *all–C* (in red), *all–D* (in blue), *k–D* (in green), cooperate with predefined probability (in yellow), LA algorithm (in pink), (d) *k–D* strategies: *0–D* (in yellow), *1–D* (in orange), not *k–D* (in black). (Color figure online)

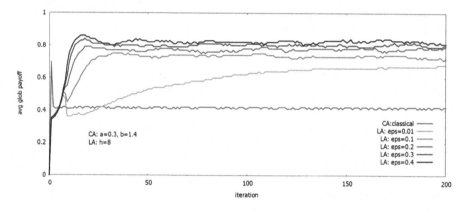

Fig. 2. Experiment #1—average global payoffs for runs of the game with classical CA vs LA–based players. (Color figure online)

4.3 Experimental Results

Experiment #1—Classical CA vs LA–based Players. Figure 2 shows typical runs of the system when either only classical CA–based players (in red) participate or only LA–based players are used. In the case of the game with CA–based players the average total payoff does not change in time, is low and equal to around 0.4. It is a result of that classical CA–based players are not aware of the payoff function, they do not have learning abilities and the value of the average payoff is a result of the initial settings. In contrast to CA, LA–based players are aware of their payoffs, they can learn, and the average total payoff depends upon a given memory size ($h = 8$) and the ϵ–value.

Experiment #2—CA–Based Players with Competition Mechanism. Figure 3 shows five typical runs of the system when only CA–based players participate but the mechanism of competition is turned *on*. In three over five runs,

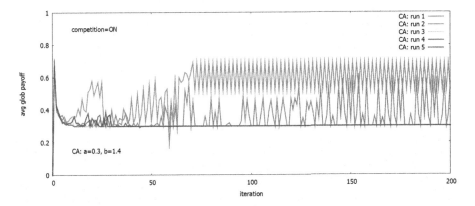

Fig. 3. Experiment #2—average global payoffs for five runs of the game with CA–based players and competition mechanism.

players reach the Nash equilibrium point providing the average total payoff equal to 0.3. In two other runs, the system is able to escape from the Nash equilibrium point what provides a higher average total payoff.

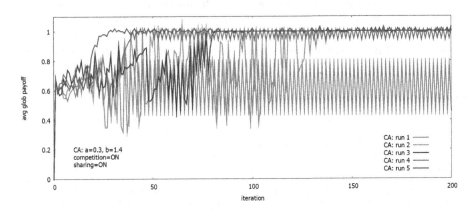

Fig. 4. Experiment #3—five runs of the game with CA–based players with competition and sharing mechanisms.

Experiment #3—CA–Based Players with Competition and Payoff Sharing Mechanism. Figure 4 shows five typical runs when only CA–based players participate, the mechanism of competition between players is turned *on*, and additionally, a possibility of local sharing exists. At this set of experiments, it was assumed that initially, each player set his flag of *"wish to share payoff"* with probability 0.7. During the game, the players could turn *off* or turn *on* this flag depending on their payoff and competition mechanism. One can see that in four over five runs, players reached the maximal level of cooperation provided the maximal value of the average total payoff is equal or close to 1.

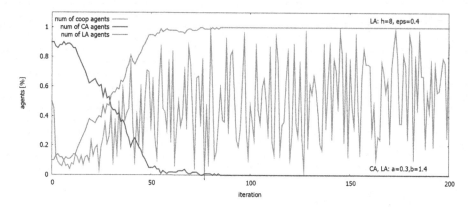

Fig. 5. Experiment #4—number of CA–based players compete with LA–based players. (Color figure online)

Experiment #4—CA and LA–Based Competing Players. Figure 5 shows a typical run when both CA-and LA–based players participate, the mechanism of competition between players is turned *on*, but now without local sharing. Initially, CA–based players were assigned with probability 0.9. LA–based players were used with parameters $h = 8$ and $\epsilon = 0.4$. Due to the competition mechanism, the type of player occupying cells could change in time. In this competition, LA–based players (in blue), despite their initial low number, fully replace CA–based players (in red). This is accompanied by an increasing number of LA–based players who wish to cooperate (in orange) but this cooperation phenomenon has a highly oscillating character.

5   Conclusions

We have presented a theoretical framework and an experimental software tool to study the behavior of heterogeneous multi-agent systems composed of two classes of automata-based agents: CA and LA agents operating in an environment described in terms of a spatial PD game. This framework was defined to solve global optimization tasks in a distributed way by the collective behavior of agents.

The essential from this point of view was to use the concept of the second–order CA and some specific mechanisms of interaction between agents were introduced. A set of conducted experiments have shown that these proposed solutions are promising building blocks enabling emergence of global collective behavior in such heterogeneous multi-agent systems. Conditions of emerging a global behavior of such systems may depend on a number of parameters and these issues will be a subject of our future work. Recently published papers show how the concept of collective behavior of automata can be used in solving some problems related to cloud computing [22] and IoT [23].

Acknowledgement. The authors wish to thank the student of UKSW Dominik Nalewajk for implementation of the simulator.

References

1. Östberg, P.-O., Byrne, J., et al.: Reliable capacity provisioning for distributed Cloud/Edge/Fog computing applications. In: European Conference on Networks and Communications, EuCNC 2017, pp. 1–6 (2017)
2. Wolfram, S.: A New Kind of Science. Wolfram-Media, Champaign (2002)
3. Narendra, K.S., Thathachar, M.A.L.: Learning Automata: An Introduction. Printice-Hall Inc., Upper Saddle River (1989)
4. Tsetlin, M.L.: Automaton Theory and Modeling of Biological Systems. Academic Press, New York (1973)
5. Osborne, M.: Introduction to Game Theory. Oxford University Press, Oxford (2009)
6. Katsumata, Y., Ishida, Y.: On a membrane formation in a spatio-temporally generalized prisoner's dilemma. In: Umeo, H., Morishita, S., Nishinari, K., Komatsuzaki, T., Bandini, S. (eds.) ACRI 2008. LNCS, vol. 5191, pp. 60–66. Springer, Heidelberg (2008). https://doi.org/10.1007/978-3-540-79992-4_8
7. Brambilla, M., Ferrante, E., Birattari, M., Dorigo, M.: Swarm robotics: a review from the swarm engineering perspective. Swarm Intell. **7**(1), 1–41 (2013)
8. Moniz Pereira, L., Lenaerts, T., Martinez-Vaquero, L.A., Anh Han, T.: Social manifestation of guilt leads to stable cooperation in multi-agent systems. In: Autonomous Agents and MultiAgent Systems, AAMAS 2017, Richland, SC, pp. 1422–1430 (2017)
9. Rossi, F., Bandyopadhyay, S., Wolf, M., Pavone, M.: Review of multi-agent algorithms for collective behavior: a structural taxonomy. IFAC-PapersOnLine **51**(12), 112–117 (2018). IFAC Workshop on Networked & Autonomous Air & Space Systems NAASS
10. Peleteiro, A., Burguillo, J.C., Bazzan, A.L.: Emerging cooperation in the spatial IPD with reinforcement learning and coalitions. In: Bouvry, P., González-Vélez, H., Kołodziej, J. (eds.) Intelligent Decision Systems in Large-Scale Distributed Environments. Studies in Computational Intelligence, vol. 362, pp. 187–206. Springer, Heidelberg (2011). https://doi.org/10.1007/978-3-642-21271-0_9
11. Axelrod, R.: The Evolution of Cooperation. Basic Books Publishing, New York (1984)
12. Axelrod, R.: The evolution of strategies in the iterated prisoner's dilemma. In: The Dynamics of Norms (1987)
13. Yao, X., Darwen, P.J.: An experimental study of N-person iterated prisoner's dilemma games. In: Yao, X. (ed.) EvoWorkshops 1993-1994. LNCS, vol. 956, pp. 90–108. Springer, Heidelberg (1995). https://doi.org/10.1007/3-540-60154-6_50
14. Seredynski, F.: Competitive coevolutionary multi-agent systems: the application to mapping and scheduling problems. J. Parallel Distrib. Comput. **47**(1), 39–57 (1997)
15. Nowak, M.A., May, R.M.: Evolutionary games and spatial chaos. Nature **359**, 826–829 (1992)
16. Nowak, M.A., May, R.M.: The spatial dilemmas of evolution. Int. J. Bifurcat. Chaos **3**(1), 35–78 (1993)
17. Nowak, M.A., Bonhoeffer, S., May, R.M.: More spatial games. Int. J. Bifurcat. Chaos **4**(1), 33–56 (1994)

18. Ishibuchi, H., Namikawa, N.: Evolution of iterated prisoner's dilemma game strategies in structured demes under random pairing in game playing. IEEE Trans. Evol. Comput. **9**(6), 552–561 (2005)
19. Howley, E., O'Riordan, C.: The emergence of cooperation among agents using simple fixed bias tagging. In: IEEE Congress on Evolutionary Computation, vol. 2, pp. 1011–1016 (2005)
20. Seredyński, F., Gąsior, J.: Emergence of collective behavior in large cellular automata-based multi-agent systems. In: Rutkowski, L., Scherer, R., Korytkowski, M., Pedrycz, W., Tadeusiewicz, R., Zurada, J.M. (eds.) ICAISC 2019. LNCS (LNAI), vol. 11509, pp. 676–688. Springer, Cham (2019). https://doi.org/10.1007/978-3-030-20915-5_60
21. Warschawski, W.I.: Kollektives Verhalten von Automaten. Academic-Verlag, Berlin (1978)
22. Gąsior, J., Seredyński, F.: Security-aware distributed job scheduling in cloud computing systems: a game-theoretic cellular automata-based approach. In: Rodrigues, J.M.F., et al. (eds.) ICCS 2019. LNCS, vol. 11537, pp. 449–462. Springer, Cham (2019). https://doi.org/10.1007/978-3-030-22741-8_32
23. Gąsior, J., Seredyński, F., Hoffmann, R.: Towards self-organizing sensor networks: game-theoretic ϵ-learning automata-based approach. In: Mauri, G., El Yacoubi, S., Dennunzio, A., Nishinari, K., Manzoni, L. (eds.) ACRI 2018. LNCS, vol. 11115, pp. 125–136. Springer, Cham (2018). https://doi.org/10.1007/978-3-319-99813-8_11

Cellular Automata Model for Crowd Behavior Management in Airports

Martha Mitsopoulou, Nikolaos Dourvas⬥, Ioakeim G. Georgoudas$^{(\boxtimes)}$⬥, and Georgios Ch. Sirakoulis⬥

Department of Electrical and Computer Engineering, Democritus University of Thrace, 67100 Xanthi, Greece
{marmits,ndourvas,igeorg,gsirak}@ee.duth.gr,
http://gsirak.ee.duth.gr

Abstract. At the airports, everything must work with remarkable precision and coordination, especially since their operational processes involve managing a large number of moving human groups in order to minimize waiting and service times of individuals, as well as to eliminate phenomena resulting from the interaction of large crowds, such as crowding and congestion around points of interest. The aim of the study is the development of an integrated automated simulation model for human behavior and traffic in the spaces of an airport. Thus, the model simulates the behavior of the human crowds in different operational areas of an airport. The area of the airport is divided into levels that are characterized by differences in the way that people move within. A fully analytical model based on the computational tool of the Cellular Automata (CA) was realised as well as an obstacle avoidance algorithm that is based on the A star (A*) algorithm. According to its structure, the model is microscopic and discrete in space and time while inherent parallelism boosts its performance. Its prominent feature is that the crowd consists of separate, autonomous or non-autonomous entities rather than a mass. During the simulation, each entity is assigned unique features that affect the person's behavior in the different areas of the airport terminal.

Keywords: Crowd modelling · Cellular Automata · Airport · A* algorithm · Obstacles · Simulation

1 Introduction

Almost recent studies on the full assessment of airports have shown that there is an imbalance between passenger terminal design and airspace planning even at major airports [1]. This stems from the fact that traditionally greater emphasis is put on the development and analysis of the airspace of the airport than on the design of the spaces used by the passengers. An immediate consequence of this potential design deficiency is the congestion problems encountered at passenger terminals in many airports around the world, a problem that is growing as the

© Springer Nature Switzerland AG 2020
R. Wyrzykowski et al. (Eds.): PPAM 2019, LNCS 12044, pp. 445–456, 2020.
https://doi.org/10.1007/978-3-030-43222-5_39

number of people using airports continues to grow at a skyrocketing rate [1]. An airport consists of three areas: (a) airspace, (b) the runway and (c) the passenger terminal(s), whereas each of these sectors is characterized by different types of flow. Airspace is the part of the airport used by different types of aircraft, the airfield is characterized by different aircraft movements in the ground and includes both landing and take-off, while the passenger terminal is the part of the airport that is occupied by flows of people, passengers and non-passengers, but also luggage. Passenger terminals are an important element of the airport structure. They are designed to serve passengers and usually consist of complex and often expensive buildings. Large airports are built to serve tens of millions of people per year [1]. Naturally, an airport's capacity is directly related to demand characteristics, operational parts, and service specifications set by the airport managing authority. Passengers travelling at the airport terminal are often forced to wait and therefore delays due to overcrowding and queues arise, usually resulting from reduced service capacity and inadequate design of the terminal facilities or terminal terminals of the airport's passengers.

An indicator of the efficiency of an airport terminal is the number of passengers served daily [2,3]. Overcrowding and congestion are major problems for hundreds of thousands of passengers. This problem has worsened over the last few years due to increased security measures at airports [2]. Therefore, capacity planning in the airport terminal planning process is more important than ever, which suggests the need for more accurate analysis methods. However, the uncertainty associated with future levels of passenger demand and the complexity of airport terminals makes this work particularly difficult. The problem of designing the service capacity of an airport terminal is concerned with identifying optimal design and capacity expansion of different terminal areas, given the uncertainty regarding both future demands and expansion costs. Analytical modeling of passenger flows at airport terminals under transitory demand is difficult due to the complex structure of the terminal. To the best of our knowledge, the airport terminal passenger capacity planning problem has not been studied holistically, meaning that the studies usually either do not take account of scalability or focus only on a specific area of the terminal [5].

One of the first models of passengers' behavior within an airport is presented in [4]. This study refers to the behavior of passengers at the airport terminal as well as to their needs. Other studies are focusing on the passengers processing times and the importance of dealing with that problem [5–7]. Studies that focus on continuously variable states indicate that such states can hardly be solved due to the complexity of the flow at an airport terminal [1]. Thus, most of them include simulations to model these random and complex flows. In these studies, simulation results are used to estimate the capacities required to make various processes more efficient [8]. In [9] the aim is to understand the dynamics of the discretionary activities of passengers. Focusing on microscopic modeling, very efficient models have been proposed that describe agents' behavioral characteristics [10–12]. A very effective model that is able to simulate the passenger behavior in situations of congestion is Cellular Automata (CA). CA describe the behavior of each person as an individual and the result of the overall system is emerging from the interactions between people that are close one to each other. CA models are widely used in crowd control [13–17], or more specifically, in controlling the disembarking or emergency evacuation of people in an airplane [18].

The main contribution of this study can be summarized as the development of a multi-parameterized, topological oriented simulation model for describing human moving behavior and traffic in the areas of an airport. The model is based on CA that combines low computational cost of a macroscopic simulation model with the focused use of separate individual microscopic features for all operational elements of the model, similar to an Agent-Based model (ABM). Moreover, an A* (A-star) based obstacle avoidance algorithm has been incorporated to the model aiming at the realistic representation of the travellers' moving tendencies. During the simulation, each entity is assigned with unique features that affect the person's behavior in the different areas of the airport terminal. It should be mentioned though that due to the fact that the density is restricted by the cell size, movement artifacts may arise because of the fixed footstep size. In Sect. 2, the proposed model is described providing all the parameters taken into consideration during the design and realisation process as well as the innovative elements that it incorporates. Section 3 presents the results of the simulation and a comparative listing of these for the various demand scenarios that may arise in the terminal of an airport during its operation. Finally, in Sect. 4 the conclusions are drawn, as well as the future perspectives of the model.

2   Model Description

This study presents a general simulation model for the final design of the airport passenger terminal using the computerized model of Cellular Automata (CA). The main and final objective is to develop an airport terminal design tool. This tool will allow the management of the terminal as well as the planning of either different designs or improvements for both existing and proposed terminals before construction. Simulation of a system of such a scale involves many complicated processes such as data collection, space modeling, experimentation, presentation and analysis of results, and proposals to be implemented according to these results. The model of an airport departure area was implemented, which is used both by passengers traveling on domestic flights and by passengers on international flights. Passengers enter the terminal after they have passed the corresponding check-in windows, depending on whether their flight is domestic or international. Then, passengers departing are characterized by freedom of movement among a number of options.

Initially, it is worth mentioning some basic principles governing the simulation model that has been developed with the usage of the MATLAB programming platform. In particular, the physical space represents the ground plan of an airport passenger terminal and is simulated by a cellular discrete mesh, each cell of which has a physical dimension of 60×60 cm, greater than 40×40 cm,

which studies have shown to be the typical area occupied by an adult stand-
ing in crowded conditions [19], as the passenger terminal does not experience
severe crowding and congestion. In addition, passengers may have to carry hand
luggage, which increases the space they occupy in total.

Also, the neighborhood selected to realize the evolution rule is the Moore
neighborhood. This means that the state $C_{i,j}^t$ of the cell (i, j) at time $t + 1$ is
affected by the states of its nine neighboring cells, including the cell i, j itself,
at this time t according to the following equation. Therefore, the evolution rule
that is applied is provided by Eq. (1):

$$C_{i,j}^{t+1} = C_{i,j}^t + C_{i-1,j}^t + C_{i+,j}^t + C_{i,j-1}^t + C_{i,j+1}^t + C_{i-1,j+1}^t \\ + C_{i-1,j-1}^t + C_{i+1,j-1}^t + C_{i+1,j+1}^t \tag{1}$$

In this way, the diagonal movement in the grid is also allowed, which represents the human movement in a more realistic way [20,21]. Consequently, each
agent can move no more than one cell within its neighborhood at each time step
of the simulation. Moreover, it was assumed that all agents entering, leaving and
moving within the airport terminal are characterized by the same speed, which is
the average walking speed of an adult, calculated at 1.3 m/s [19], corresponding
to 4.68 km/h. An initial description of the transition rule of the CA-based model
M can take place according to the following relationship:

$$M = [S, t, L, D, T] \tag{2}$$

$$S = [F, G, P, d(F, t)] \tag{3}$$

where S stands for the schedule of the flights that is created separately and it
is defined itself, by Eq. (3), with F describing the unique flight code, $G(F)$ the
corresponding gate, $P(F)$ the total number of passengers of flight F, and $d(F)$
the departure time of flight F. Continuing the description of Eq. (2), t stands for
the current time step and is the metric of time in the model. Since each agent has
to cover an average distance of approximately $\frac{1}{2} \times (0.6 + \sqrt{2 \times (0.6)^2}) \cong 0.725\,\text{m}$
at every simulation time step and the average speed of movement of persons
within the terminal equals 1.3 m/s, each of the time steps will be approximately
0.56 s [19]. Binary parameter L clarifies whether an individual is part of a group
of passengers (0), such as a family, or travels alone (1). Parameter D corresponds
to a finite set of k potential destinations that each agent can move towards,
such as gates of terminals, duty-free shops, restaurants and cafes, resting seats,
information benches, automatic cash dispensers, toilets. It can also describe the
states of an individual that wanders in the terminal area without a specific
destination, as well as the absence of movement. Finally, T describes the topology
of the terminal station that is the exact location of all possible destinations
within the terminal area.

According to the adopted modeling strategy, the services that are provided to
the agents can be divided into different levels based on certain features in order
to be more effective in managing them. In the context of this study, the first
level refers to the check-in process and includes both the check-in windows and

the queues that the passengers form when trying to approach the corresponding serving points. The generation of waiting queues in public places is a problem of great research concern [22]. In the case of airport checkpoints, it is more common to use a queue for multiple service windows, known as "snake-type" queue coupled with the so-called "fork-type" queue, where separate shorter queues are formed in front of each service window. The use of this type of queue is preferred when waiting at airports, because it allows longer queue lengths to take advantage of the space provided more effectively, and because people waiting in the queue maintain eye contact with the service windows, and thus the feeling of impatience is not increased as long as people wait [23]. Based on the airport scenario studied, the ticket control area is simulated coupling "snake-" and "fork-" type queues, ending in multiple ticket control windows [23].

In addition, the probability q of a new person to appear in the queue is adjusted by taking into account the S flight schedule. Specifically, it is inversely proportional to the time remaining until a flight departs (Eq. (4)):

$$q \propto \frac{1}{\prod_{i=1}^{n}[d(i,t) - t]} \tag{4}$$

where n is the maximum number of flights that can be served at the same time, with $n_{max} = |G|$, since n could not exceed the number of gates at the airport terminal. The model incorporates the options of increasing and/or decreasing the length of the queue, adding additional service windows, and changing the service times of each window.

As soon as the agents leave the check-points, they enter the second level. It represents the main area of the terminal and includes all the available points that an agent can visit until she/he is directed to the gate of boarding. As soon as an agent enters the main terminal area, she/he decides to move in one direction, according to the model description factors discussed previously. The factor being considered first is that of the remaining time until the departure of the flight, and whether or not it exceeds a predetermined limit. This, at a real airport terminal, is equivalent to whether the gate that corresponds to the flight to which each agent is going to fly is disclosed or not. In the event that the gateway has not yet been announced, the agent will move inside the 2nd level, choosing a certain point among all available options that are expressed by parameter D in Eq. (2). The instantaneous density $p_{AoI,t}$ of the passengers in the individual areas of the main terminal depends on the flight schedule since the total number of persons using the terminal at the airport varies not only from season to season but also during the day, and it is calculated on the basis of the following relationship:

$$p_{AoI,t} = \frac{N_{AoI,t}}{AoI} \tag{5}$$

where AoI is the area of interest and N the number of people within the AoI. The model allows the topological parameterization of the main area of the terminal station that is the topological re-location of all available visiting points within the second level. Though, it should be pointed out that the measurement area does not always coincide with the topological area.

The space around the gates, although located within the terminal, should be considered as a separate level, since individuals behave completely differently in terms of their movement when they approach the gates in order to board. Once the boarding gateway is announced, the majority of the passengers is considered to be heading towards it. The proposed system provides the potentiality for the automated calling of passengers to the gate (call for flight), which is triggered when the following relationship is verified:

$$Remainingtime(F, t) = d(f, T) - t < P(F) \times (GateDelay(t) + 0.5) + \alpha \quad (6)$$

where $\alpha = 100$ an additional time parameter for security reasons and $GateDelay(t)$, the parameterized gate delay, i.e. the average number of time steps that each agent remains at the gate from the moment she/he arrives at the gate until she/he leaves it in order to board the airplane. In case that the boarding pass check takes place automatically then the minimum $GateDelay(t) = 1$ is considered, otherwise, $GateDelay(t) > 1$. Algorithmically, the gate opens when the following relationship is satisfied:

$$Remainingtime(F, t) = d(f, T) - t < P(F) \times (GateDelay(t) + 0.5) \quad (7)$$

Then the corresponding agent tries to leave the gate as soon as possible. At these points, there are phenomena of dislocation, which are absent in both the first and second level of the terminal. Naturally, these phenomena are not particularly intense, since there are no emergency conditions under normal circumstances. Thus, there is no reason for a rapid abandonment of the site through the gate. It is worth mentioning that the model description factors are reviewed for each individual, at each time step. Therefore, the desired destination for each agent can change at any time. In the case of obstacles, agents should have the ability to avoid obstacles that may be in their route while keeping their direction to the point they want to reach. In the context of this study, an obstacle avoidance algorithm, based on the optimal path finder algorithm A* (A-star) has been developed [24] in a CA environment. The algorithm takes into account the starting position of a person, the desired destination, and the topology of the obstacles as defined by the ground plan of the airport terminal. Then it is repeatedly trying to find the optimum path to the desired point, where the optimum term is the closest route, that is, the shortest path. Taking into account that variable x represents the agent's position at time t then the fact that the distance to the destination is minimized is represented mathematically by the following equation:

$$x^{t+1} = x^t + [a, b] \quad with \quad a, b \in -1, 0, 1 \quad (8)$$

where a and b are calculated so that Euclidean distance equals to:

$$d = \sqrt{(i_g - 1 + a)^2 + (j_g - 1 + a)^2} \quad (9)$$

with (i_g, j_g) referring to the coordinates of the desired destination, to be minimised. Potential paths to the desired position are then calculated by the cells-to-extend method [24]. In case that the optimal path that each person has to follow is found, then it is stored and the person moves according to it for each next step until it reaches the desired position. The algorithm is evolved that way provided that no other obstacle appears in the calculated optimal route and there is no need or desire for the agent to move to a place other than that originally considered as desirable; for example, in the case that an agent who is moving to a vacant seat and suddenly decides to use an automatic teller machine, or another that is wandering in the terminal's premises and she/he is informed that she/he has to move to the gate of her/his flight. In the event that one of the above conditions is not met, the algorithm is called to re-calculate either the optimal path to the same desired position taking into consideration the new obstacle that has appeared or the shortest path to the new desired position. In such a manner the computational complexity of the algorithm is lowered by the implementation of the proposed method.

3   Simulation Results

The cellular grid that simulates the physical space of the airport terminal equals 150×130 cells. Therefore, the total area of the physical system is described by Eq. (10), whereas the main area of the terminal is described by Eq. (11). Then taking into account walls and set places, the space left for agents to move is given by Eq. (12). Finally, the area of interest around each gate is provided by Eq. (13).

$$A_{total} = 150 \times 130\,(cells) \times \frac{0.6\,(\text{m}) \times 0.6\,(\text{m})}{cell} = 7,020\,\text{m}^2 \qquad (10)$$

$$A_{terminal} = 150 \times 100\,(cells) \times \frac{0.6\,(\text{m}) \times 0.6\,(\text{m})}{cell} = 5,400\,\text{m}^2 \qquad (11)$$

$$Appl_{term} = A_{terminal} - seats - walls =$$
$$[148 \times 98 - 448\,(cells) \times \frac{0.6\,(\text{m}) \times 0.6\,(\text{m})}{cell}] \cong 5,060\,\text{m}^2 \qquad (12)$$

$$A_{gate} = 10 \times 10\,(cells) \times \frac{0.6\,(\text{m}) \times 0.6\,(\text{m})}{cell} = 36\,\text{m}^2 \qquad (13)$$

The scenario that is presented in the framework of this study is described by Table 1.

It is clear that the flight schedule determines how the airport terminal will operate, and any changes to it may result in various different scenarios of simulation. Figure 1 shows the evolution of the experiment based on the flight schedule of Table 1. Regarding Fig. 1(a), it is worth making two comments on that process. The first one refers to the queue that is formed before the boarding control windows; the density of people in the queue is relatively small. This is because a flight does not depart soon, thus as derived from Eq. (4), the probability q of

Table 1. The flights' schedule of the adopted scenario.

Number of flight (F)	Gate ($G(F)$)	Number of agents (Passengers) ($P(F)$)	Departure time ($d(F,t)$)
100	1	37	1,000
101	2	78	1,500
102	3	48	2,000

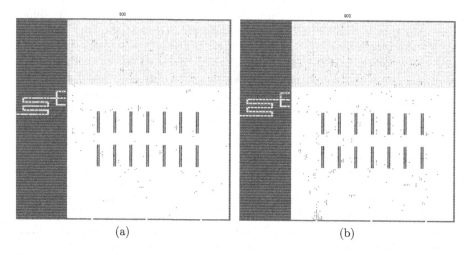

(a)   (b)

Fig. 1. Top view of the terminal as simulated by the electronic system with an emphasis on some of its separated venue; the boarding pass control windows (left), the recreation and waiting areas (centre), as well as three terminal gates that the passengers leave to board the plane (down). (a) Time step 500; no boarding (b) Time step 900; boarding from Gate 1 has commenced.

a new person to appear in the queue is relatively small. The second comment refers to the groups that form some of the agents, with a size that varies. These groups remain inseparable throughout the wandering in the terminal area until the people leave the gate. In Fig. 1(b) we can observe that the density of agents has increased significantly (900-time step) since more flights are expected. Furthermore, the process of boarding from Gate 1 has started. Besides, this fact derives from the implementation of Eq. (6), when replacing the corresponding parameters of the equation with their current values of the time step, the number of agents expected to travel on the flight served by Gate 1 and the time step that corresponds to the departure of that flight. Figure 2 shows the graphs of crowd density in relation to the time resulting from conducting this experiment. We can observe that the densities in the area around each gate initially increase, then they form a maximum and finally decrease (Fig. 2(a)). The maximum density differs for the area around each gate and it is proportional to the number of agents that will be served by the corresponding gate. The time periods that

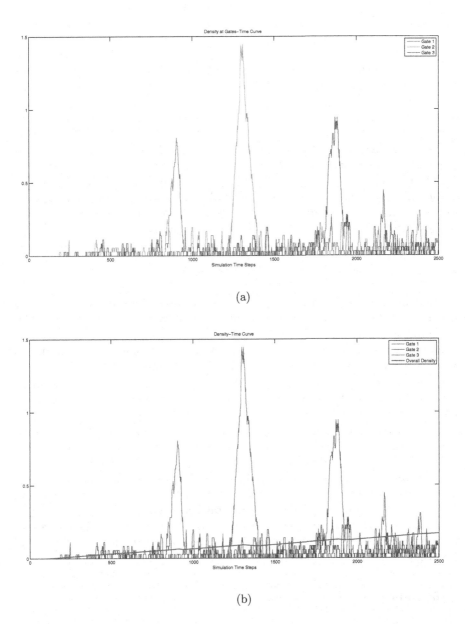

Fig. 2. (a) Recorded densities in the area around the gates as a function of time; Gate 1 (left - red), Gate 2 (center - green) and Gate 3 (right - blue). (b) The density in the terminal area (straight line with a small inclination at the bottom of the graph – red) compared to the crowd density around the gates (blue) depending on time. (Color figure online)

the density increases around the gates are identical to the periods before the scheduled departure of the flight. Finally, the moment that the density reaches its maximum value, it is the one that satisfies Eq. (6). Density is expressed in $1/m^2$ and it is obtained by multiplying Eq. (5) with $p_{max} \cong 2.77778 \frac{persons}{m^2}$, which is the maximum density for this electronic system, as the length of the side of each cell equals to 0.6 m. In Fig. 2(b), the comparisons of the densities around the gates with the densities that are observed at the rest of the terminal station take place. It is obvious that the densities reached around the gates, when agents approach them to board the planes, are much larger than those observed at the rest of the areas of the terminal station. Figure 2 highlights that the overall density has fluctuations that are strongly dependent on the flight schedule and what is happening at the terminal's gates. Initially, the total density is zero, as the first few people have not yet passed boarding documents checking. Subsequently, the density increases almost linearly with time, except for the time periods where one of the gates is evacuated, where it exhibits a downward trend.

4   Conclusions

An electronic system for the study and optimization of crowd behavior in the airport is proposed in this study. It is based on the computational tool of Cellular Automata (CA). Concerning the problem under study, CA present a number of extremely interesting features, such as local interactions, mass parallelism through the application of the rule, the flexibility of boundary conditions selection, the number of possible situations, the CA cells that form a simple structural element. Simulated experimental scenarios proved that the density of the crowd and its variations in time are directly related to the flight schedule according to which an airport operates for a given period of time. In order to avoid crowding and dissatisfaction of agents, the flight schedule must be appropriately designed so that the density does not increase beyond certain safety levels since it has a major impact on the speed at which people move of the terminal but also in the operation of the airport in general. In a physical system, both behavior and movement of people are affected by innumerable social and psychological factors. Thus, this feature could also be incorporated in the parameterization process of the proposed model. Finally, the model can be validated with the use of real data that could further enforce its efficiency.

References

1. Jim, H.K., Chang, Z.Y.: An airport passenger terminal simulator: a planning and design tool. Simul. Pract. Theor. **4**(6), 387–396 (1998)
2. Fodness, D., Murray, B.: Passengers' expectations of airport service quality. J. Serv. Mark. **7**(21), 492–506 (2007)
3. Fernandes, E., Pacheco, R.: Efficient use of airport capacity. Transp. Res. Part A Policy Pract. **3**(36), 225–238 (2002)

4. Bennetts, D., Hawkins, N., O'Leary, M., McGinity, P., Ashford, N.: Stochastic modeling of airport processing - regional airport passenger processing. National Aeronautics and Space Administration (1975)
5. McKelvey, F.: A review of airport terminal system simulation models. Transportation System Center, US Department of Transportation, November 1989
6. Hamzawi, S.G.: Lack of airport capacity: exploration of alternative solutions. Transp. Res. Part A Policy Pract. **1**(26), 47–58 (1992)
7. Schultz, M., Schulz, C., Fricke, H.: Passenger dynamics at airport terminal environment. In: Klingsch, W., Rogsch, C., Schadschneider, A., Schreckenberg, M. (eds.) Pedestrian and Evacuation Dynamics 2008, pp. 381–396. Springer, Heidelberg (2010). https://doi.org/10.1007/978-3-642-04504-2_33
8. Manataki, I.E., Zografos, K.G.: A generic system dynamics based tool for airport terminal performance analysis. Transp. Res. Part C Emerg. Technol. **4**(17), 428–443 (2009)
9. Ma, W., Fookes, C., Kleinschmidt, T., Yarlagadda, P.: Modelling passenger flow at airport terminals: individual agent decision model for stochastic passenger behaviour. In: Proceedings of the 2nd International Conference on Simulation and Modeling Methodologies, Technologies and Applications (SIMULTECH), pp. 109–113. SciTePress, Rome (2012)
10. Kirchner, A., Klüpfel, H., Nishinari, K., Schadschneider, A., Schreckenberg, M.: Simulation of competitive egress behavior: comparison with aircraft evacuation data. Physica A **324**(3–4), 689–697 (2003). https://doi.org/10.1016/S0378-4371(03)00076-1
11. Dietrich, F., Köster, G.: Gradient navigation model for pedestrian dynamics. Phys. Rev. E **89**, 062801 (2014). https://doi.org/10.1103/PhysRevE.89.062801
12. von Sivers, I., Köster, G.: Realistic stride length adaptation in the optimal steps model. In: Chraibi, M., Boltes, M., Schadschneider, A., Seyfried, A. (eds.) Traffic and Granular Flow '13, pp. 171–178. Springer, Cham (2015). https://doi.org/10.1007/978-3-319-10629-8_20
13. Tsiftsis, A., Sirakoulis, G.Ch., Lygouras, J.: FPGA processor with GPS for modelling railway traffic flow. J. Cell. Automata **12**(5), 381–400 (2017)
14. Gerakakis, I., Gavriilidis, P., Dourvas, N.I., Georgoudas, I.G., Trunfio, G.A., Sirakoulis, G.Ch.: Accelerating fuzzy cellular automata for modeling crowd dynamics. J. Comput. Sci. **32**, 125–140 (2019). https://doi.org/10.1016/j.jocs.2018.10.007
15. Mitsopoulou, M., Dourvas, N.I., Sirakoulis, G.Ch., Nishinari, K.: Spatial games and memory effects on crowd evacuation behavior with cellular automata. J. Comput. Sci. **32**, 87–98 (2019). https://doi.org/10.1016/j.jocs.2018.09.003
16. Kartalidis, N., Georgoudas, I.G., Sirakoulis, G.Ch.: Cellular automata based evacuation process triggered by indoors Wi-Fi and GPS established detection. In: Mauri, G., El Yacoubi, S., Dennunzio, A., Nishinari, K., Manzoni, L. (eds.) ACRI 2018. LNCS, vol. 11115, pp. 492–502. Springer, Cham (2018). https://doi.org/10.1007/978-3-319-99813-8_45
17. Kontou, P., Georgoudas, I.G., Trunfio, G.A., Sirakoulis, G.Ch.: Cellular automata modelling of the movement of people with disabilities during building evacuation. In: Proceedings of the 26th Euromicro International Conference on Parallel, Distributed and Network-Based Processing (PDP), pp. 550–557. IEEE, Cambridge (2018). http://doi.ieeecomputersociety.org/10.1109/PDP2018.2018.00093
18. Giitsidis, T., Dourvas, N.I., Sirakoulis, G.Ch.: Parallel implementation of aircraft disembarking and emergency evacuation based on cellular automata. Int. J. High Perform. Comput. Appl. **2**(31), 134–151 (2017). https://doi.org/10.1177/1094342015584533

19. Weidmann, U.: Transporttechnik der fussgänger. Strasse und Verkehr **3**(78), 161–169 (1992)
20. Chertkoff, J., Kushigian, R.: Don't Panic: The Psychology of Emergency Egress and Ingress. Praeger, London (1999)
21. Yang, L.Z., Zhao, D.L., Li, J., Fang, T.Y.: Simulation of the kin behaviour in building occupant evacuation based on cellular automaton. Build. Environ. **40**, 411–415 (2005)
22. Sundarapandian, V.: Probability, Statistics and Queuing Theory. PHI Learning Pvt Ltd., New Delhi (2009)
23. Yanagisawa, D., Suma, Y., Tomoeda, A., Miura, A., Ohtsuka, K., Nishinari, K.: Walking-distance introduced queueing model for pedestrian queueing system: theoretical analysis and experimental verification. Transp. Res. Part C Emerg. Technol. **37**, 238–259 (2013)
24. Hart, P.E., Nilsson, N.J., Raphael, B.: A formal basis for the heuristic determination of minimum cost paths. IEEE Trans. Syst. Sci. Cybern. **2**(4), 100–107 (1968)

A Conjunction of the Discrete-Continuous Pedestrian Dynamics Model SigmaEva with Fundamental Diagrams

Ekaterina Kirik$^{(\boxtimes)}$ ⓘ, Tatýana Vitova, Andrey Malyshev ⓘ,
and Egor Popel ⓘ

Institute of Computational Modelling, Russian Academy of Sciences, Siberian
Branch, Akademgorodok 50/44, Krasnoyarsk 660036, Russia
{kirik,vitova}@icm.krasn.ru
http://3ksigma.ru

Abstract. This article is focused on dynamics of the computer simulation module SigmaEva in connection with an unidirectional flow under periodic boundary conditions. The module SigmaEva realizes the discrete-continuous stochastic pedestrian dynamics model that is shortly presented here. A fundamental diagram (speed-density dependance) is an input for the model. Simulated specific flow rates are considered versus input ones for different diagrams. A sensitivity of the model to input diagrams is shown and discussed.

Keywords: Fundamental diagrams · Flow rate · Pedestrian dynamics model · Transition probabilities · Evacuation simulation

1 Introduction

A simulation of pedestrian dynamics is used in many fields, from entertainment (e.g., cinema and computer games) to fire safety of buildings, ships, and aircrafts. The most attractive for application is so called microscopic models, when each person is considered separately and a model determines coordinates of each person. In a model every person can have individual properties, including a free movement speed, an evacuation start time, a size of a projection, an evacuation way. These give the wider opportunities to state a simulation task and reproduce a real phenomena. Different approaches [1] from mathematically continuous models (the social force model based on differential equations [2–4]) to pure discrete models (cellular automation (CA), e.g., [5–8]) are developed already. A discrete-continuous approach combines advantages of both approaches: people move in a continuous space, but there are only fixed number of directions where a person can move. Discrete and continuous approaches are combined in models [9–11].

In the article one discrete-continuous model is considered. It was implemented in a software, and here a validation of the SigmaEva evacuation module is given

© Springer Nature Switzerland AG 2020
R. Wyrzykowski et al. (Eds.): PPAM 2019, LNCS 12044, pp. 457–466, 2020.
https://doi.org/10.1007/978-3-030-43222-5_40

with respect to a very important case study – fundamental diagrams (flow-density dependance) under periodic boundary conditions.

The model considered is designed in a way that a fundamental diagram is an input for the model. A fundamental diagram is used to calculate speed according to local density for each person. This property of the model is very convenient for practical applications because we omit a step to tune parameters to correspond desired flow-density dependance. Due to inner properties of the model, a numerical presentation of the model, restrictions which are admitted by the model (for example, constant square of projection, shape of the body) input fundamental diagram may be transformed. It means that output flow-density curve may diverge from desired (input) curve. The aim of this study is to investigate a sensitivity of the model to the input fundamental diagram and identify different manifestations of the output speed-density dependence of the model in steady movement regime.

In the next section the main concept of the model is presented. Section 3 contains the description of the case study and the results obtained. We conclude with the summary.

2   Description of the Model

2.1   Space and Initial Conditions

A continuous modeling space $\Omega \in R^2$ and an infrastructure (obstacles) are known[1]. People may move to (and on) free space only. To orient in the space particles use the static floor field S [12]. Without loss of generality a target point of each pedestrian is the nearest exit.

A shape of each particle is a disk with diameter d_i, initial positions of particles are given by coordinates of disks' centers $x_i(0) = (x_i^1(0), x_i^2(0))$, $i = \overline{1, N}$, N – number of particles (it is assumed that these are coordinates of body's mass center projection). Each particle is assigned with the free movement speed[2] v_i^0, square of projection, mobility group. It is also assumed that while moving the speed of any particular person does not exceed the maximal value (free movement speed), and a speed of each person is controlled in accordance with a local density.

Each time step t each particle i may move in one of the predetermined directions $\overrightarrow{e_i}(t) \in \{\overrightarrow{e^\alpha}(t), \alpha = \overline{1, q}\}$, q – the number of directions, model parameter (for example, a set of directions uniformly distributed around the circle will be considered here $\{\overrightarrow{e^\alpha}(t), \alpha = \overline{1, q}\} = \{(\cos \frac{2\pi}{q}\alpha, \sin \frac{2\pi}{q}\alpha), \alpha = i = \overline{1, q}\})$. Particles that cross target line leave the modeling space.

The goal is to model an individual persons movement to the target point taking into account interaction with the environment.

---

[1] Here and below under "obstacle" we mean only nonmovable obstacles (walls, furniture). People are never called "obstacle". There is unified coordinate system, and all data are given in this system.

[2] We assume that free movement speed is random normal distributed value with some mathematical expectation and dispersion [13].

2.2   Preliminary Calculations

To model directed movement a "map" that stores the information on the shortest distance to the nearest exit is used. The unit of this distance is meters, [m]. Such map is saved in static floor field S. This field increases radially from the exit; and it is zero in the exit(s) line(s). It does not change with time and is independent of the presence of the particles. The idea of the field S is imported from the floor field (FF) CA model [7].

2.3   Movement Equation

A person movement equation is derived from the finite-difference expression $v(t)\overrightarrow{e}(t) \approx \frac{\overrightarrow{x}(t)-\overrightarrow{x}(t-\Delta t)}{\Delta t}$ that is given by a velocity definition. This expression allows us to present new position of the particle as a function of a previous position and local particle's velocity. Thus for each time t coordinates of each particle i are given by the following formula:

$$\overrightarrow{x}_i(t) = \overrightarrow{x}_i(t - \Delta t) + v_i(t)\overrightarrow{e}_i(t)\Delta t, \; i = \overline{1, N}, \tag{1}$$

where $\overrightarrow{x}_i(t - \Delta t)$ – the coordinate in previous time moment; $v_i(t)$, [m/s] – the particle's current speed; $\overrightarrow{e}_i(t)$ – the unit direction vector, Δt, [s] – a length of time step that is 0,25 s.

Unknown values in (1) for each time step for each particle are speed $v_i(t)$ and direction $\overrightarrow{e}_i(t)$. A probability approach is used to find a direction for the next step. A procedure to calculate probabilities to move in each direction is adopted from a previously presented stochastic cellular automata floor field model [8]. To get speed $v_i(t)$ a local density is estimated and substituted to some speed-density dependance.

In this case, in contrast with force-based models [2,3], the task of finding the velocity vector is divided in two parts. At first, the new direction is determined; then, speed is calculated according to local density in the direction chosen. By this trick we omit the step of describing forces that act on persons in direct way, a numerical solution of N differential equations. Thereby (comparing with force-based models) computational capacity of the model is reduced and relaxation parameter Δt is allowed to be at least 10 times larger. At the same time, the modeling space is considered to be continuous which is valuable in terms of practical applications.

2.4   Choosing the Movement Direction

In this discrete-continuous model we took inspiration from our previously presented stochastic CA FF model [8]. All predetermined directions for each particle for each time step are assigned with some probabilities to move, and direction is chosen according to the probability distribution obtained.

Probabilities in the model are not static and vary dynamically and issued on the basis of the following facts. Pedestrians keep themselves at a certain distance from other people and obstacles. The tighter the people flow and the more in a hurry a pedestrian is, the smaller this distance. During movement, people follow at least two strategies: the shortest path and the shortest time.

Thus, the personal probabilities to move in each direction each time step depends on: (a) the main driven force (given by a destination point), (b) an interaction with other pedestrians, (c) an interaction with an infrastructure (non movable obstacles). The highest probability[3] is given to a direction that has most preferable conditions for movement considering other particles and obstacles and a strategy of the peoples' movement (the shortest path and/or the shortest time).

We omit here exact formulas to calculate probability for particle i to move from this position to directions $\{\overrightarrow{e_i^1(t)}, ..., \overrightarrow{e_i^q(t)}\}$, decision rules to choose direction $\overrightarrow{e_i^{\hat{\alpha}}}(t)$, and the final conflict resolution procedure. They are presented in [14].

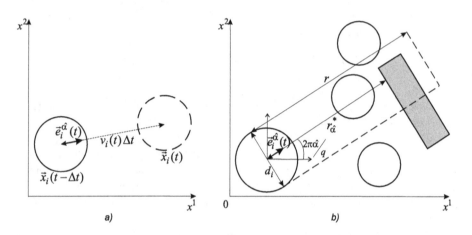

Fig. 1. Left: Movement scheme. Right: visibility area.

2.5   Speed Calculation

Person's speed is density dependent [1,13,15]. We assume that only conditions in front of the person influence on speed. It is motivated by the front line effect (that is well pronounced while flow moves in open boundary conditions) in a dense people mass, when front line people move with free movement velocity, while middle part is waiting for a free space available for movement. It results in the diffusion of the flow. Ignoring this effect leads to a simulation being slower than the real process. Thus, only density $F_i(\hat{\alpha})$ in direction chosen $\overrightarrow{e_i}(t) = \overrightarrow{e_i^{\hat{\alpha}}}(t)$ is required to determine speed.

---

[3] Mainly with value >0,9.

To calculate current speed of the particle, one of the known velocity-density dependence can be used. In the next section we present a conjunction of the model and some fundamental diagrams.

Numerical procedures which is used to estimate a local density is presented in [14]. An area where density is determined is reduced by chosen direction and visibility area which is presented in Fig. 1.

3 Numerical Experiments: Sensitivity of the Model to Input Fundamental Diagrams

3.1 Steady-State Regime, Fundamental Diagrams Considered

Manifestation of the density dependence of the velocity is implemented in the steady-state regime, when the time-spatial density is assumed to be constant and there are no conditions for transformations of the flow. People are assumed to be uniformly distributed over the entire area (e.g., in an extended corridor without narrowing) and move in one direction. Under these limitations, the speed decreases with increasing density. In terms of a specific flow, the fundamental diagram looks as follows. As the density increases, the specific flow grows, attains its maximum, and then decreases. There exist various fundamental diagrams determined by many factors, including demographics [16], which have the same basic feature.

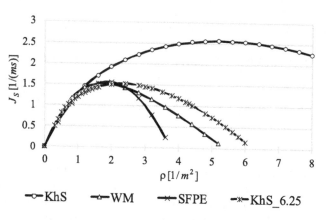

Fig. 2. Specific flows for (3)–(5), $v^0 = 1.66$ [m/s].

For example, the velocity-density dependence can be presented in the analytical form [13]:

$$v^{KhS}(\rho) = \begin{cases} v^0(1 - a_l \ln \frac{\rho}{\rho^0}), \rho > \rho^0; \\ v^0, \qquad\qquad \rho \le \rho^0, \end{cases} \qquad (2)$$

where ρ^0 is the limit people density until which people can move with a free movement speed (it means that the local density does not influence the people's speed); a_l is the parameter of adaptation of people to the current density during their movement in different ways: $\rho^0 = 0.51$ $[1/m^2]$ and $a_1 = 0.295$ for the horizontal way, $\rho^0 = 0.8$ $[1/m^2]$ and $a_2 = 0.4$, for movement downstairs and $\rho^0 = 0.64$ $[1/m^2]$ and $a_3 = 0.305$ for movement upstairs; v_0 is the unimpeded (free movement) speed of a person; and ρ is the local density for a person.

In (2) $v^{KhS}(\rho)$ goes to zero under maximum density ρ_{max}, resolution of the equation $v^{KhS}(\rho) = 0$ gives $a_l = \frac{1}{\ln \frac{\rho_{max}}{\rho^0}}$. As a result we can introduce ρ_{max} as a parameter to the formula (2):

$$v^{KhS}(\rho) = \begin{cases} v^0(1 - \frac{\ln \rho/\rho^0}{\ln \rho_{max}/\rho^0}), & \rho > \rho^0; \\ v^0, & \rho \le \rho^0, \end{cases} \tag{3}$$

In [17,18] speed versus density are given in the following way:

$$v^{WM}(\rho) = \begin{cases} v^0, & \rho = 0; \\ v^0 (1 - e^{-1.913\left(\frac{1}{\rho} - \frac{1}{\rho_{max}}\right)}), & \rho < \rho_{max}; \\ 0, & \rho \ge \rho_{max}. \end{cases} \tag{4}$$

$$v^{SFPE}(\rho) = \begin{cases} v^0 (1 - \frac{\rho}{\rho_{max}}), & 0 \le \rho < \rho_{max}; \\ 0, & \rho \ge \rho_{max}, \end{cases} \tag{5}$$

In (3)–(5), ρ_{max} is the acceptable maximum density. The original forms of the velocity-density dependencies from [17,18] were transformed to input ρ_{max} in an explicit way and to make ρ_{max} a parameter. It was done in a way presented for (2).

In Fig. 2 it was assumed that $\rho_{max} = 15$ $[1/m^2]$ (curve KhS) and $\rho_{max} = 6.25$ $[1/m^2]$ (curve $KhS\_6.25$) in (3), $\rho_{max} = 5.4$ $[1/m^2]$ in (4) and $\rho_{max} = 3.8$ $[1/m^2]$ in (5). Figure 2 shows the specific flows $\hat{J}_s = \rho v(\rho)$ $[1/(ms)]$, for (3)–(5) (curves KhS, WM, $SFPE$, and $KhS\_6.25$ respectively; $v^0 = 1.66$ $[m/s]$).

3.2   Case Study

To see the influence of the input fundamental diagram to the model, we consider the simulation experiment under the so-called periodic boundary conditions. A straight corridor 50×2 m^2 in size with the control line in the right-hand side is the modeling area, Fig. 3. People uniformly fill the entire area.

To reproduce the steady regime (periodic boundary conditions), initial number of people N should be maintained [19].

Time T required for $M = 1000$ people to cross the control line at the end of the corridor at given N is a quantity to be measured. In the stochastic model, the time should be averaged over a set of K runs under the same initial conditions.

To estimate the flow rate, the formula $J = M/T$, $[1/s]$ for each density $\rho = N/100$ is used, where $T = \sum_{j=1}^{K} T_j/K$ is the average time over K runs required for M people to cross the control line. The corresponding specific flow is

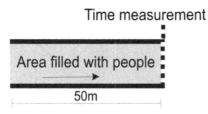

Fig. 3. Geometry set up. Corridor 50×2 m^2 in size, initial position of people (modelling area), and control line.

$J_s = M/T/2$ [1/(ms)]. This way of estimating the flow is similar to the method used in natural experiments to obtain real data.

The density $\rho = N/100$ is used to estimate the distribution of people over the modeling area (100 m^2 gray area in Fig. 3) in the simulation experiment.

When comparing the simulation and reference data and interpreting the results, it is very important to pay attention to the acceptable ρ_{max} value in the mathematical model and reference data. For example, if a square of the person's projection in the model is assumed to be 0.125 m^2, the projection has the form of a circle with a radius of 0.2 m; then, we can put closed circles with $\rho_{max} = 6.25$, [1/m^2]. Thus, it is most correct to compare the simulation and reference data with similar ρ_{max} values.

3.3   SigmaEva Simulation Results

We considered a set of numbers of people N_i, $i = \overline{1, m}$ involved in the simulation. The corresponding densities are estimated as $\rho_i = N_i/100, i = \overline{1, m}$, [1/m^2]. Each person was assigned with a free movement speed of $v^0 = 1.66$ m/s. All persons were assigned with the same square of projection, specifically, 0.125 m^2.

As far as the shape of person's projection is a solid disc, the maximum number that can be placed in an area of 100 m^2 is 625, and the maximum density is $\rho_{max} = 6.25\,1/\text{m}^2$. In accordance with the reference data, it was reduced (see below).

A set of 500 runs for each $N_i, i = \overline{1, m}$ was performed and the average times were calculated: $T(\rho_i) = \sum_{j=1}^{500} T_j(\rho_i)/500, \ i = \overline{1, m}$, where $T_j(\rho_i)$ is the time required for $M = 1000$ people to cross the control line in one run at given ρ_i.

Figures 4 and 5 show the specific flows as a function of density $\rho = N_i/100, i = \overline{1, m}$. The simulation data were obtained with (3)–(5) as a formulas to calculate speed for each person each time step[4]. The simulation data are compared with the Weidmann, SFPE, and Kholshevnikov & Samoshin fundamental diagrams correspondingly.

---

[4] All the other initial data were the same for all simulation experiments, including the model parameters. Here we do not discus them and only present values $k_P = 9$, $k_W = 2, k_S = 40$.

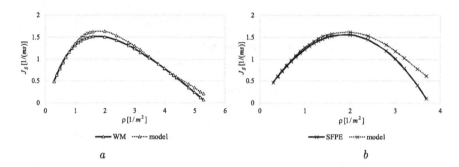

Fig. 4. (*a*) Original Weidmann data (*"WM"*) with $\rho_{max} = 5.4$ [1/m²] and simulation data (*"model"*) with (4) as an input data, $N_{max} = 540$ persons; (*b*) Original SFPE data with $\rho_{max} = 3.8$ [1/m²] and simulation data (*"model"*) with (5) as an input data and $N_{max} = 380$ persons.

It can be seen that the data in the three figures are very similar. In all cases, $\rho_{max}^{model} \geq \rho_{max}^{data}$: 6.25 [1/m²] versus 5.4 [1/m²] in Fig. 4(*a*), 6.25 [1/m²] versus 3.8 [1/m²] in Fig. 4(*b*), 6.25 [1/m²] versus 6.25 [1/m²] in Fig. 5(*b*). The other important factor is that the conditions of reference data ensure the same body size at all densities. This is consistent with the model statement that the projections of persons are solid discs with a constant radius.

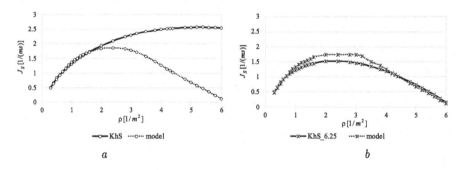

Fig. 5. (*a*) Kholshevnikov & Samoshin data (*"KhS"*) with $\rho_{max} = 15$ [1/m²] and simulation data (*"model"*) with (3) as an input data; $N_{max} = 600$ persons; (*b*) Kholshevnikov & Samoshin data (*"KhS_6.25"*) with $\rho_{max} = 6.25$ [1/m²] and simulation data (*"model"*) with (3) as an input data; $N_{max} = 600$ persons

The Kholshevnikov & Samoshin curve with $\rho_{max} \approx 15$ [1/m²] (Fig. 5(*a*)) gives the considerably higher flow at middle and high densities. Such maximum density can be obtained at smaller body sizes (square of projection) only. However, there is a lack of data on the ways of body size reduction. In Fig. 5(*a*), the model reproduces the expected behaviour of the specific flow under density variation: the flow J_s increases until a density of $\approx 2.5 - 3$ [1/m²], attains a value of 1.7–1.8 [1/(ms)], and then decreases. At the low and middle densities, the model flows agree excellently with the reference data. At the high densities, the model flow is much slower than the Kholshevnikov & Samoshin flow. This is apparently related to the strong impact of the constant square of person's projection.

Table 1. Quantitative measures for results presented in Figs. 4 and 5.

	Relative difference	Cosine	Projection coefficient
KhS	0.570195	0.852454	1.364238
WM	0.067695	0.998818	0.954949
SFPE	0.142952	0.992588	0.929509
KhS_6.25	0.068341	0.97618	0.946935

To compare curves quantitatively one can use a methods from [20]. There are three measures: Relative difference, Cosine, Projection coefficient (Table 1). The first norm provides a measure of the difference in the overall magnitude for the two curves normalized to the reference data. The norm approaches zero when the two curves are identical in magnitude. Cosine: the angle between the two vectors represents a measure of how well the shape of the two vectors match. As the cosine of the angle approaches unity, the two curves represented by the two vectors differ only by a constant multiplier. The projection coefficient provides a measure of the best possible fit of the two curves. When the projection coefficient approaches unity, remaining differences between the two curves is either due to random noise in the experimental measurements or physical effects not included in the model.

4   Conclusion

From a user point of view it is very convenient when a velocity-density dependence is input parameter of a model, and a model does not need a special tuning of model parameters to reproduce a desired velocity-density dependence as an output of the simulation. This paper shows that this is possible. The way which is used to design the presented discrete-continuous model gives such an opportunity. Of course some conditions should be taken into account. The best convergence of input and output curves is obtained if physical conditions of a simulation experiment and a reference data are close, and the main of them is condition on maximum density $\rho_{max}^{model} \geq \rho_{max}^{data}$.

References

1. Schadschneider, A., Klingsch, W., Kluepfel, H., Kretz, T., Rogsch, C., Seyfried, A.: Evacuation dynamics: empirical results, modeling and applications. In: Meyers, R. (ed.) Encyclopedia of Complexity and System Science, vol. 3, pp. 3142–3192. Springer, New York (2009). https://doi.org/10.1007/978-0-387-30440-3

2. Helbing, D., Farkas, I., Vicsek, T.: Simulating dynamical features of escape panic. Nature **407**, 487–490 (2000)
3. Chraibi, M., Seyfried, A., Schadschneider, A.: Generalized centrifugal-force model for pedestrian dynamics. Phys. Rev. E **82**, 046111 (2010)
4. Zeng, W., Nakamura, H., Chen, P.: A modified social force model for pedestrian behavior simulation at signalized crosswalks. Soc. Behav. Sci. **138**(14), 521–530 (2014)
5. Blue, V.J., Adler, J.L.: Cellular automata microsimulation for modeling bidirectional pedestrian walkways. Transp. Res. Part B **35**, 293–312 (2001)
6. Kirchner, A., Klupfel, H., Nishinari, K., Schadschneider, A., Schreckenberg, M.: Discretization effects and the influence of walking speed in cellular automata models for pedestrian dynamics. J. Stat. Mech. Theor. Exp. **10**, 10011 (2004)
7. Nishinari, K., Kirchner, A., Namazi, A., Schadschneider, A.: Extended floor field CA model for evacuation dynamics. IEICE Trans. Inf. Syst. **E87–D**, 726–732 (2004)
8. Kirik, E., Yurgel'yan, T., Krouglov, D.: On realizing the shortest time strategy in a CA FF pedestrian dynamics model. Cybern. Syst. **42**(1), 1–15 (2011)
9. Seitz, M.J., Koster, G.: Natural discretization of pedestrian movement in continuous space. Phys. Rev. E **86**(4), 046108 (2012)
10. Zeng, Y., Song, W., Huo, F., Vizzari, G.: Modeling evacuation dynamics on stairs by an extended optimal steps model. Simul. Model. Pract. Theor. **84**, 177–189 (2018)
11. Baglietto, G., Parisi, D.R.: Continuous-space automaton model for pedestrian dynamics. Phys. Rev. E **83**(5), 056117 (2011)
12. Schadschneider, A., Seyfried, A.: Validation of CA models of pedestrian dynamics with fundamental diagrams. Cybern. Syst. **40**(5), 367–389 (2009)
13. Kholshevnikov, V.: Forecast of human behavior during fire evacuation. In: Proceedings of the International Conference "Emergency Evacuation of People From Buildings - EMEVAC", pp. 139–153. Belstudio, Warsaw (2011)
14. Kirik, E., Malyshev, A., Popel, E.: Fundamental diagram as a model input: direct movement equation of pedestrian dynamics. In: Weidmann, U., Kirsch, U., Schreckenberg, M. (eds.) Pedestrian and Evacuation Dynamics 2012, pp. 691–702. Springer, Cham (2014). https://doi.org/10.1007/978-3-319-02447-9_58
15. Predtechenskii, V.M., Milinskii, A.I.: Planing for Foot Traffic Flow in Buildings. American Publishing, New Dehli (1978). Translation of Proektirovanie Zhdanii s Uchetom organizatsii Dvizheniya Lyudskikh potokov. Stroiizdat Publishers, Moscow (1969)
16. Chattaraj, U., Seyfried, A., Chakroborty, P.: Comparison of pedestrian fundamental diagram across cultures. Adv. Complex Syst. **12**, 393–405 (2009)
17. Weidmann, U.: Transporttechnik der Fussgänger. Transporttechnische Eigenschaften des Fussgängerverkehrs (Literaturauswertung). IVT, Institut für Verkehrsplanung, Transporttechnik, Strassen-und Eisenbahnbau, Zürich (1992)
18. Nelson, H.E., Mowrer, F.W.: Emergency movement. In: DiNenno, P.J. (ed.) The SFPE Handbook of Fire Protection Engineering, pp. 3-367–3-380. National Fire Protection Association, Quincy (2002)
19. Kirik, E., Vitova, T., Malyshev, A., Popel, E.: On the validation of pedestrian movement models under transition and steady-state conditions. In: Proceedings of the Ninth International Seminar on Fire and Explosion Hazards (ISFEH9), St. Peterburg, pp. 1270–1280 (2019)
20. Peacock, R.D., Reneke, P.A., Davis, W.D., Jones, W.W.: Quantifying fire model evaluation using functional analysis. Fire Saf. J. **33**(3), 167–184 (1991)

Simulating Pedestrians' Motion in Different Scenarios with Modified Social Force Model

Karolina Tytko[(✉)], Maria Mamica[(✉)], Agnieszka Pękala[(✉)],
and Jarosław Wąs[(✉)]

AGH University of Science and Technology, Krakow, Poland
{kartytko,mamica,apekala}@student.agh.edu.pl, jarek@agh.edu.pl

Abstract. A model created by Helbing, Molnar, Farkas and Vicsek [1] in the beginning of 21st century considers each agent in pedestrian movement as separate individual who obeys Newton's laws. The model has been implemented and simulated by numbers of different authors who proved its reliability through realism of agents' behaviour. To describe the motion as accurately as possible, many of them modified it by presenting their own approach of used formulas and parameters. In this work, authors consider combination of various model modifications as well as present adequate factors values, which allows to observe correct, consistent simulation of different evacuation scenarios and to track changes of Crowd Pressure in subsequent stages of visualization, depending on used exit design.

Keywords: Crowd Pressure · Social Force Model · Pedestrian dynamics

1 Introduction

Modelling and analyzing crowd motion is a crucial factor in constructing buildings, planning roads and paths locations as well as designing convenient areas for concerts, parades and many various events. Despite individual preferences of each pedestrian, which may be caused by their personality, environment, cultural background [5] or simply by their destination, we can predict crowd movement surprisingly well and accurately by making appropriate assumptions [3]. To do that, we decided to base our simulation on the Social Force Model, which was presented for the first time by Dirk Helbing and Peter Molnar in the 1990's [2]. Since then it was modified by researchers from all over the world to improve results by including new significant factors describing behaviour in specific situations (such as belonging to a group of friends or attraction to store exposition), calibrating values of model's constants and variables, or by presenting alternative ways of formula derivation.

In spite of wide acceptance of Social Force Model among researchers and within other communities that benefit from its implementation, many express

© Springer Nature Switzerland AG 2020
R. Wyrzykowski et al. (Eds.): PPAM 2019, LNCS 12044, pp. 467–477, 2020.
https://doi.org/10.1007/978-3-030-43222-5_41

the criticism towards models' 'Molecular Dynamics', due to the substantial simplification of the real world's principles. Furthermore, results of the model were not explained in consistency with psychological findings, which leads to misunderstanding it by psychologists [9]. However, modification made by Vicsek and Farkas in 2000 [1] justifies the usage of MD by aiming to create a simple but consistent mathematical model, which describes physical interactions within the crowd. Moreover, in their work, they present a universal formula for both panic and stress-free situations. This particular model was the one that we adapted the most and considered to be precise enough to become a starting point for our modifications.

The rest of this paper is organised as follows. Section 2 gives an overview of the social force model presented by Helbing et al. [1]. This is also where we introduce differences between panic and normal situations. Section 3 is where we describe our model in detail and show modifications of HMFV's model, that are visible in our approach. Section 4 shows important elements of the implementation and lists used technologies. We also focused on comparison of scenarios and pedestrians' behaviour analysis. Section 5 concludes the results of our simulation and compares it with work of other researchers.

2   Overview of Social Force Model

2.1   Panic and Normal Situations

As the model is accurate for both normal and panic situations, we present the differences between them. In normal situations pedestrians keep distances between neighbours, as well as obstacles [8]. Their movement speeds are different according to many factors like age, gender, situational context and so on. Often people do not look for the fastest route, but for the most comfortable and classical rules of proxemics are visible [10].

In competitive or panic situations the level of nervousness and stress rises. People try to move at higher speeds, often one can observe physical interactions between pedestrians and growing jams in front of bottlenecks. As a consequence, the "faster-is-slower" effect is often observed, when physical interactions and increasing pressure decrease the speed of pedestrians.

A dangerous development of this effect is the effect called "freezing-by-heating", when motion of pedestrians is completely reduced due to growing pressure of crowd. The effect can be significantly reduced by introducing appropriate obstacles in the area of accumulation of pressure and forces in front of a bottleneck.

2.2   HMFV's Approach

To simulate different evacuation scenarios and compare their results, we chose a model presented by Helbing, Molnar, Farkas and Vicsek as a starting point [1]. In the article 'Simulation of pedestrian crowds in normal and evacuation situations' authors proved that dynamics of pedestrians can be described with one consistent way regardless of the fact whether it is normal or panic situation. Since understanding mechanisms which explain people's motion is a crucial part of planing correct exit design, we decided to follow HMFV's approach which shows the similarities between pedestrians and their movement as well as points out distinctions that may be caused by environment status.

The HMFV's model for panic-free situations consists of sum of five main components, presented below [1]:

- **Acceleration Term** responsible for adaptation of current velocity vector into desired velocity vector;
- **Repulsive Social Force** $f_{ij}^{soc}(t)$ which describes 'territorial effect' and the tendency to keep certain distance to other pedestrians;
- **Attractive Social Force** $f_{ij}^{att}(t)$ which corresponds with joining behaviour of families or friends;
- **Boundary Force** $f_{ib}(t)$ which shows interaction with boundaries;
- **Attractive Force toward items** $f_{ik}^{att}(t)$ such as window displays, advertisements or other distractions.

To describe phenomenon of pedestrian movement in panic and evacuation situations, we need to simply include an additional force – **Physical Interaction Force** $f_{ib}^{ph}(t)$, which represents behaviour, when neighbours are so close to each other that they have physical contact. To do that, we have to consider **Body Force** counteracting forces that neighbours act on certain pedestrian as well as **Sliding Friction Force**. Analogically we can describe physical interaction force with boundaries.

Similar distribution of forces could be found in many studies [5,6]. However they were published years after HMFV model was presented, hence roots of solutions shown in them usually derive from articles already accepted by scientist community.

3   Presented Model

In this report, we focus our attention on the Helbing-Molnár-Farkas-Vicsek (HMFV) model [1]. Hence, the force model of pedestrian motion is represented by the equation below:

$$f_i(t) = f_i^{acc}(t) + \sum_{j(\neq i)} (f_{ij}^{soc}(t) + f_{ij}^{att}(t) + f_{ij}^{ph}) + \sum_b f_{ib}(t) + \sum_k f_{ik}^{att}(t) \qquad (1)$$

Herein, $f_i^{acc}(t)$ represents the acceleration term. In our model it is described as [1]:

$$f_i^{acc}(t) = \frac{v_i^0(t)e_i^0(t) - v_i(t)}{\tau_i} m_i \qquad (2)$$

Here, $v_i(t)$ denotes the actual velocity, $v_i^0(t)$ - desired speed and $e_i^0(t)$ is direction within a certain "relaxation time" τ_i.

$f_{ij}^{soc}(t)$ denotes repulsive social force. We use the formula from HMFV report [1], which is implemented as follows:

$$f_{ij}^{soc}(t) = A_i exp(\frac{r_{ij} - d_{ij}}{B_i})\boldsymbol{n}_{ij} \left(\lambda_i + (1 - \lambda_i)\frac{1 + cos(\varphi_{ij})}{2} \right) \tag{3}$$

Here, A_i is the interaction strength and B_i is the range of the repulsive interactions. Both are individual and culture-dependent parameters. Distance between the pedestrians i and j is denoted by d_{ij}, whereas r_{ij} is the sum of their radii r_i and r_j. Symbol n_{ij} stands for normalized vector pointing from pedestrian j to i. The angle φ_{ij} is the angle between $e_i^0(t)$ and n_{ij}. Assuming that $\lambda_i < 1$ can add an anisotropy to pedestrian's interaction. It means that with the help of this parameter we can model that the situation in front of the pedestrian has a bigger impact on his behavior than the situation behind him. In our model, we assume $\lambda_i = 0$. In this case the interaction forces are isotropic and complies with the Newton's 3rd law [1].

$f_{ij}^{att}(t)$ denotes the attractive social force. It reflects how people behave when they are surrounded by family members or friends. In this report, we use simplified version of the model and assume $f_{ij}^{att}(t) = 0$. Even though the attractive social force influences on people's behaviour, the main focus of the research is the evacuation time, not the pedestrian's trajectory of movement.

In this report, $f_{ik}^{att}(t)$ is the attractive force toward items. We assume $f_{ik}^{att}(t) = 0$, as during the evacuation the attraction toward window displays, advertisements or other distractions may be considered irrelevant.

Physical interaction forces f_{ij}^{ph} represent the interaction of an actor and his neighbours. They start to play a role when pedestrians are so close to each other that they have the physical contact. As they added up, the pedestrians' movements become unintentional. The forces are described as [1,7]:

$$f_{ij}^{ph} = f_{ij}^{body} + f_{ij}^{slidingfriction} \tag{4}$$

Here, f_{ij}^{body} stands for the body force. It represents the force that counteracts body compression.

$$f_{ij}^{body} = kg(r_{ij} - d_{ij})\boldsymbol{n}_{ij} \tag{5}$$

$f_{ij}^{slidingfriction}$ represents the "sliding friction" between pedestrian bodies.

$$f_{ij}^{slidingfriction} = \kappa g(r_{ij} - d_{ij})\Delta v_{ji}^t \boldsymbol{t}_{ij} \tag{6}$$

Herein, k and κ denotes large constants. $g(x)$ is the function, that returns x, if $x \geq 0$, otherwise 0. In this report, $t_{ij} = (-n_{ij}^2, n_{ij}^1)$ means the tangential direction to n_{ij} and $\Delta v_{ji}^t = (v_j - v_i) \cdot t_{ij}$ is the tangential velocity difference.

The interaction with walls and other obstacles is treated analogously [5]:

$$f_{ib}(t) = f_{ib}^{body} + f_{ib}^{slidingfriction} \tag{7}$$

$$f_{ib}^{body} = (A_i exp(\frac{r_i - d_{ib}}{B_i}) + kg(r_i - d_{ib}))\boldsymbol{n}_{ib} \tag{8}$$

$$f_{ib}^{sliding friction} = \kappa kg(r_i - d_{ib})\boldsymbol{t}_{ib} \tag{9}$$

Here, A_i, B_i, κ and k denote constant parameters of our model. d_{ib} stands for the distance between the boundary of the obstacle and the pedestrian's center. n_{ib}, t_{ib} denote the vector normal and tangential to the obstacle's boundary at the point where the pedestrian comes in interaction with it.

Symbol	Value	Description
A	$-mv^0/\tau$	Attraction to an exit
B	$\{1, 5\}$	Fall-off length of social repulsive force
k	$2400\,\mathrm{kg/s^2}$	Spring constant
κ	1	Coefficient of sliding friction
m	$60\,\mathrm{kg}$	Mass
r	0.35 m	Radius
τ	0.5 s	Relaxation time
v^0	1.34 m/s	Desired speed
v_{max}	1.7 m/s	Maximum speed

Values of parameters used in our simulation are presented in the table above. They were chosen and calibrated basing on solutions published in many other articles [1,5,7] as well as basing on calibration done by authors. For implementation and scene rendering purposes we needed to scale those values using proportionality factor equals to 25.

4   Simulation

We consider various scenarios of evacuation situations, where we present different 'obstructions' located in front of the exit. According to Kirchner [4] and Yanagisawa [12], such items (including barriers and tunnels) reduce both time needed for evacuation and pressure on the clogging point. To check efficiency of such solution, we compared results from three main scenarios presented on the Fig. 1.

Apart from that, the simulation allows us to compare a variety of factors which are significant for analyzing model efficiency such as actors' velocity, local density or crowd pressure, which is represented by the color-change of each pedestrian. These statistics enable us not only to plot the results but also to validate applied parameters and formulas.

Fig. 1. Frames from simulation. From left to right: no additional obstruction added to simulation; vertical barrier; tunnel.

4.1 Implementation

The simulation is fully implemented with the use of object-oriented programming language Java. Its support for *OOP* paradigm let us treat pedestrians, boundaries and obstacles as instances of previously declared classes, so it allows us to focus on general behaviour of all actors. To render each frame, we have to update the state of each pedestrian basing on their current position and environment. Below, we present a simple and schematic Algorithm 1 to achieve it.

Algorithm 1. Updating pedestrians' positions in order to render next frame

---

1: **function** UPDATEPEDESTRAINSPOSITIONS(*listOfPedestrians*)
2:    **for** *pedestrian* ∈ *listOfPedestrians* **do**
3:       **if** pedestrian.REACHEDTARGET is true **then**
4:          **continue**
5:       **end if**
6:       *accelarationForm* ← pedestrian.CALCULATEACCELARATIONFORM
7:       *neighboursInteraction* ← 0
8:       *wallsInteraction* ← 0
9:       **for** *neighbours* ∈ *listOfPedestrians* **do**
10:          **if** *neighbour* ≠ *pedestrian* **then**
11:            *neighboursInteraction* ← *neighboursInteraction*∪ pedestrian.CALCULATENEIGHBOURINTERACTION(*neighbour*)
12:          **end if**
13:       **end for**
14:       **for** *wall* ∈ *listOfWalls* **do**
15:          *wallsInteraction* ← *wallsInteraction*∪ pedestrian.CALCULATEWALLINTERACTION(*wall*)
16:       **end for**
17:       *forcesSum* ← *accelarationFrom* ∪ *neighboursInteraction* ∪ *wallsInteraction*
18:       pedestrian.*newVelocity* ← pedestrian.setNewVelocity(*forcesSum*)
19:       pedestrian.*newPosition* ← pedestrian.setNewPosition(pedestrian.*newVelocity*)
20:       pedestrian.*crowdPressure* ← pedestrian.SETCROWDPRESSURE(*forcesSum*)
21:    **end for**
22: **end function**

4.2   Comparison of Scenarios

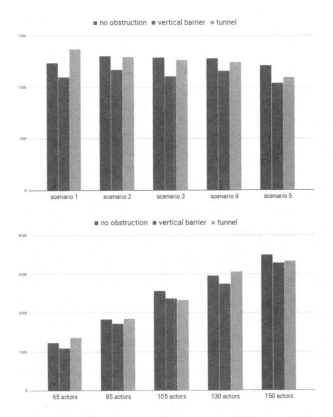

Fig. 2. Top: Comparing evacuation time for 5 groups of the same size, but different starting points. Bottom: Comparing evacuation time for 5 groups of different size. No obstruction is marked with blue colour, vertical barrier with red and tunnel with yellow. (Color figure online)

To avoid queues, clogging, 'freezing by heating' situations and crowd disasters, it is crucial to minimize the time of crowd evacuation. As mentioned before 4, locating certain obstructions next to the exit reduces pressure, which results in faster evacuation. We analyze three scenarios (see Fig. 1) and different default starting points for groups of various sizes. In the Table 1, we present results of five simulations, which describe behaviour of different groups. Duration is represented by the amount of simulation frames that were needed to evacuate the whole group. Rows marked by green colour points to the evacuation with the minimal duration.

Table 1. Comparison of evacuation times for different groups and various scenarios.

65 actors

scenario	duration
no obstruction	1235
vertical barrier	1095
tunnel	1368

85 actors

scenario	duration
no obstruction	1833
vertical barrier	1721
tunnel	1852

105 actors

scenario	duration
no obstruction	2559
vertical barrier	2367
tunnel	2328

130 actors

scenario	duration
no obstruction	2958
vertical barrier	2746
tunnel	3067

150 actors

scenario	duration
no obstruction	3493
vertical barrier	3281
tunnel	3331

Analyzing groups of various sizes and different starting point of every pedestrian, we can easily spot an improvement in evacuation time, when a vertical barrier is added (Fig. 2). However, as far as a tunnel is concerned, we cannot determine whether its presence decreases simulation duration. In most of the cases, the time needed to exit a room is slightly shorter than in a scenario where no obstruction is involved. Nevertheless, in some of the considered situations, the tunnel increases the time of evacuation, which is truly undesirable.

4.3   Behaviour Analysis

Apart from the comparison of different scenarios and results analysis, we can observe various crowd dynamics phenomena only by plotting the statistics. Accuracy of them shows that presented model is correct, therefore allows us to describe remarked behaviours.

As far as evacuation situations are concerned, we can assume that all pedestrians move towards one specific direction – the exit. This causes accumulation of actors in one spot as well as an increase of local density d for each of pedestrians in the surrounding crowds. The density is measured in pedestrian/m^2 and according to Weidmann [5,11], its maximum value is reported to equal 5.4 ped/m^2 due to high probability of being trampled by other people when d exceeds this constant.

In the Fig. 3 we present diagrams, which are based on the data collected from the simulation. They showcase local density change for certain pedestrian randomly chosen from the simulation. The left one demonstrates safe evacuation – density value increases, because of pedestrian's approach to the exit, and reaches its maximum value (which is lower than Weidmann's d_{max}) [11]. This maximum value represents the moment when an actor gets through the exit, hence density around them decreases afterwards. To collect data for the diagram on the right Fig. 3, we increased significantly the number of pedestrians in the scenario. It presents a highly unsafe situation, when global maximum exceeds d_{max} even by 30%. We decided to do that to observe 'stop-and-go flow' phenomenon [13]. The combination of extremely high density and pedestrian body collisions generates crowd turbulence as well as break down of smooth pedestrian flow, which results in 'stop-and-go' waves presented in Fig. 4. Such behaviour was observed during many crowd disasters [7].

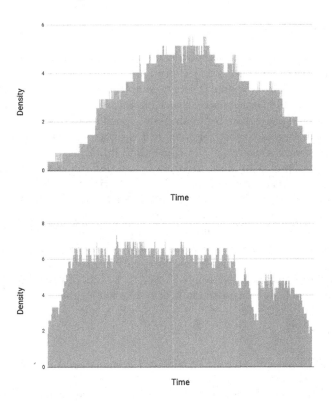

Fig. 3. Local density change in time for panic situations in a scenario with a vertical barrier. On the top, maximum value is lower than $d_{max} = 5.4$ ped/m^2, while on the bottom global maximum exceeds d_{max}.

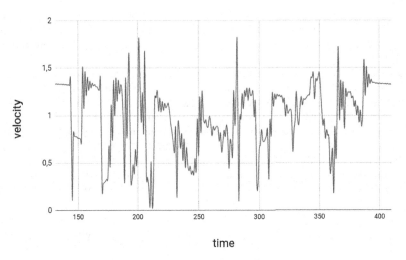

Fig. 4. Change of velocity in time in a scenario with a vertical barrier. Representation of the 'stop-and-go flow' phenomenon.

5   Conclusion

Presented model is a combination of well known and widely accepted approaches with our minor modifications and calibration. The assumption which indicates that both attraction forces equal 0, lets us simplify the model and treat actors as goal-oriented individuals, who are neither distracted by 'time-dependent attractive interactions' ($f_{ik}^{att}(t)$) or by joining any specific group of people ($f_{ij}^{att}(t)$). This solution allows us to test different exit designs as well as analyze motion and interaction of strangers, at the same time imitating events from the real world such as 'stop-and go' flow, 'freezing by heating' phenomenon, or pressure vs local density correlation. We believe that interesting behaviour noticed in this model implementation and various, potential exit designs can be an entry point for numbers of researches and modifications, mostly because of attraction forces omission, which decreases computational and time complexity of the algorithm. Testing a solution on simplified, reliable model might be a crucial part of finding the correct consisted design for perfect exit. Mostly because some of them might be proven inefficient beforehand (see the results of tunnel obstruction in Table 1) and save a significant amount of researcher's time.

References

1. Helbing, D., Farkas, I., Molnar, P., Vicsek, T.: Simulation of pedestrian crowds in normal and evacuation situations, vol. 21, pp. 21–58, January 2002
2. Helbing, D., Molnar, P.: Social force model for pedestrian dynamics. Phys. Rev. E **51** (1998). https://doi.org/10.1103/PhysRevE.51.4282
3. Helbing, D., Molnar, P., Farkas, I., Bolay, K.: Self-organizing pedestrian movement. Environ. Plann. B Plann. Design **28**, 361–383 (2001). https://doi.org/10.1068/b2697
4. Kirchner, A., Nishinari, K., Schadschneider, A.: Friction effects and clogging in a cellular automaton model for pedestrian dynamics. Phys. Rev. E Stat. Nonlinear Soft Matter Phys. **67**, 056122 (2003). https://doi.org/10.1103/PhysRevE.67.056122
5. Lakoba, T., Kaup, D., Finkelstein, N.: Modifications of the Helbing-Molnár-Farkas-Vicsek social force model for pedestrian evolution. Simulation **81**, 339–352 (2005). https://doi.org/10.1177/0037549705052772
6. Moussaïd, M., Helbing, D., Garnier, S., Johansson, A., Combe, M., Theraulaz, G.: Experimental study of the behavioural mechanisms underlying self-organization in human crowds. Proc. Biol. Sci. R. Soc. **276**, 2755–62 (2009). https://doi.org/10.1098/rspb.2009.0405
7. Moussaïd, M., Helbing, D., Theraulaz, G.: How simple rules determine pedestrian behavior and crowd disasters. Proc. Natl. Acad. Sci. U.S.A. **108**, 6884–6888 (2011). https://doi.org/10.1073/pnas.1016507108
8. Porzycki, J., Wąs, J., Hedayatifar, L., Hassanibesheli, F., Kułakowski, K.: Velocity correlations and spatial dependencies between neighbors in a unidirectional flow of pedestrians. Phys. Rev. E **96**, 022307 (2017). https://doi.org/10.1103/PhysRevE.96.022307
9. Wang, P.: Understanding social-force model in psychological principles of collective behavior. Ph.D. thesis, May 2016

10. Wąs, J., Lubaś, R., Myśliwiec, W.: Proxemics in discrete simulation of evacuation. In: Sirakoulis, G.C., Bandini, S. (eds.) ACRI 2012. LNCS, vol. 7495, pp. 768–775. Springer, Heidelberg (2012). https://doi.org/10.1007/978-3-642-33350-7_80
11. Weidmann, U.: Transporttechnik der fugänger. Schriftenreihe des Institut für Verkehrsplanung, Transporttechnik, Straen-Und Eisenbahnbau **78**, 62–64 (1993)
12. Yanagisawa, D., et al.: Introduction of frictional and turning function for pedestrian outflow with an obstacle. Phys. Rev. E Stat. Nonlinear Soft Matter Phys. **80**, 036110 (2009). https://doi.org/10.1103/PhysRevE.80.036110
13. Yu, W., Johansson, A.: Modeling crowd turbulence by many-particle simulations. Phys. Rev. E Stat. Nonlinear Soft Matter Phys. **76**, 046105 (2007). https://doi.org/10.1103/PhysRevE.76.046105

Traffic Prediction Based on Modified Nagel-Schreckenberg Model. Case Study for Traffic in the City of Darmstadt

Łukasz Gosek[(✉)], Fryderyk Muras, Przemysław Michałek, and Jarosław Wąs

AGH University of Science and Technology, al. Mickiewicza 30,
30-059 Kraków, Poland
lgosek@student.agh.edu.pl, fryderyk.muras@gmail.com, przemcom@gmail.com,
jarek@agh.edu.pl

Abstract. In this study the authors present a model for traffic flow prediction based on Nagel-Schreckenberg Cellular Automata model. The basic model was expanded to allow simulation of multi-lane roads along with representing different types of cars and drivers' psychological profiles found in urban traffic. Real traffic data from sensors in Darmstadt city (Germany) were used to perform a set of simulations on presented model.

Keywords: Traffic prediction · Cellular Automata · Nagel - Schreckenberg model

1 Introduction

Traffic draws attention of numerous scientists as traffic jams become one of the biggest problems of urban areas. Increasing number of vehicles passing through urban streets on a daily basis in connection with unadapted to its requirements infrastructure causes inconvenience not only for drivers, but also for all citizens. It should be stressed that the results of computer simulations are increasingly regarded as a very good source for planning repairs or development of communication infrastructure in given areas. Thus, one can observe an increase in interest in traffic simulations in recent years.

Cellular Automaton model created by Nagel and Schreckenberg [6] is a classical model describing flow of traffic movement in motorways. The flow depends on few coefficients (e.g. imperfect driving style, slower vehicles, accidents) which play key role in creation of traffic jams, especially in conjunction with increasing number of vehicles on streets [2]. With increasing density of cars a phase transition is observed: from fluent flow to overload. In fluent flow phase vehicles move with speed close to allowed speed limit which in urban environment increases their density. The overload phase, on the contrary to fluent flow, has negative linear correlation between traffic flow and vehicle density. It was proven that in overload phase driver's behaviour plays key role in field distribution of vehicles

© Springer Nature Switzerland AG 2020
R. Wyrzykowski et al. (Eds.): PPAM 2019, LNCS 12044, pp. 478–488, 2020.
https://doi.org/10.1007/978-3-030-43222-5_42

between each other [4]. One of solutions to traffic jam problem is a roundabout - an intersection without traffic lights offering favourable solutions in terms of flow safety. Roundabout was introduced for the first time in Great Britain and currently is widely used in majority of countries. Drivers approaching a roundabout are forced to decrease vehicle's velocity and check for potential collisions with other vehicles that already are in the roundabout before entering it.

It was conducted in many reports that risk of accident significantly decreased in comparison to conventional intersections [7]. On the other hand accidents in the roundabout are one of the most frequent causes of traffic jam creation. Research conducted on collisions of 38 roundabouts in the state of Maryland concluded that accidents happened more often at the entry than in, or at the exit of the roundabout [5].

In the current paper we propose an adaptation of Nagel - Schreckenberg (further *Na-Sch*) model that is strongly based on previous solutions [1,3,9–11]. We extended given model by randomising parameters of the model such as maximum velocity assigned to each car. Additionally, we incorporated a mechanism of intersections that allows cars to change the road they are currently on.

Our goal was to create a not very complicated adaptation of the *Na-Sch* traffic model for urban conditions, which in the future we will be a basis for large-scale simulations of traffic. In order to test the usability of the model, we test it in a confrontation with real data, namely traffic characteristics in the city of Darmstadt.

2   Mathematical Models

Cellular Automata are presented by a four (L, S, N, f), where each subsequent element stands for: space as a network of cells, collection of states, collection of given cell's neighbors and a function of configuration change in particular cells. The configuration $C:L \rightarrow S$ is a function connecting each cell with a state. Function f changes configuration C_t into a new configuration C_{t+1}.

Classic and most popular model describing vehicle movement (based on cellular automatons) is the Nagel-Schreckenberg [6] model. By default this model describes vehicle movement on a highway but after few modifications it can also be used to describe urban traffic. In *Na-Sch* cell size is stated as $d = 7.5$ m. In each cell there can be only one vehicle (or none), vehicle's velocity is described as number of cells travelled during next iteration. *Na-Sch* is specified by following rules of movement:

- Acceleration: $v(t + 1) \rightarrow \min(v(t) + 1, v_{max})$, where $v(t)$ is current velocity
- Braking: $v(t + 1) \rightarrow \min(v(t), g(t))$, where $g(t)$ is a number of empty cells between vehicles
- Random element (unexplained, illogical braking): probability p that $v(t + 1) \rightarrow \max(v(t) - 1, 0)$
- Movement: $x(t + 1) = x(t) + v(t)$

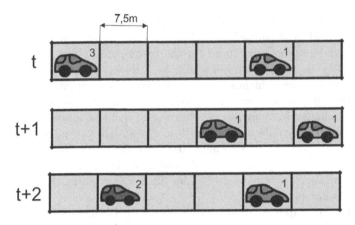

Fig. 1. Movement in *Na-Sch* for one traffic lane. Numbers at vehicles stand for current velocity → number of cells travelled during iteration (source: [11])

Amongst adaptations of Nagel-Schreckenberg model one can points out Hartman's [3] work, where author includes not only multiple traffic lanes but also roundabouts, conventional intersections, different vehicle lengths for particular vehicles (motorcycle, car, truck etc.) and modifies velocities accordingly. Publication [8] describes traffic movement in a very detailed way, replacing every intersection with a roundabout. Additionally some behaviors of drivers can be explained in terms of Agent-Based Modeling [1].

The main reason why pure *Na-Sch* model couldn't guarantee real results for urban traffic is it's primary assumption: it describes movement on a highway. Highway traffic is diametrically different than urban traffic, where vehicles alternately accelerate and brake. Almost every vehicle moves with different velocity than other close-by vehicle which requires different approach. The differences between modelled urban traffic and reality are described in the book [10].

3   Proposed Model

3.1   Model Description

Model presented by the authors is based on classic Nagel-Schreckenberg model and a generic two-lane model (more suitable for representing urban traffic) presented by Rickert, Nagel, Schreckenberg and Latour [9]. Changes introduced to those representations include:

- length of automaton cell changed to 2.5 m in order to adjust the model for real-life urban traffic speeds (the effect of this modification is that one vehicle does not fit in a single cell but the main purpose of the model was to simulate the distribution and flow of motion in general and not to simulate it accurately in micro-scale)

- maximum speed of each car is randomly generated from range [3; 5] in order to simulate different vehicle types and psychological profiles of drivers (i.e. some drivers, especially elderly ones tend to drive slower in contrast to young drivers that more often are driving in a reckless way)
- periodic boundary for conditions are no longer present – a car that exits one road is transferred to another road at the same intersection
- generic two-lane model was expanded in order to model a multi-lane road
- modified movement algorithm – in each iteration, apart from updating cars speed and position, the possibility of lane changes are considered

Single street is represented by an individual automaton for each lane with specified rules of transition between them. In each iteration a car i switches lane to another when following conditions are satisfied (from [9]):

$$gap(i) < l \tag{1a}$$
$$gap_o(i) > l \tag{1b}$$
$$gap_{o,back} > l_{o,back} \tag{1c}$$
$$rand() < p_{change} \tag{1d}$$

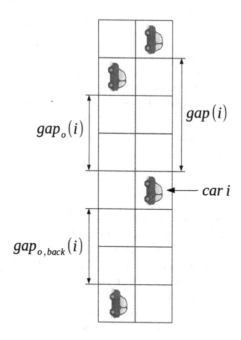

Fig. 2. Graphical presentation of the lane changing rule with "gap" parameters

$gap(i)$ denotes numbers of free cells in front of a car on current lane, $gap_o(i)$ is an analogical distance on the destination lane, $gap_{o,back}(i)$ represents distance to the nearest car behind on destination lane, l and $l_{o,back}$ are parameters defining how far the driver looks ahead and behind on appropriate lanes (both are determined by maximum speed), p_{change} denotes probability of lane change (this represents irrational behaviour of a driver – staying on current lane despite satisfaction of necessary conditions for a lane change).

Lane changes are not symmetrical – when considering switching to a lane closer to the right side of the road condition 1a does not apply. That means that a driver tries to return to the rightmost lane even if situation on his current lane does not urge him to do so. This represents the rule, enforced by traffic law, of driving as close to right edge of the road as possible (for a right-hand traffic).

In each iteration, for each lane on a road there are two steps of simulation: firstly lane changes are conducted – conditions given in 1 are checked for every car. After that, in step two, we apply regular *Na-Sch* rules for updating current speed and position of each car.

This representation of a single street is a building block for creating a model of a road network. Those blocks are joined together by a simple intersection model. When a car travels "off" the current road it is transitioned to one of the adjacent roads on the particular intersection (with certain probability distribution). On each intersection a simple traffic lights mechanism was implemented to provide collision avoidance.

The simple mechanism of traffic lights in our model is based on the assumption that cars located on only one road at the time are allowed to enter an intersection. In addition, the intervals of traffic lights changes are independent of traffic volume or time of day.

On outer edges of modelled area authors implemented a mechanism of generating new cars to populate simulation area and removing the ones leaving it. Generation of new cars happens with probability based on external data.

Cars that are generated do not have any target destination, but move around the map in a random way, associated with the probability distribution assigned to each intersection.

3.2   Model Implementation

Model described in previous section has been implemented as a Python application. Matplotlib library has been used for developing a user interface.

The input of presented application is a data set containing information about traffic volume outside represented area. Upon that data the amount of cars generated in each iteration is determined. Every minute of simulation the application returns information about traffic density at each intersection inside simulated area ("virtual sensors"). Virtual sensor is basically a counter that counts cars which have gone through the intersection in the last minute of simulation.

4   Simulation

4.1   Modelled Area

For the purpose of simulation a portion of Darmstadt city was modelled. Darmstadt is a city located in southwest Germany and has a population of about 157,400 (at the end of 2016). A company "[ui!] Urban Software Institute GMBH"[1], located in Darmstadt, provides data about traffic, based on readings from sensors distributed at about 200 intersections in the city.

Modelled area covers eight intersections connected by ten roads with various number of lanes. Chosen area is presented in Fig. 3.

Fig. 3. Modelled area with roads marked

4.2   Model Calibration

The calibration of the model was focused on finding an appropriate assignment of probability distributions for each intersection. The other parameters of the model were determined before calibration and remained unchanged.

---

[1] https://www.ui.city/de/.

4.3   Simulation Procedure

Throughout the simulation, the parameters of the intersections remain unchanged. Only the parameters responsible for the frequency of new cars are modified. Thanks to the modification of these parameters we were able to adjust the intensity of traffic flowing into the modelled part of the city. These parameters were calculated for each minute of the simulation directly from real data from sensors. At the beginning of the simulation, the grid is empty.

5   Simulation Results

5.1   Conducted Simulations

The aim of the first experiment was to examine the distribution of traffic generated upon input data from sensors outside simulated area. Output data from the simulation were later compared with actual data from sensors at considered area.

The intention of the second experiment was to check the models response for a significantly greater traffic load (in our case five-fold the normal traffic density). The results of this experiment were evaluated qualitatively, the analysed factors were: occurrence and distribution of traffic jams and general smoothness of traffic flow.

5.2   Results

In the first experiment the simulation was performed five times and average value of traffic flow for each "virtual sensor" was calculated. Figures 4, 5, 6, 7 and 8 show the comparison of those mean values and measurement data from real sensors for each minute of simulation. Results are presented for each minute separately in order to present the accuracy of the model over time. Figure 9 shows relative error between mean values and measurement data in every minute of simulation for each intersection.

Values 0 appearing for sensor A99 were present in data from real sensors and have been taken into account while establishing parameters of the model.

Results of the simulation show, that the data obtained during the simulation correlate strongly with the measurement data, except for intersection A104 in the fourth minute of the simulation. Likely cause for such non-compliance in this particular point is that the probability distribution was not selected sufficiently correctly. This issue is an area where a great deal of improvement can be achieved.

Nonetheless, we should not draw any firm conclusions about the accuracy of the model, since the simulation lasted only five minutes. However, the results obtained allow us to conclude that with appropriate modifications the *Na-Sch* model can be used for urban traffic modelling.

5.3   Numerical Indicators

In this subsection we present numerical indicators that quantify overall agreement between the model outcomes and the real data:

– Mean Absolute Error (MAE): *4.95*
– Root mean squared error (RMSE): *7.53*

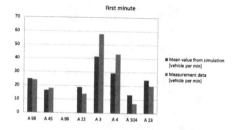

Fig. 4. The first minute of simulation

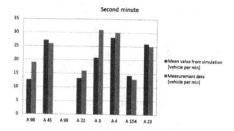

Fig. 5. The second minute of simulation

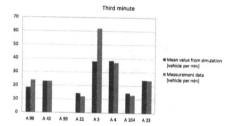

Fig. 6. The third minute of simulation

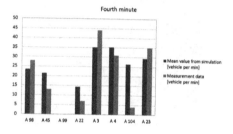

Fig. 7. The fourth minute of simulation

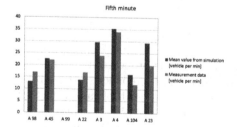

Fig. 8. Fifth minute of simulation

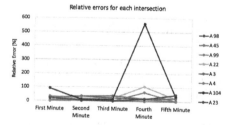

Fig. 9.   Relative   errors   for   each intersection

Results of the second experiment are presented in Figs. 10 and 11. Each red dot represents one vehicle. In presented screenshots one can observe a congestion occurring on the main roads.

Fig. 10. The second experiment (Color figure online)

Fig. 11. The second experiment (Color figure online)

6   Conclusions

The first of the experiments carried out shows that the proposed model and its parameters describe the analysed area properly. Point discrepancies between the results of the simulation and the actual data may be the result of various factors that are difficult to predict and model (e.g. blocked road due to delivery to a shop, bump, etc.). It is worth noting that there is no increase in the discrepancy between the simulation data and the measured data over time, which confirms the reliability of the model.

The second experiment assesses the ability of existing road infrastructure to accommodate a significantly higher number of cars. As can be seen in the attached simulation results, a five-fold increase in traffic volume compared to normal levels causes significant congestion on the most frequently used roads.

This is a common phenomenon and its prevention is a major challenge for road infrastructure designers. There is also an interesting phenomenon of complete congestion in Rheinstraße, one of the main arteries of the city, while there are almost empty roads parallel to it, which can unload the traffic jam. The phenomenon shows that the modelling of urban traffic presented in the project has great potential for the very desirable optimisation of urban traffic today by, for example, intelligent systems that can anticipate the possibility of congestion and prevent its formation by means of appropriate traffic light control.

Presented model can be a subject of further development in order to make it more realistic. This can be achieved by taking into account weather conditions, time of year, random events such as the passage of privileged vehicles (ambulance, fire brigade), crash or failure of traffic lights. After promising results of the simulation, the model should also be extended in the future to cover a significantly larger area than 8 junctions and during the simulation one should pay attention to the density of vehicles in particular sub-areas. However, we would like to show that using relatively simple modification of Nagel-Schreckenberg model adapted to city conditions/traffic we are able to gain reliable results. We believe that the proposed model will be a good basis for large scale models of traffic.

Acknowledgement. We would like to express our gratitude for [ui!] Urban Software Institute GMBH (https://www.ui.city/de/) for sharing their data regarding traffic in the city of Darmstadt.

References

1. Chmielewska, M., Kotlarz, M., Wąs, J.: Computer simulation of traffic flow based on cellular automata and multi-agent system. In: Wyrzykowski, R., Deelman, E., Dongarra, J., Karczewski, K., Kitowski, J., Wiatr, K. (eds.) PPAM 2015. LNCS, vol. 9574, pp. 517–527. Springer, Cham (2016). https://doi.org/10.1007/978-3-319-32152-3_48
2. Chowdhury, D., Santen, L., Schadschneider, A.: Statistical physics of vehicular traffic and some related systems. Physics reports **329**(4–6), 199–329 (2000)
3. Hartman, D.: Head leading algorithm for urban traffic modeling. Positions **2**, 1 (2004)
4. Járai-Szabó, F., Néda, Z.: Earthquake model describes traffic jams caused by imperfect driving styles. Physica A **391**(22), 5727–5738 (2012)
5. Mandavilli, S., McCartt, A.T., Retting, R.A.: Crash patterns and potential engineering countermeasures at Maryland roundabouts. Traffic Inj. Prev. **10**(1), 44–50 (2009)
6. Nagel, K., Schreckenberg, M.: A cellular automaton model for freeway traffic. J. Phys. I **2**(12), 2221–2229 (1992)
7. Persaud, B., Retting, R., Garder, P., Lord, D.: Safety effect of roundabout conversions in the United States: empirical Bayes observational before-after study. Transp. Res. Rec. J. Transp. Res. Board **1751**, 1–8 (2001)
8. Regragui, Y., Moussa, N.: A cellular automata model for urban traffic with multiple roundabouts. Chin. J. Phys. **56**(3), 1273–1285 (2018)

9. Rickert, M., Nagel, K., Schreckenberg, M., Latour, A.: Two lane traffic simulations using cellular automata. Physica A **2**, 17 (1995)
10. Wagner, P.: Traffic simulations using cellular automata: comparison with reality. Univ. (1995)
11. Wąs, J., Bieliński, R., Gajewski, B., Orzechowski, P.: Problematyka modelowania ruchu miejskiego z wykorzystaniem automatów komórkowych. Automatyka/Akademia Górniczo-Hutnicza im. Stanisława Staszica w Krakowie **13**(3/1), 1207–1217 (2009)

HPC Large-Scale Pedestrian Simulation Based on Proxemics Rules

Paweł Renc, Maciej Bielech[(✉)], Tomasz Pęcak, Piotr Morawiecki,
Mateusz Paciorek, Wojciech Turek, Aleksander Byrski, and Jarosław Wąs

AGH University of Sciences and Technology, al. Mickiewicza 30,
30-059 Kraków, Poland
maciej.bielech@gmail.com

Abstract. The problem of efficient pedestrian simulation, when large-scale environment is considered, poses a great challenge. When the simulation model size exceeds the capabilities of a single computing node or the results are expected quickly, the simulation algorithm has to use many cores and nodes. The problem considered in the presented work is the task of splitting the data-intensive computations with a common data structure into separate computational domains, while preserving the crucial features of the simulation model. We propose a new model created on the basis of some popular pedestrian models, which can be applied in parallel processing. We describe its implementation in a highly scalable simulation framework. Additionally, the preliminary results are presented and outcomes are discussed.

Keywords: HPC · Supercomputing · Pedestrian simulation · Crowd dynamics · Proxemics

1 Introduction

The purpose of simulations is to predict the behaviour of certain aspects of reality. Examples of their commercial applications are crowd dynamics during social events or city life in video games. Complex simulations, representing a wider spectrum of reality, are capable of yielding results both more precise and spanning over a larger simulation area. To satisfy increased resource requirements of such simulations, it is necessary to use computers with high computing power. Supercomputers meet this requirement due to high availability of nodes, oftentimes configurable in terms of the number of dedicated computational cores and memory.

Large-scale crowd simulations have the prospect of becoming an important tool in which one can test different crowd dynamics scenarios for public facilities and mass gatherings. Increasing availability of computing power enables the implementation of tasks in such a scale that previously was impossible to achieve/implement. However, it should be emphasized that the implementation of large-scale simulations differs significantly from standard simulations. One of

© Springer Nature Switzerland AG 2020
R. Wyrzykowski et al. (Eds.): PPAM 2019, LNCS 12044, pp. 489–499, 2020.
https://doi.org/10.1007/978-3-030-43222-5_43

the biggest challenges is the correct synchronization of individual computational domains.

In our work we present the use of the Xinuk framework[1] [2] to tackle these problems. It allows to distribute the calculations into many nodes while preserving statistical correctness of the result. It is worth noting that this solution introduces an additional load caused by the information synchronization between adjacent nodes. However, significant optimization of efficiency is achieved through unidirectional communication. This approach necessitates an appropriate modification of the model, which was created with the single-node environment in mind.

We cannot afford for the imbalance between the simplification or omission of model features and its adjustment to the solution discussed above, therefore our work focuses both on desynchronization of the distributed simulation, but also on retaining the correctness of the whole system. This introduction is followed by the presentation of the related works, then the proposed model is discussed and the implementation results are shown. Finally, the paper includes conclusions.

1.1   Related Works

In [9], the implementation of a model based on Cellular Automata with the use of the Social Distance Model was proposed. The model has been optimized for performing calculations using GPU technology. The key condition for efficient use of the GPU is to divide the calculations in such a way that the individual threads execute identical sets of functions - kernels - simultaneously. Therefore, the grid of the Cellular Automata was divided between the threads, in such a way that the one-threaded calculations were dealt with calculations related to one cell and executed the same functions in parallel with others.

According to the authors, the model is characterized by great performance and is able to realistically reproduce the movement of up to 10^8 pedestrians.

There are many simulation frameworks supporting execution in HPC environments, significant portion of which employ agent-based approach [1]. However, only a few of those are actually suited for large-scale in parallel and distributed infrastructures and capable of efficient utilization of available resources.

An example of such a tool is REPAST HPC [4], a framework operating on the Message Passing Interface (MPI) standard [11] which has been used for parallelizing a large-scale epidemiological model. Results and details of this study can be found in [5].

Another example of tool supporting large scale simulations in social context is FLAME [3] which uses a distributed memory model, Single Program Multiple Data (SPMD) with synchronization points. As in the case of REPAST, the FLAME framework also uses MPI as a means of communication between agents on different computational units. The example simulation applying FLAME

---

[1] A Scala/Akka distributed simulation framework developed at AGH University of Science and Technology.

framework can be found in [8] and includes a Carnot economic model of competing companies or modeling biological systems as ant colonies.

PDES-MAS framework, based on Parallel Discrete Event Simulation (PDES) paradigm, is also capable of running large scale social science simulations. This tool divides a simulation model into a network of concurrent logical processes, managing the shared state between agents using a tree-like network in a space-time Distributed Shared Memory (DSM) model [13].

The Pandora framework is yet another example of an MPI based tool with the possibility of using Cloud infrastructure [17] alongside traditional HPC resources.

An interesting simulation platform created specifically for complex artificial-life simulations is Framsticks [10]. As well as previously mentioned frameworks, it is capable of running in distributed environment.

All of the platforms mentioned above use standard technologies, like message passing (MPI) to create processes and perform communication in parallel and distributed environment – a well-established standard used for many years. The standard approaches employing MPI are implemented in C++, which can offer good performance, but is not particularly efficient in terms of rapid experiments development and does not provide flexibility present in more modern approaches.

A high degree of distribution provides a great magnitude of performance boost, but it also exhibits all the disadvantages of distributed computation. The performance of a distributed solution is prone to decrease as a result of inefficient data synchronization, inter-node communication and serialization. Potential instability of hardware, especially network communication, might introduce significant drawbacks on its own. Creating a solution that will maintain a consistent state throughout an entire large computational cluster is a serious challenge. It is necessary to put a lot of effort to ensure that the overhead is not outweighing the advantages of distribution, in result making it redundant and not worth utilizing the resources.

1.2 HPC Simulation Framework

The need for efficient computing in the domain of micro-scale models simulation is unquestionable. Use of High Performance Computing hardware in large scale simulations could significantly improve performance, however, implementing a scalable simulation algorithm, which can efficiently utilize hundreds or thousands of computing nodes is not straightforward. The simulation of physical systems require memory-intensive computations with a single data structure being constantly updated, while achieving scalability for highly parallel hardware imposes splitting the computational task into relatively independent parts.

Overcoming the bottleneck of data structures synchronization during model updates requires dedicated algorithms. The algorithm has to accept the unavailability of most recent data. Performing the update using slightly outdated information can lead to model constraints violation, which has to be addressed. In some cases it is possible to mitigate the influence of such desynchronization, which has been demonstrated in [14]. The presented urban traffic simulation could handle significant desynchronization of computations in different computational domains thanks to the dedicated model update algorithm.

The generalization of the concept has been presented in [2], where a universal, scalable simulation framework, called Xinuk, is presented. The framework supports parallel simulation (built according to well-known actor model of parallelism, using Akka) of physical environments, which are represented as two-dimensional uniform grids of cells. The update procedure is inspired by classic cellular automaton. The framework implements abstract information distribution model, which can represent various phenomena and allow splitting the simulated environment into separately updated computational domains.

The Xinuk framework is implemented in Scala – it uses the Java Virtual Machine. It has been successfully tested on a large-scale HPC system, the Prometheus supercomputer[2]. The results presented in [2] show that the framework can efficiently use 144 computing nodes (3456 cores) working on a single simulation task. The particular simulation models, used in the mentioned work, were very simple, used for demonstrating the scalability rather than analyze the simulation results.

In this paper we present the extensions introduced to the Xinuk firework, which make it possible to implement a variant of a well-recognized pedestrian motion model. Showing the ability of implementing its highly-scalable version is the main contribution of this work.

1.3 Crowd Dynamics Modeling

Usually two levels of modelling approaches are considered: *macroscopic* and *microscopic*. *Macroscopic* methods treat the whole crowd as a single entity, without considering the movement of single units. An algorithmic tool that can be used to implement this type of models is fluid mechanics or gas dynamics. Such approaches are developed for instance by Huges et al. [7].

Microscopic modelling analyses the behaviour of individuals. The two main approaches proposed in the literature are Social Force Model proposed in [6] and Social Distances Model (SDM) described in [16]. SDM distinguishes four social distances: *intimate distance*, where humans allow only very close relatives, *personal distance* is typical for friends, *social distance* is the most common in interpersonal interactions and appropriate for strangers, finally the last identified distance is *public distance*, also called the distance of public speeches.

2   Proposed Model

The aim of our work is to investigate the possibilities and difficulties resulting from the implementation of models designed for the environment with one computational domain, in the HPC environment. We decided to conduct our research by creating a city simulation that focuses on pedestrian traffic. Interactions between pedestrians [12] were implemented on the basis of proxemics rules

---

[2] http://www.cyfronet.krakow.pl/en/.

and especially the Social Distances Model [16]. The model has been adapted to the requirements of the Xinuk framework [2], which enables using multi-core and multi-node hardware.

2.1   Social Distances Model

The world is represented by a dedicated cellular automaton. Its cells can be in one of three states: *Movement space*, *Occupied* and *Obstacle*.

State *Occupied* means that there is a person in it. In the subsequent iterations of the simulation a pedestrian can change position and releases the cell. State *Obstacle* are cells that will never change their state and remain inaccessible.

The gray filling in the middle of the square symbolizes a pedestrian (Fig. 1b). This concept has been simplified in relation to the original SDM [16], where the pedestrians were represented by ellipses.

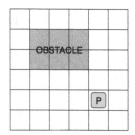

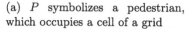

(a) *P* symbolizes a pedestrian, which occupies a cell of a grid

(b) Area calculated by one computational domain with a pedestrian

Fig. 1. Simulation schema (*a*) and GUI (*b*).

It is not possible to use the exact implementation of the SDM model in a multi-core environment. The unidirectional communication between computational domains causes difficulties which impose changes in some features. Information about the result of calculations on one node is sent to another one without getting any feedback. These places are called seams, because it expresses their nature reasonably.

The most important feature of framework Xinuk is the unique method of information propagation. Using "signal" we created social areas which are an intrinsic part of daily life according to proxemics [15].

2.2   Extended Xinuk

The Xinuk framework was a good starting point for our purposes. Its current version is used to scale the simulation, in which the reproduced world has no complicated terrain (sea or meadow), and the simulation scalability is used to give more accurate results.

The goal was to simulate pedestrian traffic in the whole city. It required a different approach than the one imposed by the Xinuk framework. Our simulation world was supposed to be a huge space. It had to be divided into many computational domains in order to acquire results in a reasonable time. We have prepared the framework to be extended with features that enable us to divide a ready-made map of a city between nodes.

In the simulation initialization process, individual computational domains are assigned an identifier, which allows to determine which part of the city a given unit calculates. We were able to initialize the grid of a CA for a specific computational unit, but it takes place independently on each node. Hence, it was impossible to set the obstacles that were located on the seams. After introducing changes we have great foundations for generating obstacles on the basis of the appropriate map fragments.

2.3   Pedestrians Movement

Persons, who participate in everyday life, move around the city from place to place. In our simulation we simulated the same process.

Each pedestrian without a predefined goal is assigned a task. Its completion concerns approaching a specific point within the simulated city. This allows missions with targets, located in the place of an obstacle (e.g. wall), to be accomplished.

In the current implementation, pedestrians do not calculate the route along they move. We came to the conclusion that the classical path finding algorithm such as $A*$ would cause a large and unnecessary increase in computing complexity. Instead, the direction is calculated in each iteration, then during the motion selection, pedestrians prioritize the cells that brings them the closest to the completion of the task. Unfortunately, this approach has a huge disadvantage, as it works correctly only if the map does not have "traps" (for example wall in shape of horseshoe). In case of the presence of mentioned "traps", pedestrians would cyclically move within a group of cells.

To make simulation credible, we introduced social areas [16]. Each pedestrian generates signal which defines his own social area. For each iteration of the pedestrian movement, there is an appropriate number of iterations of signal propagation. Therefore it spreads faster than people move.

Pedestrians should avoid obstacles and not encroach on social areas of others at the same time. We developed a function (Eq. 1) that assigns value to cells, which represents attractiveness of a cell for a pedestrian to enter.

$$f(distance, signal) = \alpha * norm(distance) - \beta * signal \tag{1}$$

A pedestrian task represented by a domain identifier and relative coordinates are converted into direction and distance. Distances are normalized so that they are within the $[0, 1]$ range. Distances are then multiplied by the coefficient ($\alpha \in \mathbb{N}_+$) for evaluating the best cell in terms of shortening the distance. The second factor taken into account is the signal emitted by other pedestrians.

The value of the repulsive signal (*signal*) including the crowd size is multiplied by a repulsion coefficient ($\beta \in \mathbb{N}_+$). Finally it is subtracted from earlier obtained value. Properly selected coefficients guarantee that the pedestrian will reach the destination quickly, while not disturbing the privacy of other participants of everyday life.

As mentioned earlier, the data is propagated unidirectionally between computational domains. When a pedestrian leaves the map fragment handled by a node, the information is sent to the adjacent one, which then takes over. Moving people are received and processed without objection, they cannot turn back, even if there is no room for them. Hence, it is necessary to take a number of actions to prevent or handle conflicts.

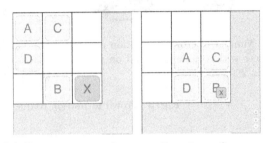

(a) X leaves node A and is sent to node B.

(b) X can not move because there is no free space for him.

Fig. 2. Cases when crowd formation is allowed. Pedestrians are marked with letters. X represents problematic one.

It may happen that a pedestrian passing between computational domains (Fig. 2a) will have to be placed in a spot that is already occupied by another one. We have solved this problem by allowing a crowd of people to form. Its graphical representation is identical to the single pedestrian, but the information is recorded and at the first opportunity the crowd separates.

We also allow it to appear outside of the border of domains. When shifting pedestrians, they are considered in an order and the decision is made on the basis of available data. Therefore, people are moved without examining the situation of others who are not considered yet. It happens (Fig. 2b) that a pedestrian is not able to make a move, because he/she is blocked by others or the terrain. Then another pedestrian takes his/her place, as at the stage of move determination it is not known that the next pedestrian will not be able to change position. Thus, a conflict occurs with two pedestrians occupying one place.

3   Results

We decided to transfer our simulation to HPC. For this purpose, we have exploited infrastructure available at AGH University of Technology in its

Academic Computer Center AGH Cyfronet. We used 'Zeus' supercomputer, which consists of 25468 computational cores.

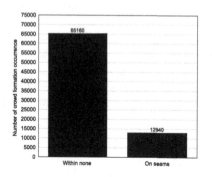
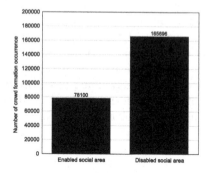

(a) Number of crowd formation occur-
rences within node area and on seams
after 200,000 iterations.

(b) Number of crowd formation occur-
rences before and after introduction of
social areas after 200,000 iterations.

Fig. 3. Crowd formation frequency statistics.

We ran the simulation in two similar configuration to compare influence of social areas effect. Both tests were conducted on 200,000 iterations, and the world was divided into 256 parts, each with 225 cells.

As crowd formation is an important difference from SDM implementation [16], we decided to keep track of how often it happens (Fig. 3a).

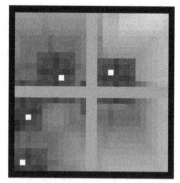

(a) Initial state of simulation.

(b) Simulation after 800 iterations.

Fig. 4. The simulation visualization with visible signal.

Presence of signal affects motion of other pedestrians, because they interpret the signal as discouraging and they rarely select cells with a high level of it.

Hence, we reproduced behaviour of preserving the distance to the strangers presented in [15], reducing the frequency of crowd formation (Fig. 3b).

To visualize social areas, the signal was presented in Fig. 4. Change of the hue of graphical representation of the cell pictures the power of signal. Social areas preserve consistency despite being located on different computing units (Fig. 4b).

At the end, we checked how big performance boost we get using many threads. Therefore we ran two 100,000-iteration tests, both with the same size of CA (240^2 cells) simulating movement of 256 pedestrians. The first one was conducted using only one processor Intel Xeon (24 cores), the latter 10 processors (240 cores). Amount of time saved due to distributed computation was significant. For many CPUs, it took only 34m 36s to finish, but one needed 15h 28m 19s.

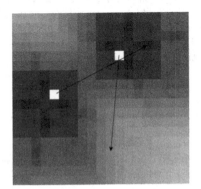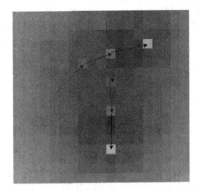

Fig. 5. Example of social area phenomenon.

Implemented movement algorithm can be presented in Fig. 5, where we also observe a behaviour of pedestrian repulsion. Two persons avoid encroaching on social areas of other pedestrians by taking other route than the shortest one (marked with the arrows).

4   Conclusions

Although the presented model seems to be relatively simple, the results obtained are satisfactory. The crowd is formed less than half (47%) as much frequently when enabling influence of social areas on the movement selection process. We successfully implemented SDM model in distributed environment.

Regarding the prepared HPC crowd simulation one can notice following, further research directions: implementation movement pedestrian on a real map, introducing obstacles and enhancement of movement algorithm.

Loading a complete map to each node creates memory lost, hence dividing it at initialization of the simulation and distributing parts through computational units seems to be the best approach.

After introducing complex environment, there is a necessity of the advanced movement algorithm, because some transition ways through the nodes may be blocked. As a solution to that problem, we propose composing movement algorithm consisted of two components: one global based on a transition graph between nodes, second one with costs of reaching adjacent computational unit using the signal mechanism.

References

1. Abar, S., Theodoropoulos, G.K., Lemarinier, P., O'Hare, G.M.: Agent based modelling and simulation tools: a review of the state-of-art software. Comput. Sci. Rev. **24**, 13–33 (2017)
2. Bujas, J., Dworak, D., Turek, W., Byrski, A.: High-performance computing framework with desynchronized information propagation for large-scale simulations. J. Comput. Sci. **32**, 70–86 (2019). https://doi.org/10.1016/j.jocs.2018.09. 004. http://www.sciencedirect.com/science/article/pii/S1877750318303776
3. Coakley, S., Gheorghe, M., Holcombe, M., Chin, S., Worth, D., Greenough, C.: Exploitation of high performance computing in the flame agent-based simulation framework. In: 2012 IEEE 14th International Conference on High Performance Computing and Communication & 2012 IEEE 9th International Conference on Embedded Software and Systems, pp. 538–545. IEEE (2012)
4. Collier, N., North, M.: Parallel agent-based simulation with repast for high performance computing. Simulation **89**(10), 1215–1235 (2013)
5. Collier, N., Ozik, J., Macal, C.M.: Large-scale agent-based modeling with repast HPC: a case study in parallelizing an agent-based model. In: Hunold, S., et al. (eds.) Euro-Par 2015. LNCS, vol. 9523, pp. 454–465. Springer, Cham (2015). https://doi. org/10.1007/978-3-319-27308-2_37
6. Helbing, D., Molnár, P.: Social force model for pedestrian dynamics. Phys. Rev. E **51**, 4282–4286 (1995). https://doi.org/10.1103/PhysRevE.51.4282
7. Hughes, R.L.: The flow of human crowds. Annu. Rev. Fluid Mech. **35**(1), 169–182 (2003). https://doi.org/10.1146/annurev.fluid.35.101101.161136
8. Kiran, M., Bicak, M., Maleki-Dizaji, S., Holcombe, M.: Flame: A platform for high performance computing of complex systems, applied for three case studies. Acta Phys. Pol., B **4**(2) (2011)
9. Kłusek, A., Topa, P., Wąs, J., Lubaś, R.: An implementation of the social distances model using multi-GPU systems. Int. J. High Perform. Comput. Appl. **32**(4), 482–495 (2018)
10. Komosinski, M., Ulatowski, S.: Framsticks. In: Komosinski, M., Adamatzky, A. (eds.) Artificial Life Models in Software, pp. 107–148. Springer, London (2009). https://doi.org/10.1007/978-1-84882-285-6_5
11. Message Passing Interface Forum: MPI: a message-passing interface standard, version 2.2. Specification, September 2009. http://www.mpi-forum.org/docs/mpi-2. 2/mpi22-report.pdf
12. Porzycki, J., Wąs, J., Hedayatifar, L., Hassanibesheli, F., Kułakowski, K.: Velocity correlations and spatial dependencies between neighbors in a unidirectional flow of pedestrians. Phys. Rev. E **96**, 022307 (2017). https://doi.org/10.1103/PhysRevE. 96.022307
13. Suryanarayanan, V., Theodoropoulos, G., Lees, M.: PDES-MAS: distributed simulation of multi-agent systems. Procedia Comput. Sci. **18**, 671–681 (2013)

14. Turek, W.: Erlang-based desynchronized urban traffic simulation for high-performance computing systems. Future Gener. Comput. Syst. **79**, 645–652 (2018). https://doi.org/10.1016/j.future.2017.06.003. http://www.sciencedirect.com/science/article/pii/S0167739X17311810

15. Wąs, J.: Crowd dynamics modeling in the light of proxemic theories. In: Rutkowski, L., Scherer, R., Tadeusiewicz, R., Zadeh, L.A., Zurada, J.M. (eds.) Artifical Intelligence and Soft Computing, pp. 683–688. Springer, Heidelberg (2010). https://doi.org/10.1007/978-3-642-13232-2_84

16. Wąs, J., Gudowski, B., Matuszyk, P.J.: Social distances model of pedestrian dynamics. In: El Yacoubi, S., Chopard, B., Bandini, S. (eds.) ACRI 2006. LNCS, vol. 4173, pp. 492–501. Springer, Heidelberg (2006). https://doi.org/10.1007/11861201_57

17. Wittek, P., Rubio-Campillo, X.: Scalable agent-based modelling with cloud HPC resources for social simulations. In: 4th IEEE International Conference on Cloud Computing Technology and Science Proceedings, pp. 355–362. IEEE (2012)

Author Index

Printed in the United States
By Bookmasters